Space Operations: Experience, Mission Systems, and Advanced Concepts

Space Operations: Experience, Mission Systems, and Advanced Concepts

EDITED BY

Michael Schmidhuber
German Aerospace Center (DLR), Oberpfaffenhofen, Germany
Craig Cruzen
NASA Marshall Space Flight Center, Huntsville, Alabama
Joachim Kehr
German Aerospace Center (DLR), Oberpfaffenhofen, Germany

Volume 242
Progress in Astronautics and Aeronautics

Timothy C. Lieuwen, Editor-in-Chief
Georgia Institute of Technology
Atlanta, Georgia

Published by
American Institute of Aeronautics and Astronautics, Inc.
1801 Alexander Bell Drive, Reston, VA 20191-4344

Front cover image design by Juliane von Geisau.

Cover Image

Top left image:	TanDEM-X. Credit: DLR
Left image:	Launch TET-1. Credit: DLR
Bottom image:	Antenna ground station Weilheim. Credit: DLR
Right image:	Mars-rover Curiosity. Credit: NASA
Top right image:	Rosetta orbiter. Credit: Astrium
Center image:	Control Room at GSOC. Credit: DLR
ISS cupola:	Credit: NASA

American Institute of Aeronautics and Astronautics, Inc., Reston, Virginia

1   2   3   4   5

ABOUT SPACEOPS

The International Committee on Technical Interchange for Space Mission Operations and Ground Data Systems (also known as the SpaceOps Organization) was formed out of the realization that the number of people involved in space mission operations was large and an organized community or technical forum was needed. Mission operations have become an increasingly large segment of space agencies' budgets. As a result, there is great interest in improving the capabilities and cost efficiencies of mission operations. It was in the spirit of providing the broadest possible managerial and technical interchange between space agencies, academia, and industry that SpaceOps was established.

Since its inception in 1990, SpaceOps has held a highly successful biennial conference hosted by various countries around the world. This international forum addresses state-of-the-art operations principles, methods, and tools and provides an excellent opportunity to foster managerial and technical interchange on all aspects of space mission operations, including such areas as robotics, human, Earth-orbiting, and deep space aspects of space operations.

As the premier organization serving the space operations community, Space-Ops encompasses the following areas:

- Mission Execution
- Data and Communication System Facilities
- Mission Design and Mission Management
- Training and Knowledge Transfer
- Cross Support, Security, Interoperability, and Standards
- Commercial Space Operations
- Launch, Sounding Rockets, and Balloon Operations

SpaceOps is managed through two committees, an Executive Committee and a Committee-at-Large. The Executive Committee consists of one member from each agency authorized to represent the entire agency delegation (with voting rights). The Committee-at-Large consists of members from each space agency installation of each country or international consortium, as well as academic or industrial representatives from the participating countries. The committees are chaired by the members who represent the most recent and next host countries for the biennial SpaceOps Conference. In addition, SpaceOps Partners are invited to participate actively in the SpaceOps Organization. Partners can be any academic or industrial organizations interested in following and promoting the activities of the SpaceOps community.

For more information on the SpaceOps Organization, visit www.spaceops.org.

TABLE OF CONTENTS

Part I   Mission Preparation and Management

Chapter 1   International Space Station: Unique In-Space Testbed as Exploration Analog **1**

Young H. Lee and Donald E. Eagles, *NASA Jet Propulsion Laboratory, California Institute of Technology, Pasadena, California*; Frank Moreno, Mike Rodriggs, Susan Beisert and Debbie Stapleton, *NASA Lyndon B. Johnson Space Center, Houston, Texas*

Chapter 2   Reengineering the Mission Operations System for the Prime and Extended Mission **27**

Joseph C. Hunt, Jr. and Leo Y. Cheng, *Jet Propulsion Laboratory, Pasadena, California*

**Chapter 3   Mission Operations Preparation Environment:
A New Approach for the Future** **41**

W. Heinen, S. Reid and S. Pearson, *Rhea System S.A., Wavre, Belgium*

Chapter 4   The Keys to Successful Extended Missions **55**

David A. Seal and Emily A. Manor-Chapman, *Jet Propulsion Laboratory, California
    Institute of Technology, Pasadena, California*

Chapter 5   Multi-Mission Operator Training Practices **77**

Jennifer M. Reiter, *Laboratory for Atmospheric and Space Physics (LASP), University of
    Colorado, Boulder, Colorado*

Chapter 6   Gamification for Astronaut Training 91

Ferdinand Cornelissen, *Nspyre B.V., Utrecht, The Netherlands*; Mark A. Neerincx and Nanja Smets, *TNO Human Factors, Soesterberg, The Netherlands*; Leo Breebaart, *Science & Technology Corporation, Delft, The Netherlands*; Paul Dujardin, *NLR, Marknesse, The Netherlands*; and Mikael Wolff, *ESA/ESTEC, Noordwijk, The Netherlands*

Chapter 7   Timeline as Unifying Concept for Spacecraft Operations 111

William K. Reinholtz, *Jet Propulsion Laboratory, California Institute of Technology, Pasadena, California*

Chapter 11 Commercial Collaboration for Collision Avoidance and Flight Operations 193

David Finkleman, *Center for Space Standards and Innovation, Analytical Graphics, Inc., Colorado Springs, Colorado*

Part II Data and Communications

Chapter 12 Overview of the Laser Communications Relay Demonstration Project 209

Bernard L. Edwards and Dave Israel, *NASA Goddard Space Flight Center, Greenbelt, Maryland*; Keith Wilson, *Jet Propulsion Laboratory, Pasadena, California*; and John Moores and Andrew Fletcher, *MIT Lincoln Laboratory, Lexington, Massachusetts*

Chapter 13 Replacing the Central Storage System 227

Bernd Holzhauer and Osvaldo L. Peinado, *DLR–GSOC, Wessling, Germany*

Chapter 14   Ten Times More Information in Your Real-Time TM 241

David Evans and Ugo Moschini, *ESA/ESOC, Darmstadt, Germany*

Chapter 15   CNES Ground Network Renewal: Challenges to Increase Capacity and to Reduce Costs 259

Jean-Marc Soula, Hélène Ruiz, Marc Palin, Fabienne Vincent Franc, Michel Recoules, and Isabelle Hernandez-Marcelin, *CNES, Toulouse, France*

Chapter 19   Intrinsic Interoperability of Services: A Dream or a Key Objective for Mission Operation Systems 353

Mehran Sarkarati, Mario Merri and Mariella Spada, *European Space Agency, ESA/ESOC, Darmstadt, Germany*; and Sam Cooper, *SciSys UK Ltd, Bristol, United Kingdom*

Chapter 20   Delay-Tolerant Networking Engineering Network: Constructing a Cross-Agency Supported Internetworking Testbed 367

Edward Birrane and Kristine Collins, *Applied Physics Laboratory, Johns Hopkins, University Laurel, Maryland*; and Keith Scott, Ph.D., *The Mitre Corporation, McLean, Virginia*

Chapter 21   CCSDS Mission Operations Services for Mission Planning 387

Roger S. Thompson, *SCISYS UK Ltd, Chippenham, United Kingdom*

Chapter 22 EDRS Precursor Systems at GSOC: Relevant Heritage and New Developments 401

Ralph Ballweg and Frank Wallrapp, *DLR/GSOC, Oberpfaffenhofen, Germany*

Part III Mission Execution

Chapter 23 Consideration of Space Debris Mitigation Requirements in the Operation of LEO Missions 413

Holger Krag, Tim Flohrer and Stijn Lemmens, *ESA/European Space Operations Centre, Darmstadt, Germany*

Chapter 24 FAST: A New Mars Express Operations Concept, Quickly 431

Daniel T. Lakey, Jonathan Schulster and Olivier Reboud, *SCISYS Deutschland, Darmstadt, Germany*; Thomas Ormston and Kees Van Der Pols, *VEGA Space, Darmstadt, Germany*; and Michel Denis and James Godfrey, *European Space Agency, ESOC, Darmstadt, Germany*

**Chapter 25   Emergency Scheduling of Multiple Imaging
Satellites with Dynamic Merging** **455**

Jianjiang Wang, Xiaomin Zhu and Dishan Qiu, *National University of Defense
   Technology, Changsha, China*

**Chapter 26   Planning and Execution of Tele-Robotic Maintenance
Operations on the ISS** **475**

M. Caron, *Canadian Space Agency, St-Hubert, Quebec, Canada*; and I. Mills, *NASA, Johnson
   Space Center, Houston, Texas*

Chapter 27   Efficacy of the Dawn Vesta Science Plan **501**

Carol A. Polanskey, *Jet Propulsion Laboratory, California Institute of Technology,
   Pasadena, California*; Steven P. Joy, *University of California, Los Angeles,
   California*; and Carol A. Raymond, *Jet Propulsion Laboratory, California
   Institute of Technology, Pasadena, California*

Chapter 28   Simplify ISS Flight Control Communications and Log Keeping via Social Tools and Techniques 517

Hugh S. Cowart and David W. Scott, *NASA—Marshall Space Flight Center, Huntsville, Alabama*; and Daniel J. Stevens, *Barrios Technology, NASA—Johnson Space Center, Houston, Texas*

Chapter 29   TDX-TSX: Onboard Autonomy and FDIR of Whispering Brothers 539

A. Schwab, C. Giese and D. Ulrich, *EADS Astrium GmbH, Friedrichshafen, Germany*

Chapter 30   Innovative Rover Operations Concepts–Autonomous Planner (IRONCAP): Supporting Rover Operations Planning on Ground 557

R. Steel, A. Hoffman and M. Niézette, *Telespazio VEGA Deutschland, Darmstadt, Germany*; A. Cimatti and M. Roveri, *Fondazione, Povo, Italy*; K. Kapellos, *TRASYS, Hoeilaart, Belgium*; and A. Donati and N. Policella, *European Space Operations Centre, Darmstadt, Germany*

Chapter 31   Integral: Investigation into Van Allen Belt and Geotail Crossings **573**

M. J. H. Walker, *SCISYS Deutschland GmbH, Darmstadt, Germany*; and J. B. Palmer, *Logica GmbH, Darmstadt, Germany*

PREFACE

Many aspects of modern life are made possible by services provided with the help of spaceflight technology. The applications span a wide range—from astronomy and planetary exploration to navigation services for everyday car travel, transport logistics, weather services, and earth observation, and to telecommunication for television and computer networks. Behind each of these systems, there is a complex world of diverse engineering disciplines that have to interact with high reliability. Besides the task to design, develop, and build the space vehicles and their payloads, there is the necessity to operate them and to provide a capable ground infrastructure. This is the world of spacecraft operations. Rockets, satellites, space stations, and science probes of all kind are operated from many locations and diverse institutions around the world. Since the beginning of spaceflight, theses centers have collected experience, developed work practices, and defined standards. Economical demands to spaceflight call for more and more cooperation between institutions and countries in order to reuse existing systems, to preserve knowledge, and to share ground infrastructure.

The SpaceOps Organization has been formed from this need to provide common approaches to system definition and to share knowledge. Since 1990, this community has been addressing interested audiences from around the world for a bi-annual conference to discuss and present the current status and newest developments of the trade. At each conference no less than several hundred excellent papers and presentations are on display.

This book presents a compilation of selected interesting and representative papers from the SpaceOps 2012 conference. Each of these manuscripts is considered to be outstanding in its group of discussion topics. They are grouped into the three main topics: Mission Preparation and Management, Data and Communications, and Mission Execution.

Mission Preparation and Management (Chapters 1–11) handles the tasks of design, development, and planning. It includes training aspects. *Data and Communications* (Chapters 12–22) describes the infrastructure needed on Earth from antennas to software in order to communicate with and retrieve data from the space-borne resources. It also includes standardizations of systems. Finally, *Mission Execution* (Chapters 23–31) focuses on aspects of specific space missions during preparations for flight and throughout operations.

For the 2012 SpaceOps conference more than 600 papers were submitted. They were grouped into eight categories or "topics": The four standards *Mission Execution*; *Data and Communications System Facilities*; *Mission Design and Mission Management*; and *Cross Support, Interoperability, and Standards* drew the most response, as usual. Additionally, four relatively new topics were included: *Training and Knowledge Transfer*; *Commercial Space Operations*; *Launcher, Rockets, and Balloon Operations*; and *University Space Operations*. While the number of papers in these newer topics was less than the four common topics, they were just as well received both from the authors and the audience participation. It can be expected that they will appear again in some form during the future SpaceOps conferences. After a detailed review by the Technical Programme Committee (TPC), 291 papers were selected for presentation. Another 38 papers were selected for poster presentation. The interesting aspect with the posters is that, other than the one-time oral presentation, the visitor comes into direct interactive dialogue with the author. Additionally, reading and later consideration is not limited. The posters were well received and in particular the dedicated poster session saw a large number of visitors.

The e-poster format, first introduced at the Huntsville SpaceOps conference in 2010, was included again in Stockholm. A limited number of presentations were given during special e-poster sessions. The main intention of this is to give authors the opportunity to present highly interactive presentations or software live-demonstrations. There is also ample time to go into various levels of detail depending on the current audience. This format has become an accepted extension and received very positive remarks.

Following the precedence of the 2006 (Rome) and 2010 (Huntsville) SpaceOps Conferences, the TPC decided to publish a printed book of interesting and representative papers from the event in Stockholm. The motivation to take the effort of printing a traditional hardcover book in the age of eBooks and electronic conference proceeding was similar to the rationale for sending a lander to the Moon or rovers to Mars...to leave a lasting and physical presence of the mission. In 2013, the most current information is available online at click of a mouse; however, the flood of information in an amorphous environment cannot compare with the impact of a prominent and thick book on your bookshelf.

With the careful selection of papers by the conference reviewer team, the reader gets a compact recommendation of the essence of the content presented at the conference. It is a reliable statement of the SpaceOps organization about issues relevant at the date of issue and the perceived outlook of spacecraft operations.

Frank Schätzing, bestselling author and invited guest speaker for the SpaceOps 2012 conference, quoted Arthur C. Clarke when he reminded us: "The best way to predict the future is to invent it – so don't hold back – invent, invent!"

Michael Schmidhuber
Craig Cruzen
Joachim Kehr
September 2013

CHAPTER 1

International Space Station: Unique In-Space Testbed as Exploration Analog

Young H. Lee* and **Donald E. Eagles**[†]
NASA Jet Propulsion Laboratory, California Institute of Technology, Pasadena, California

Frank Moreno,[‡] **Mike Rodriggs,**[§] **Susan Beisert**[¶] **and Debbie Stapleton****
NASA Lyndon B. Johnson Space Center, Houston, Texas

I.   INTRODUCTION

The International Space Station (ISS) Test Bed for Analog Research (ISTAR) project was begun in the Fall of 2010 at the National Aeronautics and Space Administration (NASA) Johnson Space Center (JSC) as a part of NASA's exploration test and risk mitigation strategy. Strategic goal 1 of the 2011 NASA Strategic Plan [1] is to "Extend and sustain human activities across the solar system." Thus, the emerging NASA vision is to launch an ambitious new initiative to enable human space exploration beyond low Earth orbit (LEO) to Lagrange points, the Moon, near-Earth asteroids (NEAs), and Mars and its environs. To accomplish this vision, it is necessary to develop and validate new and innovative exploration technologies. The 2011 NASA Strategic Plan sub-goal 1.1 is to "Sustain the operation and full use of the International Space Station (ISS) and expand efforts to utilize the ISS as a National Laboratory for scientific, technological, diplomatic, and educational purposes and for supporting future objectives in human space exploration." With the life span of the ISS extended to 2020, and possibly 2028, NASA wants to maximize the potential of the Nation's newest National Laboratory. One approach to meet the 2011 NASA Strategic Plan goal is to

---

*Program Area Manager for Operations and Member of the ISTAR Integrated Product Team, Human/Robotic Mission Systems Office.

[†]System Engineer and member of the ISTAR Integrated Product Team, Mission Operations System Engineering Group.

[‡]Deputy Manager, ISTAR Mission Integration Office, Human Exploration Development Support Office, 2101 NASA Parkway.

[§]Manager, ISTAR Mission Integration Office, Human Exploration Development Support Office.

[¶]Deputy Manager, ISTAR Mission Integration Office, Human Exploration Development Support Office.

**Former Manager, ISTAR Mission Integration Office, Human Exploration Development Support Office.

conduct sustained, full-use operations of the ISS in order to support human exploration objectives.

Exploration and ISS teams within NASA's Human Exploration and Operations Mission Directorate (HEOMD) have initiated a joint cooperative effort: the ISS Testbed for Analog Research (ISTAR), a high-fidelity operational analog that complements existing NASA terrestrial analogs in order to develop and validate innovative exploration technologies and techniques using the ISS platform. ISTAR supports and encourages investigations dubbed "exploration detailed test objectives" (xDTOs) to maximize use of the ISS platform for the evaluation of exploration technologies, capabilities, and operational concepts that mitigate human spaceflight risks.

The goal of ISTAR is to use the ISS as a test platform to reduce exploration risks for crewed NEA or Mars missions. The ISS can provide confinement and a microgravity operational environment to simulate the crew experience during long-duration transit flights and arrival activities on NEA or Mars missions. ISS preflight preparation could simulate exploration mission preparation processes, including mission management team functions, flight planning and design, crew training, flight procedure development, and certification of flight readiness. ISTAR's long-term goal is to conduct long-duration ISS Mars Analog missions on board the ISS, beginning before the end of 2015, using technologies and operational tools and concepts developed and tested during earlier ISTAR missions and Earth-based laboratory and field testing. The purpose of these ISS Mars Analog missions is to address key exploration technology and operational concept gaps before conducting human exploration missions beyond LEO. Findings from these missions will contribute to the development of a set of design criteria for spaceflight and support systems that will enable safe and affordable human exploration missions, in particular to NEAs and Mars.

II. OVERVIEW

A. ISTAR'S OBJECTIVES

ISTAR has five main objectives:

1. Identify exploration investigations that require use of the ISS to advance exploration technology and capability needs or buy-down risk.

2. Advance preparations for autonomous crew operations supporting NEA or Mars exploration.

3. Evaluate and assess new exploration technologies, operations techniques, and methods as they become available.

4. Collect lessons-learned and disseminate them to stakeholders and use them to streamline and refine subsequent ISTAR mission processes.

5. Identify effective and affordable ways to send humans beyond LEO and enable them to conduct autonomous mission operations.

ISTAR's proposed ISS Mars Analog missions could last six months or longer while performing Mars exploration mission phases and crew arrival, departure, and landing activities as realistically as possible within ISS operational constraints.

B. WHAT IS AN ANALOG?

Analog missions are remote isolated field tests in locations that are identified based on their physical similarities to the extreme space environments of a target mission. Analog missions exercise multidisciplinary activities that simulate features of human exploration missions in an integrated fashion in order to enable new capabilities for human exploration. Analog missions test robotics, vehicles, habitats, communication systems, in situ resource utilization, and human performance as it relates to these technologies. Exploration analog missions are conducted to validate architecture concepts, conduct technology demonstrations, and gain a deeper understanding of the system-wide technical and operational challenges needed to support crewed missions beyond LEO, such as to NEAs or Mars.

C. WHY ISS AS A TESTBED?

The ISS provides an in-space operational environment that cannot be completely simulated in any of the terrestrial analogs. Its environment is as close to the crewed space exploration environment as is currently possible, thereby providing a unique in-space microgravity analog opportunity not available in any of the terrestrial analogs. It is an excellent testbed with which to simulate crew activities for long-duration flights and crew arrival at a NEA or Mars, and to research 1) the effects of isolation and confinement on flight-crew autonomy, behavior, and interaction with advanced technologies, the ground, and each other; 2) the most beneficial forms of medical and psychological support; 3) the effects of microgravity and physical deconditioning on the Mars landing transition; and 4) the effects of the increasing two-way light time on crew planning, interaction with the ground, and anomaly resolution. In contrast to the current mission-control paradigm of real-time crew access to and significant reliance on mission ground control, delayed space-to-ground communications will require the flight crew to have increasing responsibility for their safety and also the safety of the flight vehicle.

D. ISTAR FIVE-YEAR STRATEGIC PLAN

ISTAR has developed and is implementing a phased approach to using the ISS as an exploration testbed and to provide a realistic exploration experience to flight-crew and ground-control personnel by 1) beginning with short analog missions to test risk-mitigation technologies and operational tools, 2) establishing baselines

TABLE 1.1   ISTAR FIVE-YEAR STRATEGIC PLAN

Phase	Major features of plan
A Evaluate ISS Capabilities (2011–2012)	Primarily current ISS operations and activities. Operational, experimental protocols to protect safety, health, efficiency of ISS crewmembers are evaluated for their applicability to Mars (and NEA) missions.
B Short-Period Simulations (2013–2014)	Discrete Mars-forward activities are inserted, such as intermittent multiday periods of different degrees of bounded autonomy by the ISS crew, including communications delays typical of Mars missions. Sets of assigned tasks would be accomplished with minimal intervention by the Mission Control Center (MCC), but few alterations would be made to onboard procedures and MCC monitoring of ISS systems. Impact on non-Mars onboard science operations would be minimized. Flight rules specify threshold at which simulation is broken in case of emergency or system malfunction. Add "exploration" tasks to post-landing timeline.
C Longer-Period Simulations (2014–2015)	More rigorous, longer periods of autonomy would be introduced. Crew procedures, MCC oversight are modified to provide a more realistic experience in autonomous operations to both crew and ground personnel. There would be some impact on onboard non-Mars science operations. Post-landing multiday exploration analogs would be conducted.
D 6-month Mission and Crew Deconditioning (post-2015)	Transits to Mars (and NEAs) would be simulated as rigorously as feasible in low Earth orbit with existing infrastructure. Progressively increasing communications delays would be introduced, reaching the maximum delay after 6 months to mimic Mars proximity. Onboard science operations would be compatible with Mars-like mission parameters. Use of post-landing exploration mission analogs would increase.

for crew performance and behavior with and without these technologies and tools, 3) developing countermeasures to the negative effects of long-duration missions, and 4) testing increasing periods of flight-crew and flight-vehicle autonomy by modifying crew procedures and mission-control operations in response to the increasing light-time communication delays. Table 1.1 describes this phased approach.

E.   APPROACHES FOR ISTAR MISSION FORMULATION, DEVELOPMENT, AND EXECUTION

It is crucial to promote collaboration and synergy where possible by fully engaging the NASA exploration community in strategy development and the execution approach for ISTAR missions. Field-tested lessons-learned from other analogs are integrated into ISTAR planning. Lessons-learned from ISTAR missions will be fed back to exploration-system planners, designers, and operations personnel to refine their processes and enhance their follow-on system development. A mission concept development process has been created that includes an integrated product team (IPT) forum to identify and vet exploration mission demonstrations that require use of the unique ISS platform. The IPT is a NASA multi-Center team, with representation from the ISS program, exploration systems, exploration analogs, the Flight Crew Office, the Human Research Program (HRP), and mission operations and engineering.

Mission planning is closely coordinated with the Human Spaceflight Architecture Team (HAT) to firmly base mission selections on exploration technology and capability needs and to buy-down risks of Mars and NEA design reference missions (DRMs). ISTAR is working with the NASA Headquarters' Strategic Analysis and Integration Division (SAID) that sponsors HAT, the Advanced Exploration Systems Division (AES) that funds AES projects (including the NASA analog missions), and the NASA Office of Chief Technologist (OCT) to establish the best path forward and develop synergistic exploration technologies and operations concepts, and to strategically plan missions that align with ISS increments. ISTAR is partnering with HRP to identify and coordinate ISTAR missions that require approval by the Committee for the Protection of Human Subjects (CPHS) review board. ISTAR also works closely with the ISS Program in mission formulation, planning, integration into ISS, and on-orbit operations.

ISTAR has established a solid working relationship with the AES projects, including the Analog Missions Project, to integrate plans for ISTAR analog testing. As a part of the NASA analog mission family, ISTAR collaborates with Earth-based analogs including the NASA Extreme Environment Mission Operations (NEEMO), Research and Technology Studies (RATS), In-Situ Resource Utilization (ISRU), and the Pavilion Lake Research Project (PLRP) in order to infuse the maturing technologies and operational tools, techniques, and concepts they have developed and the lessons they have learned into ISTAR mission designs.

III.   ADVANCED EXPLORATION SYSTEMS ANALOG MISSIONS

To prepare for the challenge of deep space exploration missions to the Moon, asteroids, Mars, or beyond, NASA conducts analog missions here on Earth, in remote locations that have physical similarities to extreme space environments. The following sections describe several of the more important NASA analogs [2].

A.   NASA EXTREME ENVIRONMENT MISSION OPERATIONS (NEEMO)

The NEEMO project (Fig. 1.1) utilizes a 45-ft-long, 13-ft-diam underwater laboratory, named Aquarius, located 62 ft below the surface within the Florida Keys National Marine Sanctuary, 3.5 miles off the Key Largo coast. A surface buoy provides laboratory connections for power, life support, and communications. Because of its isolation and real underwater hazards, this laboratory's environment makes it an excellent site for testing space exploration concepts.

NEEMO missions, lasting up to three weeks, provide astronauts with the opportunity to simulate living on a spacecraft and executing undersea extravehicular activities (EVA). During these activities they are able to test advanced navigation and communication equipment, EVA, integrated human–robotic system interactions, remote science and medical operations, and future exploration vehicles.

B.   RESEARCH AND TECHNOLOGY STUDIES (RATS)

NASA's RATS (Fig. 1.2) analog team evaluates exploration technologies, human–robotic systems, and extravehicular equipment in the high desert near Flagstaff, Arizona. RATS exercises provide information that helps scientists and engineers design, build, and operate

Fig. 1.1   An astronaut stands in front of an exploration vehicle mockup during NEEMO field tests.

Fig. 1.2   Two rovers are connected to the habitat demonstration unit during RATS field tests in Arizona.

equipment for exploration missions, and establish requirements for exploration operations and procedures.

The Arizona desert has a rough, dusty terrain and extreme temperature swings that simulate conditions that may be encountered on planetary, lunar, or asteroid surfaces. Some examples of technologies the RATS team has evaluated include high-fidelity prototype hardware, spacesuit equipment, robots, rovers, habitation modules, exploration vehicles, surface mapping and navigation techniques, and power and communication systems.

RATS objectives are to advance future human exploration capabilities by maturing operational concepts and technologies through integrated demonstrations and to reveal operational lessons-learned and technical deficiencies that enable improvements in system design.

C. IN SITU RESOURCE UTILIZATION (ISRU) DEMONSTRATIONS

ISRU is a process that harnesses local regolith (surface) or atmospheric resources at an exploration destination (Moon, asteroid, or Mars) for use in human and robotic exploration. ISRU demonstrations exercise extraction, separation, and storage of desired exploration commodities (e.g., oxygen, hydrogen, methane, and water).

NASA conducts ISRU analog demonstrations (Fig. 1.3) at Mauna Kea in Hawaii, in collaboration with partners such as the Pacific International Space Center for Exploration Systems and the Canadian Space Agency. These

Fig. 1.3   NASA and its international partners test equipment during in situ field tests at Mauna Kea volcano in Hawaii.

demonstrations are used to develop or improve systems and technologies that could be used to look for and extract desired commodities at exploration destinations.

The terrain, rock distribution, soil materials, and permafrost at Mauna Kea provide an ideal setting for testing hardware and operations not available in laboratories or NASA centers.

D.   PAVILION LAKE RESEARCH PROJECT (PLRP)

The PLRP (Fig. 1.4) is an international, multidisciplinary, science and exploration

Fig. 1.4   The Deep Worker submarine searches for microbialites during field tests at Pavilion Lake in British Columbia, Canada.

effort that seeks to explain the origin of the freshwater microbialites that grow in Pavilion and Kelly Lakes in British Columbia, Canada.

NASA conducts this analog mission because it is in a critical science research location that provides a challenging setting to test and develop research and exploration methods for future site surveys and science data collection. Scientists use submersible vehicles and methods of exploration that are similar to how robotic precursor missions would explore NEAs. The process refinements for traverse planning and science data collection will help improve techniques for future space exploration missions and scientific research.

IV. INTERNATIONAL SPACE STATION

The ISS [3] (Fig. 1.5) is the largest orbiting man-made object. It is composed of about one million pounds of hardware, brought to orbit over the course of a decade. The ISS includes 1) primary structures (the external trusses that serve as the backbone of the station and the pressurized modules that are occupied by the ISS crew) and 2) functional systems made up of replaceable units (systems that provide basic functionality such as life support and electrical power), which are composed of modular components that are replaceable by astronauts in orbit.

The ISS was constructed to support three activities: scientific research, technology development, and development of industrial applications. The facilities

Fig. 1.5   The International Space Station.

aboard the ISS allow for ongoing research in microgravity, studies of other aspects of the space environment, tests of new technology, and long-term space operations. The facilities also enable a permanent crew of up to six astronauts to maintain their physical health standards while conducting many different types of research (including experiments in biotechnology, combustion science, fluid physics, and materials science) on behalf of ground-based researchers. Furthermore, the ISS has the capability to support research on materials and other technologies to see how they react in the space environment.

Two ground facilities at the JSC in Houston are especially well suited to the preparation of flight crewmembers and ground controllers for analog operations on the ISS: the Space Station Training Facility (SSTF) (Fig. 1.6) and the Neutral Buoyancy Laboratory (NBL) (Fig. 1.7). The SSTF is a full-scale, high-fidelity mockup of the ISS module cluster. This ISS replica provides interfaces to train flight crewmembers, controllers, and instructors on ISS operations, crew systems, station maintenance, and crew healthcare. The SSTF is also used to develop and validate operating procedures planned for use on the ISS.

The NBL is an astronaut training facility consisting of the world's largest indoor pool of water, where astronauts perform simulated EVA tasks in preparation for upcoming missions. The NBL contains a full-sized mockup of the ISS.

Fig. 1.6   The SSTF at the JSC.

Fig. 1.7 The NBL at the JSC.

V. HUMAN SPACEFLIGHT ARCHITECTURAL TEAM

One of ISTAR's important stakeholders is the Human Spaceflight Architecture Team (HAT) [4], a multidisciplinary, cross-agency study team within NASA Headquarters' HEOMD that conducts strategic analysis cycles to assess integrated development approaches for architectures, systems, mission scenarios, and concepts of operation for human space exploration. During each analysis cycle, HAT iterates and refines DRM definitions to develop integrated, capability-driven approaches for systems planning to exploration destinations beyond LEO.

HAT has generated (Table 1.2) a list of risks [5] to the successful accomplishment of crewed exploration missions and a list of mission architecture questions that must be answered before completing the design for such missions. ISTAR uses these HAT-generated risks and architectural questions to influence its mission formulation, development strategy, and mission-evaluation criteria.

VI. HUMAN RESEARCH PROGRAM

All ISS experiments or activities that involve man-in-the-loop testing, including ISTAR xDTOs that require crew testing as well as planned ISTAR ISS Mars Analog missions, require coordination with and/or approval by NASA's Human Research Program (HRP) [6]. HRP, a program managed by the JSC's Space Life Sciences Directorate, seeks to perform research necessary to understand and reduce spaceflight human health and performance risks, enable development

TABLE 1.2   HAT EXPLORATION MISSION RISKS AND ARCHITECTURAL QUESTIONS

ID	Exploration Mission RISK	ID	Exploration Mission Architectural Questions
M-EDL	EDL of large Mars payloads	Q1	What is the safest way to approach a small/non-cooperative/ non-stable object? (i.e. NEA, satellite)
E-EDL	Earth re-entry at high velocities		
LV	Launch vehicle failures	Q2	What is the safest and quickest way to anchor to a NEA?
Lndr	Lander propulsion systems failure	Q3	What Earth Orbit activities are needed to reduce risk for deep space missions?
CSM	Long duration low/zero boiloff cryo-storage and management	Q4	What are the impact of the planetary protection requirements on operations and elements?
CFT	In-space cryogenic fluid transfer	Q5	What are the functional/volumetric requirements for habitation and IVA activities in zero and low − g?
ISP	In-space propulsion failures		
A-ISP	Reliability verification of advanced in-space propulsion	Q6	What is the difference in operational efficiency between crew size? (3 and 4 crew for NEA, 4 and 6 crew for Mars)
Env	Environmental risks: radiation, MMOD, dust, electromagnetic	Q7	What is the most efficient way to communicate under a long >30 sec time delay? Does this change as the time increases?
Dock	Docking/assembly failures	Q8	What improvements of logistics and packaging can be realized?
Sys	Systems failures: ECLSS, power, avionics, thermal		
EVA	EVA system/suit failure	Q9	What is the most effectives trade between level of repair and on-orbit manufacturing?
Comm	Operations under time delayed communication		
Aut	Autonomous crew/vehicle operation	Q10	How do you best reuse/repurpose disposable materials?
Health	Crew health: behavioral, health care/remote medical, micro-gravity	Q11	What is the most effective means of surface transportation? (NEA, Moon/Mars short distance, Moon/Mars long distance)
SW	Software failure	Q12	Given current robotic capabilities, what level of human/ robotic interaction provides the highest level of operational efficiency? (EVA at destination, In-space EVA, IVA, Teleoperations)
Hum	Human error		
ISRU	ISRU equipment failure: propellant, consumables	Q13	What level of IVA/EVA activities at a destination provides the most benefit?

TABLE 1.3 NASA'S HUMAN RESEARCH PROGRAM (HRP) RISKS AND CRITICALITY

HRP risks & criticality — Human Research Program (HRP) Elements:
- HHC: Human Health Countermeasures
- SHFH: Space Human Factors & Habitation
- ExMC: Exploration Medical Capability
- BHP: Behavioral Health & Performance
- SR: Space Radiation

Risk Title key — U: Unacceptable Risk, A: Acceptable Risk, C: Controlled Risk, I: Insufficient Data

HRP Elem	Risk Title (Short Title)	Lunar	NEA	Mars
HHC	Risk of Orthostatic Intolerance During Re-Exposure to Gravity (Short Title: OI)	C	A	A
HHC	Risk of Early Onset Osteoporosis Due to Spaceflight (Short Title: Osteo)	C	A	A
HHC	Risk Factor of Inadequate Nutrition (Short Title: Nutrition)	C	A	U
HHC	Risk of Compromised EVA Performance and Crew Health Due to Inadequate EVA Suit Systems (Short Title: EVA) - *Pending Human System Risk Board (HSRB) Risk Management Analysis Tool (RMAT) Approval*	A	A	A
HHC	Risk of Impaired Performance Due to Reduced Muscle Mass, Strength and Endurance (Short Title: Muscle)	A	A	U
HHC	Risk of Renal Stone Formation (Short Title: Renal)	C	C	C
HHC	Risk of Bone Fracture (Short Title: Fracture)	C	C	C
HHC	Risk of Intervertebral Disc Damage (Short Title: IVD)	C	I	I
HHC	Risk of Cardiac Rhythm Problems (Short Title: Arrhythmia) - *Pending HSRB RMAT Approval*	C	I	I
HHC	Risk of Reduced Physical Performance Capabilities Due to Reduced Aerobic Capacity (Short Title: Aerobic)	A	A	U
HHC	Risk of Crew Adverse Health Event Due to Altered Immune Response (Short Title: Immune)	C	C	I
HHC	Risk of Impaired Control of Spacecraft, Associated Systems and Immediate Vehicle Egress due to Vestibular / Sensorimotor Alterations Associated with Space Flight (Short Title: Sensorimotor)	C	A	A
HHC	Risk of Therapeutic Failure Due to Ineffectiveness of Medication (Short Title: Pharm) - *Pending HSRB RMAT Approval*	C	C	I
HHC	Risk of Microgravity-Induced Visual Impairment/Intracranial Pressure (Short Title: VIIP) - *Pending HSRB RMAT Approval*	I	I	I
HHC	Risk of Injury from Dynamic Loads (Short Title: Occupant Protection)	U	U	I
SHFH	Risk of Performance Decrement and Crew Illness Due to an Inadequate Food System (Short Title: Food)	C	C	U
SHFH	Risk of Inadequate Human-Computer Interaction (Short Title: HCI) - *Pending HSRB RMAT Approval*	C	C	A
SHFH	Risk of Performance Errors Due to Training Deficiencies (Short Title: Train) - *Pending HSRB RMAT Approval*	C	C	A
SHFH	Risk of Inadequate Design of Human and Automation/Robotic Integration (Short Title: HARI) - *Pending HSRB RMAT Approv*	C	C	A
SHFH	Risk of Poor Critical Task Design (Short Title: Task) - *Pending HSRB RMAT Approval*	C	C	A
SHFH	Risk of Adverse Health Effects of Exposure to Dust and Volatiles During Exploration of Celestial Bodies (Short Title: Dust) – *Pending HSRB RMAT Approval*	A	I	I
SHFH	Risk of an Incompatible Vehicle/Habitat Design (Short Title: Hab) - *Pending HSRB RMAT Approval*	C	C	A
SHFH	Risk of Adverse Health Effects Due to Alterations in Host-Microorganism Interactions (Short Title: Microhost)	A	I	I
ExMC	Inability to Adequately Recognize or Treat an Ill or Injured Crew Member (Short Title: ExMC)	A	A	U
BHP	Risk of Adverse Behavioral Conditions and Psychiatric Disorders (Short Title: Bmed) - Reference RMATs for Risk of Adverse Behavioral Conditions, and Risk of Psychiatric Disorders	C	A	U
BHP	Risk of Performance Errors Due to Fatigue Resulting from Sleep Loss, Circadian Desynchronization, Extended Wakefulness, and Work Overload (Short Title: Sleep)	C	C	C
BHP	Risk of Performance Decrements due to Inadequate Cooperation, Coordination, Communication, and Psychosocial Adaptation within a Team (Short Title: Team)	C	A	A
SR	Risk of Radiation Carcinogenesis (Short Title: Cancer)	A	U	U
SR	Risk of Acute Radiation Syndromes Due to Solar Particle Events (Short Title: ARS)	A	A	A
SR	Risk of Acute or Late Central Nervous System Effects from Radiation Exposure (Short Title: CNS)	A	I	I
SR	Risk of Degenerative Tissue or other Health Effects from Radiation Exposure (Short Title: Degen)	A	I	I

of human spaceflight medical and human performance standards, and develop and validate use of technologies that reduce human spaceflight medical risks.

To accomplish these goals, HRP focuses its research on establishing an evidence base on astronaut health and performance for long-duration microgravity missions, on identifying the greatest risks and developing an optimal approach to mitigate those risks, on testing space biomedical technology and medical-care procedures, and on actively collaborating with NASA's international partners on space biomedical research. Table 1.3 [7] shows HRP's list of human health and performance risks, and its assessment of the criticality of these risks.

ISTAR seeks HRP's advice, help, and collaboration when developing and executing ISTAR missions because ISTAR will exercise operations concepts that challenge flight crews to work progressively longer periods without direct assistance from ground teams, forcing them to deal with increasingly delayed communications by exercising increasingly autonomous activities. HRP and ISTAR have jointly developed clinical research investigations that assess the impact of communication delay on flight-crew performance. These investigations are being worked through the ISTAR Joint Operations Panel to establish a communication delay protocol and select specific crew tasks and procedures as part of ISTAR Mission 3. See Sec. VII.D for additional information.

VII. ISTAR MISSIONS

ISTAR's five-year strategic plan, the exploration community's exploration risks (in particular those identified by HAT, Table 1.2), HRP research objectives that mitigate exploration mission risks such as those in Table 1.3, and the rationale for using the ISS as a testbed, all guide the formulation, development, and integration of ISTAR missions into the ISS. ISTAR also heeds the following National Research Council (NRC) priorities for key technologies needed to extend and sustain human activities beyond LEO [8]: 1) radiation mitigation for human spaceflight, 2) long-duration crew health, 3) environmental control and life support systems (ECLSS), 4) guidance, navigation, and control (GN&C), 5) (nuclear) thermal propulsion, 6) lightweight and multifunctional materials and structures, 7) fission power generation, 8) entry, descent, and landing (EDL) thermal protection systems (TPS).

A. ISTAR MISSION DEVELOPMENT PROCESS

To aid in formulating its planned missions, ISTAR introduced the term *exploration detailed test objective* (xDTO) to describe the technology and operations-concept building blocks of its missions. ISTAR then developed a unique review process synchronized with the ISS payload integration template to identify, screen, score, and recommend xDTO candidates for appropriate ISS increments.

For each xDTO candidate, ISTAR documented the candidate's resource requirements (on-orbit crew time, hardware/software development cost and time, payload up-/down-mass and volume, development funding profile, projected earliest readiness date, etc.) using an xDTO Survey Form. Proposed xDTO candidates were evaluated by applying a weighting factor (using a scale of 1–3) against the following xDTO selection criteria:

1. *ISS as a testbed*: Is the ISS (or an ISS ground facility) required to test this xDTO candidate? Two points were assigned when ISS was required and no ground facility could be used to test the xDTO candidate, and one point when ISS was not mandatory because an ISS facility (e.g., NBL or SSTF), other facility, or terrestrial analog could be used to test the candidate. No points were assigned when neither the ISS nor ISS ground facility was required. ISTAR selected this criterion as one of its critical criteria with a high-value (three-point) weighting factor.

2. *Mission applicability*: What is the applicability of this xDTO candidate to an exploration mission destination? Three points were assigned when a proposed xDTO candidate technology or mitigation method(s) was applicable to a NEA, Mars, and to the ISS as a destination, two points when applicable to both NEA and Mars as destinations, and one point when applicable only to a NEA (or NEA and ISS) or to Mars (or Mars and ISS) as a destination.

No points were assigned when a proposed xDTO candidate was applicable to ISS only, to non-NEA/Mars destination(s) only, or to ISS and non-NEA/Mars destination(s) only, because exploration destinations are of high value. ISTAR selected this as another of its critical criteria with a three-point weighting factor.

3. *Safety (risk) concern*: Does this xDTO candidate introduce any risk to the ISS vehicle or crew? If so, can the risk be quantified? Three points were assigned when no risks were identified, two points when low risks or no known risks (or only acceptable risks) were identified, or one point when medium risk [or an unacceptable risk(s)] was identified. ISTAR selected this as another of its critical criteria with a three-point weighting factor.

4. *Architecture relevancy*: Does this xDTO candidate respond to HAT's assessment of Human Space Flight architecture relevance? Three points were assigned when HAT assigned the candidate a high assessed value, two points when HAT assigned a medium assessed value, and one point when HAT assigned a low assessed value. ISTAR selected this as another of its critical criteria with a three-point weighting factor.

5. *Mission-risk mitigation*: Does the knowledge gained from this xDTO candidate reduce the risks of a crewed NEA or Mars mission? Three points were assigned to Class 1 xDTO candidates (see listing of ISTAR classes immediately below), two points to Class 2 xDTO candidates, and one point to Class 3 candidates. No points were assigned to Class 4 candidates. ISTAR selected this as another of its critical criteria with a three-point weighting factor. The following are ISTAR's xDTO candidate classes:
   (a) Class 1 xDTO candidates intend to provide/improve radiation protection for flight crews (Class 1a), intend to mitigate physiological effects of long-duration microgravity (Class 1b), and intend to mitigate psychological effects of long-duration isolation (Class 1c).
   (b) Class 2 xDTO candidates support the development of technology to improve flight-crew life support (including closed loop) and/or habitation systems (Class 2d), support the development of technology to improve autonomous systems and avionics (Class 2e), and intend to improve flight-crew productivity during long-duration missions (Class 2f).
   (c) Class 3 xDTO candidates contribute to supporting flight-crew medical diagnosis and/or acute care (Class 3 g), support the development of technology to provide or improve automated rendezvous and docking (Class 3 h), and intend to improve flight-hardware maintenance/supportability (Class 3i).
   (d) Class 4 xDTO candidates include those dealing with flight operations, crew clothing, or extravehicular–intravehicular suit systems, emergency equipment, experiments/fabrication/facilities, fire detection and control, human systems, materials research, power management, and so on (Class 4j).

6. *Potential for mission-risk reduction*: If the proposed xDTO candidate is selected and succeeds, what percent of its associated mission-risk could be mitigated? Three points were assigned if significant ($>25\%$) risk could be mitigated, two points when moderate ($>10\%$ but $<25\%$) risk could be mitigated, and one point when only minimal ($<10\%$) risk could be mitigated. ISTAR selected this as the last of its critical criteria with a three-point weighting factor.

7. *Cost*: If development of this xDTO candidate is not fully funded, how much additional funding is required? Four points were assigned when US\$0.5 million or less is required, three points when \$0.5 million to \$1.5 million is required, two points when \$1.5 million to \$3.5 million is required, or one point when more than \$3.5 million is required. ISTAR assigned this selection criterion a two-point weighting factor.

8. *Crew time*: What is the total crew time (hours) needed to support operation of this xDTO candidate? This was assigned three points when 5 or fewer hours are needed, two points when 5–40 hours are needed, or one point when more than 40 hours are needed. ISTAR assigned this selection criterion a two-point weighting factor.

9. *xDTO readiness*: What is the progress of any xDTO-candidate hardware/payload-safety certification or safety and mission assurance assessment? Three points were assigned if a candidate had passed its Phase 0/I, II, and III safety reviews to date, two points if a candidate had passed its Phase 0/I and II safety reviews to date, and one point if a candidate had passed its Phase 0/I safety review to date. ISTAR also assigned this selection criterion a two-point weighting factor.

10. *ISS flight resource dependency*: What is the anticipated amount of ISS resources (e.g., power, communications, fluid/gas/atmosphere consumables, imagery, tools, crew-aids/provisioning, stowage, attitude/pointing) required to support this xDTO candidate? Three points were assigned when few or no ISS flight resources are required, two points when a moderate amount of resources is required, and one point when a large amount of resources is required. ISTAR assigned this selection criterion a one-point weighting factor.

Utilizing the ISTAR IPT forum, ISTAR conducted high-level reviews with stakeholders and management to prioritize and rank xDTO candidates for recommendation to the ISS Program. ISTAR forwarded its recommended list of xDTO candidates, denoting them for flight consideration on specific ISS increments, to the ISS Program's Research Planning Working Group (RPWG) during their ISS utilization planning for an increment period. ISTAR will monitor the on-orbit execution of its recommended xDTO candidates, conduct post-mission analyses of successfully mitigated exploration mission risk, and collect and disseminate lessons-learned.

B. ISTAR MISSION 1

Ground-based analog missions have found that time delay is an impediment to communication. An in-space operational environment is needed to validate communication delay effects on individual and team performance and behavioral health outcomes. Reduced communication is the first step toward a major cultural shift for mission flight and ground crews in the operation of exploration missions. In contrast to the current mission control paradigm of real-time crew access and significant reliance on ground mission control, delayed communication requires an increase in flight crew responsibility for the safety of the crew and space vehicle, which may cause initial discomfort for both flight and ground crews. The first ISTAR mission will study countermeasures for communication delays. HRP will sponsor a study on a later mission (planned for ISTAR Mission 3) that will look at the effects of communication delay on crew performance.

Ground-based analog missions have also found that the impact of a communication delay is lessened when autonomous procedures and text messaging are available. The primary purpose of ISTAR Mission 1, planned for ISS Increments 31/32, is to prepare the flight and ground crews for more autonomous flight operations by the execution of autonomous crew procedures and by the engineering evaluation of communication delay countermeasures (text messaging) when voice communication is not being used (but is available). The autonomous crew operations and communication delay countermeasures are separate activities and will be performed at different times so that the variables can be studied independently before the ISTAR Mission 3 test is performed.

1. *Crew procedure execution*: Communication delays will force the exploration crews and their vehicles to be more autonomous. Crewmembers will not have the ground to rely on for instant assistance, advice, and troubleshooting help while performing procedures. The objective is to prepare the flight and ground crews for more autonomous flight operations (including autonomous crew procedure execution). This will give the procedure authors experience in developing autonomous procedures, understand what extra information the flight crew would need to perform a specific procedure autonomously, and develop methods to train flight crewmembers to perform autonomous execution of procedures. This will give the crew experience in executing procedures without relying on the ground. This may also provide insight into how communication delay might affect not only procedures but also the design, building, and operation of hardware and software for future spacecraft and systems.

2. *Communication delay countermeasures*: During periods of communication delays that will be simulated in later ISTAR missions, the standard voice communication between the crew and ground is expected to be operationally ineffective. Communication delay scenarios have been simulated on Earth-based analogs and these delays have been found to make space-to-ground voice communication difficult and inefficient. The objective of ISTAR

Mission 1 is to explore other methods of space-to-ground communication so as not to sacrifice operational efficiency. As a secondary objective, the results will be compared with the results of RATS, NEEMO, and Pavilion Lake field tests.

Additional exploration-related studies will be performed on the ISS during Increment 31/32, but are not sponsored by ISTAR. These include Synchronized Position Hold Engage Reorient Experimental Satellites (SPHERES) free-flyer simulated extravehicular inspection, Robonaut 2 simulated extravehicular routine and emergency operations, and exploration-related HRP studies being performed on the ISS. The focus of these experiments will be on gathering lessons-learned for exploration-risk mitigation.

C.   ISTAR MISSION 2

ISTAR Mission 2, a continuation of ISTAR Mission 1, coincides with ISS Increments 33/34. Lessons-learned during the ISTAR Mission 1 investigation will be incorporated into the ISTAR Mission 2 study. Additional autonomous procedures will be performed, and additional variables or different countermeasures may be added to the communication delay countermeasures study, such as performing the test using more than one crewmember at the same time and/or inserting a time delay. Additional exploration-related studies that will be performed on ISS during these increments that are not sponsored by ISTAR include SPHERES, Robonaut 2, ISS Crew Control of Surface Telerobots, Radiation Environment Monitor, Microbial Growth and Control in Space Suit Assembly (SSA) Gear, and several exploration-related HRP studies. The focus of these experiments will also be on gathering lessons-learned for exploration-risk mitigation.

D.   ISTAR MISSION 3

The HRP study of the impact of communication delay on flight crew performance is the primary focus of ISTAR Mission 3. This study, starting in Increment 36, will determine whether the communication delays likely to be experienced on a long-duration mission to an asteroid or to Mars will result in clinically or operationally significant decrements in crew behavior and performance. The test will validate ground test findings and determine and evaluate 1) risks to flight crew behavioral health and performance, 2) risk of performance decrements due to inadequate cooperation, coordination, communication, and psychosocial adaptation within a team, 3) risk of psychiatric disorders, and 4) risk of adverse behavioral conditions. The crew will use the countermeasures and autonomous procedures developed for ISTAR Missions 1 and 2 in this study.

Additional exploration-related studies that will be performed on the ISS during this increment that are not sponsored by ISTAR include ISS Crew Control of Surface Telerobots, Radiation Environment Monitor, Quantification of In-flight Physical Changes–Anthropometry and Neutral Body Posture

(NBP), Microbial Growth and Control in Space Suit Assembly (SSA) Gear, and exploration-related HRP studies.

VIII. ISS MARS ANALOG MISSION

In February 2012 the Associate Administrator for HEOMD at NASA Headquarters challenged ISTAR to perform its first Mars-mission simulation on the ISS before 2016. The main goal of this ISS Mars Analog mission is to address key exploration technology and operational concept gaps before conducting human exploration missions beyond LEO. Findings from this ISS Mars Analog mission will contribute to the development of a set of design criteria for spaceflight and support systems that enable safe and affordable human exploration missions, in particular to NEAs and Mars. Discussions have begun on possible approaches to meeting this challenge. The following subsections provide a description of one possible approach to conducting such a Mars-mission simulation on the ISS.

A. MISSION OBJECTIVES (NOTIONAL)

1. Conduct Mars exploration mission launch, transit, and landing transition phases as realistically as possible within ISS operational constraints.

2. Understand the highest risks during long-duration exploration missions and learn how to mitigate them:
    (a) Understand how to mitigate risks to long-duration crew health (NRC exploration priority);
    (b) Understand how to improve exploration environmental control and life support systems (NRC exploration priority);
    (c) Understand how to mitigate risks of the deep space habitat during exploration missions;
    (d) Conduct long-term system operations [Design Reference Architecture (DRA) 5.0 key driving requirement/challenge];
    (e) Understand how to mitigate risks of operations under time-delayed communication (HAT exploration mission risk);
    (f) Gain insight into how best to plan roles and responsibilities between flight and ground for long-duration crewed missions;
    (g) Identify the critical mission preparation processes, including mission management team functions, flight design, crew training, flight procedure development, flight software needs, and certification of flight readiness, that are unique to long-duration exploration missions, including missions to Mars.
    (h) Inform customers and stakeholders of ISS analog mission results and lessons-learned.

3. Collaborate with NASA's International Partners to develop an integrated strategy for conducting joint exploration missions, including roles and responsibilities and the management model for a Mars mission.

4. Work with the International Space Exploration Coordination Group (ISECG) and HAT to ensure that the latest version of the Mars design reference mission is available, and that exploration risks and technology/capability gaps are addressed to the greatest extent possible.

5. Demonstrate and validate exploration technologies and operations concepts developed by HEOMD's Advanced Exploration Systems (AES) Division and NASA's Office of Chief Technologist (OCT) to the greatest extent possible.

6. Collaborate and synergize with NASA Headquarters' Science Mission Directorate (SMD) to infuse flight-proven robotic exploration capabilities and projected science operations to the extent possible.

B.   MISSION LEVEL 1 REQUIREMENTS (NOTIONAL)

MR1: The ISS Mars Analog mission shall mitigate the impact to ongoing nominal ISS onboard operations.

MR2: Eighty percent of ISS Mars Analog mission flight-crew activities shall support planned ISS system/experiment activities.

MR3: The ISS Mars Analog mission shall be conducted by three or more flight crewmembers.

MR4: The ISS Mars Analog mission shall be conducted for a minimum of four (to be resolved, TBR) months.

MR5: Participating flight crewmembers shall interact with only each other during 85% of the on-orbit mission phase.

MR6: If four or fewer flight crewmembers participate in the ISS Mars Analog mission, 85% of the on-orbit mission phase time shall be conducted within an ISS habitable volume of 22.5 m^3/crewmember or less.

MR7: A separate ISS Mars Analog mission ground control team shall control the ISS Mars Analog mission and report activities and simulation progress to the normal ISS ground control team.

MR8: The ISS Mars Analog mission shall be conducted with a communications delay that varies by distance from Earth and corresponds to the delay expected during an Earth-to-Mars transit mission.

MR9: The ISS Mars Analog mission communications delay shall apply to all (TBR) voice, data, and command interaction between the participating ISS Mars Analog mission ground control team and the participating ISS Mars Analog mission flight crew.

MR10: The ISS Mars Analog mission ground control team shall have representation from participating International Partners (IPs).

C. MISSION OPERATIONS CONCEPT (NOTIONAL)

The ISS Mars Analog mission timeline will be based on the latest official design reference architecture for a crewed mission to Mars [currently DRA 5.0 [9] (Fig. 1.8), created by the NASA Headquarters' Mars Architecture Working Group]. Detailed timeline activities will be based on analogous ISS operations, applicable NASA robotic exploration mission activities, and projected unique human exploration mission tasks with a special emphasis on execution of increased crew autonomous operations.

A separate ISS Mars Analog mission ground control team will control the ISS Mars Analog mission and report activities and simulation progress to the normal ISS ground control team so that the ISS ground control team can maintain overall ISS awareness and control. This will allow the ISS mission control team to monitor both ISS and Mars Analog mission systems and activities to ensure ISS safety, while the ISS Mars Analog mission ground control team controls the progress of Analog operations. Separate mission timelines will be used: the ISS mission timeline (which also contains Mars Analog mission activities) for use by the ISS ground control team, and an ISS Mars Analog mission timeline [a filtered version of the ISS mission timeline (TBR)] for use by the Mars Analog control team to govern the activities of the ISS Mars Analog mission.

At a minimum, simulation planning will include the Mars-mission phases of Earth launch/ascent (item 7 in Fig. 1.8), on-orbit (i.e., Mars transit) operations (item 8), Mars entry (represented by Earth entry) (item 9), and Mars-gravity

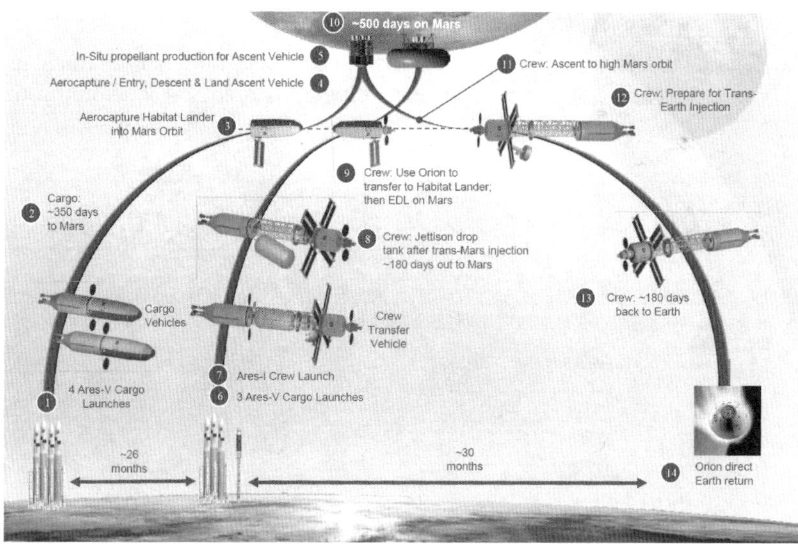

Fig. 1.8   Mars DRA 5.0 mission profile: nuclear thermal rocket (NTR) option.

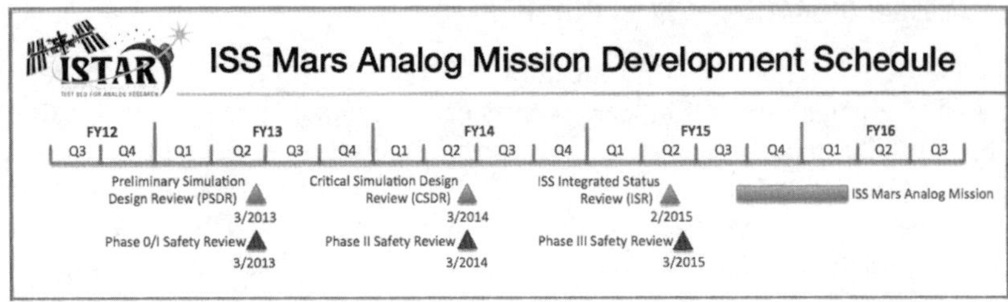

Fig. 1.9   Representative (notional) ISS Mars Analog mission development schedule.

adaptation (represented by Earth-gravity adaptation) on the Mars surface (e.g., the first three weeks of item 10).

D.   MISSION DEVELOPMENT SCHEDULE AND SIMULATED MARS-TRANSIT/ARRIVAL TIMELINE (NOTIONAL)

Figure 1.9 presents a possible ISS Mars Analog mission development schedule, and Fig. 1.10 shows a timeline of typical crew activities during the launch, transit to and arrival at Mars, and Mars-gravity adaptation periods for a crewed mission to Mars. Where possible, the simulation flight crew will conduct their ISS-experiment and ISS Mars Analog mission activities to simulate the following typical Mars-mission activities:

1. Crew launch, LEO activities, Mars transfer vehicle (MTV) checkout, and Mars-transit injection;

2. General housekeeping, food preparation and meals, equipment maintenance and repair, exercise, personal hygiene/time/recreation, and communication with family and friends;

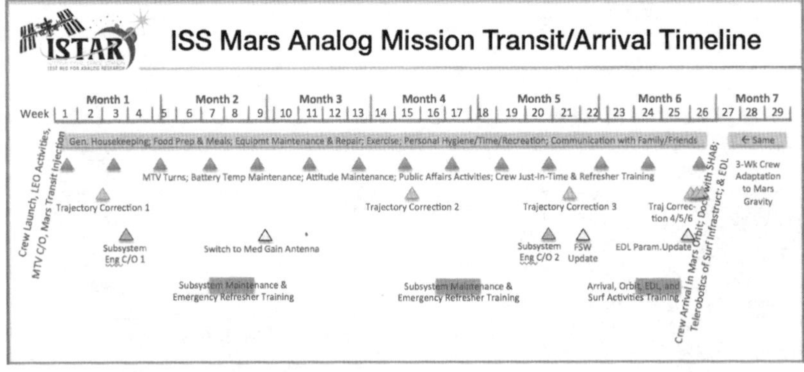

Fig. 1.10   ISS Mars Analog mission transit/arrival timeline (notional).

3. MTV turns, battery temperature maintenance, MTV attitude maintenance, public affairs activities, and crew just-in-time and refresher training;

4. Trajectory correction maneuvers 1–6;

5. Subsystem engineering checkout periods 1 and 2, MTV switch to medium-gain antenna, flight software update, and EDL-parameter update;

6. Crew subsystem maintenance and emergency refresher training;

7. Crew arrival, orbit, EDL, and surface activities training;

8. Crew arrival in Mars orbit, docking with surface habitat (SHAB), telerobotics of surface infrastructure, and EDL;

9. Three weeks of crew adaptation to Mars gravity.

IX. CHALLENGES

Several challenges must be overcome in order for ISTAR to succeed in its plan to assess and recommend critical exploration technologies, conduct ISS Mars Analog simulations, and develop operations concepts that can reduce the risks for crewed missions to exploration destinations. These challenges include the following:

1. *Building a business case for the ISS Mars Analog mission*: It will be a challenge to design an ISS Mars analog mission that will not disturb the conduct of other ISS onboard operations, in particular science-related research. Open discussions are necessary to have a better understanding of this mission's impacts on other ISS activities and to mitigate those impacts. As with the objectives of the ISTAR xDTO candidates selected to fly on the ISS, the objectives of a Mars Analog mission on ISS must be thoroughly vetted to ensure the mission makes a significant contribution to meeting NASA's Strategic Plan and, in particular, exploration goals.

2. *Exploration community buy-ins*: It is absolutely essential to work collaboratively and synergistically with all affected ISS Mars Analog mission planning and implementing organizations (including ISS teams and HRP) and stakeholders (AES and OCT) in the planning of ISTAR missions, including ISS Mars Analog missions. Obtaining adequate support from the involved organizations during mission formulation and planning, and competing with their other priorities and with today's tight budgets, will be a very daunting task. However, it will be critical to maintain their consistent participation via technical interchange meetings and workshops in order to obtain early buy-in and continuing support of ISTAR's objectives and missions.

3. *Support by HRP and NASA's Crew Office*: Conducting a Mars Analog mission on the ISS that contributes to understanding and reducing the spaceflight human health and performance risks of this exploration mission, and that also requires participating flight crewmembers to undergo the simulated

rigors of such a mission, necessitates a continuing close cooperation with HRP and the Flight Crew Office. Only with their detailed understanding and support can this ISS Mars Analog mission succeed.

4. *Resolution burn-down challenges*: During the formulation process for the ISS Mars Analog mission, adequate resources must be allocated to resolve to-be-determined (TBD) and TBR items to ensure that pre-mission preparation meets planned development timelines and that the simulation will meet planned objectives.

X.   CONCLUSION

Through its efforts to encourage use of the ISS as a test platform to reduce exploration risks for crewed NEA or Mars missions, ISTAR has made a good first step towards achieving NASA's goal to "expand efforts to utilize the ISS as a National Laboratory for ... supporting future objectives in human space exploration." Meeting the objectives of ISTAR Missions 1–3 addresses this goal by contributing to understanding the challenges and mitigating the risks of conducting crewed exploration missions. The ISTAR ISS Mars Analog mission, if recognized challenges are overcome, would support the development of design criteria for these crewed exploration missions that would enable NASA to meet its strategic goal of "extending and sustaining human activities across the solar system."

ACRONYMS

AC	Aero-capture
AES	Advanced Exploration Systems Division (or Project) (HEOMD)
A-ISP	Advanced in-space propulsion
CFT	Cryogenic fluid transfer
C/O	Checkout
CPHS	Committee for the Protection of Human Subjects (NASA JSC)
CSM	Cryogenic storage and management
DM	Descent module
DRA	Design reference architecture
DRM	Design reference mission
EDL	Entry/descent/landing (M, Mars; E, Earth)
ENV	Environmental
EVA	Extravehicular activity
FSW	Flight software
HAT	Human Spaceflight Architecture Team
HEOMD	Human Exploration and Operations Mission Directorate (NASA HQ)
HQ	Headquarters
HRP	Human Research Program (NASA JSC)

IP	International partner
IPT	Integrated Product Team
ISECG	International Space Exploration Coordination Group
ISP	In-space propulsion
ISRU	In situ resource utilization
ISS	International Space Station
ISTAR	ISS Testbed for Analog Research
IVA	Intravehicular activity
JSC	Johnson Space Center (NASA)
LEO	Low Earth orbit
LV	Launch vehicle
MAV	Mars ascent vehicle
MCC	Mission Control Center (at JSC)
MDAV	Mars descent/ascent vehicle
MMOD	Micro-meteroid orbital debris
MTV	Mars transfer vehicle
NBL	Neutral Buoyancy Laboratory (JSC)
NBP	Neutral body posture
NEA	Near Earth asteroid
NEEMO	NASA Extreme Environment Mission Operations
NRC	National Research Council
OCT	Office of Chief Technologist (NASA HQ)
PLRP	Pavilion Lake Research Project
RATS	Research and Technology Studies
RPWG	Research Planning Working Group
SAID	Strategic Analysis and Integration Division (NASA HQ)
SM	Service module
SMD	Science Mission Directorate (NASA HQ)
SPHERES	Synchronized position hold engage reorient experimental satellites
SSA	Space suit assembly
SSTF	Space Station Training Facility (JSC)
SHAB	Surface habitat
SW	Software
TBD	To be determined
TBR	To be resolved
TEI	Trans-Earth injection
TMI	Trans-Mars injection
xDTO	Exploration detailed test objective

ACKNOWLEDGMENTS

This research was carried out at the Lyndon B. Johnson Space Center and the Jet Propulsion Laboratory, California Institute of Technology, under a contract with the National Aeronautics and Space Administration.

REFERENCES

[1] "2011 NASA Strategic Plan," National Aeronautics and Space Administration
     Headquarters, NP-2011-01-699-HQ, Washington, DC, Feb. 2011, p. 5.
[2] "Analog Missions and Field Tests," National Aeronautics and Space Administration,
     NASAfacts NF-2011-04-534-HQ, 2011.
[3] "NASA: Significant Challenges Remain for Access, Use, and Sustainment of the
     International Space Station," U.S. Government Accountability Office (GAO), Report
     GAO-12-587T, 28 March 2012.
[4] Culbert, C., "Human Space Flight Architecture Team (HAT) Overview," Briefing
     to the Global Exploration Roadmap (GER) Workshop. http://www.nasa.gov/
     exploration/about/isecg/ger-workshop.html, Nov. 2011.
[5] Stegemoeller, C., "International Space Station Mars Analog Update," Briefing to
     NASA Advisory Council, 2 Aug. 2011, slides 7 and 8.
[6] Charles, J. B., "Preliminary Planning for ISS as Analog for Mars Transit," Briefing to
     Future In-Space Operations Colloquium, 29 June 2011, slide 3.
[7] "Human Research Program Requirements Document," Human Research Program,
     National Aeronautics and Space Administration, HRP-47052 Rev E, May 2011.
[8] "Restoring NASA's Technological Edge and Paving the Way for a New Era in Space,"
     Aeronautics and Space Engineering Board, National Research Council Report In
     Brief, Jan. 2012, p. 3.
[9] "Human Exploration of Mars: Design Reference Architecture 5.0," Mars Architecture
     Steering Group, National Aeronautics and Space Administration,
     NASA-SP-2009-566, July 2009.

CHAPTER 2

Reengineering the Mission Operations System for the Prime and Extended Mission

Joseph C. Hunt, Jr.[*] **and Leo Y. Cheng**[†]
Jet Propulsion Laboratory, Pasadena, California

I. INTRODUCTION

The Spitzer Space Telescope is the last of NASA's Great Observatories. Orbiting the Sun in an Earth trailing orbit, the space observatory produced images of subjects ranging from extrasolar planets to galaxies at the edge of our universe. Launched in August 2003, Spitzer is cryogenically cooled in a superfluid helium bath, allowing the primary mirror to operate at temperatures from 5.6 K to 12 K at infrared wavelengths. Spitzer's suite of instruments includes an infrared array camera (IRAC) to capture infrared light at wavelengths of 3.6 μm, 4.5 μm, 5.8 μm, and 8.0 μm, a multiband infrared photometer (MIPS) with bands at 24 μm, 70 μm, and 160 μm wavelengths, and an infrared spectrometer (IRS) with bands at 5.2–14.5 μm, 9.9–19.6 μm, 14.0–38.0 μm, and 18.7–37.2 μm wavelengths. Spitzer's primary mission began after a 90 day in-orbit checkout (IOC) and science verification (SV) period. When Spitzer's cryogen depleted in May 2009, a series of calibrations known as IRAC warm instrument characterization (IWIC) determined how one of the three science instruments, IRAC, could operate in the relatively warm temperature of 26 K. The characterization period determined IRAC could continue to operate at the warmer temperatures in two of the four wavelength bands (3.6 μm and 4.5 μm) [1].

The Jet Propulsion Laboratory (JPL) manages the overall mission to include real-time command, monitoring, and data accountability. The Spitzer Science Center (SSC) at Caltech provides science planning and instrument operations, and engineering operations and support are provided by Lockheed Martin Space Systems in Littleton, Colorado. A notable achievement is Spitzer's observational efficiency of over 90%, even well into the warm mission. This is double the efficiency of any other great observatories, and Spitzer does it with fewer people [2].

---

[*]Deputy Mission Manager/flight Director, Flight Engineering Group.
[†]Flight Control Engineer, Mission Control and Operations Engineering, Jet Propulsion Laboratory.

TABLE 2.1 SPITZER MISSION PHASES

Phase	Start and end dates
Launch	25 Aug. 2003
In-orbit checkout and science verification (IOC/SV)	25 Aug. 2003 to 1 Dec. 2003
Prime (cryogenic) and prime plus mission[a]	1 Dec. 2003 to 2 Dec. 2009
IRAC warm instrument characterization (IWIC)	16 May 2009 to 27 July 2009
Extended (warm) mission[a]	27 July 2009 to 31 Dec. 2012

[a]Reengineered.

As seen in Table 2.1, the Spitzer mission has undergone several mission phases. Although each of these phases posed unique challenges to the mission operations system (MOS), the scope of this chapter will focus only on the prime (cryogenic) mission and the extended (warm) mission.

II. SPITZER MOS

Before we begin our analysis of the reengineering processes in the primary and extended mission phase, we must define the MOS. The Spitzer MOS contains the teams, processes, and procedures required to operate the mission. This is distinct from the ground data system (GDS), which is composed of the hardware and software. The GDS not only includes computers and networks, but also distributed physical facilities such as the mission support areas, the science center, and multimission facilities. Given that people, teams, and processes are necessary to operate the GDS, it is helpful to view the GDS as a subset of the MOS in our discussion of reengineering.

We discuss the MOS reengineering using the framework of the uplink and downlink processes (Fig. 2.1). Uplink processes are the procedures and tools used to develop command products for spacecraft operations and science instrument data return. The observatory operates with preplanned command sequences, developed and uplinked at one-week intervals. These commands can take the form of preplanned stored command sequences, modules, or libraries, or they can be built, radiated, and executed in real time. The prime users of the uplink process are the science users. However, the Observatory Engineering Team (OET) is also a user, as this team is responsible for the overall health and safety at the observatory.

The downlink process begins with the Deep Space Network (DSN) receiving data downlinked by the observatory. The data are then routed to various destinations, such as navigation, science, spacecraft and instrument engineering, and real-time mission control. The function of the downlink process is not simply the return of data collected on the spacecraft, but also the validation of received data against what was planned in the uplink process.

Although the observatory is the central element in both uplink and downlink processes, according to Fig. 2.1, another function, termed *packet acknowledgment*, is also common to both processes. The packet acknowledgment process validates the receipt of science and engineering data collected by the observatory on the ground. To build commands to retransmit, or delete to free up space on the mass memory card (MMC), each packet of science data must be acknowledged and validated as being received. Therefore, packet acknowledgment is a key function in Spitzer operations. More information on Spitzer's packet acknowledgment process can be found in [3].

III. NEED FOR REENGINEERING

A. DRIVING FACTORS

There are four driving factors for reengineering. First, changes in mission capability such as Spitzer's loss of cryogen has redefined the science objectives from operating with three infrared sensors to one. Second, as NASA missions transition

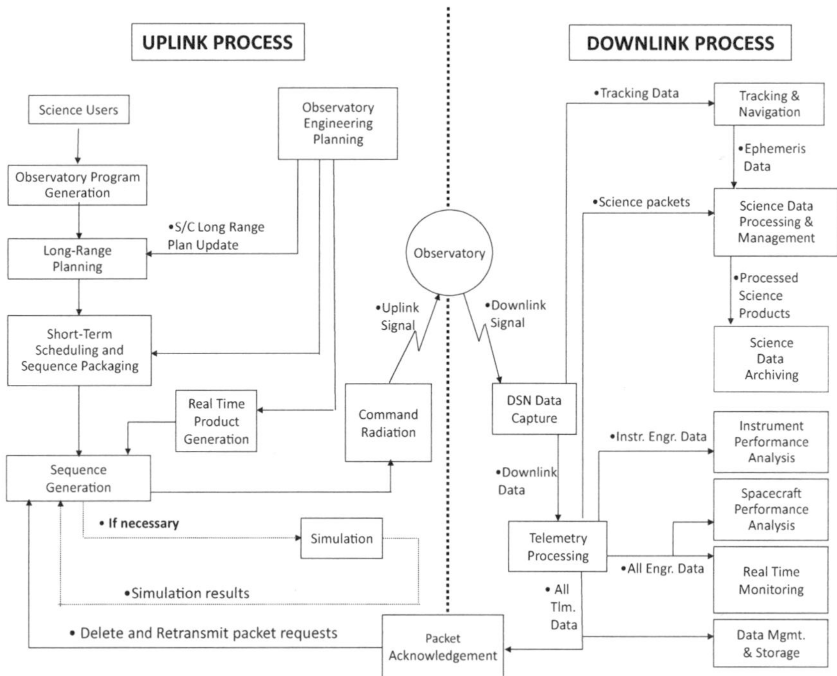

Fig. 2.1   Illustration of MOS elements using the framework of uplink and downlink processes.

from prime to extended status, most will undergo a reduction in the funding profile, so reengineering will become necessary for optimization of the MOS while resources decline. Third, missions with a long life cycle should take advantage of technological advances occurring outside the space industry. A good example may be drawn from the rapid increase in smartphone usage in our daily lives. These devices not only improve communications, but also have the computing capability of a desktop computer, but in the palm of one's hand. Finally, as the MOS progresses into a steady state after launch, lessons-learned should be incorporated to improve operational efficiency. In fact, reengineering can be thought of as a method to eliminate the unforeseen design inefficiencies that are often revealed as the MOS matures during the operations phase of the mission life cycle.

TABLE 2.2   REENGINEERED MOS ELEMENTS

MOS elements	Prime (cryogenic) mission	Primary plus	Extended
Sequence schedule and review	Paper schedule Email and fax	Automated web calendar Discussion threading with traceability/ status update (sequence tracker)	Ingesting from other databases Electronic approval
Planning products	Nonoptimized data collection	Data volume based on predicts (MMC prediction tool)	Antenna elevation angle
Uplink summary	Hard copies	Electronic forms/ approval	Editing capability
Telecom link margin	Unrestricted (34–70 m antennas)	Extrapolated telecom link Margin analysis Eliminated data dropouts due to one-way/two-way mode changes	Antenna scheduled to optimize link margin
Packet acknowledgment	Nominal PAP	No change	Express PAP
Duty roster	Laminated cards	Web-based roster with electronic notification	Smartphone interface
Workforce	Dedicated teams	Limited cross training	Multirolled staff

B. SPITZER'S REENGINEERING PATH

For the Spitzer Mission, reengineering is an evolutionary process driven by all four factors discussed in the previous section. During the primary cryogenic mission, we instituted a modest reengineering effort. The "primary plus" phase of reengineering evolved based on the steady state of the predefined MOS. Furthermore, an improved understanding of the coupling between uplink and downlink process (specifically the telecom link margin) and the MMC data volume allocation formed the basis for reengineering in the primary plus phase and continues to this day.

The loss of cryogen and the reduction in the number of science instruments marked the transition from the prime mission to the extended mission. This drove a second MOS reengineering effort in the extended mission. Also, because of Spitzer's Earth trailing orbit, the spacecraft to Earth distance gradually increases by 0.1 Astronomical Unit (AU) per year. This led to our strategy of maximizing the use of ground antenna resources. Table 2.2 summarizes each of the MOS elements affected by both the primary plus and extended mission reengineering efforts.

IV. REENGINEERING IN PRIME MISSION AND EXTENDED MISSION
A. UPLINK PROCESS

The sequence life cycle from initiation to execution onboard the observatory is a 30-day process that includes managing five sequences in various development stages across a five-day period (standard working week). Sequence scheduling for science observations begins the development phase. This phase consists of a calendar-driven timeline for tracking activities to include products, reviews, and approvals associated with command sequences. These activities are all designed to support uplink accountability.

1. SEQUENCE SCHEDULE AND REVIEW

At the start of the primary mission, the MOS tools used for sequence scheduling and review consisted of paper schedules, and e-mail and fax-based communications. As the mission progressed, Web-based communication tools were introduced to JPL, and the MOS was reengineered to take advantage of these tools. The tool Spitzer developed during the primary plus reengineering phase is known as the Sequence Tracker (Fig. 2.2).

The Sequence Tracker is a Web-based tool that provides a calendar view of deliverables and events associated with the progression from initiation to execution of a sequence activity. The tool serves as a central location for use by project members, with sequence product interfaces to post statuses as well as track a given sequence-related activity. With a click of a mouse, the Web-based calendar expands to show product delivery milestones on a given day. Clicking

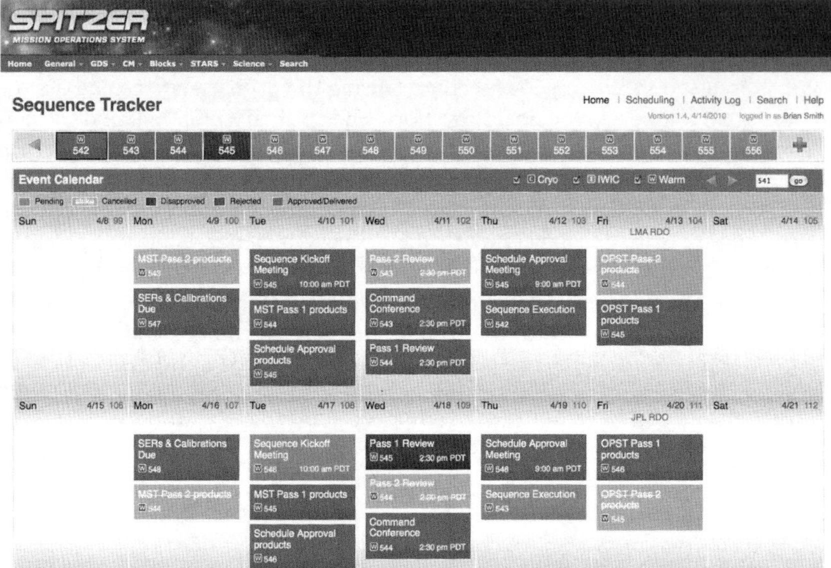

Fig. 2.2   The Spitzer sequence tracker.

on the delivery date reveals the familiar folder navigation window pop-up, which allows a file to be loaded. Moreover, threaded discussion comment fields below the delivery link allow for open collaborative discussions associated with the delivery. This eliminates disjointed e-mail discussions and the possibility of lost feedback because someone was left out of an e-mail distribution list, while adding traceability for revisions. The addition of the Sequence Tracker in primary plus has improved the level of efficiency in the review process, supported better collaboration for more remote partners, and improved searches for archived material. In the extended mission reengineering period we upgraded the Sequence Tracker with additional automation, including taking in information from other databases and electronic approvals of delivered sequence products.

2.   PLANNING PRODUCTS

In primary plus, analysis of the overall use of the MMC suggested the possibility of exceeding the capacity of the MMC after a missed downlink. The MOS was then reengineered to include the concept of operating the MMC with a single fault tolerance. A single fault is either the failure to receive data transmitted by the spacecraft or the failure to send commands to free the previous data from MMC storage. To satisfy the single fault tolerance criterion, we calculated data volume based on planned observations and engineering activities and mapped this into the antenna track allocations. This tool is called the MMC Prediction tool.

Declining telecom link margin in the extended mission created a need for planning tools that incorporate telecom performance. One tool, the Antenna Elevation Angle Predictor, was developed to help evaluate the supportable data rate based on the telecom link margin for a given antenna configuration. These antenna elevation data further aid the science planning process by ensuring there is sufficient telecom margin to support the planned downlink rate. The coupling between uplink planning process and downlink telecom margin will continue to be assessed as the observatory's orbit takes it further away from the Earth. Finally, the Antenna Elevation Angle Predictor, combined with accurate data volume predictions, give Spitzer the tools to effectively produce highly efficient sequences in the extended mission.

3. UPLINK SUMMARY

When command files are approved for uplink, they are summarized in a form known as the "uplink summary". This form is used at the command conference when approving the contents and instructions for a given uplink session. Before reengineering, a hard copy of the uplink summary was distributed, and when remote teams' signatures were required, it was transmitted via fax. After all required signatures were collected, this was faxed back to JPL for final approval. However, refaxing can sometimes result in hard-to-read copies, not to mention introducing delays while transmitting the form. After the success of Web-based tools in primary plus, the paper uplink summary was reengineered to become an electronic product. Not only does this eliminate the need for hard-to-read faxes, but it also incorporates the features of electronic signatures, traceability, and electronic archiving.

During the extended mission, one issue that arose was the problem of last-minute revisions to the uplink summary that did not impact the validity of the command itself, but instead related to the directions needed for implementation. These types of change did not require another review and signature cycle, so capability was added to allow direct editing of electronic uplink summaries during the command conference. Other changes to the uplink summary include 1) appending an automatic revision number to the uplink summary, 2) changing the status of previous versions, and 3) generating a change log for traceability. With the implementation of a Web-based uplink summary tool, hundreds if not thousands of paper products, which would require manual manipulation for retrieval, are now replaced by an electronic search facility, thereby supporting rapid response during mission operations.

B. THE DOWNLINK PROCESS

At the start of the primary mission, constraints such as antenna tracking coverage from the DSN and its effect on the telecom link margin were not an issue. Spitzer used 34 m antennas for tracking, and our maximum downlink data rate was

2.2 Mbps. During the primary plus reengineering effort, the MOS procedures were upgraded, allowing for an extension of the primary mission. This included plans to use a more diverse tracking coverage profile, for example using a 70 m antenna, and combinations of 34 m antennas arrayed to maintain higher data rates over longer durations.

The increasing spacecraft to Earth distance also reduced the telecom link margin, resulting in the need for the MOS to increase the efficiency of the use of ground antennas. One way to increase efficiency in a given downlink is to eliminate data outages during the lockup of the telemetry signal caused by one-way to two-way frequency transitions. This was done by timing the uplink signal such that the ground antenna acquired the spacecraft downlink in two-way mode, therefore eliminating the one-way to two-way transition altogether. Another increase in downlink efficiency arose from a decision to skip the "dial tone" sent by the spacecraft during acquisition of signal. The dial tone is present to allow the DSN to achieve lock, but our spacecraft has high enough data rates that the receivers lock instantaneously. Accordingly, we now have the option of sending a real-time command to initiate the early playback of science data, using the time previously allocated to the dial tone for additional science playback.

C. PACKET ACKNOWLEDGMENT

We have discussed the coupling of the uplink process to the downlink process driving the need for optimization and reengineering. However, central to the optimization of data acquisition and return is the Packet Acknowledgement Process (PAP). As illustrated in Fig. 2.1, PAP shares functions in both the uplink and downlink processes. To further explain uplink/downlink coupling, PAP can be broken down into the following steps:

1. Verify the successful downlink of science and engineering data collected by the observatory.
2. Determine if there are any missed or corrupted data packets.
3. Build commands to retransmit any missed or corrupted packets.
4. Free up space on the MMC by generating commands to delete data that have been successfully downlinked.

Steps 1 and 2 are part of the downlink process and data accountability, while steps 3 and 4 are part of the uplink and planning process. The routine execution of PAP after each downlink leads to the successful management of the MMC, which is crucial to the overall science objectives.

Beginning in primary plus, improvements to the PAP process with a view to better support single fault tolerance were investigated. The outcome was the development of an additional packet acknowledgment step before the complete data set was received. This new PAP process is called the "Express PAP". Express PAP is almost identical to Nominal PAP, except it is performed in real time rather than after the completion of a downlink. Express PAP provides for the real-time validation and deletion of a portion of received data from the MMC.

Furthermore, during the prime and beginning of prime plus, the physical ground data network limited Express PAP to a performance level ranging from 3% to 5% of what was planned. Later, during the transition from prime plus to the warm mission phase, an increase in bandwidth based on improvements in the ground network enhanced the performance further into the range 5–15%. Now, several years into the warm mission, the Nominal PAP and Express PAP combination provides a powerful toolset that is tolerant to single faults, while mitigating possible science loss and supporting the recovery of lost or degraded performance from DSN antennas.

V. HUMAN ELEMENTS

Human factors are always dynamic in any MOS, especially one containing real-time operations. Communication and coordination become key functions in multiteam environments operating in different facilities. Moreover, workforce and staffing levels can change during the mission life cycle. Improvements to the MOS must address human and team interactions that evolves with the mission life cycle.

A. DUTY ROSTER NOTIFICATION SYSTEM

There are multiple types of events during mission operations that require notification of support personnel. Mission operations both in Flight and Non-Flight environments consist of multiple layers of personnel supporting operations on different work shifts in both local and remote locations, such as the Caltech campus, Lockheed Martin, and Ball Aerospace in Colorado, and universities around the country. Spitzer therefore developed the Duty Roster Notification System to address the following problems:

1.   Who are the primary and alternate points of contact?

2.   What are their roles and responsibilities?

3.   What is their primary contact (by phone, text message, or e-mail)?

The Duty Roster Notification System provides a centralized service consolidating personnel contact information for notification. The notification system provides rapid notification to a group of roles within the roster using a variety of media devices. A text message with the time, date, and problem description is issued to alert personnel. Through a Web-based interface, the user can provide real-time updates for personnel contact and notification information. The display of information is controlled based on user privileges. Figure 2.3 illustrates how the Duty Roster Notification works.

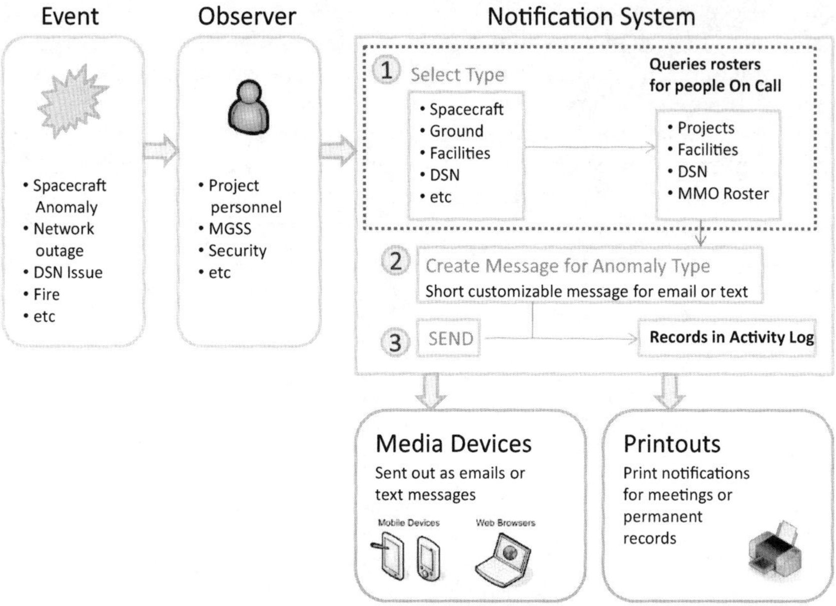

Fig. 2.3   The Spitzer Duty Roster Notification.

After its implementation in 2004, the Duty Roster Notification System helped Spitzer communicate and coordinate numerous mission-critical activities and anomalies. The elimination of laminated contact cards and paper contact lists not only reduces clutter, but also minimizes possible confusion and repetitive updates. Moreover, the Duty Roster delegates the update of contact information to the team or individual. In 2012, we updated the Duty Roster to include a mobile interface for smartphones. The success of Spitzer's Duty Roster has created a demand for similar tools for other JPL missions and services. In fact, the same team that developed Spitzer's Duty Roster has now created a version serving multiple missions and services at JPL. Future implementations of the multimission duty roster could include external missions with JPL services.

B.   CHANGES IN WORKFORCE PROFILE

Over the course of mission phases, either from higher resource demand or attrition, the staffing profile often evolves. Under these circumstances, one of the biggest challenges for the life cycle of mission operations is retaining a heritage knowledge base. The goal is to preserve heritage experience within the processes of the MOS. Spitzer achieved this by leveraging the knowledge of experienced team leads in the reengineering improvements that provided new and enhanced tools and procedures. Furthermore, co-location of some teams within the MOS proved to be a catalyst for the exchange of ideas. This was especially critical in the development of the Express PAP process as described Sec. IV.C.

VI. SUMMARY

A visual representation summarizing the reengineering processes described in this chapter is shown in Fig. 2.4. This figure superimposes Spitzer's mission phases and the reengineering efforts described in this chapter with the workforce profile. As the MOS design matured, confidence in the system allowed us to investigate modest reengineering steps. Because of the complexity of coupling between the uplink and downlink processes, unintentional changes could occur if too many changes were made at one time, or if changes were made too quickly. It is important to note that the MOS reengineering effort is occurring in parallel with nominal operations.

VII. CONCLUSION

The success of the Spitzer Space Telescope mission is a result of the systems engineering standards set out by NASA, as documented by the "NASA Systems Engineering Handbook" [4]. Focused mostly on the design and development phases of the mission life cycle (NASA phases A through D), the handbook dedicates only

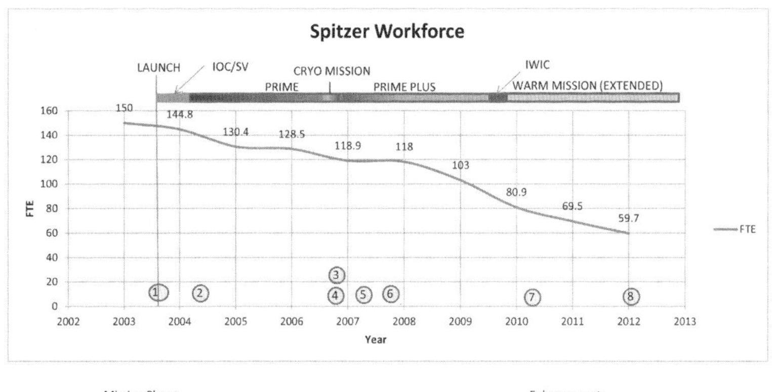

Mission Phase			Enhancements
25-08-2003		Launch (Vehicle Delta II 7920H)	1 - 2003 Electronic Contact List (Duty Roster)
25-08-2003	01-12-2003	IOC/SV	2 - 2004 Duty Roster Phase 1 (contact Info and notification via local browser)
			3 - 2006 MMC Prediction Tools (9/1/2006)
02-12-2003	15-05-2009	Prime Cryo Ops	4 - 2006 Anomaly Notification (added to Duty Roster - Sept 2006)
16-05-2009	27-07-2009	IWIC	5 - 2007 Express PAP (2/13/2007)
27-07-2009		Warm Mission	6 - 2007 Sequence Tracker (12/10/2007)
			7 - 2010 Elevation Angles - 4/14/2010
			8 - 2012 Duty Roster Phase 2 (Mobil App)

Fig. 2.4 Spitzer reengineering mapped with workforce.

33% to operations and end-of-life processes (NASA phases E and F). Our analysis demonstrates that the mission operations phase faces challenges that are often different than those encountered during design and development, but no less difficult. One way to improve the operations phase in the Handbook is the addition of a formal reengineering evaluation. Furthermore, even though extended mission are discretionary, the robustness of recent NASA missions has allowed operations to exceed beyond the prime mission phase. We therefore propose the addition of an optional "extended mission" phase between phase E (operations) and phase F (closeout).

ACRONYMS

AU	Astronomical unit
DSN	Deep space network
GDS	Ground data system
IOC	In-orbit checkout
IRAC	Infrared array camera
IRS	Infrared spectrometer
JPL	Jet Propulsion Laboratory
MGSS	Multimission ground systems and services
MIPS	Multiband infrared photometer
MMC	Mass memory card
MMO	Mission Management Office
MOS	Mission operations system
NASA	National Aeronautics and Space Administration
OET	Observatory Engineering Team
PAP	Packet Acknowledgement Process
SSC	Spitzer Science Center
SV	Science verification

ACKNOWLEDGMENTS

The research described in this chapter was carried out at the Jet Propulsion Laboratory, California Institute of Technology, under a contract with the National Aeronautics and Space Administration. The authors would like to thank the following people who helped us along the way (in alphabetical order): Eric Gillingham, Brian Smith, and Kennis Stowers.

REFERENCES

[1] Mahoney, W. A., Garcia, L. J., Hunt, J., Jr., McElroy, D. B., Mannings, V., Mittman, D. S., O'Linger, J. C., Sarrel, M., and Scire, E., "Spitzer Warm Mission Transition

and Operations," *Proceedings of the SPIE Observatory Operations: Strategies, Processes, and Systems III*, edited by Silva, D. R., Peck, A. B., and Soifer, B. T., Vol. 7737, SPIE, Bellingham, WA, 2010, 77371W.

[2]   Scott, C. P., Kahr, B., and Sarrel, M. A., "Spitzer Observatory Operations – Increasing Efficiency in Mission Operations," *Proceedings of the SPIE Observatory Operations: Strategies, Processes, and Systems*, edited by Silva, D. R., and Doxsey, R. E., Vol. 6270, SPIE, Bellingham, WA, 2010, 62701B.

[3]   Sarrel, M. A., Carrion, C., and Hunt, J. C., "Managing the On-Board Data Storage, Acknowledgement and Retransmission System for Spitzer," *Proceedings of the AIAA 9th International Conference on Space Operations (SpaceOps)*, AIAA, Washington, DC, 2006, AIAA 2006-5564.

[4]   National Aeronautics and Space Administration, "NASA Systems Engineering Handbook," NASA SP-2007-6105 Rev 1, 2007.

CHAPTER 3

Mission Operations Preparation Environment: A New Approach for the Future

W. Heinen,[*] **S. Reid**[†] **and S. Pearson**[‡]
Rhea System S.A., Wavre, Belgium

I. INTRODUCTION

The Manufacturing and Operations Information System (MOIS) operations preparation toolkit is well established throughout the European space industry and is the standard tool at the European Space Agency/European Space Operations Centre (ESA/ESOC) for the preparation and maintenance of flight operations plan(s), automated procedures, and command sequences. It is used for the entirety of the ongoing missions.

A. MOIS TOOLSET

RHEA System's MOIS is an integrated set of software tools for spacecraft mission preparation and execution. MOIS is designed to provide a practical and flexible method for handling all aspects of a mission. A key feature of MOIS is its independence from individual mission control infrastructures. The resulting flexibility in adapting to any operating environment has led to MOIS becoming the preeminent procedure development, execution, and automation platform for spacecraft missions in Europe. Indeed, it is used as standard by the ESA. The main tools that constitute MOIS are the following:

1. *Writer* enables the creation and development of procedures or timelines that may be viewed and worked on both in a linearized form (showing steps and associated statements) and as a flow-chart graphical display of the step structure (via Flowcharter).

2. *Flowcharter* enables the creation, editing, and display of procedure structures via a graphical flow chart.

---

[*]MOIS Product Manager; w.heinen@rheagroup.com.
[†]Chief Technical Officer; s.reid@rheagroup.com.
[‡]System Engineer; s.pearson@rheagroup.com.

3. *DB Editor* controls the creation, import, or editing of the MOIS operational Satellite Control and Operation System (SCOS) spacecraft database.

4. *Function Editor* provides a user-friendly interface to create and maintain functions and directives for use in procedure writing.

5. *Validator* controls validation testing of a procedure and stores the results. A Test Harness can be used optionally to emulate a control system/simulator when using with Validator to test a procedure.

6. *Scheduler* enables mission planning for satellite operations and station scheduling.

7. *Publisher* allows the production of mission documents that do not fall into the procedure category. It also makes the hard-copy production of procedures and timelines as painless as possible by automating the printing of up to thousands of documents under configuration control.

8. *Reporter* logs the definition, analysis, and solution of problems as they arise, and also performs consistency checking and analysis of sets of procedures.

9. *Library* provides a fully integrated configuration management for mission data, including procedures, schedules, operational databases, and documents, across the full suite of MOIS tools.

B.   GENERAL MODEL FOR MISSION PREPARATION PRODUCTS

The user view of Mission Operations Systems is very often through the prism of the available applications and known data structures, which are usually managed independently. This chapter will show how the lack of a single management system for all mission configuration data as well as the limitations imposed by the available views of these data can be addressed.

The principles presented here are derived from the European Cooperation for Space Standardization (ECSS) E31 Space System Model (SSM) [1]. The SSM (Fig. 3.1) defines an extensible hierarchical view of the space system, where elements such as telecommands and telemetry parameters are associated with nodes (system elements, SEs) representing components of the system such as an onboard instrument. Its objectives are to describe the system more meaningfully, to provide a means for common data transfer between systems, and to permit the isolation, duplication, and replacement of individual branches. This idea is particularly useful in Assembly Integration and Testing (AIT), where the data model can be assembled in parallel with the spacecraft. However, it has several benefits for Operations as will be demonstrated:

1. The E31 definition will be extended to cover all mission configuration data, not just the space system.

2. The spacecraft database will be subdivided into reusable "elements". The model will keep track of these elements and their evolution.

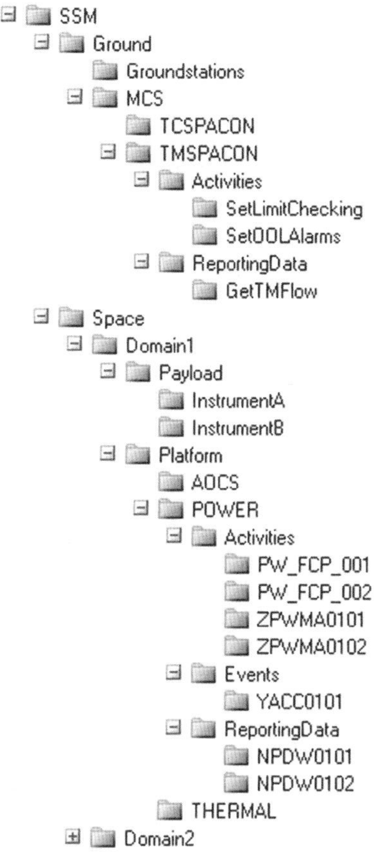

Fig. 3.1 Classification of elements within the SSM.

The intention is *not* to reformat or repackage any data. The resulting model is a view of the system, which, when created and populated, will provide a clearer view of the mission data and enable better management of it.

II. REVISIT THE DATA MODEL

A. SPACECRAFT DATABASE

The SCOS spacecraft database will be taken as a starting point for the description of the revisited data model. It is monolithic in the sense that it configures SCOS for the entire space segment. Databases for other control systems in Europe, such as Columbus Ground System (CGS) and Satellite Information System (SIS) for Open-Center, are mostly similar in this respect. The ground segment configuration is sometimes included or can be specified in a separate database.

The SCOS spacecraft database is derived from the Mission Information Base Interface Control Document (MIB ICD), which specifies a set of files that implicitly map to a set of relationally interrelated tables. These data serve two purposes: to configure SCOS and to describe the space segment in terms of telecommands, telemetry, their packet containers, and associated displays. This relational scheme minimizes duplication of data (shared digital calibration mappings for example) but by definition can have relational dependencies that make it impossible to guarantee a separation between physically separate parts of the system; for example, different onboard instruments could share a calibration curve.

The ECSS-E_ST-70-31C SSM takes a different approach. It envisages a hierarchical breakdown of the space system with telecommands and procedures (both seen as Activities), telemetry (now Reporting Data) and Events such as packet arrival and anomaly signalling, all located in the SE nodes of an SSM hierarchy reflecting the physical breakdown of the space system (Fig. 3.1). This is not just a way of managing complexity; it also facilitates a breakdown of the system that can lead to compartmentalization and therefore less validation and testing.

To realize the potential of an SSM, a mapping of SCOS elements (Telecommands, Telemetry, and their associated calibrations, packets, and monitoring data) must be made to the SE nodes where they belong in a carefully designed mission SSM hierarchy. This hierarchy is then navigated whenever the data are accessed, from the spacecraft database editor, to the procedure editors, to the execution environment.

Both procedure and spacecraft database development will therefore be based on an SSM governed by a hierarchical configuration-controlled structure. Procedures will be located in SE nodes of the tree and will be checked out from these locations in the usual way. When performing spacecraft database updates it will be possible (implicitly at least) to check out MIB elements in any branch of the tree, from individual Telecommands (TCs) or Telemetry (TMs), to whole subsystems, to the entire SSM.

This view facilitates user access control at branch level, not just for the procedures but also for the elements within the spacecraft database. If it were designed so that edits to one branch could be guaranteed not to affect the rest of the model, management of the database could be distributed and there would be no need for all database edits to go through one central authority.

B.   INTEGRATION OF THE MIB EDITOR

SCOS MIB data will not be converted to the detailed E31 form because such a two-way conversion between MIB and E31 formats may not be possible (at least, this has not been demonstrated) and would serve no real purpose. The main SCOS elements must, however, be mapped to locations in the SSM to achieve this system view.

The SCOS spacecraft database editor will need to be aware of this mapping. Elements to be edited should first be identified by their SE location, which is then reserved for editing. Because elements such as TCs and TMs have a single namespace in the MIB (they are table Primary Keys) they could also be searched for by name in the usual way to find their allocated SE.

The database editor would reserve (check out) from the SSM all element references that are related to the current edit in the spacecraft database and are therefore affected by it. This information is obviously available via SQL query, but the relationships could be complicated. It will depend on how tightly coupled the MIB data are (e.g., how many elements share the same calibration data or telecommand parameter definitions).

The MIB defines three broad table categories: Monitoring, Displays, and Commanding. There are many table dependencies, such as Packets on both Monitoring and Commanding (separately), Monitoring and Commanding on calibration (separately), and Displays on Monitoring.

1. ISSUE OF DATA DUPLICATION

The MIB relational scheme allows calibration data to be shared between two TCs or two TMs (but not between a TC and a TM). TC parameter data can be shared, as can limits and range data, but in practise this is not that common. Analog calibration data are rarely shared and often only simple digital data are shared, e.g., $1 = ON$, $0 = OFF$. It may be possible to achieve complete decoupling of these element types without much data duplication. It can be noted that this is the E31 XML schema view (hierarchical and not relational).

It should be possible to place Monitoring (TM) and Commanding (TC) elements at the SE leaf nodes, and Packets and Displays at different levels in the SSM depending on their dependencies (in general, they should be placed at the lowest-level branch that contains all their dependencies).

If, say, a calibration curve or an Out of Limits (OOL) check needs an update, it would be necessary first to locate the TM or TC that uses it. The editor would then have to check out all the TM elements in the SSM nodes that use the same calibration curve. Thus, the editor, while presenting a hierarchical SSM framework, would need to be aware of all the relational interdependencies of the MIB database.

Once the SSM hierarchal structure has been designed for the mission, a drag-and-drop editor will facilitate the location of the MIB elements (TCs, TMs, Packets, and Displays) within it. These references will then maintain their own history, like any other configuration-controlled item.

The database editing experience may not differ significantly from before. The same MIB fields will be available for editing, but on top of this the mapping of TMs, TCs, Packets, and Displays to the SSM will be visible and navigable. Importantly, no elements will be editable unless they have been mapped to the SSM. This is a fundamental requirement of the model. All the procedure editors will reference elements via their SSM references.

At the end of the editing session the spacecraft database will be checked in. At the same time, references within the SE nodes to all updated elements, together with all other elements affected relationally by these edits, will also be checked into the SSM configuration-controlled structure. This will provide a clear record of which branches and SE nodes of the SSM have been affected by each database update and hence what may eventually need to be retested.

2. FINER-GRAINED SPACECRAFT DATABASE ELEMENTS

Telecommands, Telemetry, and Event Packets can be mapped to E32 Activities, Reporting Data, and Events located at the SE nodes. However, it may well prove that the spacecraft database is too relationally coupled to be split up into these separate elements without an unacceptable amount of duplication. It is not just calibration data that can be shared between elements. For example, in the MIB it is possible to share parameter definitions between TCs. An example is the Group Repeater parameter, which is defined once and shared between all TCs with a group repeater (which is, of course, a perfectly valid thing to do in a relational scheme). It is also possible to share parameter limits and ranges between parameters.

A finer-grained division of the spacecraft database data is shown in Fig. 3.2. In addition to the main elements we have the *Parameter* element (TC or TM) and the *Calibration* and *Limits/Ranges* elements (which are both attributes of Parameter).

Dividing up the spacecraft data like this should avoid most, if not all, data duplication. It would then be possible to change a TC or TM definition without

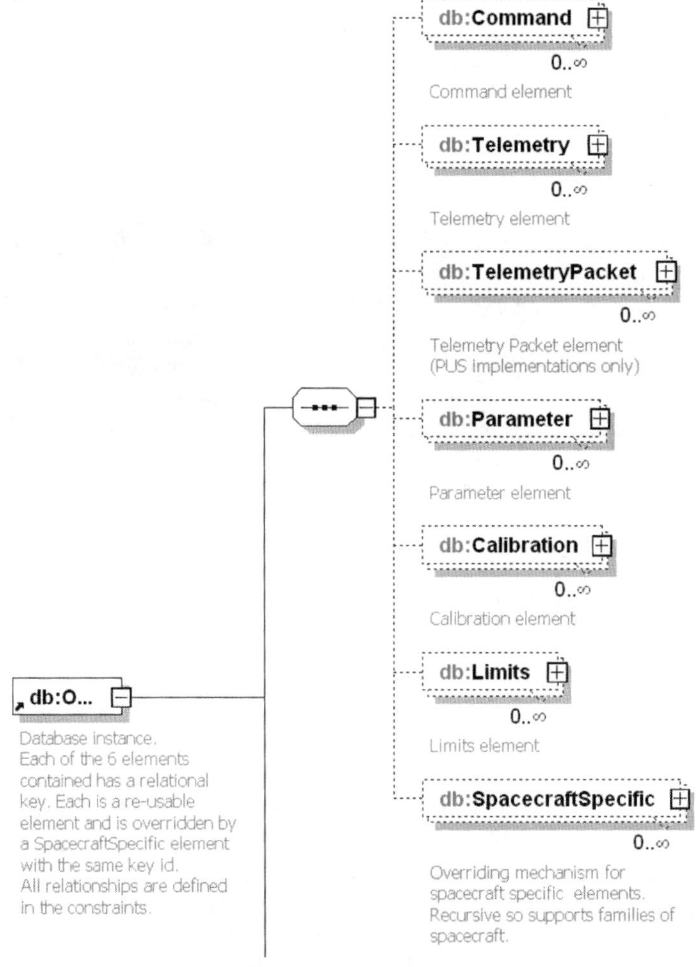

Fig. 3.2   Finer-grained spacecraft database elements.

affecting all its dependent and related data and to update a calibration curve without affecting all the TCs that use it.

The *Parameter*, *Calibration*, and *Limits* elements in the SSM hierarchy are not the same as the TC and TM elements: they never need to be referenced directly and do not map to physical parts of the system. It may therefore be possible to locate them automatically and promote them up the tree when they need to be shared more widely.

Such a categorization would greatly help the management of constellation databases.

C. INTEGRATION OF THE PROCEDURE EDITOR

Procedure editing will follow a similar but less complicated path. Procedures must be located in the SE nodes of the SSM hierarchy, and Activities (TCs or procedures), Reporting data (TM), and Events (e.g., synchronizing on packet arrival or signalled anomalies) will be referenced within procedures via this structure. It will then be possible to generate reports detailing which parts of the SSM have changed since a specified baseline was established.

This is a fundamentally different view, because the spacecraft database and procedures are now a combined dataset. The spacecraft database no longer has to be viewed as a huge interrelated single item, and the system data it contains can be modeled and displayed more clearly and logically. The hierarchical view is kept separate from the spacecraft data via element mapping. This mapping could equally apply to other database formats: for example, an OpenCenter SIS database and a SCOS database used for the same mission could use the same SSM structure. The BepiColombo mission, for example, will use both SIS and SCOS databases. Although the data are equivalent (a SIS to SCOS converter is available) the elements (TCs for example) will be named differently due to SCOS naming limitations and ESOC and Astrium conventions). It may then be useful to have a common system model as a reference.

D. ADDING SCOPE TO THE SSM

The SSM, as defined in E31, has no scope limitations. Despite its hierarchical structure, all its data are accessible everywhere. In particular, all TCs and TM parameters are available to all procedures, and a procedure can be called from any other procedure. This lack of information hiding in different contexts can lead to a lack of clarity and to greater testing needs.

There is a direct parallel in the software domain. Encapsulation and information hiding are basic tenets of object-oriented programming. The principle is that objects do not need to expose all their inner workings, so only a subset of their methods are made public. Having a public interface also means that inner mechanisms can be altered without affecting the rest of the system.

The control procedure execution (CPE) study [3] suggested a simple update to the E31 SSM to achieve this: the addition of a scope (or visibility) attribute that can take the following values (following Java and other languages): 1) Public [can be accessed by any SE (default)]; 2) Private [can only be accessed from the same SE]; 3) Protected [can only be accessed from activities in the same SE or its children (recursively)].

Such a mechanism could provide information hiding to reduce complexity and simplify integration testing, in line with standard object design principles. Figure 3.3 presents an amended branch of the E31 schema with this attribute added to an Activity (which can be a TC or a procedure).

All references in the SSM could be public initially (as at present with the flat-structured MIB). Once it becomes natural to access all database elements and procedures in the context of an SE in the SSM hierarchy, it may be considered operationally useful to reduce the scope and hence visibility of certain elements to procedures written for other SEs, both to reduce unnecessary complexity and to ensure that correct processes are always followed.

For example, if an instrument must be switched on in a certain way with specific checks performed during the process, it may be comforting to know

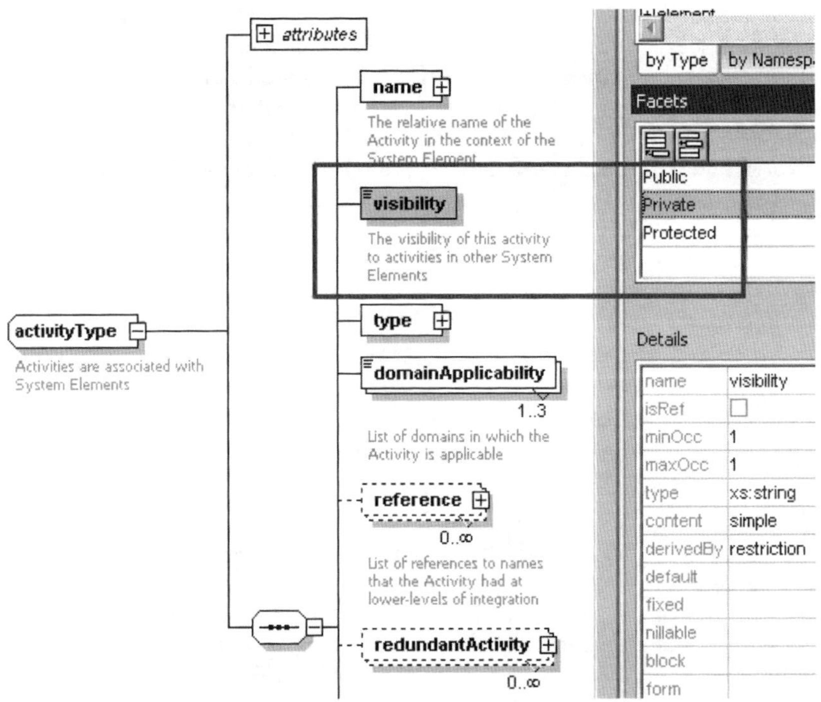

Fig. 3.3 E31 schema branch with activity visibility attribute.

that other procedures are not able to attempt to perform this procedure in another way. TCs that perform this state transition (and are therefore located in the local SE) could then be hidden from procedures written for another SE. Only the procedure (Activity) that performs the switch-on correctly would be exposed publically.

In another example, perhaps only certain state transitions for a particular instrument are legal. Its SE could be designed to expose only the procedures that perform valid mode changes between valid states, preventing, say, a switch from *dump* to *inactive* without going via the *active* state.

III. EXPANDING THE SSM DEFINITION

ECSS-E_ST-70-31C [RD5] is a delivery format for exchanging a data model of the space system including TCs, TM parameters, and procedures (but not display data). The hierarchical model should facilitate the generation of an E31 electronic database (so far by including SSM-referenced data from the MIB and from procedures).

We are now proposing to extend this model to a repository structure for all types of mission data. It should be able to hold any mission data that could benefit from being located in an extensible hierarchical structure constructed specifically for the mission, and subject to configuration control from high-level branches down to the level of individual files, procedures, and MIB element references.

As a structure it should *not* be limited to the space segment. The following would constitute the data that could be located in corresponding SEs of a ground segment model within the expanded SSM: 1) The *Spacecraft Database* elements; 2) The *Ground Configuration Data*; 3) The *Pluto Procedures*; 4) The *Flight Control Procedures* produced by MOIS Writer (note that Writer also produces a simplified Pluto output); 5) The *System Documents* that belong to the Flight Operations Plan, which are written with Publisher; 6) The *Mission Planning Rules*; 7) *Onboard Control Procedures* produced by Writer.

These data items are expressed as SSM Activities, and would naturally be stored in the SSM providing the possibility to write ground-station procedures using the same MOIS editors as for space-segment procedures. This is an important benefit of the E31 SSM's Activity abstraction. Once everything is an Activity it can be called from a procedure (which itself is an Activity) in exactly the same way. Commanding the ground station can be done via procedures in the same way as commanding the spacecraft.

IV. IMPLEMENTATION DETAILS

The objective is to redesign MOIS to create a flexible, open environment in which all types of configuration data can be managed. It will continue to provide the necessary tools for preparation for all types of procedure, but additional editors

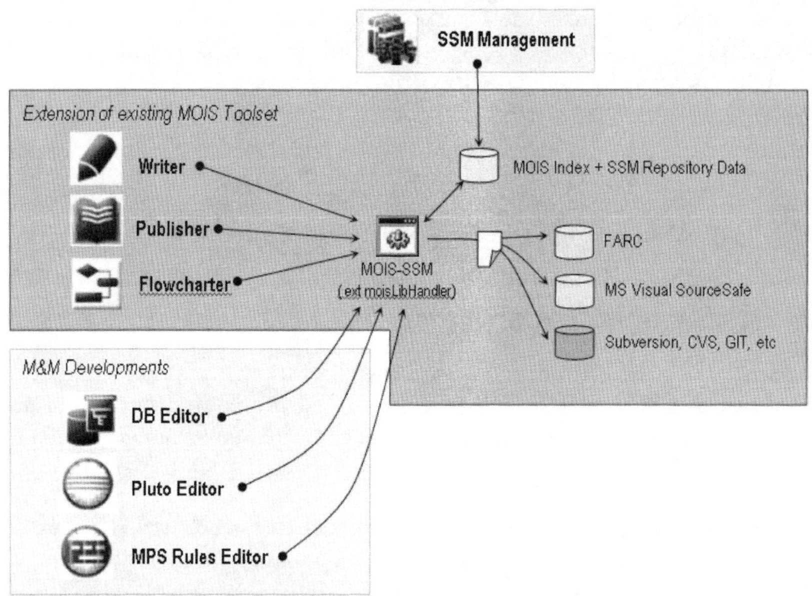

Fig. 3.4 **Diagram showing the breakdown of existing and new components and how their interaction can be envisaged.**

can be integrated and added. New editors for the TM/TC database and Mission Planning rules will be developed within this framework and at the same time the existing toolset will remain an integral part of the next MOIS product issue (Fig. 3.4).

A. SPACECRAFT DATABASE EDITOR (DB EDITOR)

The model can be introduced without necessarily having to rewrite the database editor. A data export has to be done from the database tables into an element-specific dataset. For SCOS, one can adapt the "dat/text file export & import" that is used for entire tables. The SSM data export however will not handle entire tables but just element-specific data. For example, a calibration curve with 10 points will form a dataset of one CAF.dat file with one entry and a CAP.dat file with 10 entries. It is then a matter of importing a set of dat files that are classified somewhere in the branch of the hierarchy to form the database for an editing session of that branch.

Configuration control is ensured by using that hierarchy as the repository structure, and merges and compares can be done very easily outside the specific DB Editor. The end or check-in of an editing session will regenerate the dat files and deltas will be checked in. Whether file reservation is necessary or not is questionable and indeed there are diverging opinions on this.

Reusing a database editor as-is represents a minimal solution. The study, however, intends to provide an enhanced DB Editor that takes full advantage of the model, for system element selection, filtering, check-in, and so on. ESOC have recently developed DABYS as a framework for the management of databases. This provides an environment loosely equivalent to the MS-Access tool, which allows the development of specific database applications. It stores data in a MySQL RDBMS (Relational Database Management System) and provides rudimentary configuration control. The GAIA mission implementation is a candidate for this study.

B. GROUND CONFIGURATION DATA

Ground Configuration data can be directives for the MCS (NIS) [Mission Control System (Network Interface System)] and other configuration parameter settings (e.g., MISC variables). The management of these configuration data are the subject of the Integrated Development an Validation Environment for Operations Automation (IDEA) study [6].

C. PLUTO PROCEDURE EDITOR (PLUTO EDITOR)

This Editor is compliant with the E32 procedure model and based on the established model-view design. Users can create procedures in MOIS Writer as normal, but automated procedures may also be viewed in the PLUTO text editor [2].

D. FLIGHT CONTROL PROCEDURE EDITORS (WRITER, FLOWCHARTER)

The existing MOIS Writer and Flowcharter editors can remain, but the new model offers a new way for accessing and filtering data for the design of the procedure. Similar to the Pluto Editor, the registration into the SSM determines the filter for the available procedure constructs.

E. SYSTEM DOCUMENT EDITOR (PUBLISHER)

Lightweight MSWord procedures and documents can be included in the flight operations plan (FOP) in the same way that Writer procedures are.

F. MISSION PLANNING RULES (RULES EDITOR)

Within the MOIS and Mission Operations Preparation Environment (MORE) study, a language-specific editor will be developed, initially based on the LMP (Language for Mission Planning) rule language. In a similar way to the Pluto Procedure editor, a language-sensitive editor will be developed within this study, using the LMP language specification and accessing the SSM for element insertion and rule checking.

G.   ONBOARD CONTROL PROCEDURES (WRITER AND FLOWCHARTER)

The different nature of onboard control procedures (OBCPs), compared to ground flight control procedures (FCPs), makes it necessary to extend the MOIS procedure preparation tools (Writer and Flowcharter) in a few areas. These are mainly additional statements (Set Telemetry, Send Event, Wait Event), support for OBCP parameters, and support for subprograms (subprocedures and functions) [4].

Apart from these adaptations, the procedure preparation process is the same for FCPs and OBCPs. In particular, the connection to the spacecraft database ensures overall consistency, so the editor can profit from the same additions as the Flight Control Procedure Editors. The MOIS and MORE study provides a complete OBCP solution (chosen target missions are GAIA and Bepi-Colombo) [5]). These types of procedure remain equivalent and are managed in the same way within the SSM.

H.   MANAGING THE SSM (SSM EDITOR)

Finally, the MOIS and MORE outputs will provide the tools needed to build and maintain the SSM [7].

V.   CONCLUSION

The presented new data model is a way forward to cope with the growing complexity of missions in terms of their datasets and in particular the configuration data of the software components that form a mission operations system. Complexity is controlled by classifying all operational and configuration data into one coherent dataset via compartmentalization and specialization. Existing tools will not be made obsolete; if they are not already compatible with the new data model they will be upgraded to interface to it. In either case, they will coexist with the new tools that will be specifically designed within the study to handle the increased complexity and larger data volumes more simply and more effectively.

REFERENCES

[1]   "Space Engineering, Ground Systems & Operations – Monitoring and Control Data Definition," ESA Requirements and Standards Division, ESTEC, The Netherlands, E31: ECSS-E-ST-70-31C, 31 July 2008.

[2]   "Space Engineering Test and Operations Procedure Language," E32: ECSS-E-ST-70-32C, ESA Requirements and Standards Division, ESTEC, The Netherlands, 04 August 2010.

[3]   "CPE Study TN3. ECSS Extensions and Language Conversion," ESA Requirements and Standards Division, ESTEC, The Netherlands, RHEA.CS1097.DOC.03, 2010-08-04.

[4] Heinen, W., Reid, S., and Varadarajulu, S., "Automation Through On-Board
     Control Procedures: Operational Concepts and Tools," ESA Requirements and
     Standards Division, ESTEC, The Netherlands, Spaceops, 2010.
[5] Schwab, A., and zur Borg, W., "OBCPs – an Integrated Part of BepiColombo
     Autonomy and Operations Flexibility," ESA Requirements and Standards Division,
     ESTEC, The Netherlands, Spaceops, 2012.
[6] Pearson, S., Trifin, F., Reid, S., and Heinen, W., "An Integrated Development and
     Validation Environment for Operations Automation," ESA Requirements and
     Standards Division, ESTEC, The Netherlands, Spaceops, 2012.
[7] "Manufacturing and Operations Information Systems (MOIS)," *Rhea Systems*, Sept.
     25, 2010, http://www.rheagroup.com/en/x/28/mois [retrieved May 23, 2012].

CHAPTER 4

The Keys to Successful Extended Missions

David A. Seal* and Emily A. Manor-Chapman[†]
Jet Propulsion Laboratory, California Institute of Technology, Pasadena, California

I. INTRODUCTION

Most NASA missions nearing the end of their nominal mission lifetime are candidates for extension via a structural biennial Senior Review. This process has been in place since the 1990s [Voyager, for example, has undergone 11 senior reviews, and the Mars Exploration Rover (MER) project is now preparing for its eighth], and is now formally applied to both Earth and planetary missions alike following the formation of the Science and Mission Directorate (SMD) of NASA in 2004. To be approved for a mission extension, each project must answer a call for proposals with a detailed description of the scientific benefits of extension, plans for operating the mission, health and operability of space and ground assets, and the costs of continued operations. Also relevant are the plans for continued education and public outreach (E/PO) [1]. Often, projects are supplied with guidelines (financial limits in particular), formally or informally, from their representatives at NASA headquarters, representing the expectations of the customer of the scope of the proposal based on discussions and technical interchange with the project leading up to the Senior Review process. Each proposal is thoroughly reviewed for its scientific merit and feasibility by a multidisciplinary panel whose charter is "to maximize the scientific return from these programs within finite resources" [2]. The most recent Senior Reviews have consolidated the proposal process into a single multimission senior review, the purpose of which is to "provide the best balanced science for the scarce available funding" [3].

Historically, most candidate missions (over 80%) are approved for some level of extension. However, some missions have been cancelled, and the level of support given to extensions depends on a number of factors, all of which are considered important but whose influence varies depending on the mission and circumstances. The Senior Review process must be taken as seriously as proposals for new missions, and a well-written proposal armed with strong arguments is crucial to a successful extension, regardless of the perceived success of the project during its prime mission. In many areas, the groundwork for approving an extension is laid during prime mission.

---

*Supervisor, Mission Engineering & Planning Group, Systems Engineering Section.
[†]Lead Mission Planner, Cassini Mission, Jet Propulsion Laboratory.

II. SENIOR REVIEW PROCESS

The objective of the Senior Review process is to identify those missions beyond their prime mission lifetime for which continued operation contributes cost-effectively to both NASA's goals and the nation's operational needs, and also to identify the appropriate funding levels for those missions [4]. It is clearly stated in all recent Senior Review guideline packages that the overall scientific potential of extended missions is paramount. However, acknowledgment is also made of a number of related factors, and details on a variety of issues comprise important subsections of each review proposal. These factors include:

- The importance of long-term datasets and overall data continuity through overlap; improvement in sampling via extended spatial and temporal coverage; and opportunities for synergy of multiple instruments (often via multiple spacecraft/missions);

- The contributions of mission data to advance the objectives of operational agencies such as National Oceanic and Atmospheric Administration (NOAA), the Department of Defense (DoD), and the United States Geological Survey (USGS);

- The contributions of mission data to key NASA scientific endeavors, as specified by the Decadal Survey, including fields of study acknowledged on a national or global level as being of high value (such as climate change);

- The potential for unique and unanticipated science;

- The availability and usability of scientific data produced in the Planetary Data System (including past history);

- The extent to which the scientific community beyond the mission science team may conduct research with mission data, and the adequacy of financial resources provided to support the analysis of science data;

- The uniqueness of the scientific investigation(s) compared to other sources of data regarding the same phenomena;

- The recent (last two to three years) scientific advances of the mission, as reported via refereed journal articles and other means, thereby allowing for an assessment of productivity.

Discussion of the above factors comprises the bulk of the science section requested in each proposal. Each package also requires a discussion briefing the review committee on the background of the mission, technical feasibility, budget, and (in some cases) E/PO programs. Most recently, the 2012 Planetary Science Division guidelines also required discussion on a 85% budget option, listing scientific and technical scope reductions and associated risks with a 15% cut to the proposed extended mission budget. These same guidelines also enforced a 35-page limit to the proposal (other recent Senior Review guidelines have also enforced similar page limits).

These introductory, technical, budgetary, and E/PO discussions (where requested) require commentary on the following issues:

1. The overall status and health of the spacecraft, instruments, and ground systems, including limitations as a result of degradation/aging, the use of consumables, failures, and obsolescence;

2. The robustness of the proposed mission operations plan to unexpected events [5];

3. A high-level (at least) description of end-of-life activities, in compliance with NASA planetary protection;

4. A detailed breakdown of the proposed budget, with labor, major equipment, and other expenses explained in sufficient detail to determine and justify the cost of each proposed task;

5. Parallel funding sources that are required for mission support, and status;

6. Identification and roles of international or inter-Agency partners;

7. Project management plan, including risk analysis;

8. A science traceability matrix;

9. A mission data product inventory (where applicable);

10. A publication list and historical accomplishments with respect to mission objectives;

11. Planned E/PO activities, target audience(s), and reporting process.

The Senior Review proposals for each family of missions (currently Astrophysics, Earth Science, Heliophysics, and Planetary) are formally reviewed by members of a review panel, each often with a primary reviewer and multiple secondary reviewers. Limited-time oral sessions are scheduled at NASA headquarters, and the review panel then meets, often ranking the proposals formally on bases such as a "high/medium/low utility value" [4] or more frequently a "science per dollar basis" [6, 7] based on the expected returns from each project. The panel submits this ranking in the form of budgetary recommendations and detailed discussion of strengths, weaknesses, and relevance to NASA and national priorities, upon which the highest levels of NASA management decide the missions' fates.

III. HISTORICAL EXTENDED MISSION SUPPORT AND SCIENTIFIC PRODUCTIVITY

As stated previously, historical support for extensions of NASA's mission set has been strong, with 80%+ of proposals being funded at a level sufficient to achieve some or all of the extended mission objectives. Senior Reviews to date have led to the removal of only 10–20% of the weakest extensions, some of which had partial instrument failures or significantly reduced capabilities [7]. In the past, some missions ranked low were given an opportunity to resubmit proposals with improved cost-effectiveness for evaluation in a subsequent Senior Review. However, this reclamation process is no longer in place, so projects essentially have one single opportunity to fund each extension of up to two years.

Most of the available Senior Review reports explicitly state the panel's assessments of the strengths and weaknesses of each proposal. The most frequently assessed strengths of the missions recommended for extension, with particular focus on those missions most highly rated, are as follows:

1. Uniqueness of data acquired compared to other available means, or means envisioned in the near or far future;

2. Clear science traceability to the Decadal Survey and NASA priorities, including contributions to date versus likely contributions via the mission extension;

3. Breadth of applicability of data acquired, including new applications not envisioned for prime mission;

4. Data accessibility (often quoting access statistics);

5. Synergy with other missions and/or ground-based studies;

6. Observed improvements in operations, exhibited by reduced cost and team sizes.

In contrast, the most frequently listed weaknesses, with particular focus on those proposals not recommended for extension, are as follows:

1. Inadequate quantitative demonstration of science gained, e.g., what specific improvements in understanding would result from extending the dataset(s);

2. Insufficient discussion of possible contributions to revolutionary discoveries, as opposed to incremental advances in areas already explored;

3. Insufficient distinction between the productivity of the prime mission and the added value of the extension;

4. Scientific productivity of the wider community (more from the data archive) outmatching the productivity of the mission teams (more from new observations), indicating the bulk of the current science is being done with existing rather than new measurements;

5. Observed lack of strong community interest in existing data or new results;

6. Excessively high operations costs (excessive instrument support costs are often highlighted) and no plan for cost reduction over time;

7. Insufficient budgetary detail and justification for some costs, including staffing plan, and lack of traceability of costs to science operations;

8. Lack of discussion of synergy with other missions, including operational use insufficient to serve other potential national needs.

Both lists closely trace the Senior Review guideline topics listed in Sec. II, as expected. After reviewing the many Senior Review panel reports, it seems clear that a project's ability to tell a compelling science story and justify its expenses via a well-written proposal is equally as important as the true value of the science itself. It is imperative, therefore, that projects (NASA or otherwise) devote significant resources to crafting proposals for mission extensions.

Figure 4.1 illustrates the funding profile of both prime and extended mission phases for a sample of projects. As stated in one Senior Review report, "relative to the prime phases, missions in the extended phase are expected to reduce costs quite significantly" [8], and Fig. 4.1 bears this out. The funding profile with time is shown by fiscal year (FY), with year zero containing the prime/extended mission transition (often in mid-FY). The funding profile is expressed as a percentage of the maximum prime mission phase E costs computed in real year dollars. Precise year-to-year comparisons within or between projects are not relevant here, particularly considering that some contracts are obligated on one side of a FY boundary and cause year-to-year jumps that are not an indication of base funding changes. The point of Fig. 4.1 is to observe qualitative "big picture" trends in funding profiles.

Generally speaking, mission extension funding profiles tend to fall in the range of 20–60% of peak prime mission funding, and proposals above that level can expect additional scrutiny from review panels. Cassini may be the lone exception, with its first two years of extension funded fully at the prime mission amount.

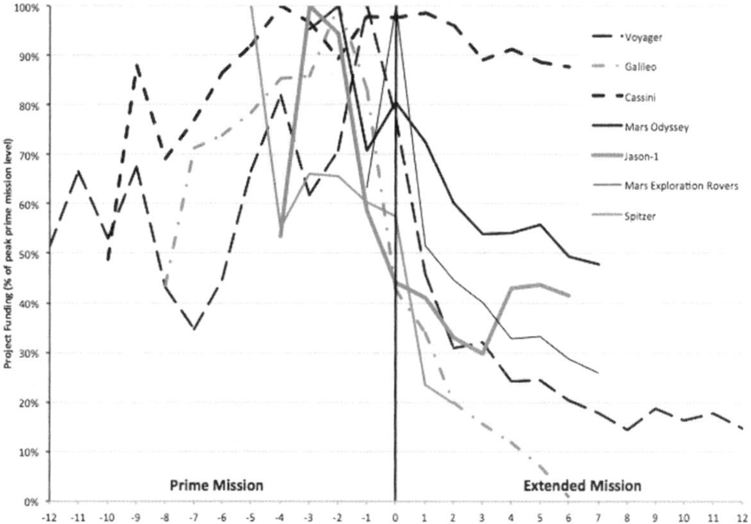

Fig. 4.1 Project funding per year, plotted as a percentage of peak annual prime mission funding.

However, these first two years (named the "Equinox Mission") were specifically pitched as a seamless, prime science extension at the same level of scientific intensity and mission optimization as the prime mission. As the Saturn environment is arguably the most target-rich in the Solar System, the additional science, both unique and repeated, was felt to merit equal funding levels. Starting in Cassini's extension year 3, and for the remainder of the second mission extension (named the "Solstice Mission"), the funding profile drops to 70–80%, with a commensurate drop in scientific intensity and optimization [9, 10].

Galileo and Spitzer are among the lowest funded extensions, in both cases because of significant limitations in their mission extensions. For Galileo, this was because of the high-gain antenna failure, problematic tape recorder, and resulting curtailments in its scientific program, and for Spitzer because of exhaustion of the cryogenic coolant, which reduced the number of instruments from three to one. Galileo's high reductions were foreseen, in fact with full agreement from the project; it was implicitly obvious, because of the spacecraft issues, that a reduction to 20–30% of prime mission funding was a foregone conclusion.

Spitzer made concerted efforts not only to reduce operations costs, but also to capitalize on a newly conceived and compelling single-instrument scientific program with significant synergy with other missions, and is one of the brightest success stories in the suite of extended missions studied. Some projects show reduced operations costs before their mission extensions; Jason-1 and Spitzer stand out, in particular, in Fig. 4.1. Note that the Voyager funding level extends off the chart to the right, continuing at approximately the 15% level for many years.

Figure 4.2 illustrates the peer-reviewed journal publication rate by year for a similar range of flight projects. Again, qualitative trends are the goal of Fig. 4.2, as an exhaustive search of all mission-related publications across all forums was not conducted. Therefore, one project's publication rate should not be compared against another assuming great precision. Also note that Spitzer year 2 and Cassini year 3 are incomplete surveys as statistics for year 2011 were not fully available as of the publication of this report.

Figure 4.2 does show the expected trend of an increasing number of publications during the prime mission as prime science data are collected. It is interesting to note that the publication rates do not noticeably slow in the extended mission phases, except after some years; it is not possible from Fig. 4.2 to distinguish between longer-term studies of prime mission observations (published after the end of the prime mission) and short-term studies and discoveries enabled primarily by extended mission science. However, common sense and the authors' and solicited project leaders' knowledge of extended-mission publications all favor the hypothesis that a significant fraction of the publications in the extended missions are enhanced or even directly enabled by extended mission measurements. It is straightforward to conclude that in terms of scientific productivity per dollar expended, extended missions represent the highest NASA return given their small incremental cost.

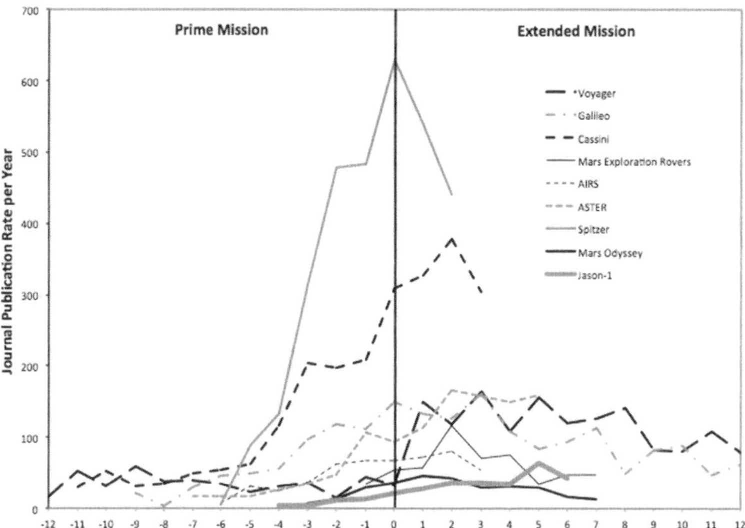

Fig. 4.2 Journal publication rate by year for a variety of projects.

IV. RECOMMENDATIONS FROM EXTENDED MISSION PROJECT LEADERS

The authors engaged in detailed interviews with a dozen project leaders (project managers, mission managers, and project scientists; see Acknowledgments) with experience spanning a wide range of NASA robotic mission extensions across four decades, with durations of only a few months to over thirty years. Their lessons-learned on the success of their extended missions, and of their experiences with the Senior Review proposal process, are summarized below in seven key categories. No priority order is implied.

A. ENSURE THE SCIENTIFIC PROGRAM IS COMPELLING

Echoed by all project leaders interviewed was the need to assemble an exciting scientific program. This is obvious both from the instructions of the Senior Review proposals and from common sense. The "science story" is often the most discussed and analyzed portion of the proposal. With a solid case, the mission will sell itself.

Many project leaders were emphatic that new and unique science is paramount, among them Dr. Edward Stone, Voyager Project Scientist and Jet Propulsion Laboratory (JPL) Director (1991–2001). "What's the discovery potential?" is a key question that must be addressed. Voyager survived two major threats to its continued program due to budget pressures, in 2000 and in 2004, after over a decade of quiet cruising beyond Neptune. The Voyager Interstellar Mission saw hints of the termination shock starting in 2002, and the fact that they were

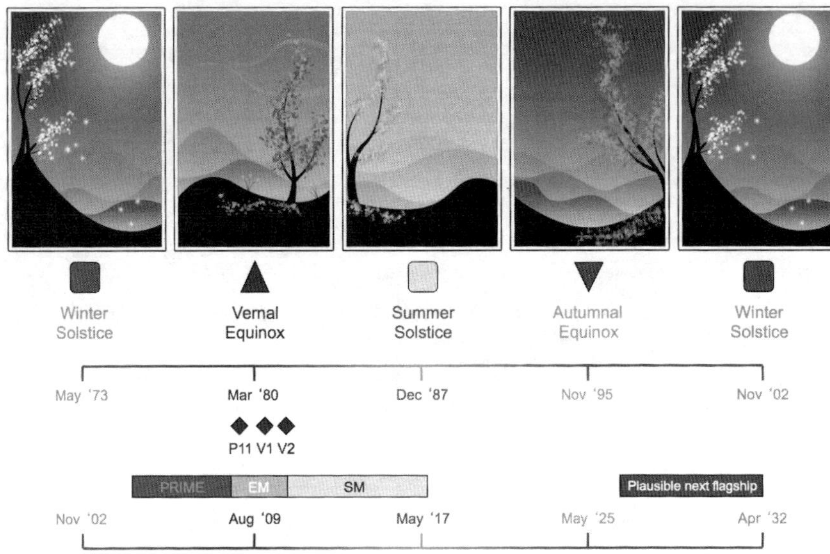

Fig. 4.3   Saturn seasons and Cassini mission coverage (bars at bottom left).

close to a wholly new environment, with high discovery potential of phenomena never before observed, was viewed as compelling and a major factor for extending the mission further. Both Deep Impact/EPOXI and Stardust/NExT also met with approval, in part by repurposing missions to study entirely different comets (Deep Impact/EPOXI from Tempel-1 to Hartley-2 and Stardust/NExT from Wild-2 back to Tempel-1).

The GRAIL (Gravity Recovery And Interior Laboratory) project, only recently extended, also knew up front that their lunar science had to be unique for an extension to gain approval. There was little case to be made that further flight around the Moon at the same altitude (55 km) would increase the knowledge of the gravitational field. They therefore opted to fly lower at half the altitude (23 km) and double the resolution of the gravity field measurements to the level of the crustal thickness, and this improvement opened up a variety of new fields of lunar study. Their proposal was arguably unique in that it was submitted before any prime mission science was even collected, because of the short prime and extended mission durations.

Spitzer may arguably be the best "dark horse" success story for new science during extension. Despite running out of cryogen, thus rendering two of the three instruments unusable, they made a compelling case for continued and new observation strategies by devoting more observing time to single objects and collaboration with other observatories (Hubble, Chandra, and Kepler). In particular, the Spitzer proposal created a niche to verify Kepler results that cannot be matched by other observatories, as Spitzer is the only infrared asset of its class currently in space. Spitzer's case was undoubtedly helped by the recognition of its target exoplanets and very distant (high-Z) objects of cosmological significance as hot topics in the Decadal Survey.

Many missions have pitched seasonal coverage as a lynchpin of a new science program, among them the Mars Reconnaissance Orbiter (MRO), the MER, Mars Odyssey, Ulysses, and Cassini. Ulysses based much of its extended science story on observations covering a full 11-year solar cycle. Similarly, Cassini devoted significant resources to arguing for an extension from the northern winter solstice (shortly before arrival in 2004) to the summer solstice in 2017, representing a full half season at Saturn, arguably covering the full range of geometric and environmental conditions. Illustration of these concepts was key to conveying such arguments. Ulysses created visualizations of the science potential plotted in a circuit around the Sun, and Cassini illustrated the seasonal coverage, including that of previous visits by Pioneer and Voyager, in charts such as those in Figs. 4.3 and 4.4. The Saturn seasons (referring to northern hemisphere conditions) are illustrated, and the prime and "Equinox Mission" (2008–2010) are illustrated by timeline bars near the bottom. The "Solstice Mission" now under way is shown by a bar labeled "SM". Note the epochs of visits by Pioneer and Voyagers 1 and 2, shown as diamonds (aligned with the previous Saturnian "year" above), indicating that only by continuing on to the Solstice mission could new environmental conditions be investigated. This illustration and others, such as Fig. 4.4, were used effectively to convey a unique environment for study with high discovery potential not likely to be achievable in the foreseeable future.

Cassini's case was made strong by tying seasonal changes to phenomena across the target-rich Saturnian system, including the evolution of lakes near Titan's poles, the variability of Enceladus's plumes, magnetospheric/solar wind

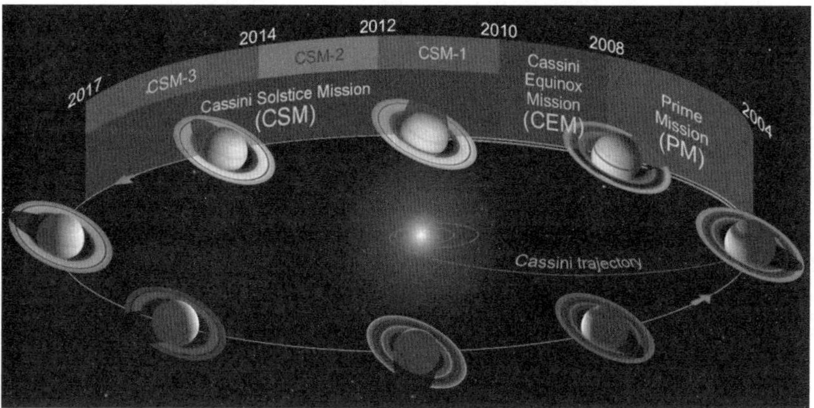

Fig. 4.4 Saturn seasons and Cassini mission coverage (another view).

interactions, Saturn's polar storms, and Saturn's ring structure (e.g., equinox offering unique illumination). Therefore, it is not simply the claim of seasonal coverage, but concrete traceability of the resulting seasonal effects to new areas of discovery potential that are scrutinized and favored by review panels.

For those missions whose new science may be a harder sell (true for some mapping missions, in particular), other arguments are available. Maintaining a continuous data record and the reliance of other agencies on the asset's continued operation are valued. Jason-1 data are used during real-time operations by both NOAA and the US Navy. Furthermore, Jason-1's applicability to climate change studies is acknowledged worldwide and by the Intergovernmental Panel on Climate Change in particular. Its continued operation during the Jason-2 mission may not overtly lead to a new data type, but increases the resolution of the combined measurement set both spatially and temporally, leading to better observability of ocean phenomena. Similarly, the Quick Scatterometer (QuickS-CAT) dataset was acknowledged by the NOAA weather community as important, and has been used to cross-calibrate ocean satellite data acquired by spacecraft of the Indian space agency, not only a synergy argument but another example of an interagency connection as well. Along those lines, Ulysses's commitment of support by the European Space Agency (ESA) in advance of NASA was also deemed likely to be a motivator for US support.

On the engineering side, Mars Odyssey represented the only UHF relay asset at Mars for some time (before MRO's arrival, at least) for missions such as MER, Phoenix, and the approaching Mars Science Laboratory, and its continued operation was deemed of critical importance as a fundamental link in the support chain to those missions. Nevertheless, Mars Odyssey also availed themselves of new science opportunities by changing their orbit geometry to map at new local solar-time lighting conditions. Also, Magellan's last mission phase included cutting-edge aerobraking techniques and was deemed an important engineering demonstration for future missions.

B.   BUILD AND MOBILIZE AN ARMY OF PROPONENTS

Preparations for extended mission proposals should begin, in most cases, at least one to two years in advance of the start of the extension. Spitzer began conceptual design three years in advance, and concerted efforts 18 months before the estimated exhaustion of cryogen. Cassini had its first focused discussions on extended mission planning a full three years before the end of its prime mission. History has shown a high approval rate for extensions, and most project leaders described high confidence that their proposal would be funded at the time of their application. However, it is important to note that all the project leaders interviewed were leaders of extensions that were successfully approved, and therefore leaders of future mission extensions that may be reading this chapter and assuming confidence for their case would be guilty of drawing a conclusion from a biased dataset. Each project should prepare for a thorough scientific and technical assessment rather than a mere coronation. Jim Erickson, project manager for MER, insisted emphatically that the entire Senior Review package must be well crafted to be successful in getting an extension approved.

Many project leaders also highlighted the need for the best writers and speakers to convey the proposal, going beyond the project office members at hand if necessary. The Spitzer project specifically brought in pundits outside its immediate management team to help pitch each of their three senior reviews. Voyager, GRAIL, Jason-1, MER, Galileo, Cassini, Mars Odyssey, and other proposals also benefitted, as described frequently during interviews, from dynamic project speakers who were known for their energy and communication abilities, including key science figures such as Drs. Ed Stone, Steve Squyres, Maria Zuber, Torrence Johnson, Tom Krimigis, William Patzert, Jonathan Lunine, Robert Pappalardo, and Jeff Cuzzi. "And that could be the difference," as stated by R. Mase, Mars Odyssey Mission Manager and Dawn Project Manager.

Spitzer also engaged its science community actively and early via workshops and even special sessions of American Astronomical Society (AAS) meetings used essentially as advertising. The project's "Science Opportunities for the Warm Spitzer Mission Workshop" [11] was held in June 2007, with 90 participants, a full two years before the exhaustion of its cryogen supply, and its charter was designed precisely to gather the information required to compile a strong extension proposal: "the goal of the workshop was to identify these [cutting edge] science opportunities and consider how to best use the observatory in an efficient manner" (P. McCarthy, Workshop Chair). This workshop, and the AAS sessions, not only equipped the project with ample ammunition to build a strong proposal, but it engaged the science community and strengthened their support for the extension by including them in its very planning, which paid dividends in perceived scientific support down the line. This approach was invaluable, and Spitzer's scientific productivity can be traced directly to these up-front efforts. The project's claim "to have one of the highest science return per dollar ratio of any of NASA's extended missions" [12] in light of its annual operating costs of one-third (or less) of the prime mission level is believable, especially considering the publication rate shown in Fig. 4.2. Other missions have adopted similar approaches, often lobbying the science community at general planetary conferences and workshops, and including Principal Investigators (PIs) and other members of the science field in the proposal writing process and oral presentations to the Senior Review panels. Letters of endorsement from the science community are also useful and were solicited by a number of missions, including Cassini in its 2012 Senior Review, which benefitted from a letter from the Outer Planets Assessment Group, which advises NASA.

Communication with headquarters stakeholders and the approval authorities is also valuable. The support for a mission extension should be built ahead of time. Program executives and scientists should be made fans of the mission, engaged, and excited. Posters and pictures of the mission and its scientific achievements should be on the walls of NASA headquarters. Project representatives should maintain a presence there and visit two to four times per year. In short, each mission, and its possible extension, should be consistently in the psyche of the stakeholders and decision makers. Furthermore, the project teams should do their homework on the target audience, the review panel specifically, to know what they are looking for and their fields of expertise. If the Senior Review panel cannot understand the science plan or budget, or be made fans of the program of observations, as is reflected by their many comments in the weakness areas of some proposals, the odds of approval drop appreciably.

Last in this category, project leaders should leverage all resources available at their home center in compiling the proposal. If there are proposal-writing institutional resources outside the project, such as the project formulation sections and teams present at JPL, their expertise and review should be brought to bear. The success of any mission extension benefits the entire center.

C.   REVOLUTIONIZE OPERATIONS PROCESSES, WITH CONTINUOUS IMPROVEMENTS

During prime science, much effort is spent optimizing science data collection. Extensions must be prepared to operate less optimally, and realize significant cost savings for approval. Dr. Stone's advice was emphatic: be prepared to redesign your operations from the ground up. Spitzer led a complete review of all of its mission and science operations functions: "What do we no longer need to do? Can we merge teams? What are we doing right that we shouldn't change? What is 'good enough?'" [12]. MER's leaders had similar experiences: "The operations process would need to transition from the high-intensity approach geared to wringing every possible bit of science out of the rovers in their presumed short lives to an approach that could be sustained indefinitely"[13]. Exceptions may include some mapping missions, which do not always revolutionize their operations, if they can demonstrate that they are already efficient and conducted at a minimum support level. However, even Jason-1 made significant operations staff reductions, starting in prime mission by implementing automation, improved software, and streamlined processes and procedures. They cut operator staff from 12 operators (supporting 24/7 shifts) to 4, and leveraged cross-training to ensure that no required expertise for managing the asset resided in one person alone.

Galileo completely redesigned their operations process to conduct targeted observations during only a small fraction of their orbit (periapsis ± 1 week) and was recast as a single-target mission (Europa). Cassini's Solstice Mission completely restructured sequencing and realized significant operational savings, in particular by reducing the number of overlapping sequencing processes that needed to operate at one time (each requiring its own support team). Voyager merged its spacecraft teams in the early 1990s and used a stepwise process to downsize, learning along the way, from 50 people at the start of the Voyager Interstellar Mission in 1990 now down to 12.5 people (since 2005). MER moved from co-located science teams at JPL to remote operations, transitioned from Mars time to Earth time, and stopped working most weekends (requiring fewer plans per week that were less complex, shortening the tactical process). They continue to make incremental improvements in procedures. MER's new operations process is enabled by separating those work days where the Mars and Earth rover times are similar (allowing the team to plan in lockstep) from those days where the downlink is later than 1300 hrs local time, where the team no longer tracks the downlink and the schedule reverts to a normal 0800 hrs start time. On these latter days, called "restricted sols," the tactical team uses data that are older, and operations are more restricted and less complex. This workshift pattern repeats every 37 Earth days and is executed by a team size of 45 (60 when both rovers are operational), down from 200 during prime mission.

Many projects change their staffing profiles and "flatten" the project structure, reducing middle management in particular and deputy positions (including even the GRAIL mission, with the shortest prime and extended mission durations of mere months). Many "worker bees" and spacecraft subsystem experts drop to part time; some depart the project altogether but are kept on call. This is best done carefully and via team sharing with other projects (conducted effectively in particular by Lockheed Martin for Odyssey spacecraft support). Often, team sizes shrink (including science teams), but the leadership, presumably with the highest level of expertise, remains the same.

Operations improvements and reductions in scientific intensity are best conveyed in proposals with quantitative visualizations where possible. During Cassini's internal review of candidate extended mission tours for the Solstice mission, the project developed "activity intensity" charts, such as Fig. 4.5, illustrating the rate of scientifically intensive events, and therefore operations workload, as a method to compare operations costs with scientific advancement. These charts clearly showed an average reduction in intensity to a 50–75% level and were used directly to scope the operations efforts. Such arguments lent credibility to the "science per dollar" related discussions during the Senior Review process and strengthened Cassini's case for delivering continued groundbreaking science at a reduced cost while remaining operationally feasible.

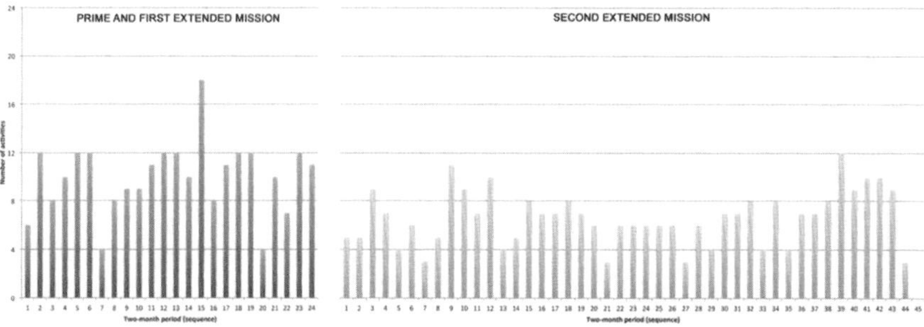

Fig. 4.5.   Cassini activity intensity, expressed as a number of intensive events per sequence.

D. HIGHLIGHT SPACECRAFT PERFORMANCE AND MAINTAIN LOW RISK TO SPACECRAFT OPERATIONS

Discussion of each project spacecraft's ability to continue operations appears nearly as often in review reports as the discussion of the scientific program. Possessing a healthy, well-behaved spacecraft is worth special mention in extension proposals. Equally as important is a detailed explanation of how limitations in resources and the management of failed or glitchy systems will be handled. Robert Mitchell, Cassini Project Manager, stated of the mission extensions that "the fact that the spacecraft had performed flawlessly was a major factor in being able to do what we did." John Callas, MER Project Manager, echoed that the quality of their vehicles was very important: they survived Martian seasons and dust storms, and climbed steep walls. Neither scenario was envisioned for the original three-month prime mission, and both were overcome despite a relative lack of redundancy compared to most JPL spacecraft. The MER rovers were designed for 1 km of surface driving and have, as of the publication of this chapter, traversed 42 km (the combined total). Similar statistics can be quoted for all missions in this chapter, again echoing the conclusion that the largest scientific and technical return can often be achieved via extended missions at a modest investment.

Each of the proposals surveyed in detail provided explanations of how the project aimed to handle its aging hardware and prove that continued operations were possible. Ulysses had to prove that their power supply was sufficient for two more orbital circuits in both best and worst cases, as well as how the unexpected nutation from the axial boom would affect data return and "science fuzziness" from the resulting reduced knowledge about spacecraft attitude. Magellan had significant anomalies in both the strings of its transmitter and tape recorder and lost the ability to transmit RADAR data altogether in 1992 (transitioning as a result to a gravity mission). GRAIL's extension proposal was met with significant skepticism in light of their original claims that their spacecraft would not survive the lunar eclipse, and was an intensely discussed topic at their Senior Review oral presentation. Voyager 2's scan platform, which jammed at Saturn (attributed to overuse at its top speed), required that rolls and turns be inserted into the science plans for Uranus and Neptune.

GRAIL's case deserves further mention. After flight trending and analysis, the project discovered that their spacecraft temperatures would not go outside of flight allowable limits during the lunar eclipse, in contrast to what was believed prelaunch and conveyed as a mission-ending event. In this and other cases, limiting spacecraft resources and circumstances analyzed preflight were found to be not so limiting in the actual mission. Cassini has benefitted significantly from only having used half of its hydrazine attitude control propellant from launch through the end of the prime mission. Frequently, significant resource margins can be found via detailed analysis by the flight team after they have operated the spacecraft, enabling mission options never before considered. In Cassini's case, the leading scientific argument for its final mission extension—in addition to seasonal coverage to the northern summer solstice mentioned previously—is the ability of the spacecraft to fly between the rings and the planet on 22 passages. This orbital geometry and the resulting unique science opportunities were never envisioned from development through even the end of prime mission, constitute an essentially wholly new Juno-like mission concept that itself would have been worthy of its own project, and were enabled primarily by the availability of significant flight resources combined with an experienced and skilled trajectory design team.

Without exception, all of the projects investigated opted to take no additional risk to their space assets, except where unavoidable (e.g., MER rover age and exposure to harsher conditions in the Martian environment, GRAIL's entrance into a thermally challenging lunar eclipse). Spitzer's number one guiding principle for its warm mission was blunt: "Don't do anything stupid" [12]. The no-additional-risk approach is not always applied to science data collection. Cassini, for example, cut back on some data-saving techniques and other approaches aimed to preserve the most important data. Science bits are slightly more likely to "hit the ground" uncollected. Spitzer likewise also accepted this fact with similar bluntness among its guiding principles: "Accept additional risk to science."

Personnel turnover is a frequent occurrence during extended mission and can incur increased risk to operations without awareness and advance planning on the part of the project. Galileo asked for statements of intent from its personnel (not signed contracts, exactly, but commitments to work for a full two years of its mission extension) and got them. The stability of staff has been a big benefit to Cassini, and the risk of loss of key personnel was tracked explicitly as a project risk. Ulysses performed cross-training and rotation of roles to ensure that the team could survive personnel losses. Deep Impact/EPOXI faced a significant challenge in losing experienced staff between its encounters, but turned in part to early career hires to fill the gaps; though inexperienced, the projects gained innovation from these staff and they, in turn, gained experience and responsibility.

MER shortened its tactical planning cycle: it released its team from working night shifts by incremental automation of previously manual steps; the tools continued to evolve as the surface mission progressed; key functions performed by each team role were refined through experience, leading to elimination of non-essential tasks and better focused communications; libraries of reusable sequences were compiled and evolved; and complexity was reduced, especially with sequences that would necessitate intricate planning with low benefit to science [13].

The loss of Mars Global Surveyor in November 2006, after operating at Mars four times longer than planned, also merits mention here. The Operations Review Board concluded, in part, that "risks associated with normal personnel turnover over time were not assessed"; "procedures and processes were inadequate to catch the errors that occurred"; and "periodic reviews should have been performed to assure that spacecraft control parameters were appropriate" [14]. These quotes relate to the systems engineering of operations and the maintenance of processes and procedures, which was also mentioned by multiple project leaders. Suzanne Dodd, Voyager Project Manager, emphasized this as a key point: "the knowledge base evaporates with time." Current Voyager project members have expended significant effort to find and maintain project documentation, particularly those memos (many of which are translated from paper to electronic form) related to spacecraft behavior, trending, commanding formats, and flight software. Ulysses maintained subsystem handbooks for each position, passed from one lead to another across its two mission extensions. These issues also bear resemblance to the breakdown in communications traced to other failures, and to the cultural problem highlighted by the Columbia Accident Investigation Board. Stephen B. Johnson, author of the article "Success, Failure, and NASA Culture" concludes that "80 to 95 percent of failures are ultimately due to human error and miscommunication. Most of these are quite simple... The mundane nature of these causes is precisely what makes them so hard to catch." "Systems management and systems engineering reduce failure rates by providing formal cross-checks that find and fix most potential mission-ending faults." [15].

The identification of a minimum operating budget has been useful in some cases as a defense against budget starvation. Longer-term missions such as Voyager, which are considered "national treasures" and seem immune to cancellation by budget pressure, still require vigilance against being squeezed to death. Cassini, too, identified levels beyond which its spacecraft could not be operated without significant additional risk to the hardware. Cutting teams equally down to this level, and then describing the more severe cuts that would be required of the science teams having to bear all of the reductions beyond this level, was an illuminating argument during the Senior Review process. Mars Odyssey also focused project discussion on cutting to the minimally comfortable spacecraft team size; the science team size, as budget cuts were absorbed, stepped down during its mission extension, and it was not considered painless.

Last in this category is the issue of conflicts with other missions. Occasionally, projects' extended mission science is deemed to be a lower priority than the prime science of others. This has been exhibited (again, infrequently) in the loss of tracking passes and skilled personnel to other organizations. Voyager loses tracking passes to other missions with some regularity and cut their data rate to 160 bits per second in part because they did not want to have to compete with primary science missions via high downlink requests. Their science collection is now a "take whatever we get" approach, purposefully designed as such, and arguably an outlier of the mission set studied. However, on several occasions, Cassini has also engaged in detailed negotiations beyond the norm with other projects in an effort to assess their mutual conflicts and redesign its own science plans to stay out of the way as much as possible. Cassini even prepared justifications and successfully defended some key tracking passes that were more critical against loss to other missions, even near some missions' (e.g., Phoenix's) critical events, where a typical result is that the critical event gets all the tracking it needs without negotiation.

Extended mission project managers should also be vigilant of their personnel and reach out to their peers on other, newer projects at the same institutions to maintain an open line of communication, with the goal of not only effectively sharing personnel (which can be a key component of low-cost extended mission operations) but to ensure that the other projects are sensitized to the impact of their luring away an extended mission's most skilled and experienced staff members, which are often the first that are considered for key positions on projects earlier in their life cycle.

E. INVEST IN PROCESS, HARDWARE, AND SOFTWARE IMPROVEMENTS

In past decades, some projects have taken the approach that both flight and ground software, processes, and procedures, and even ground hardware, mature from a development phase during phases C/D and early operations to a quiescent maintenance phase, and should even be "frozen" across the board during prime science operations. This was thought to free up resources to focus on science gathering and was simply a logical progression of ground and flight systems design. This approach is inherently flawed: projects must realize that software and process will always be evolving, and most project leaders in the current era recognize that a balanced level of development resources applied deep into prime mission is not only required but produces a clear benefit to the mission that is worth the investment.

Jason-1, in particular, made a conscious effort during prime mission to expand the capabilities of its systems, not merely for bug fixes and consistency with changes in spacecraft behavior. Jason-1 leaders recognized that the small investment of development resources in prime mission would pay dividends down the line, not only later in prime mission, but during extension. Their efforts included consolidation of their hardware integration and test laboratory (ITL) to an all-software simulator after the prime mission. Other missions, including MER, have benefitted from cost savings in extension from not having to maintain an aging ITL. One Voyager anecdote describes the project's acquisition of spare parts from military submarines for some aging but key computers used in soft simulation for sequencing, at least until after the Neptune encounter. Spitzer has improved its data compression and pointing performance via analysis, and these improvements bear benefits to science as well as operations efficiency. MER continues to devote significant effort to modernizing its ground software and hardware, the use of cloud computing, and moving towards eliminating all ground test hardware. The result is a lower-cost ground system that is more reliable, and "greener" to boot.

All project leaders agreed similarly that multimission tool, Operating System (OS), and hardware upgrades should be adopted "as they come" at some reasonable level per year, without "freezes" or waiting to take them all at once.

F. MANAGE COMPLACENCY

Of the project leaders surveyed, about half believed that complacency simply was not a concern. Of those, all agreed that this was primarily due to the presence of the stimulating analysis and replanning that was required during the mission extension; even the mapping missions have to accommodate changes in spacecraft behavior, which keeps the teams busy and stimulated. Voyager was forced to update much of its planning due to lower telemetry rates, changes in lighting conditions resulting in different exposure techniques and the need for target motion compensation, upgrades to the Deep Space Network and resulting improvements to communications, and the loss of some Voyager 2 scan platform capabilities. Similar hardware issues on Galileo and Magellan occupied many analysis hours. Spitzer teams were kept busy by solar flares and the need to reevaluate power and geometry analyses. The uniqueness of each encounter and orbit also makes planning on Cassini a consistently fresh experience; however, Cassini office managers are still reminded in quiet hours that they are responsible for keeping their teams sharp.

Other project leaders stated that complacency was a topic discussed often, with some missions having taken explicit steps against it. Jason-1 instituted regular retraining and requalification of personnel, and conducted simulated failures several times per year. MER is particularly vigilant against complacency: there are stand-down days and project retreats to address the concern, and the project deliberately puts "catch" items into plans to monitor team alertness. However, MER team excitement about the mission and science is believed to be high in general. Mars Odyssey, after 2001, began looking at commanding errors (also mandated by Lockheed Martin) to keep its staff sensitized to operations issues, and many missions now frequently track this and other related performance data types. In addition, Odyssey has a good inreach program and positive social gatherings that keep morale and motivation high. Cassini also includes frequent inreach programs as an indirect tool against complacency, and regularly posts Cassini-derived "Astronomy Pictures of the Day" and other science images in its hallways.

It is imperative that Project and Mission Assurance Managers set a tone to new and seasoned operations staff alike to "be vigilant", with phrases such as "not on my watch." Frequent inreach talks, articles, pictures, and interactions between scientists and engineers to share the exciting science being gathered can be useful motivators as reminders of the compelling reasons why the mission is being conducted.

G. VIEW EDUCATION AND PUBLIC OUTREACH PROGRAM AS EXTENDED MISSION ENABLER

NASA guidelines state that 1–2% of a project's budget be used for Education and Public Outreach (E/PO). While many view this simply as a necessary (even fun) part of a project's duties, not only as recompense for funding by taxpayer dollars but to inspire the next generation of engineers and scientists, E/PO is also an enabler for mission extensions if it is effective enough in translating the project's accomplishments into the psyche of the nation and planet. The Hubble Space Telescope is easily the prime example of this; it has been stated in many publications that its servicing missions, particularly the last in May 2009, were enabled not only by the outstanding science but as its place in the public as "the face of space science" and public outcry when its continued life was threatened. Voyager, considered a national treasure with its golden record (another outreach program altogether), is likely never to be canceled simply from budget pressure. Spitzer was at the forefront of the Myspace, podcasting, Facebook, and Twitter revolution in social media, starting as early as 2000, and their efforts to reach the younger community via these means were seen not only as part of the job but an investment in the project's future, and the quality of future matriculating space scientists. Cassini's Twitter feed now has the most followers of any NASA planetary mission in operation. Jason-1, the products of which are seen all around the world whenever there is news of El Niño/La Niña, deliberately funded its E/PO effort above the NASA guideline, to the 3–4% level. The MER rovers are very popular with the public, and NASA headquarters clearly recognizes the public value of their continued operation.

Although it is impossible to quantify the effect of E/PO on extended mission approval, there is general agreement that it is a program worth specific mention in Senior Review proposals. Each mission must find the appropriate balance in funding, but the above discussion represents a new perspective on E/PO that should be considered, especially during the early mission phases. The participation of E/PO in Quarterly and Senior Reviews alike can also serve a dual role: not only do they communicate to NASA stakeholders the extent to which the mission is reaching the public (and the public's perceived value of the mission itself), the outreach products shown are often as effective as exciting science results in converting those stakeholders into passionate supporters of the mission.

V. CONCLUSION

The Senior Review process is an effective one and the instructions to projects provide a thorough and equitable basis on which the proposals can be judged. The concepts herein, relevant not only during preparation of the extension proposal, but during the development of the prime mission priorities, represent lessons learned from project leaders across a wide range of applications and have the potential to increase a mission's lifetime or its level of financial support during extension.

The authors can think of no more fitting closing than a recommendation of the 2012 Astrophysics Mission Senior Review panel: "It may be worthwhile to have a forum to share these [extended mission] approaches, as it may benefit newer missions struggling with cost reduction and older missions searching for additional savings. NASA might consider convening an occasional workshop where missions describe their activities (e.g., Mission Operations) and the cost savings procedures that they have put in place or are considering. In such a workshop, NASA might invite additional outside experts and representatives from past missions that were particularly successful in cost reductions. Information presented at the meeting should be preserved and made publicly accessible. There may well be design choices implemented during mission development that would allow for lower-cost extended operations after the prime phase." [7]

ACKNOWLEDGMENTS

The preparation of this chapter was carried out at the Jet Propulsion Laboratory, California Institute of Technology, Pasadena, California, under a contract with the National Aeronautics and Space Administration. The authors are grateful for the time given by the project leaders in the collection of lessons-learned: Dr. Ed Stone (Voyager, and Director of JPL 1991–2001), Suzanne Dodd (Voyager, Spitzer), Robert Mitchell (Cassini, Galileo), Robert Mase (Odyssey, Dawn), Glenn Shirtliffe (Jason-1), Robert Gaston (QuickSCAT, GRACE), Shaun Standley (Ulysses), Joseph Beerer (GRAIL), Julie Webster (Magellan), John Callas (Mars Exploration Rovers), Tim Larson (Deep Impact/EPOXI and Stardust/NeXT), and Jim Erickson (Mars Reconnaissance Orbiter, Galileo, and MER). The authors also wish to thank Ann Coppin and Edward Jorgensen at the JPL for providing publication and financial data in support of this paper, and G. Mark Brown, David Nichols, and Leslie Livesay for supporting the authors' desire to bring this work to light.

REFERENCES

[1] "Extending the Effective Lifetimes of Earth Observing Research Missions," Committee on Extending the Effectiveness Lifetimes of Earth Observing Research Missions, National Research Council, NAP document 11485, 2005, http://science.nasa.gov.

[2] Swenson, C. (Chair), "Senior Review 2010 of the Mission Operations and Data Analysis Program for the Heliophysics Operating Missions," May 11, 2012 (publicly available on NASA Web site), http://science.nasa.gov.

[3] "Planetary Science Division Extended Mission Operations: Formal Guidelines for Senior Review Proposal Preparation," NASA unnumbered internal memo, 31 Jan. 2012.

[4] Freilich, M., "Call for Proposals – Senior Review 2009 and the Mission Extension for the Earth Science Operating Missions," NASA unnumbered memo (publicly available on NASA Web site), May 11, 2012, http://science.nasa.gov.

[5] "Planetary Science Division (Consolidated) Senior Review Findings 2012: Cassini Extended Mission," preliminary summary of the 2012 Planetary Senior Review board report, provided to the Cassini project Sept. 2012.

[6] Wheller, J. C. (Chair), "Report of the 2010 Senior Review of the Astrophysics Division Operating Missions," NASA unnumbered document (publicly available on NASA Web site), May 11, 2012, http://science.nasa.gov.

[7] Bregman, J. (Chair), "2012 Senior Review of Operating Missions in the NASA Astrophysics Division," NASA unnumbered document (publicly available on NASA Web site), May 11, 2012, http://science.nasa.gov.

[8] Freilich, M. (Chair), "NASA Earth Science Senior Review 2009," NASA unnumbered document (publicly available on NASA Web site), May 11, 2012, http://science.nasa.gov.

[9] Seal, D. "Cassini's Extended Mission," IAC-08-A3.6.12, International Astronautical Congress, Glasgow, Scotland, Sept. 2008.

[10] Seal, D., and Buffington, B. "The Cassini Extended Mission," *Saturn from Cassini–Huygens*, edited by Dougherty, M., Esposito, L., and Krimigis, S., Springer, 2009, Chap. 22, www.springer.com.

[11] Storrie-Lombardi, L., and Silbermann, N. (eds.), "The Science Opportunities for the Warm Spitzer Mission Workshop," *API Conference Proceedings*, Vol. 943, American Institute of Physics, College Park, MD, 2007.

[12] Storrie-Lombardi, L., and Dodd, S., "Downsizing a Great Observatory: Reinventing Spitzer in the Warm Mission," *Observatory Operations: Strategies, Processes and Systems III*, edited by Silva, D., Vol. 7737, SPIE, Bellingham, WA, 2010.

[13] Mishkin, A., Limonadi, D., Laubach, S., and Bass, D., "Working the Martian Night Shift," *IEEE Robotics & Automation*, IEEE, Washington, DC, 2006, pp. 46–53.

[14] "Mars Global Surveyor (MGS) Spacecraft Loss of Contact", preliminary report by Mars Global Surveyor Operations Review Board, NASA unnumbered document, 13 April 2007.

[15] Johnson, S. B., "Success, Failure, and NASA Culture," Academy Sharing Knowledge (ASK magazine, a NASA publication), Fall 2008, pp. 52–56.

CHAPTER 5

Multi-Mission Operator Training Practices

Jennifer M. Reiter[*]
Laboratory for Atmospheric and Space Physics (LASP), University of Colorado, Boulder, Colorado

I. INTRODUCTION

The Laboratory for Atmospheric and Space Physics (LASP) is a cradle-to-grave space sciences facility with focuses on science, engineering, mission operations, and scientific data analysis. LASP currently conducts operations for four NASA satellites and 14 science instruments from its Mission Operations and Science Operations Centers (Fig. 5.1). LASP is responsible for more than $US1.5 billion in NASA assets and processes over 100 gigabytes of data per day to support the ongoing activities of these missions. It is LASP's goal to identify and address key questions in solar influences, atmospheric, planetary, and space sciences.

The Mission Operations team at LASP is responsible for command and control of the Quick Scatterometer (QuikSCAT), Solar Radiation and Climate Experiment (SORCE), Aeronomy of Ice in the Mesosphere (AIM), and Kepler missions. These missions are operated and monitored using several customizable software packages developed in house. The Operations and Science Instrument Support–Command Control (OASIS–CC) software is used as the interface to command LASP's spacecraft and receive real-time telemetry. With OASIS–PS (Planning and Scheduling), standard command products are created daily using templates for science operations and spacecraft activities. All spacecraft data are handled by Telemetry Data Processing (TDP), which automatically processes all the real-time and post-pass data collected during spacecraft supports. These and other tools work together to allow for a streamlined operations center able to easily support a variety of missions.

As an institute of the University of Colorado at Boulder, development of the next generation of space professionals is a prime focus at LASP. To that end, the Mission Operations (Mission Ops) team comprises a mix of professional and student employees, all of whom participate in general operations and mission-specific training before undergoing formal operator certifications. Spacecraft operations are constantly evolving and, in response, training at LASP continues throughout an operator's tenure. All operators undergo yearly mission reviews

---

[*]Student Training Lead/Flight Controller, Mission Operations and Data Systems.

Fig. 5.1   The Mission Operations Center at LASP is staffed by a mixture of students and professionals.

and recertification. LASP's operations training is continuously updated and refined to provide the best resources possible for the entire ops team.

II.   STUDENT OPERATORS: COMMAND CONTROLLERS

The Mission Operations team employs approximately 20 undergraduate and graduate students from the University of Colorado (CU). These student Command Controllers (CCs) are involved in every phase of mission development, including prelaunch testing, mission simulations, launch and early orbit operations, and spacecraft decommissioning. As the title implies, CCs are the individuals who issue the actual commands to the spacecraft and are responsible for command verification during real-time uplink contacts. The student operators are also responsible for building the command products that are loaded to the spacecraft on a daily or weekly basis. In addition, CCs perform short- and long-term spacecraft telemetry trending, participate in anomaly recovery, generate anomaly reports, and interface regularly with both the missions' scientists and our partners in the aerospace industry.

CCs are the eyes and ears of the Mission Ops team, with a refined and specific knowledge base that allows them to see trends in data that might go unnoticed by the untrained eye. They are considered our first line of defense against behaviors that could indicate a near-term or future failure, and their reporting often helps to guide operations concepts in different directions. This attention to detail extends to their on-console responsibilities, where careful accounting of the commands issued to the spacecraft can prevent a commanding error and the long recovery following it. CCs are encouraged to question even the most senior of the professional staff (Flight Controllers, FCs) if the safety of the spacecraft is at stake. It is only through a sound understanding of a spacecraft's systems that undergraduate CCs become comfortable speaking up, and the knowledge they gain through LASP's Command Controller training program gives them the awareness and confidence to do just that.

Each spring the Mission Operations team selects approximately 10 undergraduate students from CU to train and work as CCs. Candidates must have a minimum cumulative Grade Point Average of 3.0 and at least two years left in their undergraduate work to be eligible for the position. CCs are selected based on an in-person interview and the recommendation of one to three professional references. Although the majority of students who apply are Engineering majors, all backgrounds are welcome as long as applicants meet the stated criteria. The Mission Ops team selects only those candidates who are responsible and mature enough to handle the job of supporting our on-orbit assets, and those who successfully complete the program are eager to meet the expectations placed upon them.

III. COMMAND CONTROLLER TRAINING

CC training takes place over 10 weeks from late May to mid-August. Students spend eight hours a day, five days a week, in training both in and out of the classroom. Because their time at LASP is limited (CCs spend two to four years in Mission Ops prior to graduation), CCs are trained on all four of LASP's spacecraft to allow for easier backfill when staffing gaps arise. It is a vast amount of knowledge to transfer in a very limited amount of time, so training is a very structured affair.

Training consists of classroom instruction, hands-on application, and small projects and assignments. The material is broken into three phases and projects are assigned throughout to support the lecture material. CCs attend lectures on basic aerospace concepts, LASP-specific operations methods, and in-depth, mission-specific training. The in-class instruction is complemented by out-of-the-classroom projects that apply the lectures to actual spacecraft behavior. CCs also learn several programming languages that make the analysis side of their jobs much easier. The hands-on component of the training allows CCs to work on actual NASA missions with close oversight by senior CCs and professionals. This aspect is heavily emphasized as it is vital to understanding how the classroom material applies to the job the CCs are hired to do.

A. PHASE I: THE BASICS

The first phase of training introduces students to LASP and the QuikSCAT, SORCE, AIM, and Kepler missions, with high-level overviews by the missions'

Flight Directors (FDs). CCs also learn how to navigate the networks of LASP and how the Mission Operations Center (MOC) fits into the NASA network as a whole. Students are immediately introduced to UNIX, the operating system used to support planning and scheduling, as well as real-time operations. The majority of the students are completely unfamiliar with UNIX and its command line interface takes some getting used to. Their first week primer gives the CCs the basic training they need to log in and run the software necessary to do their jobs every day.

Because LASP hires students from all majors it is important to equip them with a context for the job they will be doing. To that end, phase I also provides students with a basic aerospace primer. Lecturers from around the lab come into the classroom and teach topics like Command and Data Handling, Propulsion Systems, and Remote Sensing Techniques. Through these sessions students learn the foundations of spacecraft design and how subsystems work together. The mission proposal process is also explained, along with a mission's life cycle from development, to test, to launch and commissioning. Attention is placed on subsystem interactions and robust design processes to illustrate the complexities of designing and operating a space mission. These lectures are complemented by examples from LASP's missions to highlight how changes in one subsystem can be seen in trending by another (e.g., how routinely power-cycling instruments affects spacecraft heater cycling).

B.   PHASE II: OPERATIONS OVERVIEW

With a firm background in basic spacecraft design, training moves on to the specifics of spacecraft operations at LASP. This is the point at which CCs learn how commands are routed from the MOC to the spacecraft. Network protocols are revisited with an emphasis on how the MOC interacts with the various antennas of the Ground Network, Space Network, and Deep Space Network. Detailed training in spacecraft data flows complements this overview and the CCs begin to understand how data is routed to our end-users.

During the operations overview, students become familiar with the features of OASIS–CC, the interface used to command our spacecraft and receive real-time telemetry, and the Colorado System Test and Operations Language (CSTOL), OASIS–CC's control language. To do this, CCs write basic CSTOL scripts to check telemetry points, send commands based on spacecraft conditions, and respond to user inputs. CSTOL scripts are used to execute all command activities during uplink contacts, so a firm grasp of the language is essential. Eventually, CCs are expected to be able to read a CSTOL script and understand everything that will or could possibly happen during a contact based on the script's structure and the response seen at the spacecraft. The students' understanding of the CSTOL language is combined with training on OASIS–PS, the software tool used for nominal daily command planning. It is only through an understanding of these two tools that CCs become proficient command planners.

Exercises in uplink pass protocol teach the Command Controllers how to communicate with the Flight Controller during real-time contacts. Clear communication between CC and FC is vital to the prevention of command errors during both routine and non-nominal spacecraft supports. CCs learn the importance of command verification and the basics of our spacecrafts' command and telemetry systems. At this point, CCs are introduced to the many information resources available to provide more insight on current spacecraft health and safety. The operations overview flows into the next phase of training, where the basics are brought together with the specifics of the spacecraft LASP controls.

C. PHASE III: MISSION-SPECIFIC TRAINING

In the third phase of training, CCs learn how each spacecraft is designed and operated. One mission is covered per week through lectures and out-of-the-classroom exercises. The Fight Directors give in-depth lectures detailing each mission's evolution from launch to present day, highlighting the changes in operations over the life of the mission. The student subsystem team-leads teach the new CCs how their subsystem operates and any special operational considerations relating to them. Practical examples teach CCs how to spot the effects of seasonal variations on our spacecraft, such as thermal trending due to changing beta angle or the effects of a sparse starfield on the Attitude Determination and Control System (ADCS) subsystem.

During mission-specific training, special attention is placed on identifying anomalous spacecraft behavior and tracing this back to a root cause. CCs begin to work on projects that force them to use basic troubleshooting techniques to identify potential spacecraft anomalies. A common exercise begins with students looking at a set of telemetry plots that show atypical behavior for a subsystem. They must then walk through the process of determining the cause. They learn to use resources like shift reports, activity plans, and science collection timelines to make an initial assessment on whether the behavior is congruent with any special activities that may have been planned for that timeframe. Once that path has been exhausted CCs are taught to investigate trends on other subsystems at the time in question to see if the behavior is subsystem-specific or apparent across the entire spacecraft. They also compare trending from year to year to determine if the variations in the plots are repeatable during a certain season or at a particular attitude. All of this troubleshooting builds to the end goal of "telling a story" to the mission's Flight Director. When students come to a conclusion about why they see a particular trend, it is expected that they will be able to back up their claims with solid evidence. Instructors and FDs emphasize that although CCs are not always expected to understand the underlying cause of apparently anomalous behavior, they are expected to exhaust the options for explaining it away. Teaching CCs effective methods of troubleshooting allows them more independence in their job and gives FDs more confidence that their spacecraft is being watched over adequately.

D. TRAINING TOOLS

Each new topic that is introduced is often a very foreign concept for undergraduate college students lacking real-world engineering experience. Although lectures are the standard format for much of the training, they really only allow students an introduction to concepts and do not provide a sufficient understanding of how these concepts exhibit themselves on orbit. In an effort to make the connection between lecture topics and the realities of daily spacecraft operations, the Mission Ops instructors make use of several training methods that greatly enhance the CCs' comprehension of what is usually very advanced material.

One of the main ways trainers gauge the CCs' understanding of new concepts is through the use of a classroom response system. This is an interactive way to engage the students and judge their understanding of new material. At the start of each lecture day, students answer 5–15 multiple choice questions, with time for discussion and explanation following each. The entire class sees the distribution of answers selected after all responses have been registered, so the instructors can discuss with the class why one answer might be better than another. The interactive discussion highlights the intricacies of the question and gives students a better background for understanding the answer. It also exposes deficiencies in the training and provides opportunities for retraining where necessary. The classroom response system has been a popular and highly effective tool in revisiting and reinforcing new topics from day to day.

Early in the summer, the students begin shadowing the senior CCs to gain hands-on experience and familiarize themselves with the day-to-day responsibilities of their job (Fig. 5.2). At the start of the third week of training, one or two students per day will follow the CC on shift and do everything he or she does. The new students build daily command products under close supervision so they can

Fig. 5.2   New CCs prepare for certification by shadowing certified CCs.

become comfortable with the process and potential problems. Trainees observe real-time contacts alongside their CC and mimic the CC's activities whenever possible. The shadowing roles are soon reversed and trainees begin running passes themselves, with careful supervision from the CC and FC throughout the contact.

Additional hands-on training is provided as the students complete their practical checklists. At the beginning of the summer, CC trainees are issued a checklist of approximately 100 practical tasks that they must be able to perform on their own by the end of training. Each task must be explained to and watched by the trainee, and then performed independently before it can be considered mastered. These tasks include things like creating and verifying daily command products, extracting data from an engineering file, and implementation of mission rules. It takes a great deal of time to become proficient at all of the checklist items, so students must be motivated to be successful. Checklist completion also encourages interaction with the rest of the Mission Operations team, because an FC has to sign each step on the checklist once it has been mastered.

Translating in-class lectures into real-world knowledge application is a critical part of CC training. Without this step CCs will never understand the underlying causes of trends they see in their analyses. To encourage this transition in thinking the training staff regularly introduces small investigative projects that can only be completed by applying the lecture material to unfamiliar spacecraft behavior. These projects are as open-ended as possible, challenging the CCs to use the resources available to them and to think outside the box for their answer. The student trainers initiate these investigations with questions such as "Why has this temperature profile changed?" or "Explain why this data is discontinuous." These projects are usually completed in teams to mimic the way a subsystem team should work. Ultimately, these projects help trainees develop into self-sufficient workers who can attack open-ended problems in the same way as professional engineers.

IV. COMMAND CONTROLLER CERTIFICATION

At the end of the summer, CCs are formally certified on all four LASP missions. (Prior to certification, CCs are not permitted to create command products or send commands to our spacecraft without a certified CC to monitor their actions and verify correctness.) Certification requires students to pass three written exams and one oral exam over the course of their training summer. Failure to successfully complete any of these exams will result in the cancellation of a student's appointment at the end of the summer.

Certification exams each consist of 50 questions that require both short answers and calculations. The exams test a CC's ability to understand the basics of the classroom lectures and apply that information to their job in operations. They are administered every three weeks over the summer and are graded on a pass–fail basis. Exams are written from scratch each year and focus on the current state of operations as well as basic knowledge of how the spacecraft work.

If a student passes his or her written exams the oral certification is conducted at the end of the summer. The questions draw directly from the certification checklist items, so the tasks should be well practiced by the end of training. Students explain every task on their checklist and the potential risks involved with each. The FC administering the certification introduces twists in nominal scenarios to ensure the CC has a basic understanding of how to react when activities become anomalous. The trainer must be comfortable that the student can competently complete each task on their own before he or she will sign off on a CC's certification.

V. ADVANCED STUDENT TRAINING

CC training does not end with the first summer at LASP. Because a CC's career lasts for two to four years, the operations training team has implemented advanced instruction in engineering practices and applications that takes place during a CC's second summer as an operator. The training familiarizes students with the Dynamic Object Oriented Requirements System (DOORS) software tool, Earned Value Management, International Traffic in Arms Regulations (ITAR), and space policy. They are taught advanced technical topics like CCSDS standards and space weather as it relates to spacecraft operations. This is also an opportunity for advanced training on LASP's spacecraft and allows time to delve into the more complex aspects of operations that are sometimes missed by first-year students.

During this summer, CCs also prepare failure analysis reports for past NASA missions. The CCs are divided into small groups and assigned a specific project to research and present to the entire operations team. The groups describe the mission failure and explain the root cause and how it could have been avoided. The CCs extract lessons-learned from these failures and apply them to how LASP does business in order to identify areas for improvement in our operations.

VI. FLIGHT CONTROLLER TRAINING

LASP's Operations group works on missions in various stages of development on which LASP plays a variety of roles. Although Command Controllers focus primarily on on-orbit assets, the professional arm of the operations team also supports missions and instruments in development and test phases. Mission Operators use a different skill set from Instrument Operators, and both need different areas of expertise from those operating missions and instruments in the Integration and Test phase. Add to this the fact that new operators come to LASP with varying backgrounds and years of experience and it becomes very challenging to develop a structured set of training modules with which to baseline all Ops employees. In light of this, LASP has developed a set of knowledge standards that ensure all new operators have the background needed to prepare them for their role on the operations team.

A. TRAINING FORMAT

One of the challenges in achieving consistent and comprehensive training of new employees is the sporadic nature of the hiring process. Unlike Command Controllers, who are hired in bulk and trained annually as a class, Flight Controllers are usually hired on an as-needed basis, one at a time, and can start at any time during the year. This drives the Fight Controllers training to be a very individualized and operator-motivated activity. Because of this, FC training is guided primarily by a mission's Flight Director during one-on-one meetings and daily operator shadowing.

Although LASP most commonly hires one operator at a time, occasionally, several operators will be needed for one or multiple missions and will be brought on as a group. During these relatively large hiring intervals, the new operators are trained in a more structured classroom-style format similar to the Command Controller training. FCs attend training twice a week for about four weeks. Sessions can vary in length from an hour to half a day depending on the subject matter, and most classes will also include out-of-class assignments to reinforce this subject matter.

In either case, if an operator starts their employment while a Command Controller class is in training, the new FC will attend several of the CC lectures. At a minimum, this includes the modules in both the operations overview week and the mission-specific training for the spacecraft to which they are assigned. New Flight Controllers will also often attend other training classes, such as UNIX basics or Interactive Data Language (IDL) programming, depending on their proficiencies and the focus of their job. The CC classroom modules are then augmented by advanced mission training with a Flight Director in areas aimed specifically at professional operators.

B. GENERAL OPERATIONS TRAINING

Although their day-to-day responsibilities and areas of focus may differ, each operator works within the same basic infrastructure at LASP and therefore needs the same basic understanding of how LASP works as an operations center. These concepts form the foundation of every Flight Controller's training. Ideally, this begins within the first few weeks of when an operator begins work. New FCs are introduced to the basic structure of Mission Operations and Data Systems (MO&DS), with a focus on where they fit in the chain of command. They learn how the Mission Operations Center fits in the NASA network structure, as it is vital to understanding how the MOC interfaces with other NASA entities such as the Ground Network and data-routing groups like Central Standard Autonomous File Server (C-SAFS). Although not all FCs will support on-orbit missions, providing them with the understanding of how operations works makes them better equipped to develop effective test plans and ground software.

Flight Controllers generally require a greater depth of specified knowledge than their student counterparts. FCs write a great deal of LASP's operational software, so they need a keen understanding of good software development practices, which language best supports their goals, and how to release and maintain tools and scripts long term. This becomes even more challenging if a new operator does not have experience working on a UNIX platform. All FCs need a basic understanding of UNIX and those that do not have a firm background in UNIX receive training to bring them up to speed. The basic training even delves into shell scripting, and savvy FCs quickly learn to utilize their newfound knowledge to simplify their routine tasks.

One scripting language that all operators need to be comfortable with is CSTOL. From test procedures to operational scripts and file loads, LASP utilizes CSTOL exclusively for all command products. Flight Controllers in all areas of test and operations need a firm foundation in how to read, write, and understand CSTOL scripts. This is accomplished with a training package that is either presented to a group of new hires in a classroom setting or by self study of the training presentation along with reference manuals. In both cases, practice assignments are crucial to reinforcing new concepts.

C. MISSION-SPECIFIC TRAINING

In addition to their general training, each Flight Controller receives mission-specific training designed to familiarize them with one or two spacecraft, a spacecraft's instrument suite, or both. New operators are given instruction and documentation on the details of a mission, how it uses ground assets, the path data takes from the spacecraft to the end-user, and what their day-to-day responsibilities are as an operator. Much of this instruction is given by the mission's FD, and new hires work closely with the Flight Director during training to develop a detailed understanding of the mission to which they are assigned. If possible, new Flight Controllers participate in subsystem-specific CC training for their mission. As these presentations are given by LASP's subject-matter experts, it is the best opportunity for a new FC to learn and ask questions. If timing does not allow this, they are provided with the slide packages to study on their own and review with the FD.

FC training relies heavily on shadowing techniques to teach real-time or test activities. FC trainees start by shadowing the CCs during contacts to get a firm grasp of the mechanics of a real-time support or a test procedure. This allows the FC to learn from the ground up and become competent in performing all aspects of the job necessary for daily upkeep of the spacecraft. Like CCs, they progress to commanding the spacecraft themselves under the watchful eye of both certified controllers. Operators working on instrument testing take a similar path with various performance testing and data flows.

FC trainees then move on to shadowing the FCs during their shifts, checking daily telemetry plots, responding to e-mails from external entities, and troubleshooting issues with data ingestion or network connections. The new controllers support real-time contacts alongside certified controllers and learn to identify issues during uplink supports. Eventually, the trainees move into the driver's seat for a real-time contact and are responsible for giving GO/NO GOs to the Command Controllers and communicating with ground assets. This shadowing progression ensures that new FCs understand the full flow of real-time supports and are comfortable with all aspects.

D. FLIGHT CONTROLLER CERTIFICATION

FC trainees have a practical checklist similar to the CCs, with a detailed list of topics a new FC needs to understand to do his or her job. Checklists are broken into general tasks every operator needs to know, such as what to do in the event of a power outage or how to transfer files in the event of a network failure, and mission-specific items such as preparing a certain command product or downlinking data. Each task is presented to the new FC by the Flight Director or another senior FC, and any complexities are explained in detail. The operator is also expected to research these topics in depth and consult with their Flight Director to clarify complex material. Through this regular interaction, the FD easily gauges when a new FC is ready to be certified (usually after two to three months).

Once the checklist has been filled out and the Flight Controller feels confident in his or her abilities, the controller undergoes the certification process. The Flight Controller is certified when he or she has successfully completed both a written exam of short-answer questions and an oral practical exam. The practical covers activities that a certified FC must be able to complete quickly should the need arise. This includes things like scheduling a Tracking and Data Relay Satellite (TDRS) support with a quick turnaround, contacting a ground station, when and how to initiate the notification call tree for a spacecraft anomaly, or who to contact for building access if the badging system goes down. As with CC certification, practical tasks are taken directly from the FC checklist. When both the written and practical exams are complete, the results are reviewed with the FC to clarify any areas of confusion.

VII. CONTINUED TRAINING, OPERATIONAL REVIEW BOARDS, AND RECERTIFICATION

The MOC's operations concepts are adjusted routinely in response to things like the degradation of an aging spacecraft or when developing a new, better way to perform an old task. Because of this, continuous retraining is essential to keep the FCs and CCs up to date on the current method of operations. This is primarily accomplished though instruction during the weekly All-Hands meetings where the state of each mission is reviewed in detail. All members of the operations team attend this meeting and notes are issued as a record. The All-Hands meeting is the easiest place to explain changes in operations, as it reaches the entire Ops team and provides time to answer questions operators may have regarding these changes.

When significant modifications are made to an ops concept they are also reviewed in detail at the weekly Flight Controllers' meeting. These types of changes are typically accompanied by an Activity Change Request (ACR), a standard form that explains and authorizes deviations from routine operations. When ACRs cover a long-term change or direct a new routine activity they become Standing Activity Change Requests. Flight Controllers are required to initial each of these Standing ACRs to indicate they have been trained on them and understand when and how to execute them.

A.   OPERATIONAL REVIEW BOARDS

Mishap review and prevention often drives retraining not related to spacecraft degradation. When an operator-controlled incident occurs that puts a LASP spacecraft at risk (e.g., an error in commanding, a change in a telemetry trend that went unnoticed) an Operational Review Board (ORB) is held to assess the cause and determine if corrective action is needed to prevent a similar situation from occurring in the future. The ORB team comprises the QuikSCAT, SORCE, AIM, and Kepler Flight Directors, the Mission Operations Lead, the Director of the Mission Operations and Data Systems division, and the Training Lead. The FC and CC on duty at the time of the incident also participate in the ORB. The individual at the center of the incident documents what happened and how the event could have been prevented. The board provides input on the situation and attempts to identify any other areas for potential mishap relating to the event. When the board has compiled their findings an overview of what happened and any lessons learned are presented to the team at the weekly All-Hands meeting. The entire Ops team attends annual reviews of all ORBs occurring in the past year to keep these incidents fresh in everyone's minds with an aim to preventing their recurrence.

B.   RECERTIFICATION

To ensure continued proficiency, FCs and CCs are required to recertify on a yearly basis. CCs recertify during the summer by taking either the second or third exam given to CC trainees. FCs are recertified each January by completing an abbreviated version of the written FC exam for their mission(s) and performing the practical portion of the original FC certification. Questions on the exam and practical are revised annually, and special focus is placed on major changes to operations over the past year. FCs and CCs unable to successfully complete the recertification process are subject to retraining and possible termination.

VIII. CONCLUSION

In an ever-changing operations environment with a large student workforce, operator training is never finished. LASP's Command Controller program provides hands-on training and real-world experience to the students of the University of Colorado and teaches them how to think like engineers long before they enter the professional workforce. The training and development students receive during their time with LASP's Mission Operations group allows them to transition into the professional world leaps and bounds ahead of other new graduates. Students are expected to operate as junior professionals by the end of their tenure at LASP, and their continued on-the-job training ensures their success.

As new missions are launched, old missions are decommissioned, and operators transition from one mission to another, a continued level of knowledge and expertise must be maintained. The LASP Mission Operations team has implemented several techniques to train new FCs and keep veteran operators current. Because of this training, LASP's Flight Controllers possess a detailed knowledge of how our spacecraft function, and this expertise allows for a quick and effective response when presented with new challenges or anomalous situations. The annual recertification exams are a measure of the success of LASP's Operations training efforts. The results of these technical evaluations confirm operator competency, and allow a yearly opportunity to reevaluate training techniques and augment areas that may be lacking.

The continual retraining required in this dynamic environment cannot be effective without motivated team members who take pride in the work they do. The Flight Controllers' and Command Controllers' job descriptions are constantly changing and their willingness to adapt to these changes is vital to the success of the team as a whole and the missions LASP supports. An educated and willing workforce is the true key to the success of a Mission Operations Center. By frequently revisiting subjects and striving to make our processes as well-understood as possible, LASP provides the comprehensive training that is vital to both maintaining a skilled team of operators and producing the talented spacecraft engineers of the future.

ACRONYMS

ACR	Activity Change Request
ADCS	Attitude Determination and Control System
CC	Command Controller
C-SAFS	Central Standard Autonomous File Server
CSTOL	Colorado System Test and Operations Language

CU   University of Colorado
DOORS   Dynamic Object Oriented Requirements System
FC   Flight Controller
FD   Flight Director
IDL   Interactive Data Language
ITAR   International Traffic in Arms Regulations
LASP   Laboratory for Atmospheric and Space Physics
MOC   Mission Operations Center
MO & DS   Mission Operation and Data Systems
OASIS–CC   Operations and Science Instrument Support Command and
                Control
OASIS–PS   Operations and Science Instrument Support Planning and
                Scheduling
ORB   Operational Review Board
OS   Operating system
TDP   Telemetry Data Processing
TDRSS   Tracking and Data Relay Satellite System

ACKNOWLEDGMENTS

I would like to thank Katelynn McCalmont and Danielle Richey for their tireless efforts and invaluable contributions to the summer training program. To Jack Faber, thank you for your guidance and ongoing participation in CC training. Thank you to Sean Ryan, who continues to push to make each year of training better than the last. And finally to LASP's Command Controllers, past and present, thank you for providing me with the opportunity to be a part of your education. As always, I succeed when you succeed.

CHAPTER 6

Gamification for Astronaut Training

Ferdinand Cornelissen[*]
Nspyre B.V., Utrecht, The Netherlands

Mark A. Neerincx[†] **and Nanja Smets**[‡]
TNO Human Factors, Soesterberg, The Netherlands

Leo Breebaart[§]
Science & Technology Corporation, Delft, The Netherlands

Paul Dujardin[¶]
NLR, Marknesse, The Netherlands

Mikael Wolff[**]
[5]ESA/ESTEC, Noordwijk, The Netherlands

I. INTRODUCTION

Today's astronauts are on a tight schedule, both in orbit as well as during the preparation phase of a mission. The preparation of a mission includes a strict training program in a classroom environment, focusing on the maximum efficiency of the mission outcome. However, for future astronauts on manned missions beyond low Earth orbit, the duration of the mission will probably be more than two years. The scope of the actual mission and the experiments performed will subsequently exceed the scope as anticipated during mission preparations. On these missions, a shift of autonomy is expected from ground to in-flight training by the crew, and therefore astronaut training needs to be adapted accordingly. The extensive duration of the mission, the delayed communications due to distance and potentially limited line of sight, the autonomy shift, as well as ongoing changes in the mission goals, the environment, and the tools at hand, all imply a need for self-study, motivation, and a means to provide dynamic training content *during* the mission rather than only *before* the mission. A Portable Learning Application (APLA) is a study performed for the European Space Agency (ESA) to identify and fulfill these needs. The study is based on the premise that

[*]Technology Officer, Nspyre B.V.; ferdinand.cornelissen@nspyre.nl.
[†]Senior Research Scientist, TNO; mark.neerincx@tno.nl.
[‡]Research Scientist, TNO; nanja.smets@tno.nl.
[§]Software Architect, Science & Technology Corporation; breebaart@stcorp.nl.
[¶]Senior Researcher, NLR; paul.dujardin@nlr.nl.
[**]Technical Officer, ESTEC Systems, Software and Technology Department; mikael.wolff@esa.int.

gaming concepts can help to promote self-study, motivation, efficiency and effectiveness, and on-site training among astronauts.

The study has been performed in the context of a larger research program
related to the development of a mission execution crew assistant (MECA) [1].
The MECA research program focuses on crew assistance by means of an electronic partner to aid in mission operations. This also includes training and electronic learning systems. Within the MECA program, the impact on crew
operations of long-duration missions as well as how to improve human–
machine collaboration on these types of mission has been studied extensively
[1–3]. An important output of the MECA program and its related projects is to
establish a common requirements baseline for cooperative and supportive interaction between man and machine during planetary missions.

II.   GAME CHANGER: LONG-DURATION MISSIONS

All aspects of crewmember training will go through extensive evolution to incorporate changes to the training environment within future missions that foresee a
duration of more than two years (e.g., the Moon, Mars) [4]. Several changes to
training objectives, both in preparation but also on-mission in a dynamic and
unpredictable environment, are expected. Long-duration missions are a game
changer for training of astronauts. The assumption is that such missions will
require a different type of premission training, as well as refreshing training or
(re-)certification in situ. This section will go into the details of changes to training
in long-duration missions. It assumes large dynamics in mission goals, operational
context, and therefore on the training related to that. This chapter's hypothesis is
that gamification, that is, the use of gaming aspects, enhances motivation on these
types of mission.

A.   CHANGES TO TRAINING PREPARATION OF A MISSION

What makes long-duration missions different in terms of training preparation is
that there will clearly be more time between any training premission and the
execution of a typical task on site (e.g., the Moon, Mars). Training of the crewmember according to today's standards will provide the required knowledge,
skills, and attitude to execute the complete set of tasks and experiments to be performed on a short mission. It is clear that if there is a longer time between knowledge and skill training and the real performance, essential information may be
forgotten or more easily overlooked. Accordingly, during the mission, and for
the vast majority of tasks, a crewmember needs to refresh or rehearse the training
as performed premission. It might also be the case that the crewmember will see
training material for the first time during mission travel.

The entire training setup may change, because in a long-duration mission
there is a lot more information to be included in training for crewmembers,

which sets new high demands for human learning processes (such as memoriza-tion, habituation, transfer of knowledge, prevention of interfering associations). For example, in today's missions to the International Space Station (ISS), a crew-member will be onboard for typically no more than half a year. However, when crewmembers are launched on a long-duration mission, several experiments may still be under preparation and the related training material will be uplinked to the crewmembers during their travel. This therefore requires a completely different training strategy compared to short mission training, as not all training can be completed or condensed before mission commences.

One significant improvement needed in training for long-duration missions is related to skills training. A large part of the present training curriculum involves skill-based training. Skills related to operating payloads are learned using hands-on training with models of the actual hardware to be used in the mission, as well as with simulator software packages that resemble the actual software on computers and with laptops as used in-flight. If the long-duration mission training schedule of the crewmember shifts between premission and in-mission training, new skills training methods will need to be embedded into the mission. In this regard one can conceive of the utility of virtual-reality simulator training methods up to the use of advanced motion/gesture control-lers and augmented reality. Indeed, virtual reality with force feedback and 3D modeling will become part of future training capabilities. Today's motion-sensing technology will become increasingly embedded in serious gaming for professionals.

B. CHANGES TO THE MISSION

Training during long-duration missions will require a new way of scheduling tasks. It is mandatory that crewmembers are trained premission in safety-critical aspects such as handling the flight system or landing systems, knowing how to work in temperature-controlled and pressurized spacesuits, and opening hatches in a non-Earth-like atmosphere [5]. Nevertheless, it can be expected that a vast majority of day-to-day tasks and work on experiments can and will be trained only during mission travel, or even just before experiment execution. In this case, the crewmember automatic task scheduler will identify, using short-term planning tools, the tasks to be performed and will also identify the preparation path with required skills and training or the required certification levels before actually performing the experiment later in the mission.

Another difference when compared with shorter missions will be that some crewmembers may have advanced knowledge or skills on certain systems that other members lack. In a collaborative training effort, the experienced certified crewmember will train the other crewmembers for certain tasks [6, 16]. In other situations, the trainer/trainee roles may be reversed. In the MECA project, a prototype training tool (Collaborative Trainer, COLT) has been developed to support such role switching for learning payload and first-aid operations [1].

Training material will sometimes only become available during the mission; for example, exact procedures or protocol of parameters may only be uplinked during the mission. In such cases, the crewmember will need to recertify or refresh their training through courses. Access to a training database and lesson material will be required, or the crewmember may even author new procedures for training and/or later execution of the experiment. In today's missions this occurs only very rarely. A famous example is the makeshift repair of the carbon-dioxide removal system in the Apollo 13 mission. In long-duration missions these capabilities for training methods renewal will be crucial for mission success and scientific results.

C. CHALLENGES TO TRAINING CREWMEMBERS IN LONG-DURATION MISSIONS

Given the discussed changes to training of crewmembers on long-duration missions, it is clear that the human factor will become even more important when the crewmember is left to self-study, in situ, without classroom support and hands-on practice. Keeping the trainees motivated to self-study and certify themselves during such a mission become mandatory. The training tools available for use will need to facilitate this.

A long-duration mission will have an impact on the cognitive and emotional state of crewmembers. The cognitive state of the crew is impacted because they are not yet completely trained (even though they are already traveling), there is no tutor close by, and the communication delays are becoming larger as the mission evolves. The crewmembers' emotional states are affected, for example because they may feel they have been left alone with standalone training material and fellow trainees who are encountering the same issues or, more generally because they will have to cope with momentary social, cognitive, and physical stresses that will affect their mental and physical condition. A large demand on creativity and inventiveness in resolving problems will be made, requiring adequate supporting tools. Crew interaction with training databases, lessons material, and certification paths will increase when doing this in a challenging training environment. To counter these challenges, so-called serious gaming methods may provide the answers. To support crew on long-duration missions, the use of next-generation gamification training tools is envisaged; these will be outlined in the following sections.

III. GAMIFICATION: GAMING AS A MOTIVATOR

Serious gaming or applying game aspects to serious problems has been around for some time. It has been used in the military sector to train for all levels of command and operations [7, 13]. The benefits are that the use of computer systems and gaming technology can introduce situations that are normally difficult to create. These systems rely heavily on the use of simulation. The term *serious gaming*, however, does not necessarily imply the use of gaming technology, but rather the use of gaming principles, and tapping into the intrinsic drive that makes playing games fun and compelling.

Instructional design researchers are exploring the potential of serious games for education, particularly because they may help to increase motivation [1, 2]. The fact that people worldwide are spending an increasing amount of time playing computer games does suggest that there is something compelling about them. The reason for games being so compelling lies with the gaming aspects. These can also be applied to serious problems rather than the trivial obstacles that are normally pursued in entertainment games. Applying gaming aspects to solving a real-world problem is termed *gamification*, the result being serious games.

It should be noted that trainees need guidance and structure during training, which are based on didactic considerations [1–3]. Such a serious game should include realistic scenarios and dynamics that are tailored to trainees' learning objectives and progress. Building and sharing an ontology for guidance is a promising research direction [1].

A.   DRIVERS OF GAMING

Why do we want to play games? This question has often been asked and answered in the past, in some cases stating that 'playing' has been around even before we had culture [1], and is an almost primal urge. The reason we play games is because they are fun. More important, there is something compelling about games that makes you want to come back for more, either because you want to win or come back with a vengeance, or perhaps because the game provides a soothing experience that helps you to relax. Underlying these emotional qualitative expressions, or getting hard fun while playing games, are the concepts of *flow* and *fiero* [1]:

1.   The notion of *flow* is the emotional state a person can achieve when performing a task so intensely that time seems to pass without noticing [1]. These are usually moments of pure joy and happiness. The notion of flow has interesting correspondences with the didactical notion of the "Zone or Proximal Development" aiming at a balance between the offered challenge and the trainee's competencies to prevent him from getting bored or frustrated [1, 2].

2.   The notion of *fiero* is an extreme positive emotion a person can feel by the rush one gets after achieving a goal that required a considerable effort, such as winning a marathon or solving a complex puzzle. The word originates from the Italian language and means to be proud of something.

Both flow and fiero are important intrinsic drivers for continuous interaction with the game or system that causes these drivers. So, a good game designer will design a game (rules, levels, quests, achievements, etc.) around game aspects that induce these drivers.

B.   GAMING ASPECTS

Game developers everywhere try to make a game as compelling as possible, partly by providing a visually stimulating experience, but also by using gaming aspects to compel the gamer to play the game over and over, maximizing the length of the game session as much as possible. The following is a list of game aspects that we have defined as instruments to enable this by actually tapping into the flow and fiero of the gaming experience [8–10]:

1. *Personification.* By representing the player in the game, the player feels he is part of the game, rather than just a player. A good example is introducing avatars, or profiles of the player. The player feels he should care about his avatar and will subsequently return to the game. As an instrument of flow, enabling this emotional tie will help increase the time a user is exposed to the game. This effect becomes stronger when the user is able to and has invested in his character representation.

2. *Quests or missions.* Providing game experience in bite-sized chunks allows the game to give control to the player regarding how much and how long he wants to play. However, it also provides a clear overview of the game possibilities that are (currently) available to the player. These are great instruments for both flow and fiero. On the one hand, completing a quest in a manageable amount of time creates fiero soon enough to want more, thereby stimulating playing a string of quests, and thus flow.

3. *Instant rewarding.* Providing rewards immediately after successful completion of a mission will induce fiero, provided that the reward is worthwhile and can be spent within the game, for instance to unlock new and more exciting missions, or to boost character development.

4. *Epic scale.* Achieving something grand, such as completing a difficult part of the game by cooperation between players, or by stringing together the completion of smaller missions, will tap into both flow and fiero, because flow enables completing this grand mission, and the reward in terms of fiero will be huge if completion is successful.

5. *Social comparison.* Competition, as a driving force to become the best, is only achieved by trying to get better every time, either individually, or if collaborative training is allowed, as a team. By allowing players to compare their score with those of others, we expect people to strive to become the best, essentially enabling flow.

Of course, these aspects are expected to be related to one another. For instance, a good avatar system will only emotionally tie the player to the game if there are interesting missions to play and if he feels that rewarding is balanced for proper character development. Appropriately applying these aspects into a game is key to success. To evaluate the effects of gaming aspects in our situation, we translated them into an actual prototype.

IV. LET'S PLAY: A PORTABLE LEARNING APPLICATION

An important goal of the APLA project was to take the high-level user requirements for long-duration mission training support, and derive from them a generic software architecture suited to form the basis for actual support applications [11]. Figure 6.1 presents the APLA architecture for this generic APLA System, and its decomposition into functional components.

The architecture was designed in such a manner that it could conceptually and technically interface with both current crew training systems (such as that used at the European Astronaut Center, EAC) and crew operations onboard a long-duration mission platform, such as on the ISS [12].

This architecture is fully described in the APLA Technical Specification, and has some interesting overall features:

1. The use of a Knowledge Base (as opposed to a conventional database) in the Data Access Layer, which greatly improves interoperability between internal components as well as between APLA and external systems by providing a facility for storing, retrieving, augmenting, and reasoning about semantically enriched information. For APLA, knowledge representation was carried out using Web Ontology Language (OWL) ontologies to model objects and relationships. The Knowledge Base was implemented as an Resource Description Framework (RDF) triple store.

2. Loose coupling between the various architecture layers, making it easy to create APLA prototypes that experiment with different or even concurrent

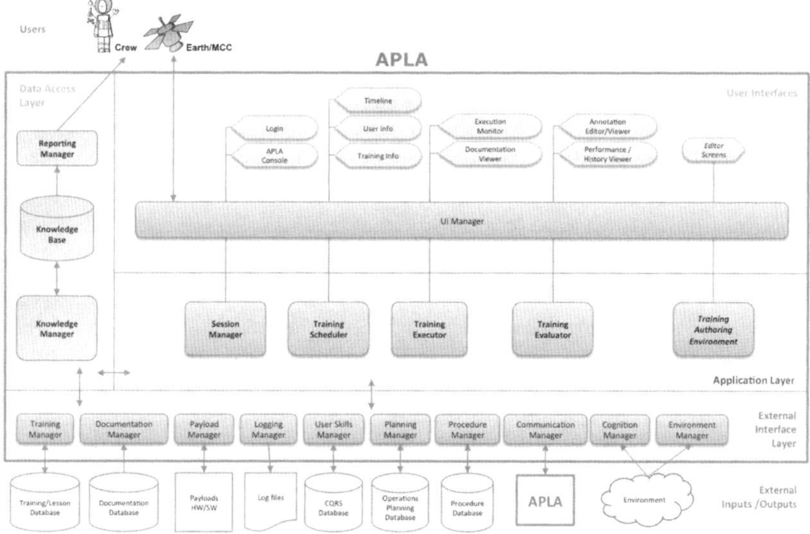

Fig. 6.1   The APLA architecture.

types of implementation (e.g., thick client vs Web application, centralized vs distributed components, classical 2-D Graphical User Interface (GUI) vs haptic interfaces, etc.) with a minimum of extra effort.

3. The possibility for separate APLA applications to communicate with each other and exchange information (e.g., synchronize Knowledge Bases). This will, for example, allow flight instructors to use the same APLA on Earth as being used by the crew.

A.　PURPOSE OF THE APLA DEMONSTRATOR

Using the APLA architecture as a starting point, we built an APLA demonstrator application for the purpose of illustrating and then evaluating our ideas on gamification as a vehicle for crew autonomy, motivation, and efficiency with respect to payload training. To keep the scope of the demonstrator manageable and within budget, and to allow us to thoroughly test a specific set of gamification ideas, we restricted the prototype (with respect to the full architecture and requirements baseline) in a number of ways:

1. The demonstrator is a prototype specifically intended for the evaluation of ideas. Although self-contained and functional, it is not a fully fledged product that can be used outside the context of a walkthrough test session, and the look-and-feel is functional, not flashy.

2. It supports training scenarios for only a single payload (the Cardiopres physiology module for measuring blood pressure, respiration rate, and ECG signals).

3. All actual lessons are implemented by external black-box applications, or by stubs containing dummy content. APLA is about the *presentation* of the training, not the training itself.

4. It implements and focuses the evaluation on the "six core principles" of gamification (discussed in the next section).

5. No authoring tools were created for either lessons or lesson metadata. Lessons themselves were stubbed (as mentioned earlier), and lesson metadata were created and maintained by manually editing RDF/OWL instance data files.

6. APLA is to be integrated as a module into the standalone research MECA Mission Execution Crew Assistant framework (rather than in an actual or simulated ISS or EAC environment).

B.　IMPLEMENTED GAMING ASPECTS

In the APLA demonstrator we have implemented a large number of requirements from the APLA requirements baseline so as to be able to study the consequences of providing personal, computer-aided training to the crew of long-duration manned missions. We particularly focused our implementation (and subsequent evaluation) on the deployment of gamification principles, as discussed in Sec. III.

The APLA demonstrator was implemented as a rich client in the Java programming language, mainly using open-source protocols and standard libraries and GUI elements. Upon startup, the application presents a login dialog. Once the user has been authenticated, APLA presents a central console window as depicted in Fig. 6.2.

We implemented six core gamification functions in the demonstrator because they support our primary goals of motivation, autonomy, and efficiency, and because they are very representative of the concepts of fiero and flow, as explained in the previous section: learning space; (procedural) training guidance; skill graph; notifications; annotations; social comparison.

1. LEARNING SPACE

The first central APLA support pillar is a transparent personal learning space that provides visualization and situational awareness to the user with respect to their current position in the training space. Its purpose is to give an overview of active, completed, and possible future learning activities, to display the trainee's current certifications and achievements, and to show the trainee's skill levels (Fig. 6.3) in various relevant categories. Learning activities are represented in the demonstrator mainly as individual, short lessons (the APLA form of quests). Certifications are formal content-based qualifications (e.g., an operator-level certification for a specific payload), and can be the result of successfully passed exams or sequences of lessons being completed. Achievements are similar, but are pure gaming elements based around playful goals such as a trainee completing their hundredth

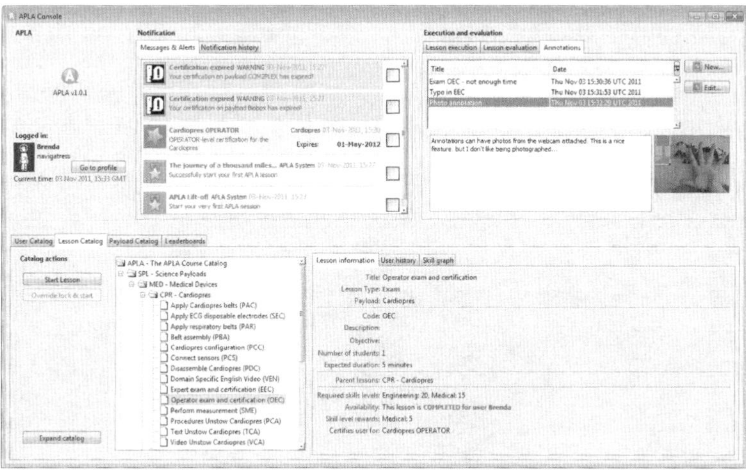

Fig. 6.2 The APLA demonstrator console.

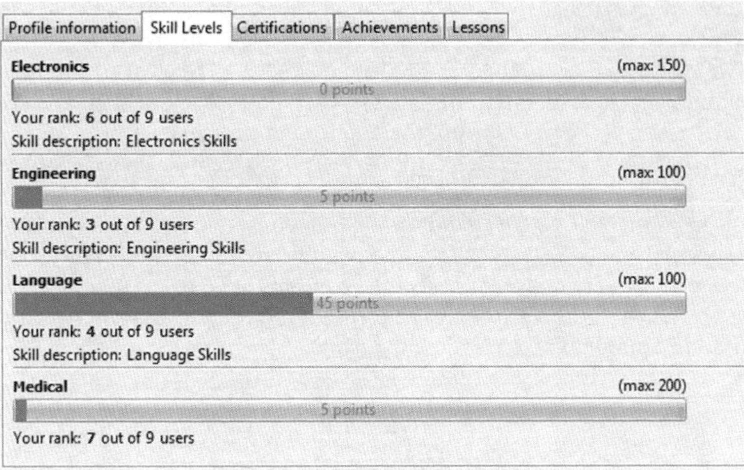

Fig. 6.3   Learning space: user skill levels.

lesson, or spending a certain amount of time in APLA. Skill categories are the key to unlocking lessons and trainee progress; every lesson requires certain skill levels before it can be attempted, and every successfully completed lesson will reward the trainee with additional skill points (Fig. 6.4).

2.   (PROCEDURAL) TRAINING GUIDANCE

Training guidance is the second central APLA support pillar. APLA supports the trainee in selecting, executing, and evaluating lessons. Every lesson in the system can be executed from within the APLA demonstrator (by starting an external application that actually runs the lesson). Currently supported in the demonstrator are video lecture lessons, viewing text or PDF-based syllabi, operational procedural training, and interactive exam lessons (Fig. 6.5). The plug-in architecture

Fig. 6.4   Learning space: user certification status.

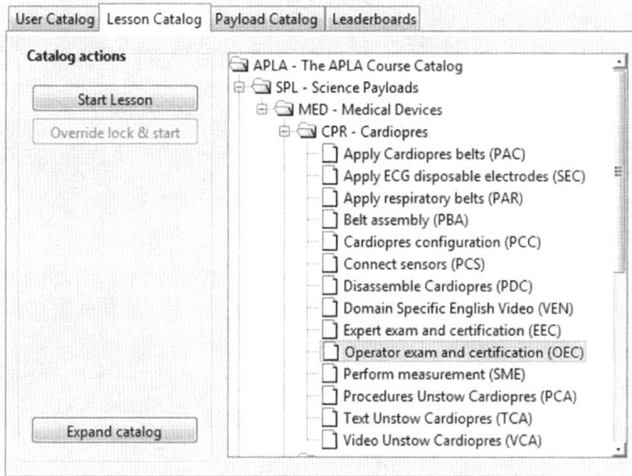

Fig. 6.5   Training guidance: lesson catalog.

is open, however, and other types of lesson can be added easily (e.g., a simulator application). Once the lesson is completed, APLA resumes control, and assists the user in evaluation, showing the results from the lesson, such as an achieved certification and increased skill levels (Fig. 6.6).

3.   SKILL GRAPH

The skill graph is an element that ties together the training guidance and learning space. It is an example of how a support application such as APLA does not just run a lesson, or show a user their status, but can also help them decide which

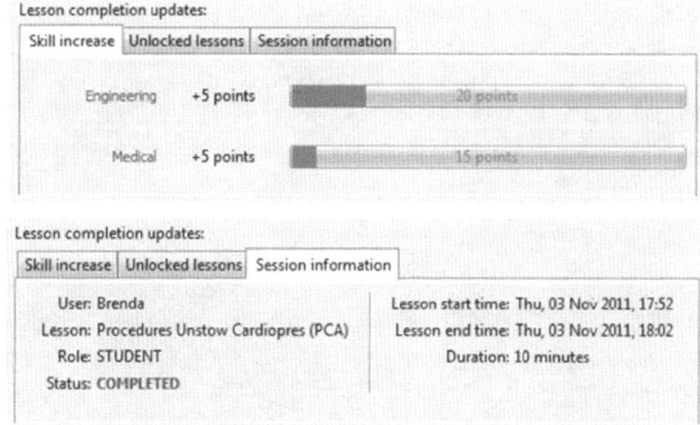

Fig. 6.6   Training guidance: lesson evaluation.

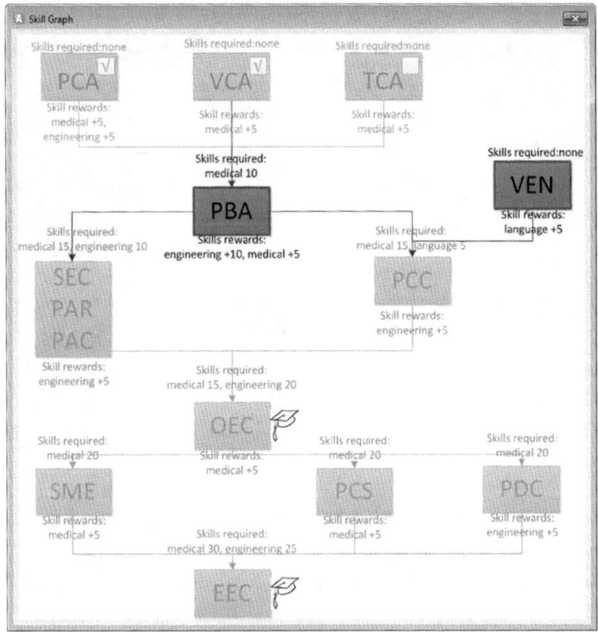

Fig. 6.7 Skill graph.

lesson to choose next; that is, it gives them the autonomy to plan their educational path through the training space. The skill graph does this by giving a 2-D graph-like representation of the dependencies between a set of available lessons (Fig. 6.7).

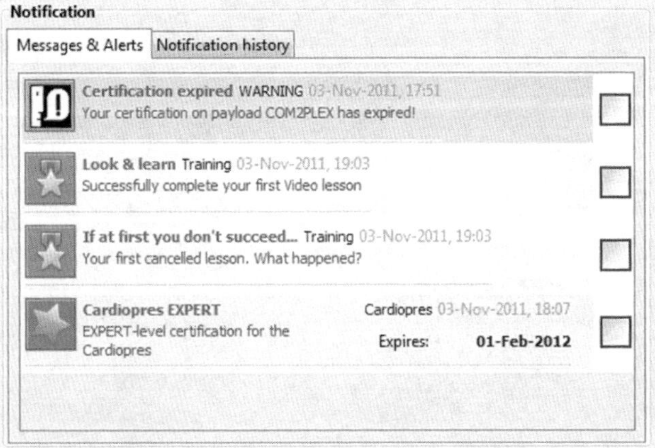

Fig. 6.8 Notifications: incoming messages.

It visualizes which lesson the user has already completed, and shows the available paths to certain goals (e.g., certification). Once an interesting lesson has been identified, the user can find detailed information about the lesson in the hierarchical training catalog (based on the one currently in use at the EAC). The skill graph also shows, at a glance, the skill requirements and rewards associated with each lesson, and whether or not that lesson is currently available to the trainee.

4. NOTIFICATIONS

The notification area of APLA is similar in concept to the notification areas found in current Smartphones and mobile devices (both for games and more general applications). The purpose is to dynamically display (and allow the user to manage) messages from the system as they come in, thus keeping the trainee in the loop and informed about what is going on at all times. Notifications can be purely informational (e.g., notifying the trainee they have achieved a goal, or that an event has been completed), but can also be warnings or other alerts (e.g., that a certification is about to, or has already, expired for that user) (Fig. 6.8). Thus, notifications are another means by which APLA can provide situational awareness and support users in choosing the best course of action, training-wise.

5. ANNOTATIONS

Annotations are a generic mechanism by which the user can actually input information into the APLA system, rather than just consume it in a read-only form (Fig. 6.9). In their simplest form, annotations are like Post-it notes: user-generated content that can be attached to almost any object (e.g., a lesson or a procedure

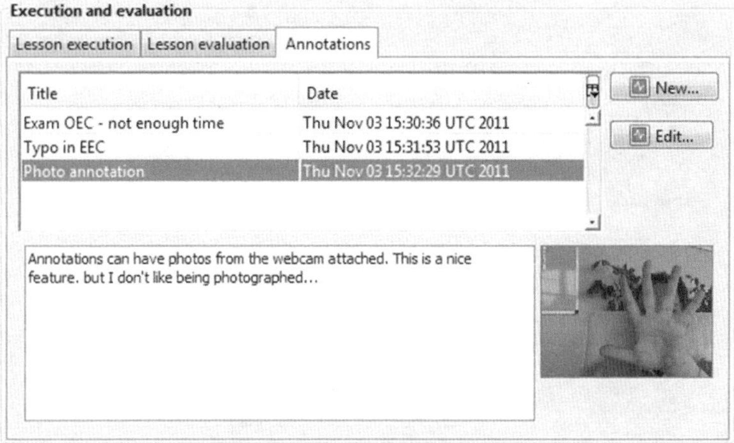

Fig. 6.9   Annotations.

step) within the APLA system. Annotations are persistent and can be viewed by other users of the system (depending on privacy settings). Annotations can be private or public, generic or specific. They can contain text, but also multimedia content (pictures, attachments, voice recordings, etc.). They can be free-form or highly structured (e.g., bug reports submission forms), and can be global or intended for a specific group of recipients. They form a simple means for the users of APLA to asynchronously communicate with training authors, with ground control, with each other, and even with their future selves. In this way, a body of knowledge will be created based on actual experience with the training system.

6. SOCIAL COMPARISON

Finally, one of the hypotheses we wanted to investigate was that a light, healthy form of competition can be an excellent means of providing motivation for a crewmember to continue and excel in their training activities. To be able to test this, APLA contains a number of social comparison elements, chief of which is the Leaderboards tab (Fig. 6.10), where each user can see how all users are ranked in terms of their skill levels for the various skill categories. Another example is the fact that the entire user catalog is initially visible to everyone, so it is also possible to see, in detail, how far other users have progressed in their training, which lessons they have executed, the position on their skill graphs, and so on. Although not implemented in the current demonstrator, it would also be possible to integrate social comparison into some of the other gamification mechanisms, such as notifying somebody when they reach (or lose) the top position in a skill category, or showing the position of multiple users within the same skill graph [15].

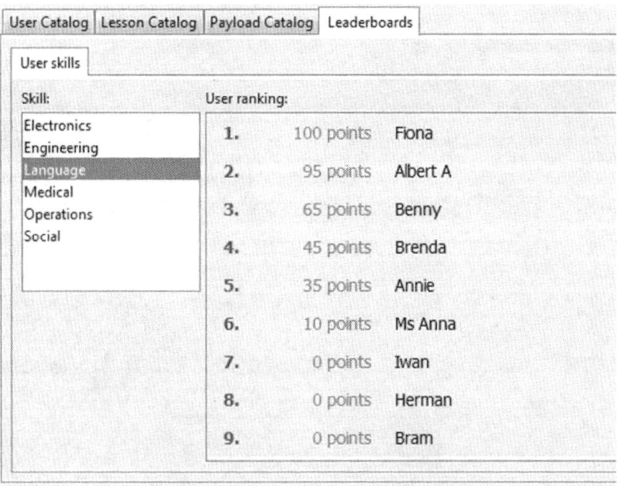

Fig. 6.10   Social comparison: leaderboards.

V. SCORE: RESULTS OF THE EVALUATION

This section briefly describes the setup of the evaluation and then the results of the evaluation.

A. STRUCTURED WALKTHROUGH

The six core functions that were implemented were justified in the evaluation by corresponding claims. Claims are concrete and testable upsides and downsides of the core function and are referred to with measures such as training effectiveness, training efficiency, satisfaction, and so on. The following are examples of some claims and their measures (tested with questions) for the core function *skill graph*:

1. + satisfaction: freedom from discomfort, and positive attitudes to the use of the product.

2. + training efficiency: resources expended in relation to the accuracy and completeness with which users achieve their learning goals.

3. − training effectiveness (superficial learning): accuracy and completeness with which users achieve specified learning goals.

The evaluation process was as follows:

1. *Short introduction*: this explained the objectives and scope.

2. *Questions*: these were related to the expectations of the participants on training support.

3. *User exploration*: the participant was given control over the demonstrator, and was asked to think aloud while using the demonstrator.

4. *Structured walkthrough*: based on a scenario, the test leader walked through the demonstrator with the participant. At times when a core function was used, the test leader asked the participant questions related to the claims.

5. *End questionnaire*: this included multiple choice as well as open questions on the activities and observations during the walkthrough.

6. *Interview with the participants*: this observed the general reception of the application.

The evaluation took about $1\frac{1}{2}$ hours. During the evaluation, a specialist was seated next to the participant, observed his actions, and, if necessary, asked questions for clarification. The experiment was performed by 15 participants, who walked through the demonstrator with the test leader. The participants were from a variety of relevant backgrounds, from engineers from ESA (space mission experience), submarine crewmembers (experience with long-duration missions in isolation), and training scientists, to trainers from EAC. For an impression of the evaluation, see Fig. 6.11.

Fig. 6.11 Participants filling in the questionnaire behind the demonstrator: on the left are the submarine crew and on the right the training experts from EAC.

VI. RESULTS

The following results were gathered:

1. For *learning space*, the mean answers show that participants were positive that APLA supports the user in his awareness about his training progress (Fig. 6.12a, question 1), that the participants are positive that APLA helps to reach certification more efficiently (question 2), and that the participants are slightly more than neutral on the motivational aspect of APLA (question 3). The variance in the answers is large.

2. *Procedural training guidance* was evaluated to be a bit more than positive on learning skill (transfer to performance in 'real' operational performance), but with a large variance in the answers (Fig. 6.12b). Feedback from participants suggested that they would like a demo with a full lesson.

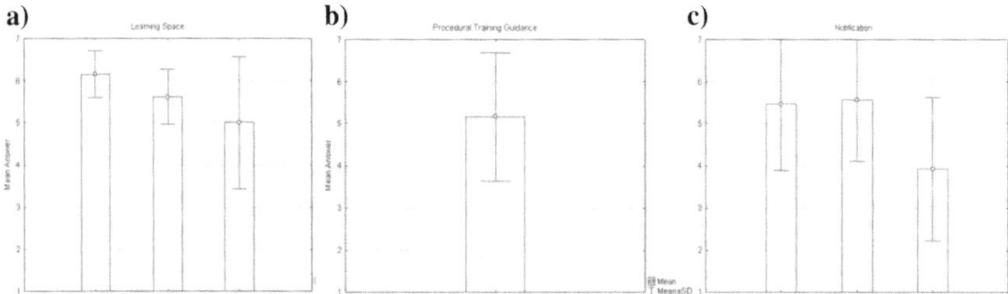

Fig. 6.12 (a) Mean answer and standard deviations to questions concerning claims for the learning space core function. (b) Mean answer and standard deviations to questions concerning claims for the procedural training guidance core function. (c) Mean answer and standard deviations to questions concerning claims for the notification core function.

3. For *notifications*, participants felt more aware about the schedule for certification (Fig. 6.12c, question 1). They trusted APLA to schedule the recertification (question 2). Notifications themselves were too interruptive (question 3).

4. *Annotations* were rated low on the claims referring to this core function (Fig. 6.13a). The function, however, was liked by participants and was labeled as important in the explanation of their answers and discussion, but not as it is now. Good suggestions were given on how to improve this function: for example, it should be possible to set visibility of annotations on for example public, private and feedback to ground. It should be possible to link annotations to a lesson and annotations can also be used as Q&A with ground or crewmembers.

5. The results to the questionnaires for *social comparison* were not that negative; see Fig. 6.13b, where question 1 asked about motivation and question 2 about insight into relative learning performance. In the comments and discussion, however, the participants made many remarks on how it was implemented. The implementation of the social comparison should be reconsidered, because, as it is, it may have a considerable negative effect on (some) users. It should probably show the relative position in crew competencies to a greater extent, rather than provide a competitive game-like ranking.

6. The results of the *skill graph* all show large variances (Fig. 6.13c). Remarks showed that the participants thought that the skill graph would not add to effectiveness (the courseware would) or motivation. Furthermore, this function should be more interactive to allow more feedback.

Several users found the demonstrator too childlike. This was a result of the avatars and colors, but also of gamification aspects such as achievements. Some participants found the achievements motivating (participants with background knowledge of training, younger representative end-users), but others found them unnecessary or even infantile (some representative end-users). An aspect

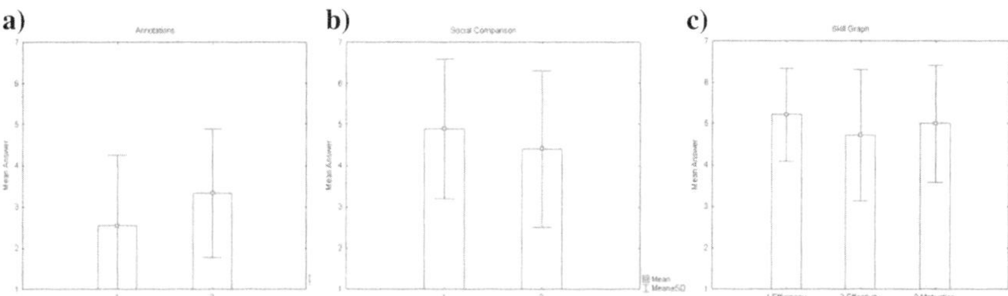

Fig. 6.13 **(a) Mean answer and standard deviations to questions concerning claims for the annotations core function. (b) Mean answer and standard deviations to questions concerning claims for the social comparison core function. (c) Mean answer and standard deviations to questions concerning claims for the skill graph core function.**

that influences the opinion on features like this can differ for users depending on age (whether they have grown up with game-like features as achievements) or background. This is something that needs to be evaluated with real end-users. Another influence is that the content was not a focus of this evaluation. If more serious content were available, this aspect might have been evaluated more positively. A recommendation is that the next demonstrator should have a more extensive and realistic content to train participants. The current prototype does not take into consideration that something that has been learned can also be forgotten over time. Imposing rehearsals or refresher training to keep the astronaut up to date should be taken into consideration.

VII.   LESSONS-LEARNED

During our study we have argued that to keep crew well prepared on long-duration missions for operational tasks in a dynamic environment, the need for training is very important. However, the emotional aspects of the long duration of the missions include a decline in motivation and a loss of interest in self-study [14]. Our initial results indicate that with the introduction of gaming aspects these may be overcome. The aspect of giving control to the trainee is perceived as very positive, in particular, as well as the overview on available lessons that the skill tree and the transparent learning space provides. As we have demonstrated, gaming aspects are not yet used in current astronaut training; however, they are a welcome addition to the training curriculum. More research should be carried out to further study the applicability of serious gaming to the training of astronauts. We have only barely scratched the surface.

ACKNOWLEDGMENTS

This research took place as part of the APLA project (ESA contract 4000102351), funded under the General Support Technology Programme (GSTP-4). We would like to thank EAC for their input during the project and as participants during the evaluation. Other parties that participated in the evaluation and we would like to extend our thanks to are the Submarine Service of the Royal Netherlands Navy, ESA-ESTEC, and TNO.

REFERENCES

[1]   Neerincx, M. A., Bos, A., Olmedo-Soler, A. Brauer, Breebaart, L., Smets, N., Lindenberg, J., Grant, T., and Wolff, M., "The Mission Execution Crew Assistant: Improving Human–Machine Team Resilience for Long Duration Missions," *Proceedings of the 59th International Astronautical Congress (IAC2008)*, IAF, Paris, France, ISSN 1995-6258, 2008.

[2]   Diggelen, J. van, and Neerincx, M. A., "Electronic Partners that Diagnose, Guide and Mediate Space Crew's Social, Cognitive and Affective Processes," *Proceedings of*

Measuring Behaviour 2010 edited by Spink, A. J., Grieco, F., Krips, O. E., Loijens, L. W. S., Noldus, L. P. J. J., and Zimmerman, P. H. (Eds.), Noldus Information Technology, Wageningen, The Netherlands, 2010, pp. 73–76.

[3] Neerincx, M. A., "Situated Cognitive Engineering for Crew Support in Space," *Personal and Ubiquitous Computing*, Vol. 15, No. 5, 2011, pp. 445–456.

[4] Baker, C., Naikar, N., and Neerincx, M. A., "Engineering Planetary Exploration Systems: Integrating Novel Technologies and the Human Element Using Work Domain Analysis," *Proceedings of the 59th International Astronautical Congress (IAC2008)*, IAF, Paris, France, ISSN 1995-6258, 2008.

[5] Smets, N. J. J. M., Neerincx, M. A., Cohen, I., and Diggelen, J., "Improving Crew Support Methods in Human–Machine Teams for Long Duration Missions," *Proceedings of the 63rd International Astronautical Congress 2012*, IAC-12-A1.1.23, IAC 2012, Naples, Italy, 2012.

[6] Dickey, M., "Engaging by Design," *Educational Technology Research and Development*, Vol. 53, No. 2, 2005, pp. 67–83.

[7] Rieber, L., "Seriously Considering Play," *Educational Technology Research and Development*, 1996, Vol. 44, No. 2, pp. 43–58.

[8] Kirschner, P. A., Sweller, J., and Clark, R. E., "Why Minimal Guidance During Instruction Does Not Work: An analysis of the failure of constructivist, discovery, problem-based experiential and inquiry-based teaching," *Educational Psychologist*, Vol. 41, No. 2, 2006, pp. 75–86.

[9] Malone, T. W., "Toward a Theory of Intrinsically Motivating Instruction," *Cognitive Science*, 1981, Vol. 5, No. 4, pp. 333–369.

[10] Peeters, M. M. M., Van den Bosch, K., Meyer, J.-J. Ch., and Neerincx, M. A., "Scenario-Based Training: Director's Cut," *Artificial Intelligence in Education*, edited by Biswas, G., Bull, S., Kay, J., and Mitrovic, A. (Eds.), Vol. 6738, Lecture Notes in Computer Science, Springer, Auckland, New Zealand, 2011, pp. 264–272.

[11] Peeters, M. M. M., Bosch, K., van den, Meyer, J.-J. Ch., and Neerincx, M., "An Ontology for Integrating Didactics into a Serious Training Game," *Proceedings of the 1st International Workshop on Pedagogically-Driven Serious Games (PDSG 2012), in conjunction with the Seventh European Conference on Technology Enhanced Learning (EC-TEL 2012)*, edited by Bocconi, S., Klamma, R., and Bachvarova, Y., (Eds.) 2012, pp. 1–10.

[12] Huizinga, J., *Homo Ludens*, Beacon Press, Boston, Massachusetts, U.S.A., 1971.

[13] McGonigal, J., *Reality Is Broken: Why Games Make Us Better and How They Can Change The World*, Penguin Press, New York, 2011.

[14] Csikszentmihalyi, M. *Flow: The Psychology of Optimal Experience*. Harper Collins, New York, 1991.

[15] Murray, T., and Arroyo, I., "Toward Measuring and Maintaining the Zone of Proximal Development in Adaptive Instructional Systems," edited by Cerri, S. A., Gouarderes, G., and Paraguacu, F. (Eds.), Springer-Verlag, Berling Heidelberg, *ITS Conference*, 2002, pp. 749–758.

[16] Peeters, M., Van Den Bosch, K., Meyer, J.-J. Ch., and Neerincx, M. A. "Situated Cognitive Engineering: The Requirements and Design of Directed Scenario-Based Training," *The Fifth International Conference on Advances in Computer–Human Interactions* (ACHI 2012), Xpert Publishing Services (XPS), Wilmington, DE, USA, 2012, pp. 266–272.

CHAPTER 7

Timeline as Unifying Concept for Spacecraft Operations

William K. Reinholtz*

Jet Propulsion Laboratory, California Institute of Technology, Pasadena, California

I. INTRODUCTION

The Advanced Multi-Mission Operations Systems (AMMOS) Mission Operations System tools and services are being reengineered to improve its capability and efficiency, and reduce operations and maintenance costs [1]. Timelines were identified as a ubiquitous data structure within the existing AMMOS, although they were noted to be implicit, informal, and non-uniform in representation and semantics, and so could convey little practical benefit to the current AMMOS [2]. Timelines were subsequently proposed as a key architectural concept and concrete data structure for the new AMMOS [3], and were in the final stages of vetting as of April 2012.

The proposed AMMOS architecture is based on the notion of the orchestrated execution of software (programs, processes, services) with the bulk of the data being of Timeline semantics and syntax, and data exchange being primarily through the Timeline database. It is an orchestrated, blackboard architectural style, with Timelines as the foundational semantics and consequent data structure. The semantics are specified a priori and are an architectural invariant.

There are several major programmatic and technical advantages to this approach:

1. It decouples the programs from each other so that they can be adapted and evolved independently (AMMOS principle: *minimize coupling*). Maintenance and adaptation costs are then reduced.

2. It makes the program interactions explicit so that it can be managed and evolved in a systematic manner. Maintenance costs are then reduced.

3. Communications, Coordination, Information Model, and software are separable so that each can be managed independently. For example, moving from software to services is facilitated. Communications can be changed (e.g., between message bus and RESTful) without rewriting the software. Exploiting the tremendous scalability of the cloud is straightforward.

---

*Principal Engineer; kirk.reinholtz@jpl.nasa.gov.

4. Incidental coupling due to ad hoc point-to-point interfaces and data structures is avoided.

5. It makes it relatively easy to add new programs to the ensemble, which reduces the temptation to construct large monolithic applications. Adaptation and maintenance costs are then reduced.

6. It allows smaller, more cohesive programs to be constructed and integrated (AMMOS principle: *maximize cohesiveness*). Adaptation and maintenance costs are then reduced.

7. It facilitates effective program interactions across organizational boundaries because of the common language of timelines, and comprehensive closed-loop control of the spacecraft at various levels: Integrated Planning and Sequencing, comprehensive Accountability. All are made practical.

8. Programs can be orchestrated in new ways, to provide new capabilities.

The following are some of the many spacecraft operations that are organized around timelines in AMMOS: 1) recording telemetry channel values as received; 2) computing spacecraft state timelines from telemetry; 3) creating command timelines from plan/activity timelines; 4) updating expected spacecraft state timelines based on commands to be sent to the spacecraft; 5) comparing expected spacecraft state with actual spacecraft state so as to adjust future plans and activities; 6) graphical and textual real-time display of timeline values to monitor spacecraft health and performance; 7) graphical and textual display of timeline historical values to monitor spacecraft health and performance; 8) automated flight rule checking; 9) science planning; 10) spacecraft engineering activities; 11) subsystem monitoring and trending and health analysis; 12) science product availability prediction and notification.

All these operations manipulate timelines, although, at present, the timelines are tacitly defined and the definitions have not heretofore been formalized (or even expressed, for that matter) or shared among the capabilities. That lack of sharing has led to stove-piping that has hindered cost-effective interoperability.

This chapter defines the key timeline concepts (immutability, versioning, timeline type, timeline name, timeline instance, and timeline value), key types (state, measurement, event, and activity), and several mission operations software architectures that exploit timelines to reduce costs and improve spacecraft operability.

II. RELATED WORK

The work described here is one of several tasks related to the modernization of processes and software used within the NASA-funded AMMOS program. The Operations Revitalization task (OPSR) [1] is developing an AMMOS architectural framework and set of Mission Services that include adaptable multimission engineering specifications. These specifications include essential documentation such as operational concepts and scenarios, requirements, interfaces and agreements, information models, and mission operations processes. Those products will provide clear and rigorous specification of operations that inform the work described here.

The second task, called Sequence Revitalization (SEQR) and the subject of this chapter, is developing detailed timeline semantics and data structure specifications, and most of the software architecture and implementation that uses timelines.

Both tasks use timelines as a key architectural concept, but treat them from somewhat different perspectives. The OPSR task is concerned with the use of timeline information by operations personnel—the human need for information. The SEQR task focuses on the detailed structure of information and how it can be manipulated using software to service human needs for information. This has led to different levels of detail in the two tasks' respective timeline information models and implementation. The timeline information models are joined into a unified and coherent whole in [2] as part of the OPSR task.

It was recognized that realization of the full potential of these related tasks required that they work to a common set of architectural principles and objectives, as otherwise there would doubtless be clashing assumptions and implementation, as well as semantic gaps and conflicts that would be costly to resolve (and would probably damage the integrity of the architecture until resolved). The teams therefore participated in the development of a common architectural vision, the results of which are described in [3].

III. INTRODUCTION TO TIMELINES

The timeline [4] is abstractly defined as a container of items indexed by time, or of items related by time. The abstract definition is intentionally rather open, and the edge between timeline and not-timeline is fuzzy. The abstract definition does not need to be formalized, because it is the types of timelines that are actually defined that have practical impact. Timelines are made practical by creating concrete types of timelines that can be precisely defined, stored in databases, manipulated in software, and so on. An informal example follows to illustrate the concept of timeline. The definition of new broad types of timelines is a matter of detailed system engineering and engineering judgment and is not discussed further here.

1. *Record measurements.* Many telemetry channels are measurements; for example, a periodic sampling of the spacecraft bus voltage. Such measurements are recorded on timelines (usually one timeline per sensor). The timeline is conceptually a list of timestamp and value tuples, where timestamps must have values and must be unique within a given timeline. The timestamp is defined to be the primary key of the tuple. In this case the timestamp is the time at which the sensor was read, and the value contains the sensor reading.

2. *Record state.* "... state variables are the smallest possible subset of system variables that can represent the entire state of the system at any given time. ("State space (controls)," Wikipedia, The Free Encyclopedia). Measurements are sampled in discrete time. State in principle has a value at all points in time. A state timeline is typically implemented as piecewise continuous interpolators. We require that all state timelines return a value for any possible time $0, \ldots, \infty$; otherwise individual software elements would hard-code the meaning of values (probably conflicting with other elements and probably not logically reasonable for all timelines) outside the domain of the timeline. Measurement and state timelines may well share a common structure (tuple indexed by time yielding sample, or tuple indexed by time yielding interpolator). We have found it best to architect around their mathematical distinctions, rather than implementation similarities.

3. *Record events.* These are a labeled instant in time. Events have zero duration. "The switch turned on" for example, or entries in a time-stamped error log. ("Event (Relativity)," Wikipedia, The Free Encyclopedia).

4. *Record commands.* This is a timeline that represents the information received (or may be received) by the spacecraft [or more generally, the system being controlled, e.g., a Deep Space Network (DSN) antenna] for the purpose of changing its behavior.

5. *Record intent.* Intent timelines (often expressed today as activities or plans) indicate an acceptable envelope of spacecraft operation. Intent is not a single state trajectory. It is, at least abstractly, a set of possibilities. The general act of operating a spacecraft uses intent to determine specific commands to impose on the spacecraft, and reconciles the resulting state with the intent to determine what adjustments must be made to the intent and what further commands must be uplinked. It is basically a closed loop control between intent and state.

IV. KEY TIMELINE CONCEPTS

A *timeline* is informally and most generally defined as a mathematical construct that contains items that are strongly temporally related (e.g., the estimated battery voltage as a function of time, or the discrete finite list of bus voltage measurements indexed by time of sample acquisition, or "this and that must not both happen at the same time" or "this must happen no more than five minutes after that"). We prefer to describe timelines in mathematical terms, rather than those of software engineering, to keep focus on the engineering use of the timeline (e.g., a function continuous in time that represents the battery voltage at any point in time) rather than on the implementation details of the timeline (e.g., a series of linear interpolators). In particular, we do not want mere implementation details to leak into all the software that uses timelines, as this would make it very costly to change those details (they are not mere details once that leak happens; they become tacit aspects of the architecture).

Several mathematical timeline types may use the same data structures (e.g., a list of events indexed by time and a list of interpolators indexed by time may use the same data structure, but are quite distinct mathematical concepts). Such sharing is simply an implementation detail and must not leak past the Application Programming Interface (API), because we must to be able to evolve the data structure without performing costly changes to all the software that uses the timeline.

Not everything that relates to time is a timeline. If time is the dominant index and/or relationship in the structure, it is probably a timeline. If not, it may not be a timeline. For example, you could store all science images received from the spacecraft on a simple timeline that is indexed by time and yields the image that was acquired at that time (we call such a timeline an event timeline). However, that is probably not the dominant access pattern and there are many other likely forms of query and relationships, so this is probably not the most appropriate use of a timeline.

Best practices suggest that you use the simplest timeline type that will work for your application. For example, you can do everything with a temporal constraint network timeline that you can do with an event timeline, so in theory the event timeline is unnecessary. However, practical considerations (performance, simplicity, robustness, etc.) bring great utility to the linear timeline: "In theory there is no difference between theory and practice. In practice, there is."

Various implementation mechanisms are used to map an instance of a mathematical/system-engineering timeline into a software data structure for storage and transmission. The architecture is flexible on this point so that the mapping can be selected based upon the use cases for the particular timeline. For example, an estimate of the spacecraft bus voltage is, mathematically, a function continuous in time, and may be mapped into software as piecewise continuous polynomial interpolators.

The values of a timeline are stored in a *timeline instance* (typically but not necessarily in a database of some sort). Each timeline instance contains multiple *timeline versions*. In principle, every mutation of the timeline instance may form a distinct version, although in practice the programmer may choose to perform a number of mutations as an atomic operation. The timeline instance is, in effect, an L-value ["Value (computer science)," Wikipedia, The Free Encyclopedia] that has a name, and provides a name for each mutation (e.g., write or assignment) to the instance. A reference to every mutation of any instance can thus be created and dereferenced in a uniform manner, providing referential transparency.

Every timeline instance is assigned a unique Timeline ID number (TLID) by the system when the instance is created. Two timelines are defined as the same timeline if they have the same TLID; otherwise, they are not the same timeline (identical copies perhaps, but not the same timeline). Renaming a timeline is defined as associating the new name with the ID that the old name previously referenced. Copying a timeline is defined as creating a new ID with the new name, and performing, in effect, a deep copy (for performance reasons one may choose to be clever in sharing structure, although the effect will always be that of a deep copy) of data into the new ID.

Every timeline instance is given a unique *timeline name*. Names are *namespaced*; for example, each mission would probably have its own namespace, so that timeline names need only be managed at the project level. Namespaces basically work like Linux directories. There is a root, which has directories, the directories have directories, and directories may contain timeline names. The timelines within a namespace need not reside in the same database and may be moved among databases without changing the name; migration and tiered storage is thus supported. Immutability means that they may also be mirrored for replication purposes. Names are assigned to timelines by the users, and may be changed. The architecture per se does not depend on the structure of the namespaces, but software and processes no doubt will, so it will be a (very important) matter of system engineering to design an appropriate namespace scheme. The scope of the namespace in effect defines the scope of the system boundary; if two independent namespaces are created, then the same name may refer to two different timelines, which means that those two namespaces must be understood to define nonoverlapping system boundaries.

A version of a timeline instance, once created, is at once and forever *immutable*. It follows that the timeline instance full name and timeline version, taken together, will always reference exactly the same value, assuming that the indicated version exists. The story is a bit more complicated, but still sound, when renaming is allowed. The story is also a bit more complicated, but still sound, when physical deletion is allowed. More on these complications later.

When a reference to a particular version of a timeline instance is dereferenced, the result is a *timeline value*. A timeline value is an R-value. Most of the AMMOS software will probably manipulate either versioned instance references, or values. Explicit manipulation of the versions will probably be the domain of configuration management.

Immutability of timeline versions means that timeline values are referentially transparent. The value can always be recreated from a reference to the version, so no timeline value can exist that is not a version of a timeline instance. It is never necessary to store a file containing a timeline value, because the value can always be recreated from the versioned timeline name. This is the primary reason for specifying the immutability principle at the architectural level.

Timeline versioning (in fact all versioning, including mutable timeline metadata, and even nontimeline information) is modeled in terms of the system change number (SCN). The SCN is, in effect, an integer that is incremented by at least 1 within each data-mutating transaction. The SCN must be allocated in strictly increasing order so that the mutations are strictly ordered by SCN, at least in a quiescent system (although they may not be exposed in the same order due to transaction semantics). Holes are allowed so that certain performance optimizations may be applied, and duplicates are never allowed. The SCN is used to label each change made within each transaction, where transactions are in effect serialized (we only literally serialize them where semantics depend on serialization, and otherwise reserve the option for nonserialized execution for performance reasons). The SCN represents the instant in time at which the resulting database state is defined to have been created or logically deleted. The SCN therefore precisely labels each state of the database, and no state exists that is not labeled by an SCN. Each record is, in effect, tagged with the SCN at which it was created and the SCN at which it was logically deleted. The record is defined as logically existing for all SCN: $\text{SCNCreated} \leq \text{SCN} < \text{SCNdeleted}$. The creation SCN must not be changed after initial creation of the record. The deletion SCN is given the value of infinity at record creation, and can be mutated exactly once thereafter to assign a deletion SCN. An implementation may choose to log each SCN along with pertinent details as to who did it, why, when, and so on, to provide a detailed change log of each change to the state of the Database (DB).

Physical deletion is allowed. Any operation that references a deleted SCN will return an appropriate error indication. An explicit error must be indicated, because a deletion would otherwise cause violations of immutability and so referential transparency. The architecture depends upon immutability, and physical deletion is a practical necessity, and so an error is thrown. Physical deletion is only allowed when it is proven that the deletion cannot alter the results of any query (other than to throw the error noted above), for that would violate immutability. For example, if there were a "how many records in the database as of SCN x"? query, then any physical deletion would alter that query for all time into the future. Such queries must not be provided for other than administrative purposes, otherwise physical deletion would be prohibited.

Transactions are only required to be serialized where the system depends upon the serialization: for example, mutations to a given timeline instance would generally be serialized so that a long-lived transaction cannot cause violations of immutability, but mutations to distinct instances probably do not need to be serialized. System engineering determines what serialization models should be used.

Every timeline instance has *timeline metadata* that describe the static (i.e., cannot be mutated once the instance is created) and dynamic (can be changed over the life of the instance) properties of the timeline. Mutable metadata (e.g., the timeline name) are subject to SCN semantics. The metadata are used make explicit the information that is needed to find and use the timeline (contrasted with imbedding the information in the tools that use the timeline). For example, the schema (or "type") of the items in the timeline should be in the metadata, so a timeline visualization tool could in principle read the schema and display the timeline using only that schema.

Because the timeline instance name may be changed and is therefore under SCN semantics, access via name is a two-step process: 1) the name is dereferenced to a TLID with respect to a given SCN; and 2) the TLID is dereferenced to data potentially in another SCN. We expect that in the common case, the same SCN will be used for both operations, although the architecture supports the use of distinct SCNs.

Every item in a timeline instance must have a key that uniquely identifies that item within version(s) of the instance. Often the index is a time value, but it may be more complicated. It follows that name, version, and item name form a durable and immutable *timeline item reference*. That reference forms the basis of data structures that track relationships between items. For timeline structures where there is a well-formed notion of the location of, or index of, an item in the timeline, that location or index should serve as the item name; indeed, no point in making up new names when existing ones will do. For example, the time index of an event timeline is a perfectly good name for the item at that time index.

Every timeline instance is immutably associated with the time system (e.g., International Atomic Time (TAI)) of the time values within the instance, and the time format (e.g., large integer fixed point offset from epoch in nanoseconds) in which those values are stored within the instance. The time system must include the location of the clock, so there is no confusion when relativistic effects are significant. For example, an atomic clock onboard a particular spacecraft, even a perfect one, forms its own time system. Various mechanisms for conversion between commensurate time systems, and presentation via display formats, are of course provided.

V. TIMELINE CATEGORIES

The major types (or classes or categories) of timelines comprise an open, but highly controlled, list. It is open because we expect that as new technologies and techniques come along we will need to extend the list of supported timeline types. It is highly controlled to counter the apparently natural human tendency to define a new type with semantics very specific to the application at hand, rather than use a previously defined type. If the list is too easily extended, then type proliferation will dilute the key advantage of having common timeline semantics standardized and shared among many applications.

Timeline categories have two distinct perspectives: system engineering perspective, which is primarily concerned with the semantics of and operations upon the various types, and implementation types, which encode the semantics into computer data structures that can be serialized and versioned and stored.

A. SYSTEM ENGINEERING CATEGORIES

1. *Measurement*: A measurement timeline contains sampled measurements, usually obtained from a sensor of some sort, that are recorded as a totally ordered time-indexed sequence of data points. Typically, there is one timeline per sensor. Multiple data points at the same time index are not allowed, because of the confusion that typically follows when various tacit, implicit,

and/or ad hoc mechanisms are inevitably used to force an order onto the duplicates. If multiple points can occur at the same time instant then a composite data point type is defined that can contain multiple samples, and any required ordering semantics if required are made explicit as part of the definition of that composite type. Measurements are generally input to estimation processes, which result in state timelines. Processes other than estimation should generally operate on state timelines, rather than raw measurement timelines.

2. *Event*: An event is a labeled instant in time. Event timelines are often used to label points in time at which events did or are intended to occur. For example, an event timeline might be used to mark the instant in time at which each telemetry pass may first expect to receive a signal. As with measurement timelines, the events are totally ordered and time indexed. Events are defined in this way to prevent their creep (via the addition of durations, or allowing them to consume resources or directly cause things to occur) into the semantic domain of state or intent timelines.

3. *Command*: A timeline that represents the information that has been or may be received by the spacecraft (or more generally, the system being controlled, e.g., a DSN antenna) for the purpose of changing its behavior.

4. *State*: The engineering concept of system state (for example the state of the spacecraft) is that the state at some point of time is sufficient to predict how the system will respond to inputs (gravitational forces, commands, etc.) without reference to the past history of the system inputs and outputs. "The state" is obviously a very large and generally unknowable value. Much of the art of system engineering is determining a useful subset of that platonic state that can be known (often estimated from measurements) and is required to operate the spacecraft. That subset is known as the state variables, where one variable may be the attitude of the spacecraft, another may be the velocity, yet another the acceleration, another for camera articulator angles, available fuel, battery state of charge, commands queued up for execution, and so on. The spacecraft state is represented by state timelines (probably tens of thousands), usually one timeline per state variable. There are also state variables for other systems of interest to the spacecraft operator, for example, rover surface assets, the configuration of the DSN antennas, and telemetry relay assets. A state timeline is abstractly a function of time that yields a state variable value at that time. The key distinction between a measurement timeline and a state timeline is that the domain of the measurement timeline is a finite list of timepoints, and the domain of a state timeline is the infinite set of all possible timepoints. The other distinctions noted here are a matter of sound engineering practice but not intrinsic to the architecture. If you interpolate between measurements, you are using the measurement timeline as a state timeline and really should use a state timeline instead. State timelines are total in time (yield a value for any possible time input, from birth of

universe to infinity beyond), to avoid the mess that would come if applications themselves encoded the meaning of times not covered by the state timeline. It is better for the timeline to return an explicit "spacecraft did not exist at that time" than to encode that interpretation into all software that uses state timelines.

5. *Intent*: Spacecraft operations often involves creating activities that represent higher-level desired spacecraft behavior, which are then decomposed into commands that can be executed by the spacecraft and that accomplish the activities. Telemetry is monitored, perhaps by comparing actual state timelines (computed in turn from measurement timelines) to predicted state timelines (perhaps generated via simulation) so as to assess and manage progress in executing the activity. An activity timeline contains activity instances. An activity instance is a named and parameterized interval of time, where the start and end time of the interval may be assigned values, or may be variables or allowed ranges of times. The actual semantics of the activity name and parameters is outside the scope of the timeline itself. We expect that several intent timeline types will be developed to support spacecraft operations automated planning and scheduling systems. A survey of such systems and the possibilities of developing common timeline semantics for them is covered in [5].

B.   IMPLEMENTATION TYPES

The system engineering timeline categories are implemented as shown here. There is no one-to-one relationship between the system engineering category and the implementation mechanism, because in some cases a single mechanism can support several implementation types without compromise.

Every timeline instance ("timeline instance" as defined herein and elaborated upon in [4]) must have the following associated information (not just the slots; values must be assigned): 1) timeline type: immutable; 2) TLID: immutable, assigned by system at the time of creation; 3) timeline name: mutable, may be changed after timeline created; 4) time system: immutable; 5) time format: immutable; 6) physical schema: immutable; 7) type-specific information as specified for the timeline type.

Every timeline version must be associated with the information necessary to allow the receiving entity to interpret the bits, and must contain the information necessary to retrieve the bits from the timeline instance that contained the version. It is not necessary to serialize this information with each version: it is acceptable to use the timeline name or TLID as a primary key ("PK") for this information where that makes sense, for example, 1) all information specified as common to every timeline instance; 2) the query that extracted the information from the instance; and 3) the instance SCN to which the query was applied.

The *physical schema* describes how the value is serialized into the database. It may indicate an External Data Representation (XDR) encoding of a data structure containing two floating point numbers, for instance. The physical schema information must be sufficient to allow software to convert the bits in the database into a data structure (e.g., a Java object), but does not indicate the interpretation of that object. This is so that a struct of two floats can be used for several purposes, e.g., xy coordinates or a linear interpolator.

Every timeline instance is assigned a name (where the name consists of a namespace prefix followed by a short name). The name uses path syntax and semantics ("Path," Wikipedia, The Free Encyclopedia). The name does not define the location of the timeline (e.g., it is not a URL to a particular database), although an important property of the syntax is that it is designed to be used as part of an HTTP URL without the need for character encoding should one wish to construct a URL from the name. Specifically, the names are constructed with reference to RFC 2616 and RFC 2396 (RFCs are available from the official location www.ietf.org/rfc).

The name syntax is intentionally rather restricted. This is so that the name is not likely to collide with symbols used in other protocols and languages.

The timeline name has the following syntax as specified in Augmented Backus-Naur Form (BNF) as defined in RFC 2616:

```
TLNAME = *1 ( "/" ALPHA * (ALPHA | DIGIT | "_" | "-") )
```

1. GROUNDED LINEAR TIMELINE (GLTL)

The GLTL is used to store system engineering timelines of type Measurement, Event, and State. The GLTL is the simplest timeline type. It is purposefully restricted to a simple form, for two reasons. First, a large number and percentage of timelines used in spacecraft operations are of this simple type. Second, we want a simple, theoretically sound, easily explained, easily implemented timeline for the ubiquitous case. Where more complexity is required, different timeline types are used.

A GLTL is defined as a sequence of {timepoint, item} tuples where 1) all timepoints are represented in the same time system and in the same time format; 2) all timepoints are grounded, meaning they are assigned specific absolute values (not variables, not offsets relative to previous timepoints); and 3) the timepoints have a total order, meaning in particular that there are no duplicate values.

State Timeline discretized values (e.g., ON or OFF) are stored directly, with the common semantic that the value holds over the interval from the timepoint of that item in the timeline to, but not including, the timepoint of the following item.

State Timeline Interpolated values may be stored in two ways:

1. Store discrete values in the timeline, then compute and evaluate the interpolation function as needed. In this case the interpolation function is computed and discarded every time it is used. The NAIF Type 8 ephemeris[†] is an example of this approach.

[†]http://naif.jpl.nasa.gov/pub/naif/toolkit_docs/C/req/spk.html#Supported%20Data%20Types

2. Store the interpolation function in the timeline, where each function is defined as valid over the interval from the timepoint of that item in the timeline to, but not including, the timepoint of the following item. The NAIF Type 2 ephemeris is an example of this approach.

The purpose of this specification is to define the minimal syntax, semantics, and attributes of grounded linear timelines such that they may be exchanged between systems without semantic clashes or ambiguity, and with a minimum of syntactic conversion machinery. We expect that concrete bindings will be developed (e.g., XML schemas and documented CSV formats) but those bindings are not part of this specification.

The purpose of a GLTL is to store and communicate things that happen at an instant in time (an event) and things that have values that vary with time. The definition is very broad, as is the application of the GLTL. An event might be a telemetry frame and the time might be the Earth Received Time (ERT) of the first bit of the frame. It might be an Event Record (EVR), indexed by Spacecraft Clock as determined by a physical clock onboard the spacecraft (SCLK) of the EVR. The notion of an event is that it is most naturally considered to "happen at a time". The other category of GLTL information stored in a GLTL is that which varies with time: for example, the voltage on a spacecraft bus, the position of the Sun relative to the spacecraft at a given time, the number of people on the Jet Propulsion Laboratory (JPL) campus at any time. The notion here is that the information is most naturally considered as having a value at all times.

For time-varying values the GLTL tuple value for a given timestamp is applicable from that timestamp to, but not including, the following timestamp. There is no separate "duration" value; that is strictly implied by the timestamps. If you find yourself wanting to describe things that have overlapping durations, you are probably into a nonlinear timeline representation or you are looking at the problem the wrong way. You could force it into the linear timeline with ad hoc information in the TLVALUE field, but that is probably an indication that the GLTL is the wrong representation for whatever you are trying to describe, or that you have defined your timelines incorrectly. As an example of the latter point, consider the mechanical pointing of all DSN antennas. If you try to put it all on a single timeline, well, that is wrong and leads to the overlapping durations problem. If you have a timeline for each antenna that records where it is mechanically pointing at any time, that is a classic GLTL representation.

A GLTL is defined as a list of tuples <TIMESTAMP, TLVALUE> where

1. TIMESTAMP is in the same time system and time format (as defined by NAIF) for each tuple in the GLTL.

2. The time system and time format are specified.

3. TIMESTAMP is assigned a literal value ("grounded").

4. The list is totally ordered by TIMESTAMP (no duplicate values of TIMESTAMP).

5. The "type" of TLVALUE, which is defined herein as its structure in mathematical terms of integers, reals, enumerations, sets, lists, . . . , is specified.

6. The "physical schema" of TLVALUE, which is defined herein as the representation of the mathematical structure in terms of serialized computer constructs such as ints, floats, lists, . . . , is specified.

7. The "interpretation" of TLVALUE, which is defined herein as the semantics of the "type" (e.g., the list of floats is interpreted as coefficients of an interpolation function), is specified.

8. If the GLTL represents a value over an interval of time, the interval is defined as starting from TIMESTAMP and continuing to, but not including, the following TIMESTAMP (there must never be a separate "duration" concept).

9. If the GLTL represents a value over an interval of time, the first interval in the timeline must start at TIMESTAMP=0 and the last interval is defined as covering TIMESTAMP..Infinity.

10. If the GLTL represents an event at an instant in time, the time at which the event occurred is defined as TIMESTAMP (there must never be a separate "occurred at" concept).

Some design notes:

1. "Type" and "Interpretation" are distinct concepts so that we can reuse a type (e.g., list of floats) via different interpretations (polynomial coefficients, Chebyshev coefficients).

2. "Physical Schema" and "Type" are distinct concepts so that the same Type (e.g., list of floats) may be serialized into different physical schemas (XDR, ASN.1, Protocol Buffer, Thrift) as needed. The physical schema can then be evolved to suit size/performance requirements at hand without altering the software that uses the timeline.

3. The interpretation should include units if the notion makes sense for the GLTL at hand. There should be a standard for expressing interpretations so that things that are common to many (but not all, for otherwise we would specify it outside of the interpretation) GLTL types (such as units) can be extracted by software in a standard manner.

2. ACTIVITY

An activity timeline is a type of intent timeline that is in common use in the current AMMOS system.

An activity instance has activity type; activity instance ID (unique, never recycled); parameter values; start timepoint; end timepoint [where a timepoint is minimum time, maximum time, expected time (between min and max)].

Time slots must all have assigned values: Use 0..Infinity to indicate wide-open values.

VI. ARCHITECTURE PRINCIPLES

A number of principles, outlined here, guided the development of our architecture. Our objective in developing and maintaining this list was to describe the "culture" of our architectural work, so that the architecture might not rot as team members come and go over the life of the project.

1. *A domain-specific architecture is good.* The purpose of AMMOS is to operate spacecraft. The architecture does not need to extend to other domains (e.g., AMMOS will not be used to implement a banking system). This principle allows us to avoid over-generalization of the architecture by making commitments as to what the architecture will not do, which leads to lower adaptation costs by reducing the number of decisions that must be made for every adaptation and reducing the volume of infrequently used concepts and code.

2. *Timelines are the central organizing concept and data structure.* Timelines are ubiquitous in spacecraft operations, as described earlier in this chapter as well as [1] and [2]. A key factor of our architecture is that the semantics of timelines are specified in detail a priori. Any data structure that is reasonably directly mapped to those semantics can then be converted to another representation, which makes the details of the data structure something of an implementation detail because easy conversion is a priori known to be straightforward. That said, sound engineering suggests that data structures should be reused where reasonable so as to avoid the expense of writing and maintaining low-value conversion code.

3. *Immutability: strive towards immutable data structures.* Immutability means that once a particular version of a particular data structure is created, that version is never further mutated. A reference to that version thus forms a reliable reference to the bits in the version. In the case of timelines, the architecture itself then supports queries of the general form "As of last Friday noon, what was the predicted bus voltage midnight Saturday?" Tracking data provenance becomes practical, because every version of every data item has a well-formed reference that can be used to track the many important data relationships found in spacecraft operations systems:
    (a) Immutability applies across the system: information about or relating timelines must follow the same principle in order for the architecture as a whole to provide immutability;
    (b) It eliminates a vast range of potential bugs, most of which are basically pointer errors writ large;
    (c) It eliminates much ad hoc code that provides localized immutability;

(d) It provides first-class names for every version of every data structure instance;

(e) It orms a solid basis for modern stateless Internet protocols such as REST ("Representational State Transfer," Wikipedia, The Free Encyclopedia).

4. *Good theoretical foundation to the architecture.* A sound implementation is unlikely to result from unsound architecture.

5. *Minimally constraining architecture: but the rules that are there are hard and fast.* Architectural mandates become the foundation of the implemented system, so changing the mandates is naturally very expensive. We strive to mandate only things that are essential to the integrity of the architecture. So, for example, the architecture does not require a particular language (Java, say) or a particular communications style (pub/sub for example). Those are important design decisions, of course, but they are not architectural. We strive to constrain only where the "physics" of the situation makes such constraints a good long-term bet.

6. *Data structure, not data stream.* Spacecraft generally emit a telemetry stream, which tends to lead to systems that are architected around that data stream. The stream is parsed, processed, displayed, and archived. Focus on the stream in turn leads to a tendency to leave access to historical data as an implementation detail. Each endpoint on the telemetry stream must deal with archiving the data for historical access, recovering historical information on startup, and so on. Our focus is on the overall timeline, with uniform access to the past and present. A focus on data structure also simplifies the system design because it becomes easier and more natural to exchange data custody and perform flow control across interfaces.

7. *Exploit but isolate vendor-specific capabilities.* We want to exploit best-of-breed technologies (e.g., replication services in high-performance databases), so that we do not have to build such capabilities ourselves on a less powerful foundation. However, we must also be prepared to replace the product with another that does not have the same capabilities. We meet these seemingly conflicting needs by using the vendor capabilities, but not allowing the vendor-specific details to "leak" across the system as a whole. We strive to move such advanced capabilities into "'ilities", rather than operational necessity, so there's a good trade between cost and capability. Then, if an advanced feature is needed (say real-time transactional replication), we can either use a product that implements it (an expensive commercial database product) or implement the advanced feature ourselves on top of a "free" database. As another example, we use SQL databases, but the SQL is not available across the system. Queries are supported via a neutral interface, and the query is executed behind that interface in whatever terms the underlying database technology requires.

8. *Generally prefer solutions with a minimum of "machinery."*

9.   *Evolvable and scalable.*

10.   *Security levels "adjustable," and the architecture does not limit how far you turn the knob either way.*

VII.   ARCHITECTURE

The AMMOS System structure will have six major elements, described below. This is true no matter how the system is designed (e.g., changing database technologies or using Web-based tools does not change this structure). The system is structured in this manner so that the system elements have interfaces that are more a consequence of the architecture than the design, so that the design and implementation of each element can evolve without disrupting the overall architecture.

1. *Timeline semantics and data structure.* Much of the value of the AMMOS architecture will come from the unified manner in which the many time-oriented data structures used in spacecraft operations are represented. The timeline is that unifying data structure.

2. *Timeline database.* The architecture is based on the execution of coordinated processes that operate on timelines that are stored in one or more timeline databases (TLDBs). The database, in general, stores the timelines such that time-related aspects are reflected in the DB schema or structure, but the values related to the times are opaque to the database. The database structure thus forms a stable foundation for the system, because it does not need to be changed as new timeline types are introduced. In general, the TLDB can only perform time-oriented lookups. That said, it is plausible that if extreme Timeline Database (TLDB) performance is necessary and if that performance depends upon value-based operations (not just time-based), then the TLDB will need to exploit the structure of the timeline values themselves. This will not break the architecture but it will probably increase adaptation and maintenance costs.

3. *Timeline database library.* There is executable code in front of the database and executing on the server side (assuming a server side) that converts higher-level technology-neutral queries into DB-specific primitives, and also processes the results. For example, the server code might perform a bulk query to minimize the number of DB interactions. The client code cannot perform such operations itself, because the DB product or technology choice would then leak into the client code and make it difficult to change DB products or technologies. The server side may also support plug-ins to present project-specific views of some timelines in a uniform manner. That could be done on the client side, but may be less costly and easier to maintain if done once and for all on the server side.

4. *Client library.* The client library element is a library that is compiled with the application program (e.g., compiled with APGEN) that provides an

application-oriented interface to the timelines. The architecture constrains this library much less than the other elements, so that the library can rapidly evolve to contain functions that are found to be useful to the application programmer. The client library contains the bulk of the code that requires knowledge of the semantics of a particular timeline type or instance. There may be multiple libraries, each for a different purpose, for example analytics vs real-time vs modeling.

5. *Orchestrator.* The orchestrator coordinates the execution of the various programs that are run to perform mission operations. A project may choose to use more or less orchestration. The orchestrator may interact with workflow if a project uses formal workflow mechanisms. Program execution can also be initiated manually, or triggered by timeline database events. The orchestrator may be explicit (e.g., a Business Process Execution Language (BPEL) engine), or may be encoded into pipe-and-filter, dataflow, and so on.

6. *Name server.* The name server returns the current physical location(s) of a timeline, given the name of the timeline. Its function is analogous to the Internet domain name system (DNS). Note that the scope of the nameserver effectively defines the boundary of the system, because two independent nameservers allow the same name to refer to two different timelines and therefore must be independent, distinct systems.

A. TLDB

The database interface is a key architectural invariant, along with the timeline. It is designed to allow the database technology to be selected to meet mission needs. For example, a small mission may choose to use a free database. A larger mission may choose to use a commercial database that provides robust hot backups, offsite mirroring, local caching, cloud scaling to infinitely, and the like. A mission may choose to put the data in a commercial cloud. It may even change DB technologies over the life of the mission, using something cheap and light in formulation, heavy in operations, and optimized for archival access in perpetuity. The interface stays the same no matter what technology is used by a project, so that the spacecraft operations software suite will operate the same regardless of the DB technology used.

The point is that the database technology itself (SQL, NOSQL cloud, whatever) is not an architectural invariant or "load-bearing wall," and so can be changed as needed as long as it meets the interface specifications. The interface is architectural, and much of the upfront engineering rigor in the definition of timelines and the operations they support is there so that the interface will not need to be modified later. "Interface" as used here does not mean the details of the communication fabric (RESTful, message bus, etc.). Those must remain design choices. "Interface" means the basic protocols: what functions are performed, the data types that go in and out, that sort of thing. It has been done right when it is easy to switch from one communication style to another. It has been done wrong if doing so becomes a big deal. Do not let communication fabric details leak into (or even become) the architecture.

The TLDB interface provides fairly direct access to the DB data structures. Client-side libraries perform further conversions into application-specific forms. A thin server-side library is between the interface and the database and converts the raw timeline information in the database into the forms that are used by the client software (and vice versa). The idea is that the server-side library is an "adaptor" between the fixed API behavior on one side, and the implementation specifics of the chosen database technology on the other side. It thus isolates the database technology from the client software, so that the database technology can be selected based on its 'ilities (e.g., performance, cost, comprehensiveness, and so on) without having to do a costly and risky rewrite of the client software. Basically, it avoids a lock into the DB technology. In contrast, if we exposed Structured Query Language (SQL) or Java Database Connectivity (JDBC) at the interface level that would lock into at least relational databases, and quite likely would lock into a particular vendor. Timelines were designed so that such a commitment is not necessary.

The server-side interfaces will be the minimal set that provides a computationally complete set of operations on the timelines, so that the interface is likely to prove stable and support all desired abstractions as they evolve. "Code comes and goes, Data is forever"; the timeline schema is the most strongly vetted, followed by the database interface, then the official library code, and finally the abstracted services. Orderly evolution is thus supported, with room for ad hoc extensions at the client level and an evolutionary path for appropriate extensions to be made available for more general use or even inclusion into the architecture itself.

B. LIBRARIES

The client-side library computes the runtime and application-specific presentation of the timelines. For example, on the database representation of a given timeline (say the state timeline on the battery state of charge) there might be stored a time-tagged list of interpolators with parameters encoded into XDR and sampled at a fixed interval, say every hour. The client-side library would provide several views of that timeline. It would provide a floating-point value of the battery SOC for any time in $0, \ldots, \infty$ and yielding either a voltage or other information, for example that the spacecraft did not exist at the selected time. It may also provide a more detailed, type-specific interface that gives access to detailed uncertainty information and the like.

The client library provides mathematical operations that can be performed on timelines, such as differencing and integration. Such operations are provided in a functional programming style so that the caller (typically application-specific code in a component) need not express incidental machinery (e.g., a for loop) to perform computations on timeline. That isolation allows the library to operate in several modes (bulk computation, real-time as new data arrives, lazy evaluation) without modification of the user code.

We expect that several variations of the library will exist, each presenting a different "personality": perhaps one for extremely high performance and fairly direction interaction with the TLDB, and another that presents a mathematically oriented interface for analytics and simulation (as described above).

C.   COMPONENTS

A large number of the computations that are performed by the current AMMOS can be easily reinterpreted as one of several types of computations performed on timelines, and yielding timelines. For example, many derivations are performed, such as computing power as the product of voltage and amperage, and many checks are performed, such as checking that the power does not exceed some limit (where the limit is either a constant, or itself a timeline). Other computations, such as simulation-based predictions, are more complicated. In any case, such computations are performed by components that have several important properties:

1.   The component is fairly isolated from its execution environment, so it can be repurposed (e.g., computing the power product as a one-shot computation, or performing it in real time as new voltage and amperage values become available).

2.   The component generally does one thing (check that a timeline does not exceed a value), rather than many things (check it does not exceed a value, send out an SMS alert to subscribers if that limit is exceeded) so as to maximize reusability. The components are then composed into higher-level capabilities (e.g., compose the limit check component that generates a Boolean timeline with another that sends a text if the Boolean timeline ever becomes True).

D.   TIMELINE MANAGEMENT SYSTEM

The Timeline Management System (TMS) is the name for the system comprising one or more TLDBs, one or more libraries, many components, the orchestration mechanism for executing the components to carry out mission operations, and a few other elements like the timeline name registry. The TMS is a Web highly scalable and cloud-compatible server with a RESTful API, so it is easy to manipulate via modern Web-aware scripting and programming languages.

E.   CLIENTS

A client is defined as any system outside of the TMS that is using the TMS to perform mission operations. For example, a sophisticated simulator may read its initial state from the TMS timelines and write its computed state timelines back into the TMS. That simulator would be defined as a client because its execution is not orchestrated from within the TMS.

F.   NAME SERVER

The programmatic scope in which names are managed in a single rooted namespace (in other words, the level at which uniqueness is guaranteed) is an important decision. It is likely that the scope will be much larger than a single database instance, and so there will be a name server that covers many instances. Within that scope, names, TLIDs, and SCNs are known to be unique, so the timelines will play well together no matter what database they are in. The nameserver is very much like the Internet DNS ("Domain Name System," Wikipedia, The Free Encyclopedia) in that it associates various information (e.g., the TLID and location) with the timeline name. The location can thus be changed without recoding the users of the relocated timeline.

VIII.   CM OPERATIONS

Many processes within the MOS involve the application of Configuration Management (CM) processes on timelines [2]. The architecture, via the semantics of the SCN and Namespaces, can implement a number of CM models in a simple and straightforward manner:

1. *First write wins/optimistic locking.* The SCN of the version to be edited is recorded. Upon write, the recorded SCN is compared to the SCN of the last mutation. The write succeeds only if they match. If the write fails, the caller must in effect redo the complete read/edit/write operation, based on the current version of the timeline. This model basically mimics what happens when two people edit the same file, and the editor says "file on disk has been modified."

2. *Last write wins.* The SCN of the version to be edited is recorded. Upon write, any logical insertions performed since the recorded SCN are logically deleted, and any logical deletions performed since the recorded SCN are reinserted. The write is then performed. The write cannot fail in this model. This model basically mimics what happens if two people edit the same file. The last person to write the file completely obliterates whatever the other people wrote.

3. *Pessimistic lock.* The timeline is "locked" before editing is started, and the lock is only released after the edits have been written to the timeline. Other editors that attempt to acquire the lock during that time will fail to acquire the lock. This model mimics various source code control systems (SCCS, RCS, SVN, etc.) that provide for locking a file to force serialized edits.

4. *Branch/merge.* A set of timelines, or segments thereof, are deep copied from the parent timelines into new timelines for editing. The SCN of each parent timeline is recorded at the time of the copy. The user can read/write the new timelines without concern for conflicts, because the user is the only one that can read or write those copied timelines. When the user is done with the edits, any changes applied to the parent timelines since the deep copy are merged into the workspace, and the merges are then posted back into the parent timelines. If any parent was updated since the merge, the write fails. The user repeats the merge/write cycle until the write succeeds. The merge does not need to preserve the total semantic correctness of the parent; errors will be caught later during consistency checks. The object of the merge is to minimize the odds of errors being found, but it need not guarantee there are no errors. If the full set of consistency checks are quickly and easily performed, they may be applied at the time of the merge to greatly reduce the odds of a later error, but again, that is only an optimization, not a requirement.

IX. ADMINISTRATIVE OPERATIONS

1. *Physical deletion of certain SCNs.* The immutability principle, in its pure form, implies that no data can ever be physically deleted, because the SCN in which the data did exist may be queried at any time in the future. However, physical deletion is a practical necessity in order to manage disk space. The immutability principle is thus extended to require that any possible query against any past SCN always returns the same result (this is the "pure" clause) or it returns an error indicating that the query is no longer possible due to physical deletion (this is the extension to the definition). Physically deleting SCNa, . . . , SCNb is defined to mean that any query outside that range will function normally, and any query within that range will return an error. Physical deletion is implemented thus:

    (a) SCNa, . . . , SCNb are entered into a "Deleted SCNs" table.
    (b) All queries raise an error if the queried SCN is in the Deleted SCNs table.
    (c) An extended SCN range is constructed that is the given range plus any contiguous previously deleted ranges either side of the given range.
    (d) Any row that was both created and deleted within the extended range is physically deleted.

    It should be noted that administrative database operations are by their nature not immutable.

2. *Physical deletion of timeline instance.* All rows related to the given timeline are physically deleted using normal DB deletion mechanisms, except that the TLID is marked as physically deleted rather than physically deleted from the TLID table. This marking is not strictly necessary, but does provide

more robust error messages and could be the basis of automatic forwarding of requests to the proper database.

3. *Moving a timeline.* The new location of the database is entered into the Nameserver. Note that this type of move is administrative and so by definition not subject to SCN semantics.

4. *Splitting a database.* Create the new database, then use "move a timeline" administrative operation to populate the new instance and "physical deletion of timeline instance" to remove it from the source database.

5. *Merging databases.* All rows in the source database are copied into the destination database. This "simply works" because the TLID, SCN, and name are global across all instances. Physical deletion is then performed on the source database.

X. FUTURE WORK

Much of the theory and implementation proposed here can be applied to repositories in general, not just timelines. The generalization is fairly straightforward: consider timelines a type of repository, add an association to each TLID [which should be renamed Repository ID ("REPID") indicating the type of repository it references], and use the immutability and SCN mechanisms to store other repository-type information.

For example, we are exploring the development of a file repository, where the REPID indexes a row containing, among other things, a binary column of essentially unlimited size in which information usually stored in literal files can be placed. The existing SCN and immutability mechanisms would provide robust and rigorous versioning and naming and security and audit logging (and so on) far beyond what a file system can provide, for only a few days of additional effort.

ACKNOWLEDGMENTS

K. Reinholtz thanks the many people who devoted time and intellect to the constructive critique of these concepts in various presentations and reviews over the last three years. This work is greatly improved thanks to your contributions.

REFERENCES

[1] Bindschadler, D. L., and Delp, C. L., "Principles to Products: Toward Realizing MOS 2.0," *SpaceOps 2012 Conference*, Stockholm, Sweden. Available from: http://www.spaceops2012.org/proceedings/proceedings.html.

[2] Chung, S. E., "Timeline-Based Mission Operations Architecture," *SpaceOps 2012 Conference*, Stockholm, Sweden. Available from: http://www.spaceops2012.org/proceedings/proceedings.html.

[3] Estefan, J., "AMMOS Architecture Vision," NASA/JPL MGSS DOC-000780, 2012.

[4] Reinholtz, W. K., "Timeline Central Concepts," JPL Technical Rept. D-71055, 2011.

[5] Chien, S. A., Knight, R., Johnston, M., Cesta, A., Frank, J., Giuliano, M., Kevalaars, A., Lenzen, C., Policella, N., and Verfaille, G., "A Generalized Timeline Representation, Services, and Interface for Automating Space Mission Operations," *SpaceOps 2012 Conference*, Stockholm, Sweden. Available from: http://www.spaceops2012.org/proceedings/proceedings.html.

CHAPTER 8

Pools: A More Efficient Way to Support Spacecraft Operations

A. Codazzi[*] and M. Kim[†]
German Aerospace Center, Munich, Germany

I. INTRODUCTION

The DLR (Deutsches Zentrum für Luft- und Raumfahrt) is Germany's national research centre for aeronautics and space, with main research fields in aeronautics, space, transportation, and energy. There are a total of 33 institutes and facilities all over Germany with a total work force of approximately 6900. Spacecraft mission operations are carried out in one of these institutes, the German Space Operations Center (GSOC), located in Oberpfaffenhofen. At the GSOC, the mission operations department performs the project's operational tasks.

This department can be divided into three main subunits. The first focuses on manned missions, the second takes care of communication satellites, and the third supports technology and Earth observation projects. The latter in turn consists of three groups: the projects management group (PMG), the software engineering group (SEG), and the operations engineering group (OEG).

The OEG supports all the tasks related to operations engineering, as defined by the ECSS standards [1]:

1. During phase A (mission and operational analysis, feasibility study, and conceptual design), the OEG contributes to the mission requirements analysis and to concept development. It supports the definition of the operational aspects of the mission concept at the space system level. Using the experience of over 40 years in spacecraft operations, the OEG can analyze subsystem-related tasks, contributing to the project effort estimation.

2. During phase B (preliminary design), the OEG supports the specification of the ground systems requirements. It contributes to the operational analysis and to the development of the operational concept.

3. During phase C (detailed design), the OEG contributes to the mission operations plan.

4. During phase D (production, assembly, integration, test and verification), the OEG supports system integration and its operational validation. The group's

---

[*]Operations Engineering Group Leader, Mission Operations Department; Alessandro.Codazzi@dlr.de.
[†]Operations Engineer, Mission Operations Department; Mischa.Kim@dlr.de.

engineers develop and validate the flight operations procedures starting from the command sequences delivered by the spacecraft manufacturer. The operations engineers also prepare the displays that are used for monitoring spacecraft telemetry. The group takes care of the administration, monitoring, and maintenance of the spacecraft "failure detection, isolation, and recovery" routines. The preparation of training and simulation sessions and participation in those sessions is also an OEG task.

5. During phase E (mission operations), the OEG contributes to the planning of subsystem operations. The group supports subsystem operations during the launch and early orbit phase (LEOP), and the commissioning and routine phases. The group engineers analyze the performance of the spacecraft subsystems and handle anomalies. The OEG also delivers on-call support, and takes care of subsystem configuration management.

6. During phase F (disposal), the group supports the preparation of mission termination and space segment disposal operations.

This chapter focuses on the OEG and the measures applied to fully capitalize on its human capital. It is described how, to reach this goal, the OEG works not only on the organizational structure and on the processes applied to carry out the daily business (the "hard" factors), but also considers the "soft factors." These include leadership styles, shared values, and the actual skills and competencies of group members.

Typically, hard elements are easier to define, and can be directly influenced by management. Soft elements, on the other hand, can be more difficult to describe, are less tangible, and are influenced more by the group culture. Nevertheless, there is clear evidence to show that these soft elements are as important as the hard elements for the success of an organization [2].

In the following section, the multimission support strategy applied by the mission operations department is introduced.

II. POOLS STRATEGY

The resources necessary to support a project throughout its lifetime vary. The typical project's course is outlined in Fig. 8.1. Workload fluctuations are normal for any project, especially those creating "something new" [3]. This is the case with most spacecraft missions.

More often than not, spacecraft are designed and built as prototypes. Therefore, the operations systems often have to be built from scratch or, in the best case, need to be strongly adapted.

Spacecraft missions present further challenges for managers. After launch, anomalies and contingencies support and recovery lead to workload peaks that cannot be predicted. Even before the start of the mission, launch delays, difficulties in meeting the delivery deadlines, and other factors can strongly affect the project

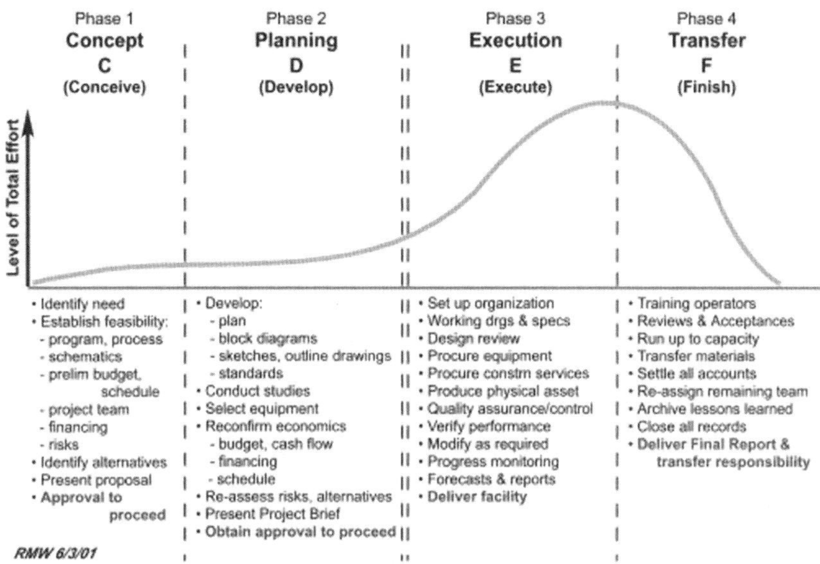

Fig. 8.1 Typical project course, showing how the resources necessary to support a project vary throughout its lifetime. (From http://www.maxwideman.com)

schedule. The effects of launch delays are shown for a project in Fig. 8.2, with the required support increasing as the launch comes closer. The launch delay then causes a drop in support needs, and, typically, a pause is introduced until the next launch preparation phase. Efficiency in coping with such variable support requirements is a very important factor for company success. Team flexibility and scalability are therefore core assets when dealing with spacecraft operations.

In the past, the GSOC mission operations department supported each mission with a dedicated team. As the number of missions increased, the department faced a challenge in that this strategy could no longer be supported with the available resources. Instead of hiring new operations engineers, the decision was taken to create a structure that, by coordinating the sharing of resources between projects, would allow a more efficient utilization of personnel.

In 2009, three groups of engineers were created by considering the department's core competencies: the software engineering group (SEG), which was to take care of the development and implementation of software solutions, the project management group (PMG), which grouped together the managers of the spacecraft missions operations, and the operations engineering group (OEG). A lead was assigned to each group. The SEG and OEG leads were assigned with the task of controlling project efforts with an adequate resources allocation plan, to keep project managers informed, and to make suitable adjustments as necessary.

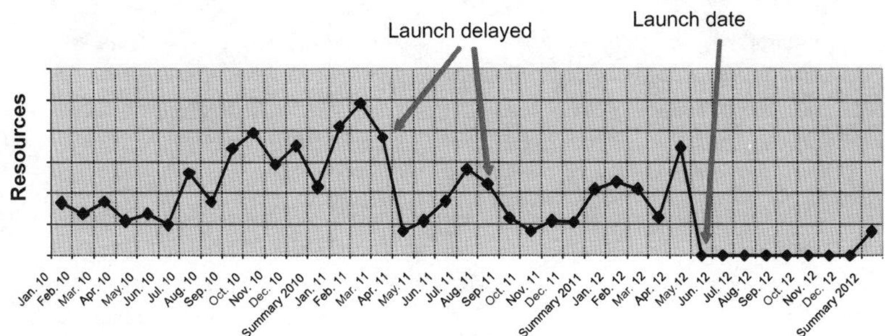

Fig. 8.2   Consequences of launch delays on resources utilization in the period from January 2010 to April 2012.

The three groups were supposed to cooperate closely, and share information, tasks, and some resources. Intense communication between the group leads was therefore encouraged by the department management introducing the pool concept.

III.   OPERATIONS ENGINEERING GROUP

There are approximately 25 engineers in the OEG. Of these, roughly 40% are core members, 15% are contractors, and the remaining 45% are engineers from other groups who dedicate part of their time to operations engineering. In total, the OEG can provide a workforce of roughly 17 men, of which 80% are DLR employees and 20% contractors (obeying all German Work Agency Act regulations).

The OEG is divided into four subgroups according to core competencies: onboard computer, telemetry and telecommand (OBC/TMTC); attitude and orbit control (AOCS); power and thermal (PTS); and operations system engineering. The first three correspond to the main subsystems of a spacecraft (excluding the payload), while the fourth relates mainly to coordination tasks and the establishment of interfaces with other systems not operated by the OEG (e.g., the mission planning system, flight dynamic system, or payload calibration system).

Each subgroup has a coordinator, who is the prime point of contact the OEG lead refers to for subgroup-related issues. In addition, the subgroup coordinators support the OEG lead in planning tasks assignments, estimating the projects' related efforts, evaluating employee performance, training and developing of (new) employees, and staffing. However, the group coordinator spends the larger part of the nominal working day providing technical support to the missions.

Thanks to the support of the subgroup coordinator in fulfiling management tasks, the OEG lead can focus on leadership-related aspects.

IV. CHALLENGES

The establishment of the pool concept was a significant change in the GSOC mission operations department. New players (i.e., the group leads) in charge of allocating scarce resources have been introduced into the organizational structure. Established routines, for example, those related to the assignment of resources, have been challenged. As a consequence of this strategy change, operations engineers have moved to a multimission environment, where most have to support several projects in parallel.

The McKinsey 7 s model (Fig. 8.3) states that for a change to be successful there are seven elements that need to be adjusted (or readjusted). These elements, divided into hard (strategy, skills, structure) and soft (systems, staff, style, shared values) factors, are mutually interconnected.

A survey conducted in large German companies showed that only 20% of the changes undertaken fully succeeded in reaching their objectives. The results were used to create an analytical model for explaining successful change. The model identifies four key success factors: leadership, experience in coping with uncertainties, motivation, and congruence [2]. The leadership factor consists of top-management engagement, the clarity of the objectives, the consensus on top-management level, and the support of middle management. The experience in coping with uncertainties has to be seen from different perspectives. On the one side, employees must have enough expertise to understand and cope with uncertainty. On the other side, managers have to carefully consider employees' feelings and intervene whenever necessary.

Actors' motivation is a factor that can strongly affect the change process results. The more the employees are motivated to support the change, the higher the probability of success. According to [4], almost half of employees withdraw support for the change process. In many cases, they resist or even fight it. Resistance to change is not always explicit and easy to detect. The behaviors aimed at avoiding the change process, without giving the impression of resisting it, can be grouped under the term "BOHICA," which translates to "bend over, here it comes again" (the wind of change). BOHICA is in fact the most common behavior following changes that overstrain employees.

According to the literature, a change has to take into account as many organizational aspects as possible, thus increasing the congruency of the process. The more factors are considered when

Fig. 8.3   The McKinsey 7s model.

performing a change, the higher the probability of success. Often, the agents of change (i.e., the employees directing the change) concentrate on the hard factors, leaving the soft unchanged. It is worth stressing the importance of organizational culture: according to [5–7], the most frequently cited reason for change process failure is neglect of the organizational culture.

A. FROM HIERARCHICAL TO MATRIX

The department structure has changed from being hierarchical to a matrix structure (Fig. 8.4). In the hierarchical structure, every operations engineer was subordinate to a single project manager. In the matrix, all operations engineers report to the OEG leader. At the same time, they are assigned to different projects and therefore report also to the corresponding project managers.

The OEG serves as the interface to coordinate the allocation of operations engineers. Project managers, who no longer have direct access to resources, may perceive this change as working against their interests. The OEG is allowed to codetermine the level of resources to be assigned to a project, with possibly negative impacts on project results and achievements.

A feeling of uncertainty may arise, triggering actions that can be interpreted as resistance to enforcement of the pool strategy. However, resisting loss of status, loss of pay, or loss of comfort are not the same as resisting change [4].

The OEG lead decided to use an "adaption strategy" instead of an "overcoming strategy" and scheduled a number of meetings with the different project managers. The purpose was to listen to and understand their concerns, adapting the OEG goals, plans, and features as a consequence. The OEG group lead did not

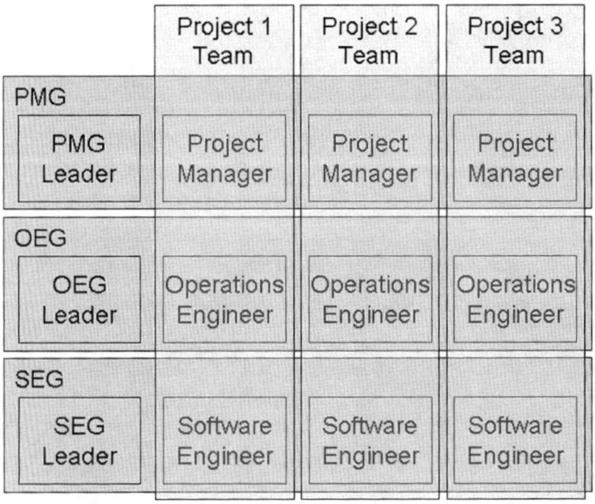

	Project 1 Team	Project 2 Team	Project 3 Team
PMG			
PMG Leader	Project Manager	Project Manager	Project Manager
OEG			
OEG Leader	Operations Engineer	Operations Engineer	Operations Engineer
SEG			
SEG Leader	Software Engineer	Software Engineer	Software Engineer

Fig. 8.4 The department matrix structure.

expect the project managers to be enthusiastic about the new strategy, because it reduced their influence on resource allocation. Instead, the group lead focused on getting the project managers' acceptance, defined as "approval despite diverging interests." In fact, the critical factor for success is acceptance of the change process, not enthusiasm [4].

To properly distribute the tasks between the group lead and the project manager, thereby avoiding possible conflicts, it is important to understand the difference between "management" and "leadership." James Kotterman suggests that these two terms are often used interchangeably in companies, thus creating confusion [8].

In the literature, "Leadership" is defined as a "process of social influence in which one person can enlist the aid and support of others in the accomplishment of a common task" [9], whereas "Management" is the "act of getting people together to accomplish desired goals and objectives using available resources efficiently and effectively" (see http://www.radford.edu).

Kotterman states that "there is a general acceptance that the functions of leaders and managers are conceptually different, but no universal acceptance of what those functional differences are is apparent". In his work, he compares leadership and management as shown in Table 8.1.

In the end, the input from project managers turned out to be very constructive and useful. The role of the group lead and project managers, as well as their respective tasks, were discussed and better defined. The skepticism of project managers progressively disappeared as they felt their input was considered and integrated into the pool concept.

It was agreed that the OEG lead has to make sure that the employees within the group recognize the value of the supported projects, and has to motivate them to actively engage in it. The OEG lead also takes care of aspects such as networking between group members, creativity and innovation management, employees' career development, (work) culture, and feedback. The assignment of project-related tasks, decisions about priorities inside the project, and planning of project activities fall under the responsibility of the project manager.

In the area of allocating resources (Fig. 8.5), where the competencies of the group lead and project manager overlap, the following process is enforced: the project manager is free to define not only the support of project needs, but also the amount of personnel and staffing assignments. The function of the OEG lead is to satisfy and coordinate the requests coming from the different projects. In the case of conflict, an alternative is suggested that has to be approved by the project manager. In other words, the OEG lead does not force the project to accept a solution. Potential conflicts between the projects are discussed and solved with the department lead.

The task of controlling is assigned to the group lead to allow a better comparison between projects. Thanks to a controlling system based on monthly reports, the OEG lead is able to give feedback to the project manager about the amount of resources spent in supporting the project's main activities. In the case of

TABLE 8.1 COMPARISON OF MANAGEMENT AND LEADERSHIP PROCESS DIFFERENCES IN THE WORKPLACE

Process	Management	Leadership
Vision establishment	Plans and budgets Develops process steps and sets timelines Displays impersonal attitude about the vision and goals	Sets the direction and develops the vision Develops strategic plans to achieve the vision Displays very passionate attitude about the vision and goals
Human development and networking	Organizes and staffs Maintains structure Delegates responsibility Delegates authority Implements the vision Establishes policy and procedures to implement vision Displays low emotion Limits employee choices	Aligns organization Communicates the vision, mission, and direction Influences creation of coalitions, teams, and partnerships that understand and accept the vision Displays driven, high emotion Increases choices
Vision execution	Controls processes Identifies problems Solves problems Monitors results Takes low-risk approach to problem solving	Motivates and inspires Energizes employees to overcome barriers to change Satisfies basic human needs Takes high-risk approach to problem solving
Vision outcome	Manages vision order and predictability Provides expected results consistently to leadership and other stakeholders	Promotes useful and dramatic changes, such as new products or approaches to improving labor relations

Source: Kotterman [8].

discrepancies between the plan and reality, a dedicated planning update meeting is called by the OEG lead.

In general, it was agreed that project-related tasks have a higher priority than those related to the OEG, which are normally not associated with fixed milestones or delivery dates. Priorities between projects are assigned by the department lead.

Discussions between the OEG lead and the project managers has helped to establish an intense and open communication that became one of the crucial factors in the success of the pool concept. Also, clear decision responsibility assignments are defined for each project for nominal and emergency situations.

B. LOYALTY AND IDENTIFICATION

With the former strategy, each project was supported by a dedicated team. The current multimission concept implies that the operations engineers often support several projects in parallel. However, some questions arise: can a person be loyal to more than one project? Can an employee identify with the different missions supported?

"Identification" (e.g., with a company or a team) is defined as "an active process by which individuals link themselves to elements in a social scene" [10]. Organizational identification can also be viewed as an alignment of individual and organizational values [11]. As a consequence, decision makers desire to choose the alternative that best promotes the perceived interests of the organization [12]. Organizational identification is an important aspect because of its relationship with commitment to the organization [13], which in turn positively affects motivation, job performance, job satisfaction, employee interaction and retention (with positive effects on productivity, efficiency, effectiveness), and individual decision making [14].

"Loyalty" can be defined as "the willing and practical and thorough going devotion of a person to a cause" [15]. Scholars have observed that, in general, multiple loyalties do not cause disloyalty. This would imply that an operations engineer can be loyal to different projects at the same time. "Organizational culture" refers to the "taken-for-granted values, underlying assumptions, expectations, collective memories and definitions present in a company. It reflects what an employee expects from others, and constitutes the guidelines to interpret a particular situation" [16].

Project managers often have very different characteristics, talents, and values, and these influence their management style. Furthermore, the values of the project often reflect those of its manager. Therefore, each project team may develop its own "culture."

Because OEG members often work for several projects at the same time, they have had to be able to adapt to and ideally identify (e.g., align their values) with different projects marked by different sets of values. The issue was discussed, and it has been concluded that, while adaptation is possible, identification might

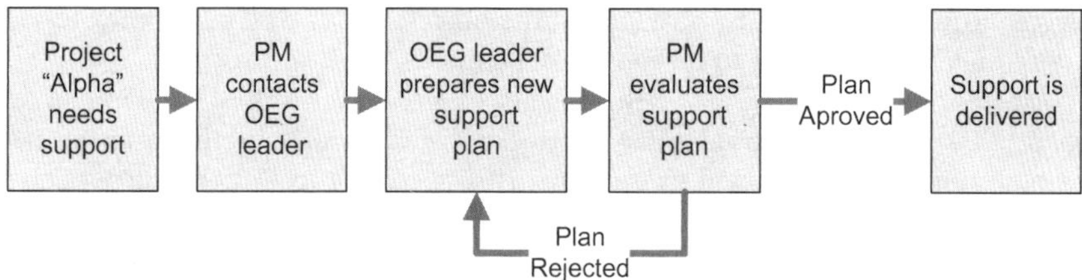

Fig. 8.5   The process of allocating operations engineering resources.

become problematic. This discussion also leads to another very important aspect: OEG culture harmonization.

OEG members agreed that a common culture would improve the efficiency of the group. In fact, a common culture provides an "unproblematic background consensus." In cross-cultural communication, this background is reduced by an amount that is proportional to the cultural differences. Scholars believe that difficulties in coordinating and integrating processes or organizational activities, for example, are often a result of culture clashes among different subunits [16].

Furthermore, by establishing a set of values, rules, and beliefs agreed by everyone in the group, group members were provided with a guideline with which they could identify, thereby helping their performance, independent of the culture of the projects on which they were working. The diagnosis of the organizational culture was made according to literature [16].

As a first step, a survey was provided to the engineers. The results were used as a starting point for several meetings involving all OEG members. The purpose of the meetings was to talk about the current cultural situation, the goals, and the measures to be applied to achieve them. This resulted in the "OEG culture" text, which is included in the Appendix.

The establishment of a group culture has some side effects that have to be considered carefully. In fact, a strong group culture may emphasize subunit cultural differences, fostering alienation and conflict. For the particular case described in this chapter, this would mean clashes between the OEG and other groups, or conflicts between operations engineers and project managers. To guard against these effects, the OEG group lead constantly encourages flexibility in dealing with different cultures and encourages communication with employees outside the OEG.

C. MOTIVATION

Employee motivation is considered to be a crucial factor for the success of a company. Several scholars have engaged themselves in understanding the triggers of motivation. In his "acquired-needs" theory [17], David McClelland proposed that individual needs develop over time and are shaped by life experiences. According to McClelland, most needs can be classified as either achievements, affiliation, or power. Motivation is influenced by these three ingredients. The degree to which one need prevails over the others, and the way to motivate as a consequence, is assumed to be subject-dependent.

Steven Reiss, starting from the results of previous studies involving more than 6000 people, proposed a list of 16 basic desires that motivate all human beings' actions [18]. Among those are curiosity (desire to learn), independence (need for individuality), order (need for organized and predictable environment), power (need for influence of will), and status (the need for social standing). The degree to which one need prevails over another depends on the employee's characteristics.

Herzber's two-factor theory [19] states that there are aspects in the workplace that result in job satisfaction. They can be divided into motivators and hygiene factors. Examples of motivators are responsibility, recognition, or the possibility to work on challenging tasks. The hygiene factors are not motivators, but result in demotivation if not present. Examples of hygiene factors are salary and job security.

The goal of the OEG lead is to enhance the intrinsic motivation of group members, capitalizing on the possibilities offered by the pool strategy. Intrinsic motivation refers to motivation that is driven by an interest or enjoyment in the task itself, and is based on taking pleasure in an activity rather than working towards an external reward (see http://www.tuition.com.hk/psychology/i.htm).

The OEG lead fosters intense and open communication with group members to obtain information about their specific characteristics, understanding their ambitions, strengths, and weaknesses. In doing so, the lead acquires the knowledge necessary to distribute tasks according to the talent of employees, their potential, and their preferences. The goal is to satisfy the projects' requirements in the most effective and efficient way, fully capitalizing on the group's human capital by assigning to each employee the most suitable tasks portfolio.

The flexibility of the pools strategy constitutes a big advantage, because the employee no longer has to focus on a single project. This factor enhances the possibilities the OEG lead has in assigning tasks to employees. In fact, the employee can support different projects in parallel, thus developing a broader spectrum of competencies. Still, this is not a must. Employees who have difficulty in staying tuned with more missions in parallel can ask to focus on a single project.

The intense communication between the lead and group members is also necessary for the lead to optimize tasks distribution on a regular basis. In fact, some iteration is necessary in order to fine-tune the tasks portfolio so that it is challenging without overstressing or overloading employees. Furthermore, the interests and ambitions of employees change with time and experience, and new and interesting tasks arise from the project that naturally attract OEG group members and have to be assigned. Although the distribution of tasks is subject to a constant optimization process, employees are assured that nothing is changed without listening to their opinion first. This leads to a feeling of working in a dynamic, yet predictable environment.

The scenario is made even more attractive by the courses and training offered by the DLR, giving employees a further chance to extend their knowledge and learn about new topics.

Keeping the Reiss theory in mind, the OEG lead allowed the subgroups to work as self-managed teams, in order to satisfy the desire for independence of the engineers. The lead leaves as much autonomy as possible to the coordinators, and they are instructed to do the same with their engineers. The lead avoids acting in the area of competence of subgroup coordinators; overruling a coordinator decision is applied only in extreme cases and only after talking with the coordinator.

Another initiative that had a positive impact on employee motivation was to encourage creativity and innovation, as explained in the following section.

D. CREATIVITY AND INNOVATION

Creativity is defined as "a mental characteristic that allows a person to think outside of the box, which results in innovative or different approaches to a particular task" (see http://www.businessdictionary.com/definition/creativity.html). Innovation is defined as "the process by which an idea or invention is translated into a good or service for which people will pay, or something that results from this process. Innovation involves deliberate application of information, imagination, and initiative in deriving greater or different value from resources, and encompasses all processes by which new ideas are generated and converted into useful products" (see http://www.businessdictionary.com/definition/innovation.html).

Because the main goal of the OEG is to use its resources in the most effective and efficient way, creativity and innovation are considered important aspects. On the other hand, mission operations must follow strict rules and validated procedures, with the purpose of avoiding mistakes that may lead to contingency situations and, in extreme cases, to the loss of a mission.

To encourage creativity and innovation, while maintaining the mentality necessary to perform mission operations in the most safe and reliable way, the OEG decided to create a "separated" environment, where the rules that are necessary to carry out daily business no longer apply.

For one afternoon every week, operations engineers are free to engage themselves in topics different from their normal activities. A power and thermal engineer can, for example, read about team management, or have a talk about orbital dynamics with one of the experts working in the GSOC.

The group's members are also free to call meetings or brainstorming sessions to discuss their ideas, and organize presentations to get feedback from colleagues. It has been observed that many ideas that seem good at first fail during this process, but discussion with group members represents a sort of decision board rating the new proposal. This gives employees the feeling that their ideas have been heard and considered, which is not always the case when the idea is evaluated by a single person (typically the boss) and not pursued further where there is a negative opinion.

During these times reserved for creative work, mistakes are tolerated, and all ideas can be brought up for discussion, no matter how unconventional they might be. People cultivate their creativity and learn to deal with the resistance that is typical when proposals challenge routines that have been in place for a while. While discussing innovation, group members exercise communication with other group members, sharing useful experiences that can help in supporting satellite missions.

The possibility of working on an idea that may improve the way operations are supported in the GSOC, as well as the feeling that opinions can be expressed and are considered by colleagues, are factors that have a positive effect on employee motivation. When talking about innovation, no pressure is applied to come up with results. In fact, it has been demonstrated that pressure and incentives in the form of money do not help innovation. On the contrary, they may hinder it. Quoting the career analyst Dan Pink, for example, "If you want people to perform better, you reward them, right? [...] But that's not happening here. You've got an incentive designed to sharpen thinking and accelerate creativity, and it does just the opposite. It dulls thinking and blocks creativity."

When a group decides to foster innovation, the crucial role of the "product champion," defined as "a senior individual or executive at a company that takes charge of furthering the internal development and external promotion of a certain good or service" (see http://www.businessdictionary.com/definition/product-champion.html), has to be considered. In fact, the product champion significantly increases the probability of success of an innovation inside a market (in this case, the company). Therefore, the OEG lead takes care in presenting promising ideas to the department lead for further evaluation.

While fostering innovation, the OEG lead considers the possible conflicts with the defined group culture. In fact, corporate culture can rule the group, and the pressure to conform can become overbearingly high. Innovation has often to do with breaking rules, and a strong culture may therefore hinder it.

One may state that creativity does not come by request on a defined day. However, the establishment of the free-thinking afternoon is a constant reminder of the importance of innovation, and of the decision of the OEG to strive for performance at the leading edge.

V. ACHIEVEMENTS

The mission operations team supporting a generic mission for Earth observation typically includes the following members: one AOCS engineer, one PTS engineer, one OBC/TMTC engineer, and one mission operations coordinator. For redundancy purposes, a backup should be provided for each position. In total, eight engineers are required.

The OEG is currently offering a similar service for ten satellite projects, among them two missions with two flying spacecraft each. Two of these four satellites are far beyond their expected lifetime and require demanding battery operation. One mission is currently close to launch, three projects are in the design phase, and four others are on "standby."

Low management costs are achieved thanks to the establishment of a common group culture, which provides a basic understanding that helps to minimize discussions and simplify the internal decision processes.

The group engineers declared themselves satisfied with the way the OEG operates and the tasks assigned. All perceived an increase in work efficiency and a decrease in uncertainty. Some reported difficulties in working on several projects in parallel. However, these problems are normally resolved by communication with the group lead. On the other hand, the multicolored task portfolio is highly appreciated.

Positive feedback was given about the campaign to harmonize the group culture, especially from new employees, who had the chance to learn about the basics of the cooperation inside the OEG from the very beginning of their experience at the DLR. Particularly positive was the feedback about the time dedicated to innovation and creativity. According to group members, this program enhanced employee motivation and improved communication between the different engineers.

VI. CONCLUSIONS AND FURTHER DEVELOPMENTS

This chapter describes the mission operations support strategy implemented and applied since 2009 at the GSOC, focusing on the OEG. The structure of the group is described, as well as the way engineers have transitioned from supporting a single project to working for more projects in parallel. The advantages of this approach are presented in terms of efficiency, job satisfaction, and employee motivation.

Together with these advantages, the establishment and maintenance of the new support strategy imply different challenges to be faced. The issue related to the role of the group lead and the possible conflicts with project managers are presented, together with the possible problems of loyalty and identification.

The positive experience of the OEG convinced the GSOC management to extend the area of application of the pool strategy. So far, operations engineers supporting the communication mission have also been integrated in the group. The synergies between Earth observation and communication missions have still to be proven advantageous in practice. Nevertheless, the networking and services offered by the OEG were considered a strong argument in favor of integration.

With the increasing size of the group, communication will become more challenging. This aspect is particularly important considering the crucial role an intense and open communication plays for the performance of the group. At the moment, several possibilities are under investigation to cope with this issue.

The improvement in terms of employee motivation and performance achieved after harmonization of the OEG culture convinced the lead of the SEG to go through the same campaign. This campaign will take advantage of the experiences gained by the OEG lead. The harmonization of the SEG culture will also consider the OEG culture with the purpose of achieving not only the maximum efficiency and effectiveness inside the SEG, but also the best possible cooperation between the SEG and the OEG.

APPENDIX: OEG CULTURE

The Operations Engineering Group's main task is to support the satellite projects.

We are aware of the importance of this duty, and we fulfill it with particular focus on the quality of our output and the adherence to the due dates.

We are convinced the best results are achieved cooperating with the colleagues. We believe competition and rivalry must not be part of our Group's culture.

We support each project in a proactive way; we take care to report to the project manager about the possible improvement area we observe. On the other side, we are aware of the role of the project leader, and we are ready to accept his decisions.

Any Group's member knows that his performance and behavior has an impact not only on his reputation, but also on the reputation of the whole Group.

We consider the internal and external communication as a crucial factor to enable knowledge transfer, avoid single-points-of-failure, and improve the Group's achievements.

We take particular care of having a meetings concept which enables communication at operations subsystem level, as well as cross communication between the different subsystems.

The feedback culture is strongly encouraged. This is true inside the Group, but also in working together with the projects team. We provide (and expect from each other) a polite, honest and direct feedback.

We consider the development of the operations engineering as one of our top priorities. Our goal is to be on the leading edge! Therefore, we are constantly looking for management solutions enabling the implementation of new concepts, vision, and possible improvements. There is no idea which is too crazy to be discussed!

With this respect, we are always willing to expand our field of view, through the participation to conferences, seminar, and department meetings, as well as reading the information available on the DLR media instruments.

We commit to integrate any new Group member in the fastest way possible, and we expect the same commitment from any person joining our team.

ACRONYMS

AOCS	Attitude and orbit control system
DLR	Deutsches Zentrum für Luft- und Rauhmfahrt
GSOC	German Space Operations Center
LEOP	Launch and early orbit phase
OBC/TMTC	Onboard computer/telemetry and telecommand
OEG	Operations Engineering Group

PMG   Project Management Group
PTS   Power and thermal system
SEG   Software Engineering Group

GLOSSARY

Spacecraft subsystem (or subsystem): a part of the spacecraft system; typical subsystems are the AOCS, the OBC/TMTC, and the PTS.

Space system: the spacecraft.

REFERENCES

[1]   European Cooperation for Space Administration (ECSS), "ECSS-E-ST-70C, Space Engineering, Ground Systems and Operations," ESA-ESTEC Requirements & Standards division, 2008, Chap. 5.

[2]   Houben, A., Frigge, C., Trinczek, R., and Pongrats, H. J., *Veränderung erfolgreich gestalten*, C4 Consulting, Düsseldorf, Germany, 2007.

[3]   Baglieri, E., Biffi, A., Coffetti, E., Ondoli, C., Pecchiari, N., Pilati, M., Poli, M., and Sampietro, M., *Organizzare e gestire progetti*, Etas, Perugia, Italy, 2004, Chap. 3.

[4]   Pongratz, H. J., and Trinczek, R., "BOHICA, Change zwischen Akzeptanz und Wiederstand," München, 2005.

[5]   Caldwell, B., "Missteps, Miscues," *Information Week*, 20 June 1994.

[6]   Gross, T., Pascale, R., and Athos, A., "The Reinvention Roller Coaster: Risking the Present for a Powerful Future," *Harvard Business Review*, November–December 1993, pp. 97–107.

[7]   Kotter, J. P., and Heskett, J. L., *Corporate Culture and Performance*, Free Press, New York, 1992.

[8]   Kotterman, J., "Leadership Versus Management: What's the Difference?," *Journal for Quality & Participation*, Vol. 29, No. 2, 2006, pp. 13–17.

[9]   Chemers, M., *An Integrative Theory of Leadership*, Lawrence Erlbaum Associates, ISBN 978-0-8058-2679-1, Mahwah, NJ, 1997.

[10] Kleinig, J, "Loyalty," *The Stanford Encyclopedia of Philosophy*, http://plato.stanford.edu/entries/loyalty/. [last accessed April 2012].

[11] Pratt, M. G., "To Be or Not to Be: Central Questions in Organizational Identification," *Identity in Organization*, edited by Whetten, D. A., and Godfrey, P. C., Sage, Thousand Oaks, CA, 1998, pp. 171–207.

[12] Cheney, G., and Tompkins, P., "Coming to Terms with Organizational Identification and Commitment," *Central States Speech Journal*, Vol. 38, No. 1, 1987, pp. 1–15.

[13] Tompkins, P., *Apollo, Challenger, Columbia: The Decline of the Space Program, A Study in Organizational Communication*, Roxbury, Los Angeles, CA, 2005.

[14] Scott, C., Corman, S., and Cheney, G. "Development of a Structurational Model of Identification in the Organization," *Communication Theory*, Vol. 8, No. 3, 1998, pp. 298–336.

[15] Cheney, G., *On the Various Changing Meanings of Organization Membership: A Field Study of Organizational Identification*, Communication Monographs, Vol. 50, No. 4, 1983, pp. 342–362.

[16] Cameron, K. S., and Quinn, R. E., *Diagnosis and Changing Organizational Culture*, Addison-Wesley, U.S.A., 1999.

[17] McClelland, D. C., *Power: The Inner Experience.* Irvington, New York, 1975.

[18] Reiss, S., *Who Am I? The 16 Basic Desires That Motivate Our Actions and Define Our Personalities*, Berkley Trade, New York, 2002.

[19] Herzberg, F., Mausner, B., and Snyderman, B. B., *The Motivation to Work*, Wiley, New York, 1959.

CHAPTER 9

NASA Space Launch System Operations Strategy

Joan A. Singer* and **Jerry R. Cook**[†]
Space Launch System Program Office, NASA Marshall Space Flight Center, Alabama

Christopher E. Singer[‡]
Engineering Directorate, NASA Marshall Space Flight Center, Alabama

I. INTRODUCTION

NASA's Space Launch System (SLS) will be a unique infrastructure asset for missions of national and international importance as the Space Age continues to unfold (Fig. 9.1). This super-heavy-lift human-rated rocket will be the first exploration-class launch vehicle since the Apollo Program's Saturn V Moon rocket in the 1960s and 1970s. The SLS tenets of safety, affordability, and sustainability are guideposts for developing and fielding a system that will be ready to fly in 2017, and that is optimized for streamlined, efficient operations, so that the societal, economic, technological, and other benefits it empowers will far surpass the cost of operating the world's most capable launch vehicle.

The SLS will transport astronauts, cargo (including habitation modules and surface rovers), as well as science payloads (such as astronomical telescopes and planetary probes) beyond Earth's orbit. It will provide the capability to perform national and international missions that foster the pursuit of mutual objectives, as outlined in the National Space Policy of the United States of America (June 2010), in NASA's 2011 Strategic Plan, and in the Global Exploration Roadmap [1–3]. Potential destinations for human exploration include asteroids, Lagrange points, the Moon, and Mars.

Delivering a cost-effective capability is imperative to advancing the global space agenda. Future generations are the ultimate beneficiary of the knowledge that space exploration affords and the associated emerging markets upon which it depends. As summarized below, the SLS Program has developed an operations strategy commensurate with its goals of safety, affordability, and sustainability for entirely new missions beyond Earth's orbit that may benefit all Earth's people.

---

*Deputy Program Manager, SLS Program.
†Deputy Director, NASA Stennis Space Center.
‡Director, Engineering Directorate.

Fig. 9.1   NASA's Space Launch System is designed for human and scientific exploration beyond Earth's orbit (artist's concept).

II.   SLS TEAM, PARTNERS, AND STAKEHOLDERS

The SLS Program is managed from the Marshall Space Flight Center, where its partnerships with the Safety and Mission Assurance Office and the Engineering Directorate provide foundational skills and the experience base to develop the next heavy-lift human-rated space transportation system. The SLS team includes design talent from the Constellation Program and operations experts from the Space Shuttle Program, among others. The SLS Chief Engineer directs systems engineering and integration activities and provides technical management for this nationwide effort. Aerospace prime contractors and the industrial supplier/vendor base are key partners on which the success of this major venture depends.

The SLS Program closely collaborates with its partners, the Orion Program at the Johnson Space Center (JSC) and the Ground Systems Development and Operations Program at the Kennedy Space Center (KSC), through technical interchange meetings, cross-program reviews, and other formal and informal avenues. These important lines of communication foster real-time integration and problem solving. SLS also works closely with its parent organization—the Exploration Systems Development division of the Human Exploration and Operations Mission Directorate at NASA Headquarters—to make informed, integrated decisions that optimize design solutions and support safe, affordable, sustainable operations.

Affordability in this context is defined as the ability to develop and operate the SLS within national means to sustain funding for the Program. With this in mind, SLS stakeholders provide schedule and technical requirements, as well as funding conveyed in public law (e.g., the NASA 2010 Authorization Act), and guidance contained in policy documents (e.g., the 2011 NASA Strategic Plan). Providing maximum value for the U.S. taxpayers' investment demands an accountable, disciplined approach to solving the challenges that come with an engineering feat of this magnitude.

The 2011 NASA Strategic Plan identifies expanding international partnerships as an overarching strategy for managing and achieving NASA objectives. The importance of SLS international partners has been confirmed by the recent announcement by the European Space Agency (ESA), following discussions with NASA, that it is ready to build a service module for the Orion Multi-Purpose Crew Vehicle. This potential synergy would allow NASA–ESA cooperation to expand beyond the International Space Station (ISS), and provide for a mutually beneficial arrangement for optimizing space exploration technology.

The SLS operations strategy addresses managing stakeholder requirements and incentivizing contractors to find and implement actions that lead to fixed and recurring cost reductions. The SLS Concept of Operations [4] articulates the expectation for a "culture shift," which places emphasis on cost-as-an-independent-variable and personal responsibility for keeping costs within bounds. It also provides a framework for defining the outcome of work being done today to fly Orion in 2017 and 2021, while positioning the system for streamlined, efficient operations for additional missions that will be added to the manifest as the Agency solidifies its plans in the months ahead.

The summary below includes nonrecurring design, development, testing, and evaluation (DDT&E) affordability tactics and results, as well as fixed and recurring cost-reduction initiatives that will be implemented in the production and operations (P&O) phase. Achieving a safe, affordable, and sustainable SLS will be made possible through the ideas, actions, and commitments of the SLS team, partners, and stakeholders.

III. SLS CONCEPT OF OPERATIONS

The SLS Program's Concept of Operations describes the system's attributes, as well as how it will be built, tested, shipped, processed, and launched. This document is an objective tool to calibrate and validate stakeholder requirements and expectations as the vehicle design is successively refined, manufactured, tested, operated, and retired, according to NASA's life cycle regulations and requirements (Fig. 9.2).

The SLS Program successfully completed its Mission Concept Review (MCR) in March 2011, and the SLS architecture was announced by the NASA Administrator in September 2011. The System Requirements Review/System Definition Review (SRR/SDR) technical objectives were completed in March 2012, with the programmatic baseline completed in June. The SLS Preliminary Design Review (PDR) is scheduled for 2013. Throughout this process, changes to the vehicle's baseline, including affordability approaches and implementation, will be captured in the Concept of Operations.

NASA Life Cycle Phases	Approval for Formulation ▼	FORMULATION		▽ Approval for Implementation	IMPLEMENTATION		
Program Life Cycle Phases	Pre-Phase A: Concept Studies	Phase A: Concept & Technology Development	Phase B: Preliminary Design & Technology Completion	Phase C: Final Design & Fabrication	Phase D: System Assembly, Int. & Test, Launch & Checkout	Phase E: Operations & Sustainment	Phase F: Closeout
Program Life Cycle Gates and Major Events	KDP A ▼	KDP B ▼	KDP C ▽	KDP D ▽	KDP E ▽ EM-1 Launch ▽		KDP F ▽
SLS Reviews	MCR ▼	SRR/SDR ▼ 2012	PDR ▽ 2013	CDR ▽ 2015	▽ SIR FRR ▽ 2016 2017		

Legend

CDR: Critical Design Review	KDP: Key Decision Point
FAD: Formulation Authorization Document	PCA: Program Commitment Agreement
FRR: Flight Readiness Review	SIR: System Integration Review

Fig. 9.2 SLS is in the design and development stage.

Fielding the world's most capable rocket is a commitment of limited resources, so the Agency called on some of the top Government and industry experts to gather the information upon which it based this important architecture decision. Throughout the process, the Agency sought input from internal and external stakeholders, using the affordability figure of merit as a key driver for hundreds of potential architectures that were studied during the selection process.

From the beginning, internal requirements analysis cycle studies and independent analyses performed through a Heavy-Lift Propulsion Technology Broad Agency Announcement identified potential avenues for major cost reductions that are now being implemented. The SLS Program has adopted best practices from industries ranging from automotive to space transportation. These operations and affordability drivers, which are catalogued and referenced in the SLS Concept of Operations, include principles related to leadership focus and direction; cost management techniques and tactics; people costs; supply chain management; and exploring new business models. The Agency has had great success with lean developments and manufacturing, and is applying these approaches to both DDT&E and P&O to reduce waste and save money.

A. SLS DESCRIPTION: SIMPLE, FLEXIBLE, EVOLVABLE DESIGN

Three principles that have been successfully driven into the SLS operations concept and related analyses are 1) simplicity of design and implementation, 2) evolvable, flexible design to meet specific mission requirements, and 3) margin in performance, cost, and schedule. The current SLS configuration is the most cost-effective choice, while offering unprecedented performance (Fig. 9.3).

The NASA Authorization Act of 2010 directs NASA to develop the SLS as a follow-on to the Space Shuttle, with the capability of accessing cis-lunar space and the regions of space beyond low Earth orbit (LEO) to enable the United States to participate in global efforts to develop this increasingly strategic region [5]. The Act also provides a series of minimum capabilities that the SLS vehicle must achieve: 1) initially lift 70 metric tons (t) to LEO and be evolvable to 130 t or more; 2) lift the Orion spacecraft; and 3) serve as a backup system for supplying and supporting cargo and crew delivery requirements for the ISS in the event that such requirements are not met by available commercial or partner-supplied vehicles. These requirements drive detailed technical trade studies and resource planning as the SLS concept is refined through the development process, in response to the Agency's specific design reference missions (e.g., high Earth orbit and geosynchronous Earth orbit) and figures of merit (e.g., safety, affordability, and reliability).

To reach the goal of first flight in six years from authority to proceed, the Agency decided to leverage the existing core stage engines from NASA's Space Shuttle inventory and to complete propulsion hardware that is already well into the development phase. This approach will keep the DDT&E budget flat, rather than result in the funding escalation normally experienced by developmental

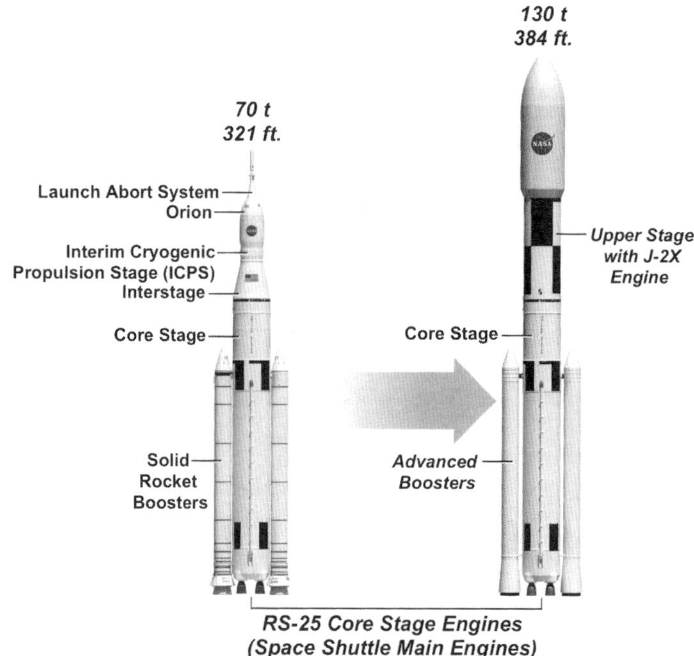

Fig. 9.3 The SLS block-upgrade approach delivers near-term and long-range capabilities.

Fig. 9.4 SLS RS-25 core stage engine being transported for storage and green-run testing.

programs. The SLS offers a relatively simple architecture, with competitive opportunities for affordable performance upgrades as the vehicle evolves over time.

The basic SLS design comprises a common 27.5 ft (8.4 m) metallic core stage that will contain liquid oxygen/liquid hydrogen (LOX/LH$_2$) tanks to fuel four RS-25 engines (Space Shuttle Main Engines). This takes advantage of 30 years of U.S. experience with LOX/LH$_2$ propellants, including manufacturing and launching facilities, and uses existing flight-ready core stage engines and qualification engines currently in stock, which will support the first several missions (Fig. 9.4) [6]. These human-rated engines built by Pratt & Whitney Rocketdyne had a 100% mission success rate in support of 135 total Shuttle flights. In a recent decision regarding how to prepare these valuable assets for flight, the SLS Program descoped the Main Propulsion Test Article and added green-run testing at the Stennis Space Center before mating with the core stage structure and shipping the integrated first stage to the Kennedy Space Center to save almost $300 million in development costs. Such decisions are analyzed by engineers and business support personnel, recommended by the SLS Chief Engineer (along with supporting cost, schedule, budget, and risk rationale), and approved by the SLS Program Manager.

The core stage also will form the SLS structural backbone. The upper stage for the SLS evolved capability will share common attributes with the core stage, such as its outer diameter, material composition, subsystem components, and tooling, to capitalize on synergies in those areas and positively affect the balance sheet as the vehicle is evolved. The same is true with payload adapter hardware. Various fairings will provide a modular, flexible platform for launching a variety of high-priority missions.

The SLS initial capability will use the world's most powerful solid rocket boosters (SRBs), now in development. The solid rocket motor is designed to generate up to 3.6 million pounds of thrust at launch. Although similar to the Space Shuttle SRBs, the five-segment SRB includes upgrades such as a larger nozzle throat, as well as upgraded insulation and liner [7]. To date, the upgraded motor has completed ground testing in the development phase and has begun the qualification phase, with testing scheduled for 2013 (Fig. 9.5). Prime contractor ATK has conducted extensive lean value-stream mapping (VSM) to streamline the SRB process flow and yield significant savings. Hundreds of changes were identified by the contractor that are projected to improve SRB processing times by nearly 50% and reduce project costs by millions of dollars. VSM has been applied to the SRB in all major production areas, including metal refurbishment, insulation, propellant, nozzle, and final assembly. These process improvements received final approval

Fig. 9.5 Solid rocket booster development test series.

from the SLS Boosters Office in March, and they are being implemented during the fabrication and assembly of the first qualification motor. In October 2012, ATK presented VSM improvements for booster separation motor and test area processing.

The SLS evolved capability will require even more advanced boosters that exceed the current limits of today's technology, providing a competitive opportunity for industry to deliver cost-effective, innovative hardware for deep-space missions post-2021. The engineering demonstration and risk-reduction phase for advanced boosters began in late 2012, followed by full-and-open competition for DDT&E [8].

As the SLS is evolved through planned block upgrades, the J-2X upper stage engine, now in the testing phase, will provide mission flexibility right-sized for performance requirements (Fig. 9.6). An advanced version of the Saturn's IVB and Saturn V launch vehicle upper stage engines, the J-2X will generate approximately 294,000 pounds of thrust to LEO, or 242,000 pounds of thrust from LEO into deep space, again using LOX/LH_2 [9]. The J-2X is designed to start and restart while on orbit, unlike the RS-25, which is designed to start while on the ground.

While the SLS operations concept defines surge capacity and a robust system that can sit on the launch pad in a state of readiness for months at a time, the SLS architecture is being developed around a low rate of missions (nominally, one per year), so any additional missions will make it even more affordable.

Fig. 9.6   J-2X upper stage engine test, May 2012.

NASA is using modern manufacturing and vehicle processing techniques, implementing risk-based insight/oversight practices, and streamlining contractor deliverables. For example, applying a lesson from the Constellation Program, SLS has streamlined its decision-making process, including reducing the number of working groups and formal boards, and clarifying lines of authority and each team member's roles and responsibilities. As another example, the amount of formal deliverables required of contractors has been reduced. There are no Type 1 documents that must be approved by the Government (versus hundreds in the Constellation Program), and deliverables are accepted electronically in the contractor's preferred format to save money on production and reproduction costs. Documentation reviews focus on how requirements are addressed, versus editorial comments. Cumulatively, simple improvements such as these add up to significant savings.

In the area of robust margins, the SLS Program holds reserves to be able to trade performance for cost and schedule. From a programmatic perspective, each hardware element office holds management reserves, allowing decisions to be made at a subsystem level and approved at the system level. SLS employees and partners will be continually challenged to strike a balance between having performance margin to spare and keeping adequate budget reserves to address the unexpected challenges that arise during an endeavor of this magnitude.

B.   PLANNED AND POTENTIAL MISSIONS

The SLS changes the paradigm of what can be launched because its performance and fairing size will far exceed that of any current or planned vehicle. In addition to launching the Orion spacecraft, the SLS will provide a heavy-lift capacity, allowing satellites and spacecraft with much higher masses to be launched. Dramatically larger fairings will be attractive for multi-element systems, science instruments with greater mass fraction, larger electrical power supplies, and more physical mass for radiation shielding. This translates into a positive return on investment for the SLS user community through reduced mission times and the ability to design more straightforward, less complex payloads.

To give context for the operations flow that follows, the first two SLS missions will send Orion all the way to the Moon. Exploration Mission One (EM-1), scheduled for 2017, will send an uncrewed Orion MPCV on a high-angle lunar trajectory to test the spacecraft's systems (especially the heat shield and reentry parachutes) as well as the SLS rocket. EM-2, scheduled to fly in 2021, will send a crew into orbit around the Moon and back for the first time since 1972. The Orion spacecraft will provide emergency abort during the launch ascent phase, sustain the crew during space travel, and provide safe reentry from deep space return velocities (Fig. 9.7) [10].

The SLS will be designed to support a surge to a maximum of three launches in a 12-month period to support NASA's capability-driven framework, which scopes one to three launches per mission campaign. The SLS will be interchangeable

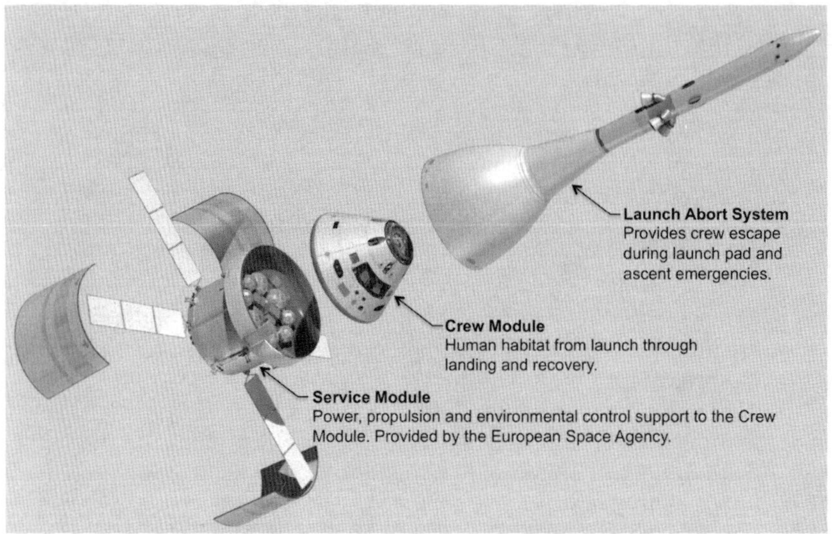

Fig. 9.7   Expanded view of the Orion Multi-Purpose Crew Vehicle.

within a block configuration with any payload type (Orion, cargo, Orion plus cargo) such that no significant changes in processing flow will be required.

In addition, SLS elements and line replaceable units (LRUs) below the payload adapter (refer to Fig. 9.3) will be interchangeable at the launch processing facility at KSC. The payload adapter will provide a common interface to the vehicle's core stage or upper stage, while supporting various configurations of Orion and other payloads. The launch infrastructure is being built such that it provides a common frame, with the option of adding access platforms and moving umbilical interfaces up for the Block 1 (70 t) to Block 2 (130 t) transitions (Fig. 9.8). These and other modular/interchangeable hardware and infrastructure augmentations are key to reducing costly touch-labor and keeping SLS affordable in the out-years.

C.   LAUNCH VEHICLE PROCESSING

The SLS mission operations function consists of several phases conducted by a cadre with clearly defined roles and responsibilities, including the mission planning phase, the mission training phase, the flight operations phase, and the post-flight activities phase. The SLS Flight Evaluation Task Team will coordinate with the SLS Ground Operations Task Team to ensure that all element- and vehicle-level objectives (from vehicle assembly through ascent and beyond) are addressed. The facilities required to implement the mission operations function include, but are not limited to, the JSC Mission Control Center, the MSFC Huntsville Operations Support Center operated by the Engineering Directorate, and various KSC Ground Systems Development and Operations facilities.

There are a number of steps involved in readying the launch vehicle for the first two Orion missions, as well as for others that will soon be manifested (Fig. 9.9). Given in the following sections is a top-level synopsis of the steps that will be taken to ready the SLS for its initial missions.

1. MANUFACTURING AND LOGISTICS

The SLS elements will be designed and managed to minimize open manufacturing work at the launch site. Stages will be manufactured by the contractor at NASA's Michoud Assembly Facility (MAF) in New Orleans, Louisiana, where Orion also is being fabricated. The solid rocket boosters will be built at the prime contractor's plant in Promontory, Utah. The core stage engines will be tested at the Stennis Space Center in southern Mississippi, and shipped to MAF for integration. The integrated stage will travel by barge to KSC, while the solid rocket boosters will be transported by rail. Any task that can be done at the contractor's facility (including test and checkout activities) will be completed there to minimize processing costs.

Fig. 9.8   The SLS 130 t vehicle is ready for launch (artist's concept).

2. GROUND OPERATIONS

At KSC, the Ground Systems Development and Operations Program manages and performs ground operations, which involve assembly, integration, and testing of the vehicle, as well as launch-pad operations. A single-string architecture will be used [one high bay in the Vehicle Assembly Building (VAB), one Mobile Launcher (ML), and one launch pad at Launch Complex 39B] to reduce integration complexity. The SLS will be assembled and integrated vertically on the ML in the VAB, then rolled out to the launch pad for final checkout. The SLS will be made ready to launch within a minimal timeframe to reduce the need for limited-life items and consumable servicing during pad operations.

Block 1

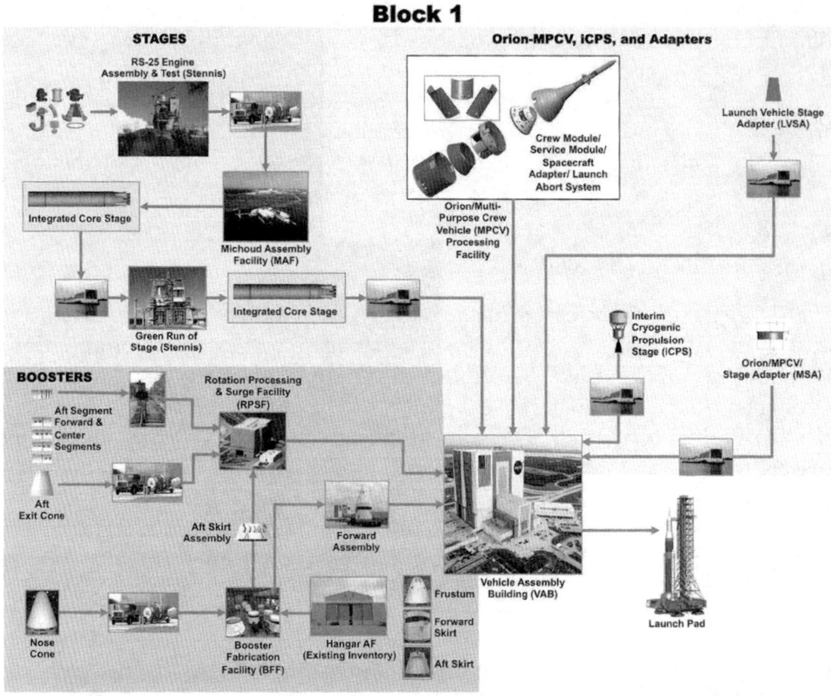

Fig. 9.9 **A typical process flow for the SLS Block 1 (70 t) configuration.**

3. LAUNCH

Countdown operations will be conducted in the KSC Launch Control Complex through first motion of the SLS rocket. Vehicle health and status data will be monitored. Cryogenic propellants will be loaded, and all propellant replenishing and/or pressurization activities completed. Launch holds will be accommodated to minimize battery depletion and the chance of a scrubbed launch. If present, the crew will board the Orion spacecraft via the access arm late in the countdown to limit their waiting time while suited up for launch.

Final configuration, checkout, and monitoring will be performed remotely. The SLS will switch to internal power. Autonomous handover to the rocket will occur. It will accept ground-computed mission parameters (verified before launch) via a ground system hard line. Remaining final configuration and automated verification of systems will be completed.

Launch control and mission managers will then make a go/no-go decision for launch, based upon their evaluation of compliance with launch commit criteria and flight and range safety rules. During nominal terminal countdown, the main engines will be started shortly before Time Zero (T-0) when the SLS flight computer will be sent the signal to ignite the boosters. Shortly after booster ignition, the integrated stack will lift off from the launch pad.

4. ASCENT AND DISPOSAL

Ascent begins at first motion as the SLS rocket lifts off the pad and ends with disposal of the core stage. Flight operations include disposal of the other SLS elements, as described below. The SLS rocket provides automated and autonomous flight operations during launch and ascent. It manages the states and modes of the SLS integrated vehicle systems as the Orion spacecraft and/or payload are transported safely to the ascent target. The SLS rocket provides power to all integrated subsystems and elements; performs guidance, navigation, and control; and provides propulsion to the integrated stack. It monitors vehicle health and status and controls the integrated vehicle trajectory, as well as providing attitude control of the stack during ascent.

Experience has shown that expendable hardware is less costly to design, build, and operate than reusable systems. Disposal begins when the SLS boosters separate, sinking in the Atlantic Ocean after impact, which saves money on recovery and refurbishment. The SLS core stage engines will continue to burn until orbit is achieved and engine cutoff is internally commanded. During cargo missions, the SLS payload fairing will be jettisoned, sinking in the Atlantic Ocean after impact. During crewed missions, the Orion spacecraft's service module fairing will be jettisoned in a similar manner. The SLS core stage will separate after the Orion spacecraft and/or payload is inserted into the ascent target. The core stage will follow a trajectory designed to ensure safe disposal in an uninhabited area of the ocean, where it will break up and sink upon impact.

D. SUSTAINING ENGINEERING

Sustaining engineering operations will include post-flight analysis and performance assessment for each SLS flight. This analysis will include performance trends, failure causes and effects, root cause analysis, imagery analysis, reliability and maintainability trends, and safety hazards. Safety issues will be addressed immediately in the production flow and take priority. Other issues will be forwarded to block-upgrade planning.

Sustaining engineering operations will be focused on vehicle production and engineering services, block upgrades, infrastructure maintenance, launch operations, and post-flight analysis. The block upgrade approach will be periodically applied to address design changes to improve operability, understand previous flight anomalies and marginal performance trends, improve manufacturing affordability, cover obsolescence issues, and so on. Engineering services will also focus on maintaining the vehicle's production schedule in support of the SLS manifest. These services will address supply chain management, manufacturing and transportation infrastructure maintenance, and anomaly resolutions.

E.   SLS BENEFITS FROM ORION'S 2014 TEST FLIGHT

As a final operations strategy example, the SLS and Orion Programs have signed a bilateral exchange agreement that allows the MPCV Stage Adapter (MSA) interface to be used during Orion's upcoming Exploration Flight Test (EFT-1) in 2014 (Fig. 9.10). It will be part of the stack when the Orion test article is launched on a Delta IV rocket during the EFT-1 Earth-orbit mission, as well as on the SLS rocket during its beyond-Earth-orbit missions.

The MSA is being designed once for both applications, as part of NASA's aggressive pursuit of affordable solutions for the human exploration of space. This is an example of commonality of hardware that supports flexibility, while reducing costs by minimizing the number of unique hardware items that must be developed and processed for flight. SLS and its partner programs will also use this opportunity to meet several objectives related to ground processing and mission operations.

IV.   CONCLUSION

NASA's vision is "To reach for new heights and reveal the unknown, so that what we do and learn will benefit all humankind." The first plank in NASA's strategic goals is to "Extend and sustain human activities across the solar system." The SLS (Fig. 9.11) is integral to establishing a capability-driven framework that will open a frontier filled with possibilities by building on the Apollo, Space Shuttle, and ISS experience, and moving out into the solar system. See www.nasa.gov/sls for more information.

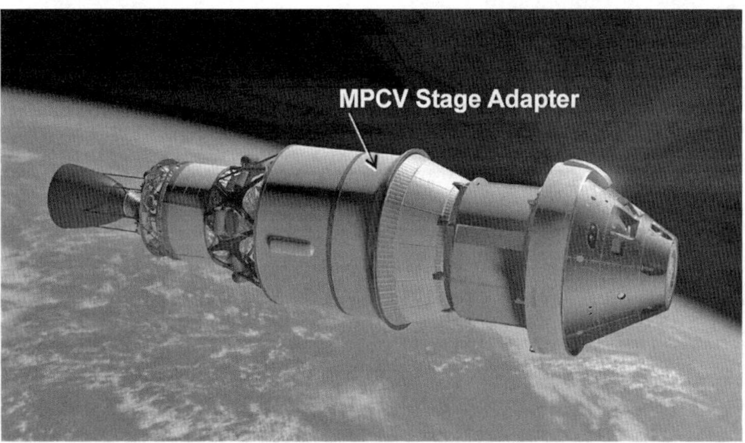

Fig. 9.10   Orion's first flight test will include the SLS stage adapter (artist's concept).

Fig. 9.11 The SLS is scheduled for first flight in 2017 (artist's concept).

Creating a sound technical and programmatic climate for sustainable exploration has far-reaching ramifications. Guided by NASA's 2011 Strategic Plan and validated by numerous internal and external and stakeholder organizations, the SLS gives missions defined by NASA, the International Space Exploration Coordination Group, the National Research Council, and others a mode of transportation out of Earth's orbit for astronauts to explore new worlds and for science instruments to rewrite textbooks with new knowledge. The SLS will provide the capacity required for entirely new flagship missions and will be an impressive asset with which to advance global exploration and international partnerships.

The SLS planning strategy uses existing hardware and finishes developments in progress, as well as incorporating advanced technologies if the return on investment supports affordability goals. The overall development strategy provides incremental stakeholder value by fielding an initial capability that will be successively refined in line with budget realities. Using common elements, the simplicity of an evolvable, flexible design, and adequate performance margins and funding reserves supports affordability principles that extend from design to operations. The ultimate benefit of the SLS operations strategy will be to decrease the price of flying the Agency's launch vehicle fleet so that more resources can be dedicated to expanding the boundaries of science and technology.

REFERENCES

[1]   The White House, "National Space Policy of the United States of America," http://www.whitehouse.gov/sites/default/files/national_space_policy_6-28-10.pdf [last accessed 28 June 2010].

[2]   National Aeronautics and Space Administration, "2011 NASA Strategic Plan," http://www.nasa.gov/pdf/516579main_NASA2011StrategicPlan.pdf [last accessed 14 Feb. 2011].

[3]   International Space Exploration Coordination Group, "The Global Exploration Roadmap," http://www.nasa.gov/pdf/591067main_GER_2011_small_single.pdf [last accessed 22 Sept. 2011].

[4]   National Aeronautics and Space Administration, "Space Launch System (SLS) Program Concept of Operations (Con Ops) Document," NASA Marshall Space Flight Center, 27 Oct. 2011.

[5]   111th U.S. Congress, "Public Law 267: NASA Authorization Act of 2010," http://www.gpo.gov/fdsys/pkg/PLAW111publ267/html/PLAW-111publ267.htm [last accessed 11 Oct. 2010].

[6]   National Aeronautics and Space Administration, "NASA Moves Shuttle Engines From Kennedy To Stennis," http://www.nasa.gov/home/hqnews/2012/jan/HQ_12-018_Shuttle_Engines_Move.html [last accessed 12 Jan. 2012].

[7]   National Aeronautics and Space Administration, "NASA Successfully Tests Five-Segment Solid Rocket Motor," www.nasa.gov/exploration/features/dm3.html [last accessed 8 Sept. 2011].

[8]   National Aeronautics and Space Administration, "NASA Awards Space Launch System Advanced Booster Contracts," http://www.nasa.gov/home/hqnews/2012/oct/HQ_12-339_SLS_Awards_Contract.html [last accessed 1 Oct. 2012].

[9]   National Aeronautics and Space Administration, "NASA Begins Testing of Next-Generation J-2X Rocket Engine," http://www.nasa.gov/mission_pages/j2x/j2x_ignition.html [last accessed 14 July 2011].

[10] National Aeronautics and Space Administration, "Explore the Exploration Vehicle," http://www.nasa.gov/exploration/systems/mpcv/explore.html [last accessed 16 May 2011].

CHAPTER 10

A University-Developed COSMOS to Operate Multiple Space Vehicles

Trevor C. Sorensen,[*] Eric J. Pilger,[†] Mark S. Wood[‡] and Miguel A. Nunes[§]
Hawaii Space Flight Laboratory, University of Hawaii, Honolulu, Hawaii

Bruce D. Yost[¶]
NASA Ames Research Center, Moffett Field, California

I. INTRODUCTION

The Hawaii Space Flight Laboratory (HSFL) was established at the University of Hawaii at Manoa in 2007 for two primary purposes: 1) to educate students and help prepare them to enter the technical workforce, and 2) to help establish a viable space industry that will benefit the State of Hawaii. HSFL is currently leading the effort to develop a solid-propellant launch vehicle, called Super Strypi, which is capable of placing a small satellite ($<$300 kg) into low Earth orbit (LEO), and various satellites from CubeSat (1 kg) to microsatellite ($\sim$100 kg) size [1]. HSFL is installing the infrastructure and facilities to support space missions, such as clean rooms for integration and testing, ground stations, mission operations center, and simulators/test beds. It is the goal of HSFL to provide full life cycle mission operations support for its space missions. This requires the use of specialized software to develop and sustain mission operations. Based on a trade study and analysis of the various software packages and systems available, we determined that none met the full functionality and flexibility we desired while staying within our budget constraints. The primary author has experience with mission operations from many space missions and has developed some tools that have been successfully used in an LEO mission and a lunar mission [2] and would be well suited to HSFL's needs (with some modifications and improvements). The idea of developing a comprehensive system of software and hardware to efficiently support mission operations for multiple spacecraft, especially small satellites, was thus born. We decided that such a system should have a framework for "plugging in" external applications and tools that would

---

*Professor and Project Manager, AIAA Fellow.
†Flight Software Engineer.
‡Instrumentation Engineer.
§Graduate Student, AIAA Student Member.
¶Program Manager.

enable them to work with the whole system (i.e., plug-and-play attribute). There was also an obvious advantage to the system being open architecture, including software, which would enable a large community of users to help rapidly develop new features and enhancements to the system that would be difficult in a closed system. This new system and architecture was named the Comprehensive Open-architecture Space Mission Operations System (COSMOS) and was the basis for a successful proposal to NASA in 2010, resulting in a three-year NASA EPSCoR grant to develop and deploy COSMOS. The grant started in 2010 and ends in 2013. NASA Ames Research Center is the primary collaborative NASA center. We are past the mid-point of this project and although the design of COSMOS has been presented in previous publications [3, 4], this chapter looks at the evolution and development of COSMOS and how it has been adapted to support different types of mission, including some types that were not foreseen when the COSMOS project was proposed and initiated.

II. COSMOS OVERVIEW

COSMOS is a framework of software and hardware elements that addresses all phases of a spacecraft life cycle: design, development, implementation, and operations. It provides elements for the creation of simulators, test beds, and flight and operations software, all fully interactive. Sorensen et al. [3] describes the functional architecture and operational concept of COSMOS. The guiding principle of the COSMOS suite is that it will be easy to port to different locations, and to configure for different spacecraft. The following tools are used to perform the major functions of mission operations for COSMOS: 1) Mission Planning and Scheduling Tool (MPST); 2) Mission Operations and Support Tool (MOST); 3) Test Bed Control Tool (TBCT); 4) Ground Segment Control Tool (GSCT); 5) Data Management Tool (DMT); 6) Flight Dynamics Tool (FDT); 7) COSMOS Executive Operator (CEO); and 8) various analysis and support tools.

To achieve this end, COSMOS is based on a limited number of key design elements, put together in a layered approach (Fig. 10.1). These key elements are based as much as possible on existing protocols and approaches. The foundation of COSMOS consists of a set of libraries supporting the various functionalities available in the suite. This includes mathematical functions, orbital and coordinate calculations, protocol support, and hardware and simulation support for the Operations Test Bed (which will help test and verify command loads before they are uploaded to the satellites).

The COSMOS project is an effort to create a complete open environment for the design, development, and operation of small spacecraft. The broad philosophical approach can be found in [3], although the concepts and implementation are continuing to mature and evolve as more of COSMOS becomes operational. What follows is a synopsis of the major fundamental software elements (or building blocks) of COSMOS, with specific details that will be important to discussions in later sections.

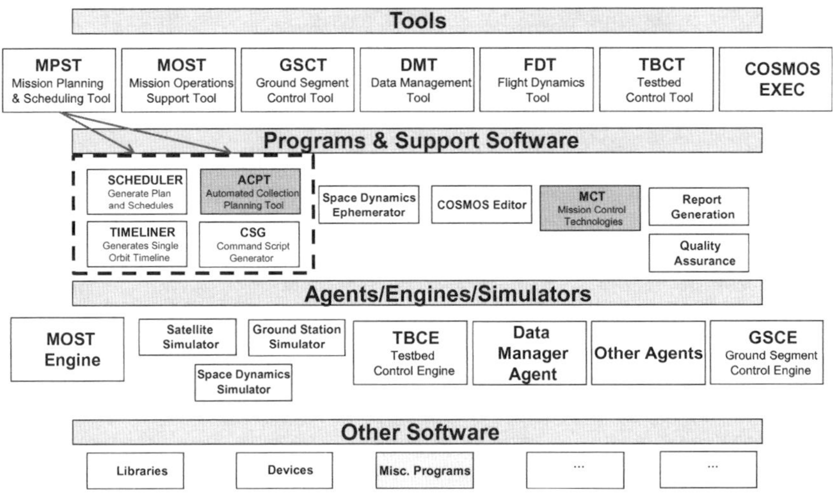

Fig. 10.1   COSMOS software.

A.   PROTOCOLS

1.   UNIFIED NAME SPACE

As originally implemented, the COSMOS Name Space was a completely flat mapping of names to single values. In this approach, each name represented a single string or number. The meaning, units, and data type of each value were predefined. An attempt was made to further simplify things by assuming that values such as position, which could be defined in multiple systems, were always represented in one predefined system. As no arrays were used, column (and row) positions were indicated with digits embedded in the name. A name like "panel_01_point_01" represented point \#1 of panel \#1.

Each member of the Name Space was then mapped to a unique location in memory; the memory locations being drawn from a globally accessible storage space. This Name Space Map was stored as a simple list of entries, allowing either forward or backward translation between COSMOS Name and memory storage.

This approach has proven quite flexible. However, it has also caused a number of problems, especially as the space has grown in size. This led to a number of changes, which are detailed in Sec. III.

2.   COSMOS SUBSET OF JAVASCRIPT OBJECT NOTATION (COSMOS JSON)

In keeping with the original definition of the Name Space, the original COSMOS JSON included only strings and numbers. Arrays and complex objects were not supported. The purpose behind this was to support the completely flat Name Space, and to simplify the mapping of names to variables in memory. It has subsequently become necessary to expand this usage, both in support of the Name Space changes noted above, and to more accurately represent values in memory.

3. AGENT COMMUNICATIONS PROTOCOL

The original implementation of Agents in COSMOS called for them to send out a *Heartbeat* packet of information at regular intervals to a predefined COSMOS Multicast address. This packet would contain the IP address of the machine on which the Agent was running; a unique *Request* port, on which the Agent would listen for incoming requests; the size of the buffer available for Requests, and any resulting Responses; the name of the Agent; and the period of the heartbeat (in seconds). While this basic approach has proven successful and remains mostly unchanged, certain supporting elements are still in flux. The best mechanism for determining the IP address and Port of the Agent, the relative merits of Broadcast versus Multicast, and the possible use of protocols other than IP are still being discussed.

4. NETWORK PROTOCOLS

The original design for COSMOS specified the use of IPv4 UDP protocols in both a Unicast and Multicast mode. The Lightweight Communications and Marshaling (LCM) protocol was then used on top of this to provide communications and data translation between Agents. It soon became clear that LCM incurred a large amount of programming overhead, and replicated many features already inherent in our use of JSON and the Name Space. We are still also trying to define the best protocol for use between the satellite and the ground. Although the NORM protocol is still in the running, we are also giving serious consideration to the Cubesat Space Protocol (CSP). At this juncture, we are designing in support for multiple protocols, thereby allowing the user to choose the one that makes the most sense for a particular connection. We are also continuing to search for a robust File Transfer Protocol, investigating such things as the Saratoga File System and the CCSDS File Delivery Protocol (CFDP).

5. CONFIGURATION FILES

As originally conceived, a number of configuration files were used to communicate with MOST, or COSMOS in general, allowing them to be fine-tuned to a specific mission. Foremost of these is the Satellite Description File, named "satellite.ini," which is a JSON description of all the static elements of the vehicle. It starts with a description of the physical Parts, followed by a listing of any Components that are then tied to specific Parts, followed by specific Devices that are tied to Components. Parts include any physical qualities. Components describe general electrical qualities. Device for each Device Type describes qualities unique to that device. The goal of the Satellite Description File is to provide sufficient detail to reasonably simulate all aspects of the vehicle. A separate Ground

Station File, named "groundstation.ini," is also created for each mission so that the Ground Segment for that mission can be simulated. This basic concept has since been simplified into one JSON description string. At the same time, the contents have been greatly enhanced to meet our new concept of COSMOS "Nodes," described in more detail in Sec. III.C.7.

6. SELF-GENERATING USER INTERFACE FILES

The original user interface (UI) files for MOST were designed by hand to demonstrate the sort of functionality we wished to see in the finished product. However, this was not consistent with the long-term goals of having MOST, and all other COSMOS tools, easily configurable for a wide range of different applications. We have subsequently developed an infrastructure for coding UIs that create themselves on the fly, based on the configuration information described above.

B. SUPPORT HARDWARE

It has been the ongoing goal of the COSMOS project to develop hardware simulators for all the components to be found in a spacecraft. Extensive progress has been made on the groundwork for this effort with basic off-the-shelf hardware and processors.

C. SUPPORT SOFTWARE

The suite of libraries, as initially laid out, has not changed significantly, other than to have numerous bugs identified and eliminated. Two additional libraries proved necessary, one in support of enhanced mechanisms for satellite description, called *satlib*, and a second for support of OpenGL graphics, named *graphlib*.

D. AGENTS

Only a cursory description of Agents was provided in [3], so a more detailed explanation will be given here. As described above, Agents make their presence known through a *Heartbeat*. Any client wishing to communicate with an Agent listens to the COSMOS Multicast address until it either receives a *Heartbeat* from the Agent it is waiting for, or it times out. Once it acquires the desired Agent, it sends a *Request* to the IP Address and Port taken from the *Heartbeat*. It then waits for a *Response* on that same IP address and Port until it receives one, or times out. All *Requests* and *Responses* are in plain ASCII.

E. TOOLS

Extensive work has been done on MOST since it was first described [4]. The latest version set up for a 3-U CubeSat is shown in Fig. 10.2. Quite a few features that were merely described have now been implemented, and significant changes to

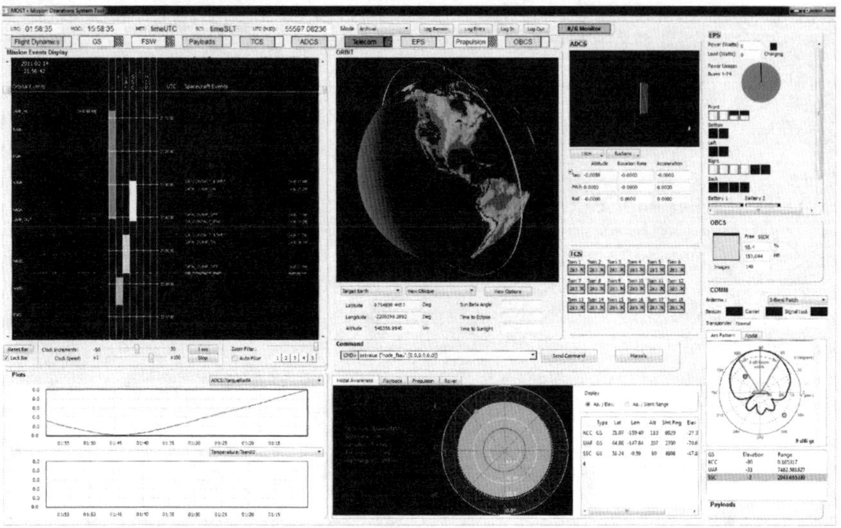

Fig. 10.2 MOST overview display for 3-U CubeSat.

both internal workings and features have been driven by our work on the various Test Cases that are listed in Sec. III.A. The need for an overarching Tool to control all the Entities functioning in a COSMOS system has also become clear. We have dubbed this the COSMOS Executive Operator (CEO).

MOST is written in C++ using the Qt (pronounced "cute") compiler [5]. A modular approach using Qt "widgets" has been used with idea of making MOST easily reconfigurable using configuration files. Qt compiles on Windows, Linux, and Mac, so MOST will be able to run on any of these systems. The Mission Operations Support Tool (i.e., MOST) displays information about a spacecraft (or other node). It is the main interface between a spacecraft and an operator. It shows position and attitude related to an object it is orbiting. It shows telemetry data and past and present orbital and spacecraft events. Each subsection of MOST is contained in a "Widget." Each widget contains information about a specific aspect of the spacecraft. The widgets presently used in MOST are widgets that contain the time, Caution & Warning Buttons, Mission Events Display (MED), orbit, subsystem data (e.g., ADCS, EPS, OBCS, Comm), Payloads, TimeLine Plots, and a command line to send commands to the spacecraft.

III. MODIFYING MOST FOR NEW SPACECRAFT

A. TEST CASES

Since 2010 we have had a number of opportunities to fit MOST, and the underlying COSMOS framework, to a variety of real-world examples. This has allowed us to 1) identify problems with our existing code base and make appropriate corrections; 2) spur the creation of additions and enhancements to both the existing tools, like MOST, and to the whole philosophy of COSMOS, especially making it easy to adapt to new vehicles (nodes); 3) identify areas of future concern that will need solutions, and spur the invention of future features.

This subsection provides detail on these test cases, while the following subsections describe the changes that were brought about in both our code and our overall approach to COSMOS development as a result.

1. SUMMER HIGH SCHOOL INTERNSHIPS

We were host to six high-school juniors through a UH College of Engineering intern program. With this group, we were able to start putting to use some of the basic components of the Operations Test Bed (OTB) and COSMOS software. For the OTB, we laid down the foundational use of Arduino processors, coupled with XBee radios. For COSMOS software, we developed the first test version of an Interpreter Agent to interface a non-COSMOS data stream, which we then displayed in MOST in real time.

2. NASA ARC PHONESAT

In collaboration with the NASA ARC, we are developing a version of MOST that can be used with a personal receive-only ground station to monitor passes of PhoneSat from any location. We are currently generating the description of the satellite. We will then import data from recent sounding rocket and balloon flights so that it can be visualized with MOST.

3. NASA ARC EDISON DEMONSTRATION OF SMALLSAT NETWORKS (EDSN) MISSION

We are planning to provide COSMOS to NASA Ames Research Center to support their Edison Demonstration of Smallsat Networks (EDSN) mission, which is planned to launch an interconnected group of eight 1.5-U CubeSats during the ORS-4 launch in 2013 (the same one that is carrying the HSFL Hawaii-Sat 1 HiakaSat microsatellite). This application is driving the development and initial deployment of the CEO tool.

4. NASA ARC MISSION CONTROL TOOLS (MCT)

We are studying the similarities and differences between MCT and COSMOS, with the intention of creating a support framework between these two environments. Interesting possibilities exist both for giving access to the COSMOS structure to MCT, and for embedding elements of MCT in MOST and other tools. We are currently working on an MCT program that will first capture COSMOS data in real time, and then eventually communicate with the Data Management Engine.

5. SDL/USU DICE MISSION

We have agreed to modify MOST to provide shadow operations for this mission, which in late 2011 launched a pair of CubeSats developed by the Space Dynamics Laboratory at Utah State University (USU). We are converting their mission telemetry to a form that MOST can understand, thus allowing the mission operators to view it within MOST.

6. CABLE MISSION

Our most detailed test case, and the one that has driven the greatest number of advances to date, has been the Canadian American British Lunar Expedition (CABLE). This collaborative effort [6, 7] allowed us to extend MOST to demonstrate its utility in support of a multivehicle, planetary mission that also included a rover. This mission includes three vehicles: a Lander with thrusters, a Rover with a drivetrain, and a free orbiting Intersatellite adapter. It includes elements of ground-based operations, with multiple vehicles on multiple planetary bodies. This proved quite challenging to the COSMOS team, and drove numerous enhancements and improvements in the software. We created a number of enhancements to MOST, as well as defining features we would like to have in later versions of COSMOS. Examples of the work done on MOST in support of CABLE are shown in Sec. III.C.

7. UH UNDERGRADUATE OUTREACH

As part of the educational aspect of HSFL in general and COSMOS specifically, we have created a number of projects for UH students who are NASA Space Grant Fellows, and for a Mechanical Engineering class. The Space Grant Fellows have allowed us to push forward a variety of projects related to OTB hardware and flight-related concepts. The Mechanical Engineering class provided much needed field experience.

8. COSMOS ITSELF

The next big Tool-level project in the COSMOS suite is the CEO. Always part of the planning for later stages, development of this software has risen in priority by funding of the EDSN mission by NASA (Sec. III.A.3). Consideration of what this software will need has driven a number of changes to the existing software and plans for future enhancements.

B. PROBLEM AREAS IDENTIFIED

Perhaps the most important outcome of our various test cases has been the identification of problems that were going to hold us back in either the short or long term. These problems are detailed below along with any immediate response.

These are provided as an aid to software programmers developing similar types of system. Any long-range changes to our philosophy are then covered in subsequent subsections.

1. CODE ERRORS

Some of the first things to be identified, from the very first test case, were the various errors that existed in the code we had already written. In some cases the errors were subtle issues, but in others it was impossible to move ahead until flaws in either logic or implementation had been fixed. This subsection details these various problem areas.

a. Basic Functionality

An immediate benefit of the work one of our summer interns did with the Interpreter Agent was the discovery of numerous bugs and logical flaws in our support libraries. To deal with this in a systematic way, we set our first student hire to work on developing exhaustive test suites for the various libraries. He has completed work on the Math Library, and has begun work on the Conversion Library. Our long-range plans are to have a similar test program for each library in the COSMOS suite.

b. Thread Safe Code

As we continued support for Hawaiisat-1, and moved on to developing support for missions like PhoneSat and DICE, we started creating fully functioning agents. It became clear at this point that more care had to be taken with the thread-safe nature of the support libraries. This was especially true for the JSON library, where multiple threads for communication with the outside world used the same engine. In both the creation and parsing of JSON strings, and in the reading and writing of Name Space memory, multiple threads were found to be interfering with each other. This has been resolved by allowing the assignment of different buffers for use by the different threads.

In the long term we will analyze this issue in more detail, and also consider other areas in the code that could be open to conflict. The end goal will be to relieve end-users from having to spend too much time worrying about system-programming types of issue.

c. Transfer Buffer Size

The Agent and JSON support libraries are all about the creation and transfer of buffers. As we moved into extensive use of these libraries through development work for the CABLE mission, we started running into problems with both relative buffer sizes and absolute buffer limits. The JSON support routines were coded with dynamically expanding buffers, and both sets of routines keep careful track of the sizes of buffers. However, it is still an issue when one agent that can only handle a 4 kB buffer is handed a 40 kB buffer. This problem only became worse as we started making use of agents dispersed over the network and started running into problems with routers and bridges.

In the short term we juggled increased buffer sizes and decreased data transmission to stay within the workable limits of our immediate environment. In the long term we are considering changes that will allow the fragmentation of buffers that will decrease the magnitude of this problem and increase the scope of networks over which we can travel.

2.   HARD-CODED BEHAVIORS

As with any software project, the requirement to get some initial functionality in place (i.e., prototype), combined with deadlines, led to a version of MOST that was hand-tailored to our original mission, HawaiiSat-1. As we began to adapt this to other missions, especially the multivehicle CABLE mission, we found ourselves rewriting the same program multiple times.

[1] Integrating the code of several different software developers, while keeping all the functionality, can be an issue. Some sort of version control was essential, so we adopted code development using TortoiseSVN (Subversion) (see http://tortoi sesvn.net/). Simply adopting a version control solution was not enough, however, and we have subsequently adopted conventions for usage that optimize the effectiveness of this approach. This approach saw its greatest challenge during our work for CABLE, and we now feel confident working multiple versions of the software with multiple programmers simultaneously.

a.   Hand-Crafted UIs

We have modified MOST several times to support different missions. For each mission we have identified parts of the code that have previously been hard-coded, and have modified the code to allow the UI to be created "on the fly" using a configuration file. The goal is to allow MOST to be configurable for many different missions with little or no modification of the code. For example the Mission Events Display (MED) had some mission-specific subsections hard-coded. During our last update, we removed any mission-specific coding from the MED, and the MED is now configured by the mission's configuration file.

b.   Hard-Coded Name Space

The COSMOS Name Space is proving to be quite useful, and will prove to be a critical element in the openness and flexibility of COSMOS. At the same time, it has had to go through a number of iterations since it was first imagined, with more changes in the offing. Chief among these was a major modification of the naming rules. As originally defined, all numerical values were to have their own names. However, this soon proved to be quite cumbersome, as well as causing problems with associated but disconnected values. We therefore made a major rule change to the naming convention, and now assign names to larger structures, such as *position*.

This major change of naming rules, as well as some other minor changes, has also emphasized the problems with hard-coding Name Space support. As a result, we are formulating some future mechanism by which we can define the Name Space in some configuration file from which the support code would be automatically generated.

3. PERFORMANCE ISSUES

Part of our last update included changing the way MOST handled orbital events and spacecraft events for the MED. Previously, we had two files, and correspondingly two software routines that handled them: one routine for each type of event. We combined these two into one events file, and now have one routine to handle these data. This speeds up the process of displaying the data and reduces the amount of code needed.

a. Name Space Mapping

Another aspect of Name Space mapping that we soon discovered needed work was related to its performance. Setting up, and then using the map, was taking far too much time. As a result, we modified the code from a straight linear search to a hashed database approach. This has dramatically decreased setup time, as well as increased access speed.

For a Universal Name Space to work properly, there must always be a physical structure in memory to which to map each name. The strict requirement of having a physical location for each name can lead to a number of unfortunate consequences. Chief among these is the large amount of space that must be reserved for every possible instance of a name. This can be mitigated somewhat by limiting the number of instances of a name, but this then leads to the second problem of a limited Name Space. We have mitigated this problem somewhat in the current version of MOST by being more efficient in how we define things. In the long run we will have to do something more dramatic. Our ideas for this will be covered in Sec. III.D.2.

b. Graphical Widget Reuse

As we introduced more graphical elements into MOST, it became clear that we needed to be far more efficient in both our coding and use of plots. We found ourselves being inefficient both during coding (through rewriting essentially the same code in different parts of MOST) and during execution (through unnecessary recalculations of drawing commands). We have taken a step back and are rethinking our graphics environment to identify common elements and views. We have already implemented draw lists for commonly used elements such as satellite models and planetary bodies, as well as using standard widgets whose behavior is modified through parameters, thereby making the same code reusable. In future we hope to create a simpler, more efficient, yet flexible approach to all our graphical views.

C.   CURRENT ENHANCEMENTS AND FEATURES

1.   IMPROVED NAME SPACE AND COSMOS JSON

We have relaxed some of the limitations on both the Name Space and the JSON that represents it. The use of a flat Name Space proved to be too limiting, so we have expanded it to allow complex structures built upon simple elements. To support the various types already defined in COSMOS, we have had to provide support for arrays and complex objects in JSON. We now support most of the vector and matrix type defined in mathlib, as well as a variety of position and attitude types defined in convertlib. At the same time we have embraced a broader range of names, allowing names that reflect the exact system in which a set of values, such as position, are defined. The mapping of JSON to memory now includes automatic conversions to other systems where it makes sense.

We have extensively reworked our use of the Name Space, mapping, and JSON. As an additional enhancement, we have added support for equations. Multiple Names pointing to simple scalars can be combined with basic arithmetic or Boolean operators to return a double precision answer. This result can be used anywhere a simple name would, such as displays and strip charts. It will be integral to the event-driven command queue. To support this feature, while at the same time addressing the concern for thread-safe code mentioned above, we have reworked the approach to mapping the Name Space to memory. The user is now required to supply a memory location for storage; at the same time, mapping is done as an offset, allowing one map to serve multiple memory locations. Separate memory locations can be provided for Static and Dynamic data, in support of the Static/Dynamic split as described in Sec. III.C.4. To support the various new features we have added, mapping is now done for both complex objects and their components. Additionally, mapping is now accessible from either the name or the offset. Finally, in support of "fragmentable JSON" described later, we have slightly altered our use of the JSON specification. Our JSON strings now consist of what is effectively multiple JSON strings with a single JSON object in each.

2.   SATELLITE DESCRIPTION GENERATOR

The more vehicles we embraced, the clearer it became that we could not continue generating the Satellite Description File, satellite.ini, by hand. In support of the CABLE mission, we developed a number of tools to help us with this process. The list of Parts, Components, and Devices is nothing more than a simple database, and we have developed a set of databases using FileMaker Pro^{TM} that help us manage the information. This can then be dumped to a series of files that are read by a COSMOS program that spits them back out as a valid JSON strings and writes them to node.ini. A greater challenge has been the generation of the Parts to put in the database. We have developed a process using a simple model developed in SolidWorks 3D CAD that is then exported as VMRL, and finally converted with a second COSMOS program to a Parts description. This is all being integrated into our Data Management System framework, which will consist of a set of libraries and Tools that allow monitoring and manipulation of all data in a COSMOS environment.

3. ADDITIONAL VIEWS

To support the variety of vehicles and different planetary bodies represented by the CABLE mission, we were required to create some new variations on our existing Vehicle and Orbital views. New Vehicle views include a Chase view that follows close behind the vehicle, either in orbit, or in the case of the Rover on the ground, as well as a Lander view, which remains at a fixed point on the ground and follows the vehicle. The Orbital view was modified to support different planetary bodies as its center of focus. Finally, we developed a two-dimensional Map view showing the horizontal placement of the vehicle. Examples of these various views of MOST running the CABLE mission are shown in Figs. 10.3 to 10.7.

4. STATIC VERSUS DYNAMIC DATA

As we started using full sets of telemetry data, and to consider what maximum frequency of data we would like to handle in the future, we began to search for ways to be more memory-efficient. We quickly realized that the arrangement of values in memory that made sense for a simple simulation was quite wasteful of memory when transferred to an archival array of thousands of values. As a result, we split the storage into static and dynamic values and now store them in different parts of the global data structure.

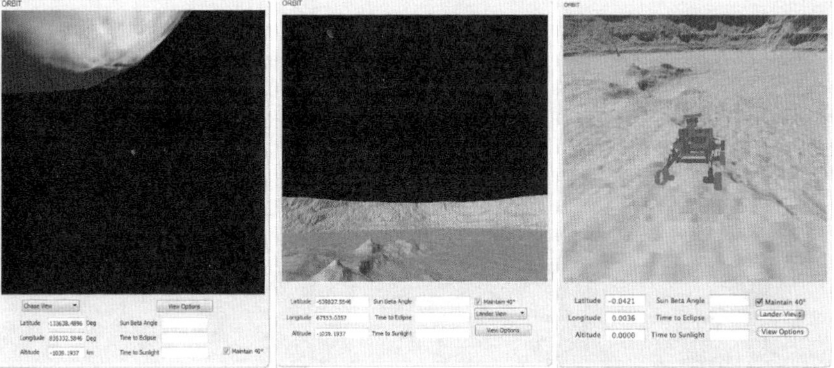

Fig. 10.3   Chase view.   **Fig. 10.4   Lander view.**   **Fig. 10.5   Lander view: Rover.**

5. MOST FEATURES ENABLED

It has always been a goal of MOST development to support not only archival data, but also instantaneous near-real-time (NRT) data. We also planned to allow simple spacecraft commanding, also in NRT. In support of the class project of a Mechanical Engineering class, we were able to develop both these features. MOST can now receive state-of-health (SOH) updates dynamically, as transmitted by a COSMOS SOH Agent. The commanding feature of MOST now interfaces via the network to an Agent that can accept Requests. Several predefined Requests are shown on a dropdown menu, or a Request may be typed in manually. It is then sent to the appropriate Agent via the Agent Communications Protocol.

MOST caution and warning lights have also been enabled to indicate the warning level. Normal operation is Green, the cautionary condition is Yellow, and an anomalous condition is Red. The indicator lights also now show a pattern indicator to the right of each button that changes in accordance with the warning level. This allows a color-blind person to easily determine the warning level.

6. MOST ENGINE

Through development of the CEO it has been clear that we needed to separate the data management functions of MOST from the graphical display and interface. We are therefore in the process of separating MOST into a UI component and a second Agent (the "MOST Engine") that runs as a background task. The engine constantly monitors telemetry data and does any required calculations. The UI portion of MOST communicates with the Engine via the Agent Communications Protocol, and receives and displays the data. The "MOST Engine" can constantly monitor the mission, even if no UI is active. It will generate alerts and have the capability to e-mail or text important developments or problems to key personnel.

Fig. 10.6   Cis lunar map.

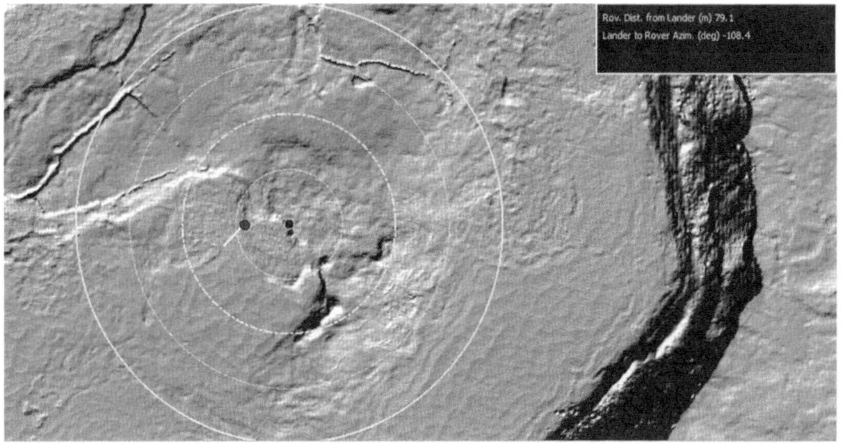

Fig. 10.7   Surface map for Rover operations.

Separating the computationally intensive aspects of MOST into its own Agent will allow the CEO much better access to, and control of, multiple missions. As the number of missions grows, the number of MOST Engines can easily be scaled up through Grid computing solutions, such as Condor. It also allows more flexible access to specific mission data from multiple graphical interfaces simultaneously.

7. NODAL CONCEPT

As we have stretched the limits of COSMOS, it has become clear that it has potential for far more objects and systems than just spacecraft, or even vehicles. In an effort to develop a broader concept in which spacecraft are just a subset, we have adopted the idea of the COSMOS "Node." In this new context, Nodes are locales of action where assemblages of related Agents tie together functionality on pieces of related hardware. Actions occur as the result of communications between Agents, using the COSMOS protocols, and supported by the COSMOS Data Management System. The Node Description, or "node.ini," has become the information that ties together all elements of a Node. Every Agent or Tool that represents a Node will be configured by the same "node.ini." Tools that work with multiple Nodes will access multiple Node descriptions. As a result of this change, COSMOS is setting itself up to be operations software that can be tailored to operate anything you choose: spacecraft, ground system, rover, robot, or even factory floor. In its adaptation to spacecraft usage, it can be configured to tie together spacecraft, ground stations, and mission operations in one unified environment.

8. SELF-GENERATING GRAPHICAL UI ELEMENTS

To meet the needs of the Nodal concept, it has become an absolute requirement that the display and control elements of COSMOS not be tied to specific usages. Everything must be able to adapt itself to both the data and the systems that the particular Node supports, as indicated in the Node Description. This must be automatic to some extent, so that a new Node is easy to configure, but still allow for modifications to be made if some sort of hand-tailoring is desired.

To solve this problem, we have developed a set of routines to self-create generic UIs based on the "node.ini" file. These UI files only use names from the Name Space, plus a set of specially developed widgets capable of loading the values represented by these names. Routines can then be written that will automatically create a UI combining Name Space names with the specialized widgets. A generic loader routine is then capable of implementing any of these UIs. Once the UIs have been created, they are stored on disk to be automatically loaded the next time the Tool is loaded. The end-user can enhance these UIs as much as desired, as long as the Name Space names are not altered. New Nodes can be supported simply by defining a "node.ini" and then starting MOST or some other Tool in an empty directory.

D. FUTURE ENHANCEMENTS AND FEATURES

For every improvement added to COSMOS, our work with the previously described test cases has generated new ideas for future improvements. The following sections are a short list of the ones we plan to work on first.

1. MEMORY-FRIENDLY COSMOS

We currently assign the global data structure statically, which has the dual disadvantage of reserving large amounts of memory, even when it is not needed, and at the same time limiting the number of various items to preset maximum values. We are working on a mechanism that will allow us to assign the space dynamically, while retaining some of the current convenience of being able to set one fixed structure equal to another.

2. FRAGMENTABLE JSON

Upper limits to the size of packets, and their fragmenting as they pass through different networks, has become a problem as COSMOS has grown. We are developing a modification to our use of JSON that will allow JSON strings to be as long as required and yet survive over any network link.

The proposed approach makes what, in standard JSON, will actually be a series of single-element JSON strings. This will allow the string to be broken between any two elements if, at any time during its passage, it comes across a bottleneck smaller than its current size. When split, each new element will be given its own Timestamp so that all will arrive at their destination with proper time information.

3. SELF-CONTAINED MOST

The PhoneSat team has asked for a version of MOST that could be used in combination with a small portable ground station for receive-only monitoring. We are combining a copy of MOST and a MOST Engine in a virtual Linux machine that could run on any other operating system (OS).

4. MULTIPLE NODES PER MISSION AND NODAL AWARENESS

COSMOS uses a powerful feature to support more than one Vehicle (or Node) per mission. We are building in the ability to recognize multiple Nodes within a single mission, allowing rovers to be aware of their landers, or satellites in a swarm to be aware of each other.

5. CHANGING MOST CLOCK

The MOST "master clock" determines what time is displayed as the "present" time. It has been using the built-in Qt QDateTime structure to keep track of time. This has led to some confusing issues, and has limited the resolution of the master clock to one second. This will be changed from the QDateTime structure to fractional Julian Day values, allowing higher resolution of time and fixing some of the issues that have caused confusion.

IV. OPERATIONS TESTBED AND SIMULATORS

The COSMOS OTB is coherent with the open-source system architecture that integrates hardware and software components to operate a low-cost Satellite System Simulator (e.g. FlatSat) that can be integrated into the MOC setup for command scripting testing, personnel training, mission rehearsals, and anomaly resolution. The OTB has tools for satellite technology integration and development that allows for relatively cheap satellite subsystem integration and testing (as would be required by universities developing CubeSats). Most of the OTB tools are based on Custom-Off-The-Shelf (COTS) parts that are affordable to university laboratories, while some tools are being developed under the COSMOS project using proven standards and made available to the small satellite community. The OTB is part of the four major processes in mission operations that are supported by COSMOS, namely Mission Planning and Scheduling, Real-Time Contact Operations, Mission Analysis, and Data Processing and Management. This testbed is being designed to accommodate multiple spacecraft testing by reutilization of the same tool, which is highly configurable and therefore generalizable for different satellites. Another important design feature of the OTB is that it is capable of scaling up for testing satellite constellations.

One important aspect of the OTB is that it makes it possible to provide an interface with the different satellite hardware and simulators that are needed for the global testing procedure for different missions. This platform also allows mission segment functional simulation and mission rehearsals from the command sequence to software and hardware performance. One more important aspect to note is that the OTB is being designed so that it may be remotely operated, allowing people from different remote locations to use this same setup to help in their satellite development or mission operations.

To completely operate the OTB, its setup must integrate six main constituents: 1) the actual MOC control tool, or MOST; 2) the Ground Station Simulator (GSS); 3) the Satellite System and Subsystem Simulator (SSS); 4) the Test Bed Engine (TBE); 5) The Test Bed Controller Tool (TBCT); and 6) the Test Bed Controller User Interface. This segmentation is shown in Fig. 10.8.

The MOC System Simulator allows the end-user to conduct NRT spacecraft system and subsystems testing and operational activities, including mission planning; assessment and maintenance; instrument health monitoring; and communications, command and control function. The integral part of the MOC System Simulator is MOST, which is one of the two interface tools between the OTB and the end-user. Sorensen et al. [3] expands on some of these functionalities of the OTB.

Open source frameworks used in real-time systems are considered primary resources for the development of the OTB (including YARP, ACE, LCM, etc.).

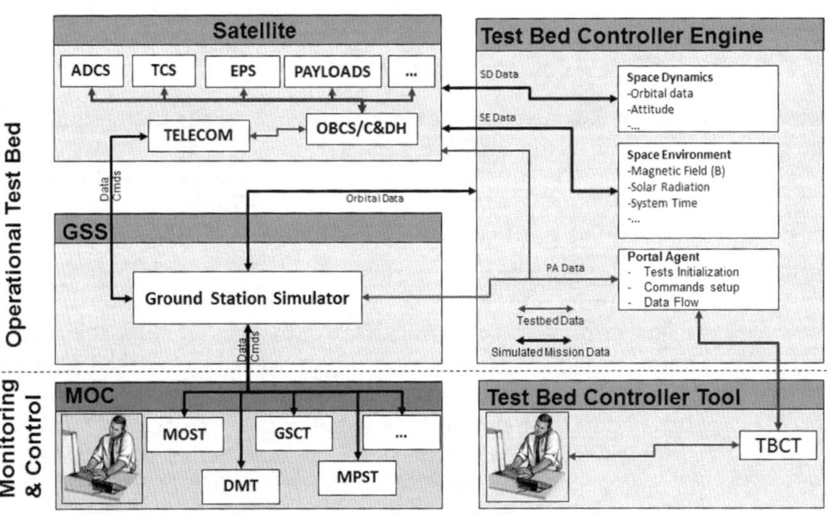

Fig. 10.8   OTB architecture.

Fig. 10.9   OTB being developed in HSFL's laboratory.

The SSS platform integrates all the satellite subsystems to be operated (e.g. ADCS, TCS, EPS, Telecom). These can be either fully operational with the engineering model hardware components or else software-simulated if the hardware components are not readily available. Based on the TBE, it supports full propagation of the test satellite's conditions, in both real and faster than real time. Figure 10.9 shows the OTB hardware platform being developed for the HiakaSat-1 microsatellite. The OTB replicates the HiakaSat hardware subsystems and it is placed on top of an air-bearing testbed setup. This testbed was developed with the help of Daniel Wukelic, a Space Grant fellowship student working with HSFL.

The COSMOS OTB is acquiring the Astro-und-Feinwerktechnik-Adlershof (Astrofein) Attitude Control System (ACS) testbed, which will be installed at HSFL in 2013. HSFL has been awarded a grant from the Air Force Office of Scientific Research for the FY 2012 Defense University Research Instrumentation Program (DURIP) towards the purchase and installation of this state-of-the-art ACS testbed. This testbed simulates the space environment using different simulators with a high degree of precision for sensor inputs: load-free (equivalent to zero gravity testing), Sun luminescence, GPS, and Earth's magnetic field. This will greatly increase the capabilities of HSFL and the COSMOS project. Figure 10.10 shows the ACS testbed setup with an air-bearing table inside the Earth-magnetic-field, GPS, and Sun simulator system.

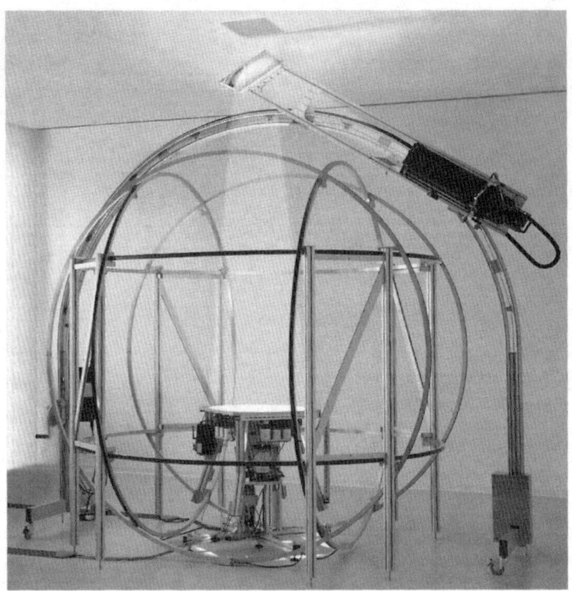

Fig. 10.10   Attitude Control Testbed from Astrofein.

This OTB setup will initially be used to test the Attitude Determination and Control Subsystem for HiakaSat-1. Using COSMOS MOST, software commands can be sent through a GSS to the mockup. This part of the OTB is comprehensive because it is the culmination of many previous OTB subsystem tests, after which the system was integrated and then tested before launch.

The Test Bed Dynamics Engine provides a software-simulated space environment to the OTB to allow more realistic operation of the whole platform. The dynamics engine also controls the different hardware and software configurations in the satellite system simulator and allows the tuning and mixing of signals and interrupts, adding noise and possible failure modes. All this is either controlled by the controller user interface or by a scripting sequence. The TBCT is an application to support the experimental setup for the OTB architecture. The TBCT interfaces with the GSS, the satellite system, the TBE, and the end-user. It allows initializing and controlling the satellite system platform and the TBE according to user decisions or scripting. The UI control tool is software, like MOST, for operating and changing the OTB parameters and testing sequences.

The COSMOS OTB can incorporate different hardware parts that are made available for testing and experimentation. These components can include common sensors, actuators, and other hardware systems that are common for satellite integration. Other specific features of the OTB include calibration and testing of hardware components; integration of software tools for hardware simulation; subsystem validation & monitoring; subsystems interaction & dynamics monitoring; pseudo-environment input; anomaly resolution; measurable performance, such as pointing, timing, speed, fast, power, etc.; remote control of the OTB using scripts; NRT testing and simulations; mission training and rehearsal; trending and analysis; system operation rehearsals and simulations with statistical analysis (e.g., Monte Carlo, dead reckoning); operability with different standard software development tools and languages, including MATLAB, LabView, Python, C/C++, and/or other engineering COTS software utility tools; and supporting the development and operational testing of different satellites.

V. COSMOS EXECUTIVE OPERATOR

COSMOS is being designed to handle multiple satellites on multiple missions. When there are only a few (maybe less than a couple of dozen), COSMOS can handle multiple missions in a single MOC, with each satellite having its own session with the major COSMOS tools, either on the same or on different consoles. If the facility resources permit, it would be simpler and easier to dedicate one console per satellite, with another to host the GSCT, one for the DMT, and a top-level coordinating console running the COSMOS Executive program.

The CEO program provides situational awareness of multiple spacecraft or simulated spacecraft simultaneously (Fig. 10.11). In its initial implementation,

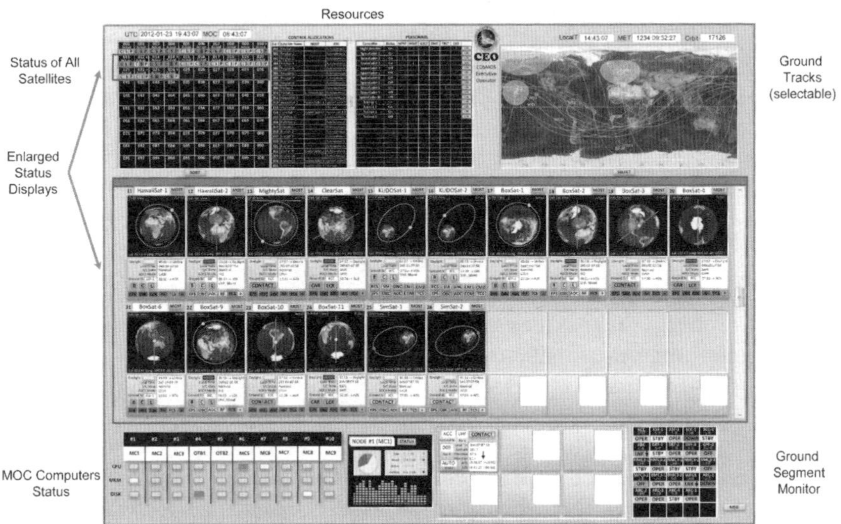

Fig. 10.11   Mockup CEO display, showing 20 satellites and 24 ground station antennas.

CEO can handle 100 spacecraft, but there is no reason why it cannot be modified to handle more. There are three different selectable levels of monitoring:

1. Low: spacecraft identification, status of spacecraft, status of payload, ground station contact status.

2. Medium: shows orbit position and data, day/umbra status, ground station contact status, status of subsystems, spacecraft or attitude/pointing modes, etc.

3. High: similar to main display of MOST, giving detailed information.

The CEO collects information from the MOST Engines running with the Data Management System (one MOST Engine is running for each satellite being monitored). These MOST Engines send their data out through the DMS to any application that wants it, such as a session of MOST or the CEO.

CEO also provides situational awareness of the Ground Segment, including the status and operations of all the ground stations in the network. It can launch the GSCT for more detailed information or commanding of the Ground Segment.

Besides the external elements of COSMOS outside the MOC, the CEO also provides management of MOC operations. It monitors allocation of COSMOS tools to spacecraft, monitors personnel utilization, access console logs (current or archived), and communicates with one or more MOC positions directly. The CEO monitors COSMOS system performance, such as console computers' performance and utilization, and the status of the COSMOS tools and the data flow between the COSMOS elements. The CEO can also launch any of the COSMOS tools, such as the MPST, MOST, GSCT, DMT, TBCT, and analysis tools.

The design of the CEO was completed, and coding started by the end of 2012. It is expected that a prototype version will be running by April 2013, and a basic operational version of CEO finished in mid-2013.

VI. CONCLUSION

The COSMOS project is two-thirds of the way through the three grants provided by NASA. Some of the major elements have been developed and demonstrated in a prototype form, and it will soon be deployed to support various current or upcoming space missions, initially in the monitor-only mode, but eventually with full command capability as well.

During the first half of the project, the software building blocks needed to develop the high-level operations tools were developed and mostly completed, although improvements are being made as our experience with the system grows. The lynchpin of COSMOS and the tool most developed, MOST, has shown its versatility in being adapted from supporting just Earth-orbiting missions to prototyping a version to support lunar missions, even extending its capabilities to supporting control of rovers. This was beyond the original range of applications that were considered at the outset of the project, but once the flexibility and capabilities of MOST were demonstrated, we came to realize that its application potential was much larger than anticipated. The Canadian firm MPB Communications has a contract with the Canadian Space Agency to build and demonstrate a miniature planetary rover using Earth-analogous terrain. They have requested a version of MOST to support this effort.

The OTB is the other major element of COSMOS that is well under development. This is currently being used in the development and testing of HSFL satellites.

During the remainder of the project period our primary goals are to deploy operational versions of MOST at various MOCs, including at NASA Ames Research Center and the USU Space Dynamics Laboratory. We will also be working on the development of the remaining tools, especially MPST and GSCT. The CEO is vital to controlling multiple spacecraft or objects simultaneously, and development of it has started but is in the early stages. A fully functional basic version of COSMOS should be operational by the end of the NASA grant, at the end of August 2013. However, COSMOS will continue to be developed within HSFL and with the aid of our government, university, and industrial partners who constitute the COSMOS community.

ACKNOWLEDGMENTS

We would like to acknowledge the hard work put into this project by the following members of the COSMOS team: (faculty and staff) Harold Garbeil, Daniel Watanabe; (undergraduates) Erik Wessel, April Vogt; (high school summer interns) Mathew Esporas, Max Dylan Matsuda-Hirata, Daron Lee, Erin Main, Jennifer Nishida, and Grant Takara.

REFERENCES

[1] Sorensen, T. C., French, L., Chan, J. K., Doi, W. K., Gregory, E. D., Kobyashi, M. H., Lee-Ho, Z. K., Nunes, M. A., Pilger, E. J., Yamura, R. A., and Yoneshige, L. K., "Hawai'iSat-1: Development of a University Microsatellite for Testing a Thermal Hyperspectral Imager," AIAA, Anaheim, CA, 2010, AIAA-2010-8922.

[2] Sorensen, T. C., Tran, T. T., Geldzahler, B. J., Horan, D. M., and Prescott, R. J., "Effective Science Mission Planning and Operations – The Clementine Approach," *1st Annual Reducing the Cost of Space Ground Systems and Operations Symposium*, Rutherford-Appleton Laboratories, Chilton, Oxfordshire, UK, 1995, Paper RAL.GS.31.

[3] Sorensen, T. C., Pilger, E. J., Wood, M. S., and Nunes, M. A., "Development of a Comprehensive Mission Operations System Designed to Operate Multiple Small

Satellites," *25th Annual AIAA/USU Conference on Small Satellites*, Utah State University, Logan, UT, 2011, SSC11-IX-3.

[4]   Sorensen, T. C., Pilger, E. J., Wood, M. S., Gregory, E. D., and Nunes, M. A., "Development of the Mission Operations Support Tool (MOST)," *SpaceOps 2010 Conference*, AIAA, Huntsville, AL, 2010, AIAA 2010-2230.

[5]   Qt Technical Overview whitepaper, http://cosmos-project.org/documents/qt4_whitepaper.pdf.

[6]   Kruzelecky, R. V., Haddad, E., Nakhaei, A., Jamroz, W., Cloutis, E., Sorensen, T., Mouginis-Mark, P., Shan, J., Hamel, J., de, Lafontaine, Barnet, M., Teti, F., and Ghafoor, N., "CABLE Canadian American British Lunar Explorer," *Global Space Exploration Conference*, Co-sponsored by AIAA, Washington, DC, 2012, GLEX-2012.03.1.5x12366.

[7]   Sorensen, T. C., Pilger, E. J., Wood, M. S., Nunes, M. A., Garbeil, H. M., Wessel, E. K., Kruzelecky, R. V., and Jamroz, W., "Adapting an Open-Architecture Mission Operations System for a Lunar Rover Mission," *63rd International Astronautical Congress*, International Astronautical Federation, Naples, Italy, 2012, IAC-12.A3.2C.7.

CHAPTER 11

Commercial Collaboration for Collision Avoidance and Flight Operations

David Finkleman[*]
*Center for Space Standards and Innovation, Analytical Graphics, Inc.,
Colorado Springs, Colorado*

I. INTRODUCTION

The investment, utility, and productive capacity of man-made satellites in Earth orbit demand extremely safe and reliable operations. At present, there are no rules, agreements, or legal instruments that facilitate or enforce operational safety among satellites or operators from different nations or organizations. It is traffic as existed in cities at the turn of the nineteenth century, except that the elements of transportation travel at speeds of kilometers per second and cannot easily arrest their motion nor divert their courses responsively to avoid imminent catastrophe. Most are juggernauts with massive, uncontrolled inertia.

Space traffic cannot be controlled. It can only be planned, trusting that plans can be executed as intended and that the parties involved develop and execute maneuvers collaboratively to avoid exacerbating an already dangerous situation.

Voluntary collaboration for the common good is currently the most proactive and realistic approach to collision threat mitigation. Not all operators will be willing to participate because knowledge of a competitor's movements and capabilities can be a significant competitive advantage. Benefit for one is the other's disadvantage. National technical means of treaty verification from space are the most obvious example of bearing collision risk to preserve important knowledge. Collision risk is contagious. It involves at least two parties and many satellites suffer the consequences. An operator cannot in good conscience decide the risk to be imposed on others or ignore the impact (pun intended) on others. A collision benefits no-one.

The inevitable inability to know where all satellites are at all times precisely compounds the problem. No observation network could ever keep track of everything, at any cost. All observation measurements are imprecise. This imprecision and unavoidable approximation of the forces that govern satellite motion makes estimates of satellite trajectories quantifiably uncertain and extrapolation into the future for the purpose of avoiding collisions arguable. Risk mitigation must be extremely conservative to accommodate the uncertainty. Only a few satellites

---

[*]Senior Scientist; dfinkleman@centerforspace.com.

have their motion almost continuously monitored and have dedicated avoidance processes in place, so only for these can we claim great confidence in relation to collision avoidance. The International Space Station (ISS) is the best example.

Fortunately, what is perceived as a crowded environment is still reasonably sparse by terrestrial standards. The environment is also not uniformly congested. Some orbit regimes are much more densely populated than others, such as Sun-synchronous or geosynchronous orbits. There are hundreds of perceived approaches, within a few kilometers, every day. The greatest cumulative likelihood of actual physical contact is about one chance in ten per year in the most crowded orbit regimes. The instantaneous probabilities of actual physical contact for a single close approach are extremely small by terrestrial standards [1].

The consequences of collision are unacceptably high, so even low probabilities of its occurrence still justify extreme caution. The threshold established by most owner/operators is one in ten thousand, or even one in a million in some cases [2]. Low probabilities are deceptive. The probability of physical contact is low when orbits are known very precisely, even though satellites might approach each other very closely. The probability of physical contact is also low in the more common situation when we think we know the orbits very precisely but really do not because measurements are very imprecise and models of forces are insufficient. We may be very close to a hypothesized orbit that itself is wrong. However, the probabilities are still low even when we do not know clearly where satellites are, particularly when it appears that the satellites are far apart (although that might only be a perception) [3].

Accidental collisions are very rare. Unintentional collisions during active operations are as likely as nonoperational accidents; that is satellites maneuvering in a manner that inadvertently places them and others in jeopardy or satellite trajectories that deviate from what was expected leading to inadvertently close approaches.

No entity has even a marginally sufficient volume of timely data. The U.S. Air Force has the most comprehensive data, particularly regarding the orbits of inactive and natural debris in Earth orbit. However, by doctrine and practice, Air Force orbit determination is derived only from U.S. Government (USG) observations. There is very little capacity to keep track of satellites during maneuvers, and orbits derived only from USG sensors can in fact differ significantly from owner/operator information [4].

The Space Data Center (SDC) is executing a phased approach to mitigating, if not overcoming, these issues by using owner/operator knowledge of their own operations and plans and by exploiting diverse and pervasive sources of observations.

II. SDC EVOLUTION

Collision avoidance practices were established almost simultaneously with the first Space Shuttle mission. U.S. Space Command directed these operations from

Cheyenne Mountain using Air Force resources and orbit estimates. Because the Shuttle and the ISS were NASA missions, the Department of Defense (DoD) supported NASA decisions with more intense observation schedules and more timely orbit estimates. NASA manned missions naturally have different threat criteria and procedures than unmanned (robotic) missions. Unique mechanisms were established for data transfer (Conjunction Summary Messages) and collaboration. The fullness of even classified data available to NASA operators is an essential element of the process. Where established data exchange content and format are insufficient for the mission, NASA and DoD users can delve more deeply. Commercial owner/operators do not enjoy that privilege.

Commercial operators were so concerned that some of them established financial relationships with organizations that could use (but not release) privileged data to monitor risks to specific satellites or constellations. Intelsat arrangements with the Massachusetts Institute of Technology (MIT) Lincoln Laboratory and the Aerospace Corporation were most well advertised. MIT could gather independent observations of geostationary satellites within the field of view of its radars in Massachusetts, while Aerospace Corporation operated under the aegis of the Air Force Space and Missile Systems Center. Neither of these was cost-effective for the operators or could meet their needs, and neither was continued.

In 2005, the Center for Space Standards and Innovation (CSSI) began a free, Web-based conjunction assessment service. SOCRATES has provided at least twice-daily assessments of man-made objects in Earth orbit based on publicly available Two Line Element set (TLE). SOCRATES evolved from TS Kelso's Celes-Trak Web site, which has been widely used for more than 25 years (see http://celestrak.com/). CSSI also developed advanced probabilistic conjunction assessment techniques, debris and other capabilities for addressing outcomes models, and techniques for extracting greater value from TLEs (see http://celestrak.com/NORAD/elements/supplemental/). These techniques enabled analyses of the Fengyun 1-C, Iridium 33 – Cosmos 2251, and other very important dangerous

Action	NORAD Catalog Number	Name	Days Since Epoch	Max Probability	Dilution Threshold (km)	Min Range (km)	Relative Velocity (km/sec)
				Start (UTC)	TCA (UTC)	Stop (UTC)	
Analysis	38036	SOYUZ-TMA 03M [+]	1.982	1.000E+00	0.000	0.000	0.000
	38222	PROGRESS-M 15M [+]	1.982	2012 May 03 12:00:00.000	2012 May 03 12:00:00.000	2012 May 10 12:00:00.000	
Analysis	38096	ATV-3 (EDOARDO AMALDI) [+]	1.982	1.000E+00	0.000	0.000	0.000
	38222	PROGRESS-M 15M [+]	1.982	2012 May 03 12:00:00.000	2012 May 03 12:00:00.000	2012 May 10 12:00:00.000	
Analysis	06659	METEOR 1-15 [?]	5.259	3.710E-02	0.009	0.036	14.773
	26654	CZ-4B DEB [-]	6.338	2012 May 07 16:15:42.490	2012 May 07 16:15:42.829	2012 May 07 16:15:43.167	
Analysis	09481	METEOR 1-26 [?]	1.110	2.620E-03	0.053	0.089	3.397
	38124	FENGYUN 1C DEB [-]	5.950	2012 May 03 13:10:57.312	2012 May 03 13:10:58.784	2012 May 03 13:11:00.255	
Analysis	25394	RESURS O1-N4 [?]	3.952	2.604E-03	0.035	0.136	13.714
	33865	IRIDIUM 33 DEB [-]	4.752	2012 May 06 19:57:10.016	2012 May 06 19:57:10.381	2012 May 06 19:57:10.745	

Fig. 11.1   Sample SOCRATES report.

encounters among satellites. Figure 11.1 presents examples of a near-real-time SOCRATES report.

Collegial relationships and research collaboration with major communication satellite operators fostered extrapolating the SOCRATES capability to owner/ operator trajectories and much more refined orbit data in general. Presaged by discussions between Intelsat and NASA and commercial operator discussions with the Aerospace Corporation, meetings between CSSI and Intelsat flight dynamics principals in 2007 and 2008 led to a seminal gathering at Telesat in Ottawa in December 2009. This led to preliminary terms of reference and organizational concepts for the Space Data Association (SDA), incorporated on the Isle of Man in 2010.

The SDA objectives may be summarized as follows:

1. Seek and facilitate improvements in the safety and integrity of satellite operations through wider and improved coordination between satellite operators.

2. Seek and facilitate improved management of the shared resources of the space environment and the radio-frequency spectrum.

The SDA founding members were Intelsat, SES, and INMARSAT, joined later by Eutelsat, and these comprise the executive members and board members.

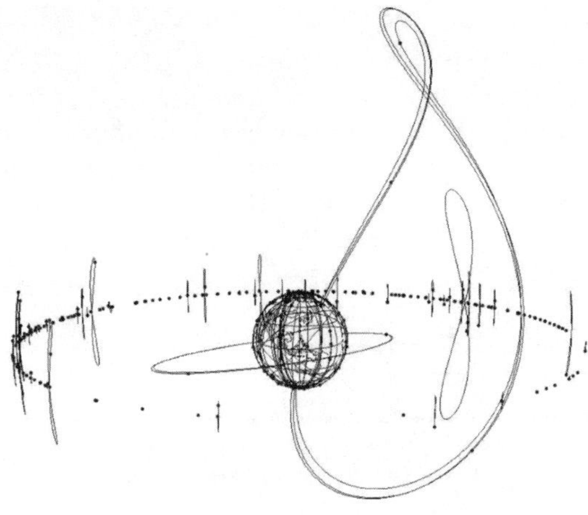

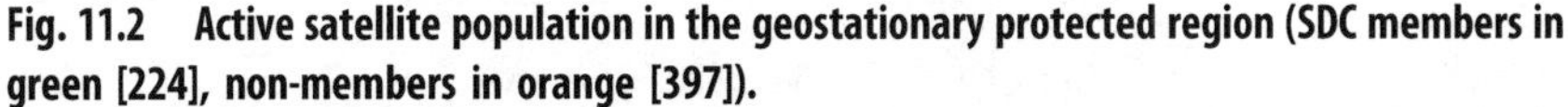

Fig. 11.2   Active satellite population in the geostationary protected region (SDC members in green [224], non-members in orange [397]).

Operational concepts and doctrine were deemed exceptionally important. The principals are competitors, so their collaboration was a milestone in industrial cooperation to meet a common challenge. Data integrity and security, actionability, and reliability were major considerations. Neutrality among claimants from different nations and jurisdictions was critical. The SDA conceived requirements for a conjunction assessment center with these criteria and doctrines. The SDA solicited operational and technical capabilities, and a contract was awarded to AGI in the spring of 2010. Initial operational capability was achieved in July 2010 based on SOCRATES capabilities and technology. Final operational capability followed in September 2011 based on modern Web services, flexible software architecture, and state-of-the-art security and reliability.

SDA membership, or a commercial contract or agreement, is required to contribute data to the SDC and receive conjunction assessments. At present, there are 17 active member organizations and over 300 geostationary and lower orbit satellites in the scheme. These include several governmental organizations. Membership continues to grow.

Figure 11.2 demonstrates the current active population of the geostationary protected region, indicating SDA subscribers and those who are not yet subscribers.

III. SDC CAPABILITIES

The SDC is based on commercial off-the-shelf elements and state-of-the-art service-oriented software architecture and Web services. The capabilities are illustrated in the work flow shown in Fig. 11.3.

A. HIGHLY ACCURATE AND PRECISE SATELLITE ORBITS

Owner/operator observations and ephemerides are the best representations of the states of their satellites. Operators use techniques they trust and that have enabled reliable resource management. These techniques have evolved to be unique to constellations satellites, or even individual satellites. They are not uniform, even among constellations controlled by a single operator. Operators acquire satellites on orbit that were previously controlled by others. Some organizations, such as Intelsat, operate satellites for other service providers.

The quality of orbits determined by operator ranging on downlinks or even onboard global positioning system (GPS) is variable. Although most operator orbit estimates are very good, based on frequent and diverse observation geometries, some estimates are degraded by single station ranging or unfavorable viewing geometries. It is still inarguable that operators know best where their satellites are than anyone else could.

There are many techniques for developing orbit estimates from observations. Most are described in ISO DIS 1123 [5]. They are also enumerated in Vallado's

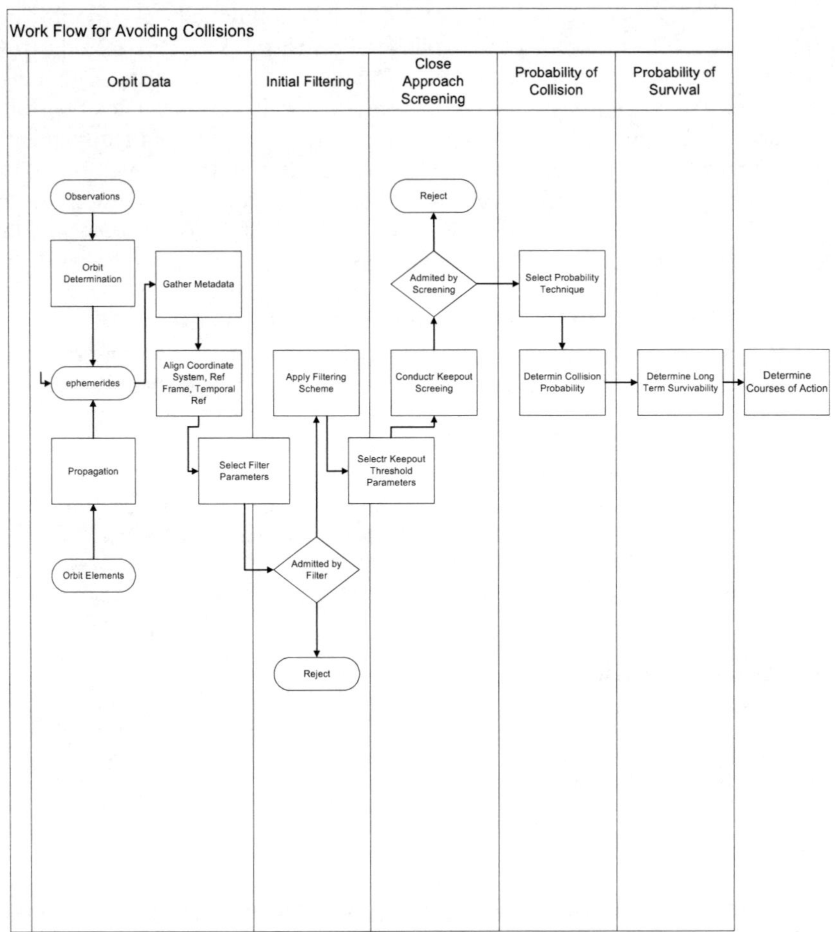

Fig. 11.3   Representative collision avoidance work flow.

text [6]. They include a variety of statistical filters, least squares, and differential correction of recent orbit estimates. The SDC can use most of the alternatives, but trustworthy conjunction estimates require that output ephemeris products from both conjunction partners be consistent and compatible.

The SDC has demonstrated the ability to ingest and consistently apply ephemerides in many forms and to produce consistent representations of orbit estimates and uncertainties. The SDC has also demonstrated the advantages of predictive filters, but it can accommodate other orbit determination approaches should users prefer.

B. ESTIMATE AND REPORT CONJUNCTIONS

SDC principals have developed internationally recognized requirements for Conjunction Assessment (CA) from their SDC experience. Conjunctions are close approaches among satellites. They are not collisions. Different operators have different judgments regarding what constitutes a close approach. The SDC accommodates criteria the operators establish as opposed to what might be convenient for the SDC. The most important requirement is that conjunctions perceived include as many high-probability collisions as possible. The SDC has established a mathematically rigorous approach to determining close approach thresholds sufficient for individual operators' risk tolerance.

No-one can predict conjunctions. Analysts can only estimate what conjunctions might occur in the future. Estimates are not predictions.

Conjunction assessment can be computationally intensive. Starting with the entire catalog, the possibilities are first thinned with a filtering technique, which eliminates satellite pairs where it is virtually impossible to lead to close approaches. The surviving pairs are screened more precisely, reducing the number that must be examined intensively to estimate collision probability. ISO CD 16158 [7] enumerates the widely used filtering and screening techniques. The SDC can use all of them. SDC principals have developed very effective techniques, well described in the literature, but these are not given preference [8]. Any approach that a user prefers can be used.

The SDC exercises arguably unique diligence in assuring actionable conjunction assessments. CSSI experts have defined requirements for actionability [9]. These include considering the age of the orbit estimates included, the reference frames and coordinate systems of contributing data, the time span over which observations were gathered, the force models used in owner/operator generated ephemerides, the reliability of the owner/operator maneuver plane, and many other factors.

All seriously involved in conjunction assessment now recognize that the task demands orbit estimates that are much more accurate and have uncertainty much better characterized than any other satellite operational task. Even proximity operations can bear reasonable uncertainty because the end game is closed-loop on real-time, extremely precise local observations.

One criterion for response is that there is sufficient time to plan and execute evasive maneuvers with minimal expenditure of stored energy. Kelso and Oltrogge have confirmed that accurate orbit estimates about two days into the future are required to facilitate operator action [10]. CSSI has developed maneuver planning tools that estimate the energy required, the thrust vector, and the time required to reduce collision probability to a safe value [11].

Owner/operators also plan maneuvers and estimate ephemerides for different time spans in the future. These are operational imperatives unique to the satellite and mission. The SDC has developed procedures to bring orbit estimation time spans for conjunction pairs into a common time span. There are several ways to do this, for example, by truncating all ephemerides to the shortest time span among them or using ephemerides provided to extend orbit estimates to a longer time span using trusted orbit determination techniques.

Data alignment is important because different owner/operators conduct operations in a variety of reference frames and potentially different timescales. Absolute time from a common past reference is critical as it determines the Earth orientation parameters (EOPs) necessary to correlate observations from an Earth-fixed, rotating reference frame to the inertial reference frame in which satellite orbits are determined and in which stellar references reside. The correlation between coordinated universal time and time expressed in Earth orientation and rotation angle is key. Therefore, the timescale and reference frame are interdependent. The variety of reference frames and alternative orbit element formulations in operational use today is astounding. Sometimes operators themselves do not realize what reference frame their software uses. Even if they do, reference frame descriptions change with time and using an old version can mislead the interpretation of ephemerides. SDC principals have devoted significant effort to reference frame-related forensics data analysis before including data into the SDC.

Reporting conjunction perceptions is much more than sending an e-mail. Essential elements of information must be included, both data and metadata. Operators whose satellites are in jeopardy must be able to trust and confirm conjunction assessments. The SDC accomplishes this through data provided by SDC subscribers on their own satellites, quality controlled and timely use of the public catalog, databases of subscriber satellite characteristics, and very active contribution to international standards for exchanges such as orbit data messages [12] and conjunction data messages [13]. The SDC maintains continuous vigilance for close approaches to all subscriber satellites, screening with user-desired thresholds in a neighborhood watch, and immediately transmits conjunctions with a notable probability of collision. The usual format is that provided in the SOCRATES free Web service. Prudently, the SDC does not presume to manage evasive maneuvers, which are best dealt with collaboratively by the parties involved. The timing of close approaches, relative velocities, and close approach distances in the SOCRATES format are ultimately verifiable, because orbit data, force models, propagators, and other essential elements of analysis are well described and easily accessible.

Our diligence in characterizing orbit determination and propagation techniques is widely recognized and well documented. We were principal authors of the only normative standards and best practices [14, 15]. We can use almost any widely used schemes and force models. We have examined the effects of different force models, representations of the atmosphere, multibody effects, and influences of solar and other radiative fields [16]. The SDC examines, quarterly, the accuracy and precision of owner/operator ephemerides [17]. We are zealots for proper use of the terms accuracy and precision. Accuracy is how well orbits match a reference. Do we determine the satellite's state consistently with independent ground truth? Precision is the dispersion of repeated measurements or estimates. We have demonstrated that TLE sets are often inaccurate and imprecise [18]. Often there is no truth. Observations themselves are imprecise. It is very important that the imprecision be understood and characterized. Imprecision is represented by dynamic variances and covariances. As previously stated, covariances are the most important element of conjunction and collision assessment.

The issue of a realistic covariance matrix cannot be overemphasized; in fact, it has received considerable discussion over recent years. Batch least-squares techniques produce a covariance matrix as a result of the estimation process and the additional state parameters included – solve for parameters, track weighting, and so on. This technique has been used for many years, but may experience difficulties when nonconservative forces introduce significant error into the solutions. Using a Kalman filter that incorporates mathematically derived process noise has the advantage of not being constrained by the fit span, and the limitations that imposes on dynamic variables in the solution. However, the Kalman filter solution may not model the mean long-term behavior in prediction as well as batch least-squares methods, although various fading memory approaches can be used to adjust bias and drift uncertainty [19].

Figure 11.4 presents the geometry of a conjunction encounter. The figure shows a generic close approach scenario. Each satellite has an associated velocity and covariance. The combined error ellipse is simply the summation of each individual covariance and the relative velocity vector is used to define the encounter plane. The sigmas (n) should be the same for each error ellipse.

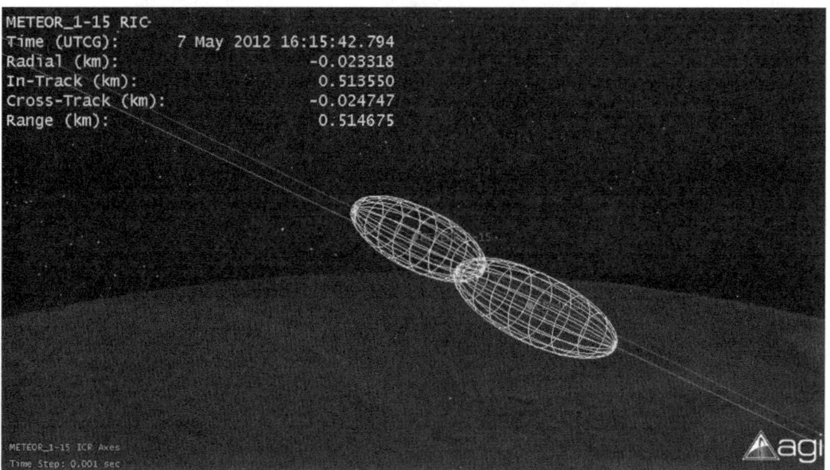

Fig. 11.4 Conjunction encounter geometry. (STK image courtesy of AGI.)

For orbit propagation, maneuver detection, and tracking, the covariances must be a consistent product of orbit propagation. Covariances that are scaled or otherwise manipulated independent of mathematical orbit determination or propagation processes should not be used to estimate collision probabilities. We are very serious about covariance realism and consistency [20].

C.   SENSOR CHARACTERIZATION AND TASKING

Several sensor networks are not well enough characterized for emerging SSA requirements. The U.S. Air Force space surveillance network radars were developed for Cold War missile warning. Although they have been upgraded several times and are very capable, operational concepts and procedures do not meet the needs of maneuver detection and collision assessment. Vallado and Griesbach [21] analyzed the locations of US Air Force sensors. Consider Fig. 11.5, which shows USG SSN, ESA, and ISON sensors. None has a complete view of the satellite population! Sensors are located around the world, but coverage is not particularly good in the Southern Hemisphere and over the oceans. The US SSN (circles), ESSS (stars), and ISON (icons) systems are shown. More important, [20] examined the combinations and permutations of observation opportunities for many satellites with the measure of merit being the quality of resulting orbit estimates. Finkleman used similar techniques to demonstrate the capability of existing sensors to perceive maneuvers and support international codes of conduct and contribute to space traffic control [22]. These tools can support efficient observation tasking based on the contribution of each observation opportunity to the quality of orbit determination. This is often called "covariance-based tasking" [23].

Fig. 11.5   Representative worldwide sensors. (Image courtesy of AGI/STK.)

It is as important that we characterize the measurement uncertainties as it is to have the most precise observations possible. The networks can be characterized through intercomparison and calibration against sensors and satellites that are themselves well characterized and precise. The International Satellite Laser Ranging Network [24] is invaluable, but its precise orbit estimates (POEs) are created only for a small set of satellites in very stable orbits, such as LAGEOS. Measurement precision is situational. Satellites in less stable orbits, with less opportunity for observation, and less favorable viewing geometries introduce different error sources and magnitudes. The SDC can infer sensor measurement uncertainties from network observations of the same satellites by several different sensors. Similar techniques have been used for many years to ascertain minute but important variances among atomic clocks that are coordinated to produce coordinated universal time (UTC).

D.   MANEUVER DETECTION AND FORENSICS

The best way to deal with maneuvers is to know operator maneuver plans. The SDC is arguably unique in being a trusted agent for orbit and maneuver data from operators who compete with one another. This trust was earned by demonstrating exquisite system and process integrity and security. This is the only nongovernmental capability to advise operators of potentially dangerous consequences of otherwise uncoordinated maneuver plans.

When maneuvers are not communicated in advance, preferably in the Orbit Data Message (ODM) format, there are several forensic techniques to infer the actions and estimate outcomes. Complete and accurate force models are essential to propagate orbits and covariances through gaps in observations. Figure 11.6 shows the relationship between force models, Orbit Determination (OD) accuracy, and prediction accuracy. Note the importance of force models during the OD phase. For filter applications, force models are often reduced during the observation interval; however, data gaps necessitate additional force models during this time. (In the figure, notice the two-day gap in the data during which OD force models become important; in the prediction phase, force models are important, as well as knowledge of future maneuvers.) For beyond line of sight systems, the general approach is to keep the force models the same between the OD and prediction phases. The presence of unknown maneuvers complicates the propagation accuracy considerably for any application having no knowledge of those maneuvers.

When maneuvers must be detected without any advance notice of information, the time required to recover a sufficient orbit estimate is the best performance measure. The recovery process has three elements: perceiving that the satellite is not behaving as expected, assessing where the satellite might be, and acquiring data for estimating a correct orbit. The first two depend on consistent and realistic covariances. For example, a satellite's perceived trajectory being very different from excursions consistent with the covariances of the most

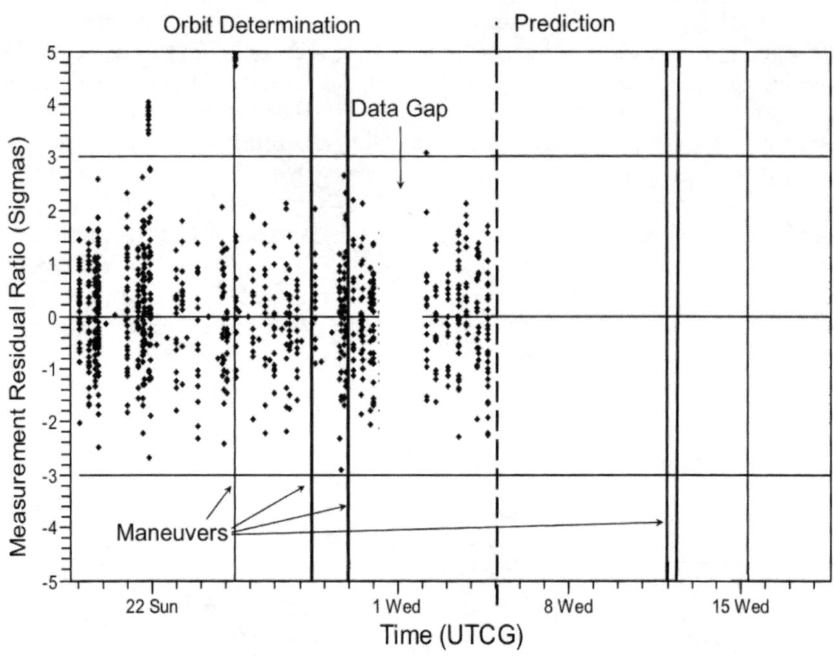

Fig. 11.6   An example residual ratio plot is shown where a period of observations is processed (orbit determination).

recent trustworthy orbit estimate is a good indication of maneuver only if the covariances are realistic. Using observations acquired during a maneuver is extremely misleading unless the maneuver process model is included in orbit estimation.

E.   QUALITY CONTROL AND DATABASE MAINTENANCE AND ASSURANCE

For nearly 30 years, the CelesTrak service has been the most widely used and trusted value-added source of orbit data, including extensive and searchable historical records. CelesTrak is the paradigm for SDC operations. The SDC scrutinizes publicly available orbit data, usually the first and only detection of cross-tagging (associating an orbit estimate with the wrong satellite, missing data, and other anomalies). The SDC maintains current databases of points of contact among subscribed operators and the satellite characteristics that are important for orbit estimation, such as satellite mass and ballistic coefficients. The integrity of sensitive orbit and satellite data is exquisite and spotless, with state-of-the-art distributed server reliability and Web services, including the physical security required by critical revenue-sensitive operations.

IV. OPERATIONAL EXPERIENCE AND VALUE ADDED

The SDC is also the source of new standards and operational practices. It is pioneering fusing owner/operator data with independent observations. It has enabled advances in track association, choosing safe keepout volumes, and understanding the advantages and deficiencies of existing public satellite catalogs. There is no substitute for doing the job. That experience was gained over eight years, beginning with SOCRATES.

This experience has contributed much to recent international and national standards. For example, the diversity of reference frames, coordinate systems, and timescales among satellite operations was astounding. However, operators should control what coordinate systems, reference frames, and timescales are best for their operations. To communicate essential data to those who must collaborate, for example, to avoid collisions, these physical constructs must be completely described to others or they should conform to standard approaches that are widely accepted and documented. Orbit and conjunction data message international standards facilitate this. Such standards are based in great measure on SDC experience and direct participation of SDC principals.

The utmost importance of trust among collaborators and the ability to verify mission and revenue-critical data are among the most important of SDC legacies. Conjunction reports and assessments are estimates based on past information. The time of closest approach, minimum separation between satellites, and relative velocity are imprecise to a degree determined by the imprecision and inaccuracy of contributing orbit estimates. The uncertainty can be such that dangerous situations might not be identified, or safe passage is confused with danger. Evasion and mitigation are planned and executed by affected operators based on their best mutual understanding of each other's kinematic states. This is generally not the same data used to detect danger or issue warnings. The risk of accepting another's estimate of the risk to a commercial operator's property and revenue is unacceptable. Those with sufficient resources must have information for them to verify conjunction warnings. Those who operate few satellites or are disadvantaged must have ultimate confidence in those they rely on. The SDC provides such fullness of information to subscribers and potentially others who are at risk. The SDC has earned the confidence of many who operate few satellites or who lack resources or skills to verify independently. These principles are reflected in the international best practices for avoiding collisions currently being developed with SDC leadership [25].

It is important to separate operations from maintenance and development, and maintaining close connection is very important. The SDC has demonstrated the ability to sustain reliable operations while developing new capabilities and resolving operational problems. Having separate but identical servers has been critical. The SDC has an independent "sand box" for offline development.

V. PLANS AND EMERGING CAPABILITIES

One should use the best orbit data, data transfer, and supporting databases to help perceive, locate, and mitigate electromagnetic interference with critical Comsat operations. The SDC is well along the development of this capability. We have demonstrated the value of more precise orbit data in classical Time Difference of Arrival (TDOA), Frequency Difference of Arrival (FDOA), and cross-correlation geolocation. The SDC can host member data spanning operational contacts, satellite electromagnetic characteristics, reference emitters, and transponder coverages for subscribers. It is charged by the SDA to enable the commercial Radio Frequency Interference (RFI) warning process.

No space surveillance capability is ubiquitous. None are required to be because the requirement is to obtain a sufficient number and distribution of observations to estimate orbits consistent with satellite missions and avoiding collisions among them. This can be approached with a spectrum of observation locations and capabilities from which the most appropriate set can be created. Presently, the SDC has only subscriber observations and orbit data and U.S. Air Force TLE sets. Although the U.S. Air Force has putatively the best orbit estimates for debris, which is the greatest threat in LEO, it does not have competition-sensitive subscriber orbits, which the SDC has. We support and hope for an industry–government collaboration.

Lacking such collaboration, the SDA seeks independent sources of trustworthy observations and independently reliable orbits of debris objects in orbit in order to minimize the risks to subscribers. An initial announcement in Madrid in 2011 was followed by a more substantial set of requirements and solicitation at the first SDA Users' Meeting during *Satellite 2012*. Several institutional and industrial observation organizations have responded. The U.S. National Oceanic and Atmospheric Agency uses SDC services. The diligence of SDC orbit determination, quality control, data assurance, and security makes incorporating these sources of information straightforward.

New valued-added capabilities are developed continuously. One of the most important is a mathematically and physically sound way to balance the probability of correctly identifying a potential collision with the probability of incorrectly including false alarms among pairs of satellites potentially at risk. The technique guides determining filtering consistently and keep out/screening criteria. This is described in recent conference papers [26] and practice documents [27].

VI. CONCLUSION

Nearly 300 geostationary satellites participate in SDC workflows. These include satellites owned by the major Geosynchronous Earth Orbit communication satellite owner/operators and as well as those who operate smaller constellations. Their operations span numerous operational software suites and operational concepts. Combining these was a monumental task. However, the entire operation and capability cost a very small fraction of member operational expenditures or what a comparable Government-sponsored project would cost.

The SDC demonstrates that competitors who face a common risk can collaborate to mitigate the threat to all. Competitors and other participants reside in many nations, industries, and governments. While policy and diplomatic communities deliberate confidence building and transparency, civil and commercial cohorts have already accomplished them. This demonstrates the feasibility and wisdom of national space policies that encourage or even mandate furtherance of commercial capabilities.

ACKNOWLEDGMENTS

The author of this chapter dedicates his efforts to his co-workers: Daniel Oltrogge, T. S. Kelso, James Wilson, Adam Himes, David Vallado, Salvatore Alfano, Paul Welsh, and many others.

REFERENCES

[1] Oltrogge, D. L., "Getting to Know Our Space Population from The Public Catalog," *AAS 11-416, AAS/AIAA Astrodynamics Specialists Conference*, Girdwood, Alaska, July 2011.

[2] Foster, J., and Frisbee, J., "Comparison of the Exclusion Volume and Probability Threshold Methods for Debris Avoidance for the STS Orbiter and the International Space Station," NASA Technical Paper 2007–214751, May 2007.

[3] Kelso, T. S., and Alfano, S., "Satellite Orbital Conjunction Reports Assessing Threatening Encounters in Space (SOCRATES)," *AAS 05-124, AIAA/AAS Space Flight Mechanics Conference*, Copper Mountain, Colorado, January 2005.

[4] Kelso, T. S., "Improved Conjunction Analysis via Collaborative Space Situational Awareness," *AAS 08-235, AAS/AIAA Spaceflight Mechanics Meeting*, Galveston, Texas, January 2008.

[5] ISO S-1123, 2012.

[6] Vallado, D. A. *Fundamentals of Astrodynamics and Applications*, 3rd ed., Microcosm Press, Portland, Oregon, 2007.

[7] ISO CD 16158, "Avoiding Collisions Among Orbiting Objects: Best Practices, Data Requirements, and Operational Concepts," 2012.

[8] Alfano, S., "Toroidal Path Filter," *AAS 11-438, AAS/AIAA Astrodynamics Specialist Conference*, Girdwood, Alaska, July 2011.

[9] Oltrogge, D. L., "Space Data Actionability Metrics for SSA," *Proceedings of Improving Our Vision Conference V*, Chateau de Betzdorf, Luxembourg, June 2011.

[10] Oltrogge, D. L., "Ephemeris Requirements for Space Situational Awareness," *AAS 11-151, AAS/AIAA Astrodynamics Specialist Conference*, Girdwood, Alaska, July 2011.

[11]   Alfano, S., "Collision Avoidance Maneuver Planning Tool," *AAS Paper 2005-08, AAS/AIAA Astrodynamics Specialist Conference*, Lake Tahoe, California, July 2005.

[12]   ISO 26900, "Orbit Data Messages," 2009.

[13]   CCSDS 508.0-W-8, "Conjunction Data Message," 2012.

[14]   "Astrodynamics – Propagation Specifications, Technical Definitions, and Recommended Practices," ANSI/AIAA Standard # S-131-2010.

[15]   ISO S-1123, "Space Systems – Orbit Determination and Estimation – Process for Describing Techniques," 2012.

[16]   Vallado, D. A., "An Analysis of State Vector Propagation Using Differing Flight Dynamics Programs," *AAS 05-199, AIAA/AAS Space Flight Mechanics Conference*, Copper Mountain, Colorado, January 2005.

[17]   Vallado, D. A., "Verifying Observational Data for Real World Space Situational Awareness," *AAS 11-439, AAS/AIAA Astrodynamics Specialist Conference*, Girdwood, Alaska, July 2011.

[18]   Vallado, D. A., "Covariance Realism," *AAS 09-304, AAS/AIAA Astrodynamics Specialist Conference*, Pittsburgh, Pennsylvania, August, 2009.

[19]   Sharma, J., "Toward Operational Space Surveillance," *Lincoln Laboratory Journal*, Vol. 13, No. 2, 2002, pp. 309–333.

[20]   Vallado, D. A., and Seago, J. H., "Covariance Realism," *AAS 09-334, 19th AAS/AIAA Space Flight Mechanics Meeting*, Savannah, Georgia, February 2009.

[21]   Vallado, D. A., and Griesbach, J. D., "Simulating Space Surveillance Networks," *AAS 11-580, AAS/AIAA Astrodynamics Specialist Meeting*, Girdwood, Alaska, July 2011.

[22]   Finkleman, D., "Space and Verification, Part II, Technical Aspects," Lecture, U.S. Air Force Academy, Colorado Springs, Colorado, 2010.

[23]   Vallado, D. A., and Alfano, S., "A Future Look at Space Surveillance Operations," *AAS 99-193, AAS/AIAA Space Flight Mechanics Meeting*, Breckenridge, Colorado, February 1999.

[24]   Pearlman, M. R., Degnan, J. J., and Bosworth, J. M., "The International Laser Ranging Service," *Advances in Space Research*, Vol. 30, No. 2, 2002, pp. 135–143.

[25]   ISO CD 16158, "Avoiding Collisions Among Orbiting Objects: Best Practices, Data Requirements, and Operational Concepts," 2012.

[26]   Finkleman, D., "Requirements and Guidance for Conjunction Assessment," *AAS 11-434, AAS/AIAA Astrodynamics Specialists Meeting*, Girdwood, Alaska, July 2011.

[27]   Finkleman, D., and Berry, D., "Cross-Agency Collaboration and Standards For Conjunction Assessment," *U.S. Air Force Ground Systems Architecture Workshop*, 2011.

CHAPTER 12

Overview of the Laser Communications Relay Demonstration Project

Bernard L. Edwards[*] **and Dave Israel**[†]
NASA Goddard Space Flight Center, Greenbelt, Maryland

Keith Wilson[‡]
Jet Propulsion Laboratory, Pasadena, California

John Moores[§] **and Andrew Fletcher**[¶]
MIT Lincoln Laboratory, Lexington, Massachusetts

I.   INTRODUCTION

The communications link between space-borne observatories and the Earth has long been a critical mission systems driver. The information from a scientific or exploration discovery has to get back to Earth; the more data that can be sent back, the better the probability that the mission will produce more valuable science. Several technologies, such as radio-frequency (RF) communications with higher data bandwidth and lossless data compression have improved the capability over time, but are failing to keep pace with the needs of the advanced instrumentation that can be flown in space today.

Optical communications (or laser communication, "lasercom") is a revolutionary technology that will enable NASA to undertake more complex missions in the future that require even greater increases in data rates or decreased mass, size, and power burden on the spacecraft. For approximately the same mass, power, and volume, an optical communications system will provide significantly higher data rates than a comparable RF system, and for the same data rate (e.g., 1 Gbps of output), an optical communications system will require less mass, power, and volume than a comparable RF system.

The near-term demand for high-bandwidth communications services is driven by NASA's Science Mission Directorate, which wishes to deploy more capable instruments onboard spacecraft. Longer term, there will be requirements for very high data rates to support subsurface exploration of outer planets, bidirectional high-definition television transmission to and from the Moon or

---

[*]Chief Communications Systems Engineer, NASA Goddard Space Flight Center.
[†]Principal Investigator, NASA Laser Communications Relay Demonstration Project.
[‡]Principal Member of the Technical Staff, Jet Propulsion Laboratory, Pasadena, California.
[§]Assistant Group Leader, Optical Communications Technology Group, MIT/LL.
[¶]Technical Staff, Optical Communications Technology Group, MIT/LL.

other near-Earth exploration objectives, and eventually to support astronauts at Mars. Near-Earth (including lunar) spacecraft will need bidirectional links supporting hundreds of Mbps to Gbps. Deep-space missions will need tens to hundreds of Mbps from destinations such as Mars and Jupiter. An image from the Mars Reconnaissance Orbiter (MRO) currently takes 1.5 h to transmit back to Earth at the MRO maximum data rate of 6 Mbps. This bottleneck becomes the limitation on the return of science. The Lunar Reconnaissance Orbiter has been able to transmit more data than all planetary missions combined with its downlink of 100 Mbps. Order of magnitude (or more) increases in data rate over these current mission capabilities is possible using optical communications.

Beyond the potential savings in mass, weight, and power, the higher data bandwidth will allow missions with the present-day data volumes to operate while requiring less time for communications activities. This savings in communication time will reduce operational constraints for both the spacecraft and the supporting communications network infrastructure.

There are some differences between the technological approaches to optical communications specifically designed for near-Earth missions and those designed for deep-space missions. For example, the vastly differing ranges and data rates for near-Earth and deep-space missions mean that some of the technologies applicable to each domain differ profoundly; however, there are also many technologies that are similar to both! Coordination of system development for these two domains will maximize NASA's return on investment. The Laser Communications Relay Demonstration (LCRD) flight payload will demonstrate critical technologies relevant to both near-Earth and deep-space optical communications systems, including modulations, codes, pointing and tracking techniques, and so on. LCRD will also demonstrate network-based relay operations in both near-Earth and deep-space applications.

Recent developments in optical communication technology have demonstrated the ability to achieve bidirectional near-Earth data links beyond 10 Gbps utilizing differential phase shift keying modulation (DPSK). Similarly, deep-space links with downlinks up to 1 Gbps and uplinks up to 100 Mbps can be achieved using photon counting and pulse position monitoring (PPM) modulation techniques. Photon counting and PPM are also critical technologies for very small low-Earth orbit (LEO) spacecraft, which need high-performance, but low-power and low-mass communications systems. The LCRD mission will provide a space-based technology demonstration of optical communications, using both DPSK and PPM modulated signals. LCRD will use photon counting on the PPM downlink and will use existing systems and minimal modifications to other flight systems to fully characterize high-data-rate optical communications in a space flight environment.

II. LEVERAGING NASA'S LUNAR LASER COMMUNICATION DEMONSTRATION

NASA is currently making a large investment in the Lunar Laser Communication Demonstration (LLCD) [1], which is scheduled to launch in 2013 as a secondary payload on the Lunar Atmosphere and Dust Environment Explorer (LADEE). LLCD will prove the feasibility of optical communications, but, owing to the very limited operating time (potentially less than 16 h over the life of the mission), it will not provide the necessary operational knowledge to allow optical communications to be mission-critical on future missions. However, LLCD will ably demonstrate pulse position modulation; photon counting on the downlink; inertial stabilization; high-efficiency transmission and reception of pulse position modulation; very low size, weight, and power space terminal; integrating an optical communications terminal to a spacecraft; link operation under some conditions (limited due to the very limited operating time); and a scalable array ground receiver.

Unfortunately, however, LLCD does not go far enough. To make optical communications useful to future projects, long-mission-life space terminals must be developed and proven. Operational concepts for reliable, high-rate data delivery in the face of terrestrial weather variations and real NASA mission constraints need to be developed and demonstrated. To increase the availability of an optical communications link and to handle cloud covering a ground terminal, there needs to be a demonstration of handovers among multiple ground sites. For near-Earth applications, a demonstration needs to show the relaying of an optical communications signal in space. There also needs to be a demonstration of the modulation and coding suitable for very high rate links.

NASA's new LCRD optical communications project will answer the remaining questions for near-Earth applications. LCRD's flight payload will have two optical communications terminals in space and two optical communications terminals on Earth to allow the mission to demonstrate

1. High-rate bidirectional communications between Earth and geostationary Earth orbit (GEO);

2. Real-time optical relay from Ground Station 1 on Earth through the GEO flight payload to Ground Station 2 on Earth;

3. Pulse position modulations suitable for deep-space communications or other power-limited users, such as small near-Earth missions;

4. DPSK modulations suitable for near-Earth high-data-rate communications;

5. Demonstration of various mission scenarios through spacecraft simulations at the Earth ground station;

6. Performance testing and demonstrations of coding, link layer, and network layer protocols over optical links over an orbiting testbed.

The LCRD Project Office is also working closely with NASA headquarters (HQ) to possibly demonstrate LCRD with an optical communications terminal flying on a LEO spacecraft, such as the International Space Station (ISS). Thus the flight payload on the GEO spacecraft has a requirement to be able to support high-rate bidirectional communications between LEO and GEO as well as between Earth and GEO.

III. BASIC CONCEPT OF OPERATIONS

The basic optical communications concept of operations assumes there is a specific amount of data at the source that must be transferred to the destination in a specific amount of time, not unlike a typical RF scenario. The high-level process assumes a scheduled approach where the communications process starts at a specific time, with a link establishment process between the source spacecraft and the destination (ground station or relay satellite). If the link is successfully established, data are transferred during the specified time and the link is terminated according to plan. If link establishment is not successful or the link cannot be maintained for the required duration due to clouds or other link impairment, then an alternate process is required. At a minimum, the data not transferred must be stored until they can be transferred later (to another ground station or relay) or deleted.

The first consideration in link establishment is whether there is a line of sight between the source and destination. Geometric line of sight is calculated based on the source spacecraft trajectory, location of the ground station(s), and any local terrain considerations (e.g., mountains, buildings, trees). Free-space optical communications through Earth's atmosphere is nearly impossible in the presence of most types of clouds. Typical clouds have deep optical fades so it is not feasible to include enough link margin in the link budget to prevent a link outage. Thus, a key parameter when analyzing free-space optical communications through the atmosphere is the probability of a cloud-free line of sight (CFLOS) channel. A mitigation technique ensuring a high likelihood of a CFLOS between source and destination is therefore needed to maximize the transfer of data and overall availability of the network. Such mitigation techniques include the following:

1. Using several optical communications terminals on the relay spacecraft, each with its own dedicated ground station, to simultaneously transmit the same data to multiple locations on Earth (hopefully one will be free of clouds and allow the transmission to succeed). Unfortunately, this mitigation technique can be cost prohibitive.

2. A single optical communications terminal in space utilizing multiple ground stations that are geographically diverse, so that there is a high probability of CFLOS to a ground station from the spacecraft at any given point in time.

3. Storing data until communications with a ground station can be initiated. Of course a key issue is making sure all the stored onboard data can be transmitted within the allocated time.

4. Having a dual RF/optical communications systems onboard the spacecraft. The RF system could support low-data-rate, low-latency communications, and the optical communications could support high-data-rate communication of data that can be stored for some time.

NASA has studied various concepts and architectures for a future optical communications network. The analysis indicates ground segment solutions are possible for all scenarios, but usually require multiple, geographically diverse ground stations in view of the spacecraft. Thus, optical communications handover strategies will have to be developed. Those strategies may differ depending on the transmission time required; for example, handover strategies that work with an optical communications terminal in LEO may not work for an optical communications terminal located at Mars.

The availability of a communication link between a spacecraft and a ground station network depends on many factors, including the number and location of the sites in the network and the orbit of the spacecraft, which together determine the elevation angle of the link and the path length of transmission through the atmosphere. In our concept, a ground station is considered available for communication when it has a CFLOS at an elevation angle to the spacecraft terminal of approximately $20°$ or more. The network is available for communication when at least one of its sites is available. Typical meteorological patterns cause the cloud cover at stations within a few hundred kilometers of each other to be correlated. Consequently, stations within the network should be placed far enough apart to minimize these correlations, thus maximizing the probability of CFLOS. This requirement may lead to the selection of a station that has a lower CFLOS than sites not selected, but is less correlated with other network sites. The stations also need to be close enough to each other to maintain continuous access with the spacecraft as its position with respect to the ground changes with time.

Depending on the scenario, free-space optical communications operations can take advantage of cloud prediction at each ground site to maintain CFLOS, and thus maximize availability. Maintenance of CFLOS can be accomplished by knowing whether the line of sight to each ground site is cloud-free at a given time, and knowing how many minutes into the future each site is expected to remain cloud-free. Having local weather and atmospheric instrumentation at each site and making a simple cloud forecast can significantly reduce the amount of time the space laser communications terminal requires to repoint and acquire with a new ground station. Useful categories of instrumentation include the following:

1. *Weather.* Weather information is gathered locally by standard meteorological packages that monitor temperature, humidity, barometric pressure, and wind speed and direction.

2. *Clouds.* A thermal infrared cloud camera is used to monitor the extent of cloud coverage. These sensors indicate not only whether there are clouds or

no clouds at very high temporal resolution, but also the sky temperature and emission.

3. *Daytime sky radiance.* A Sun photometer provides this measurement.

4. *Atmospheric loss.* During the day, a Sun photometer is also used to measure atmospheric loss. At night, a calibrated photometric system that tracks stars of stable emission, e.g., Polaris, can be used.

5. *Clear air optical turbulence.* A differential image motion monitor (DIMM) is the predominant method of measuring seeing. During the night this instrument tracks stars, and during the day, the Sun.

In addition to outages or blockages due to weather, an optical communications link also has to be safe. As well as needing to be safe for humans, some satellites have sensitive detectors that can be damaged by intense light sources originating from Earth. Safe laser beam propagation to near-Earth and deep-space spacecraft starts at the transmitter facility, extends through navigable air space and the near-Earth region, and eventually into deep space. In the United States, regions of concern for safe beam propagation are under the purview of the Occupational Safety and Health Administration (OSHA), the Federal Aviation Administration (FAA), and the Laser Clearing House (LCH).

As a federal government agency, NASA generally coordinates its outdoor laser activities with the LCH, a Department of Defense component within U.S. Strategic Command. The American National Standards Institute (ANSI) standard for outdoor laser propagation (ANSI Z136.6) also recommends that lasers that meet certain criteria also coordinate with the LCH.

The LCH provides predictive avoidance analysis and deconfliction with the United States, allies, and other space operations due to the potential to affect satellites or humans in space. It can determine whether a laser has the potential for interference or damage to particular satellites. Lasers with very high peak powers and very small beam divergences (10 μrad or less) probably have the greatest potential for causing damage to these very expensive systems in space. Rather than give detailed guidance on the levels that could produce damage to satellites, criteria are provided as a starting point where coordination with the LCH is desirable. Lasers that can produce an instantaneous irradiance exceeding 1 mW/cm^2 at an elevation of 60,000 ft (18 km), even for 1 ns or less, should contact the LCH. Based on laser propagation details, the LCH can either issue a blanket approval of transmission at a facility or require coordination of all laser beam propagation activity.

Thus, a laser safety system and operational procedures have to be in place to ensure safe operations. Those procedures may lead to periodic outages of the link just as if a cloud had blocked a station.

As optical communications link through the atmosphere will be intermittently interrupted, the use of data return acknowledgement and retransmission protocols, such as delay-tolerant networking (DTN), help improve data throughput.

Such protocols cope with atmospheric disturbances to the optical communications link and also prevent loss of data during redirection of the downlink to a second ground station if the first should be obscured by clouds. Data acknowledgment requires sending information back to the spacecraft either through an optical uplink, or alternatively through a RF uplink. Because the data rates on optical communication links will typically be higher than those of RF links, research should be conducted on the best data return acknowledgement and retransmission protocols for use with optical links.

IV. FLIGHT PAYLOAD

The LCRD flight payload will be flown on a GEO spacecraft and consists of the following:

1.   Two optical communications modules (heads);

2.   Two optical module controllers;

3.   Two DPSK modems;

4.   Two PPM modems;

5.   High-speed electronics to interconnect the two optical modules, perform network and data processing, and to interface to the host spacecraft.

An optical communications terminal on LCRD consists of an optical module, a DPSK modem, a PPM modem, and an optical module controller.

V. FLIGHT OPTICAL COMMUNICATIONS MODULE

Each of the two optical communications terminals to be flown on the GEO spacecraft will transmit and receive optical signals. When transmitting, the primary functions of the GEO optical communications terminal are to efficiently generate optical power that can have data modulated onto it; transmit this optical power through efficient optics; and aim the very narrow beam at the ground station on Earth, despite platform vibrations, motions, and distortions. When receiving, the GEO optical communications terminal must provide a collector large enough to capture adequate power to support the data rate; couple this light onto low-noise, efficient detectors while minimizing the coupled background light; and perform synchronization, demodulation, and decoding of the received waveform.

Each optical module, shown in Fig. 12.1, is a 4 in. reflective telescope that produces an approximately 15 μrad downlink beam. It also houses a spatial acquisition detector, which is a simple quadrant detector with a field of view of approximately 2 μrad. It is used both for detection of a scanned uplink signal, and as a tracking sensor for initial pull-in of the signal. The telescope is

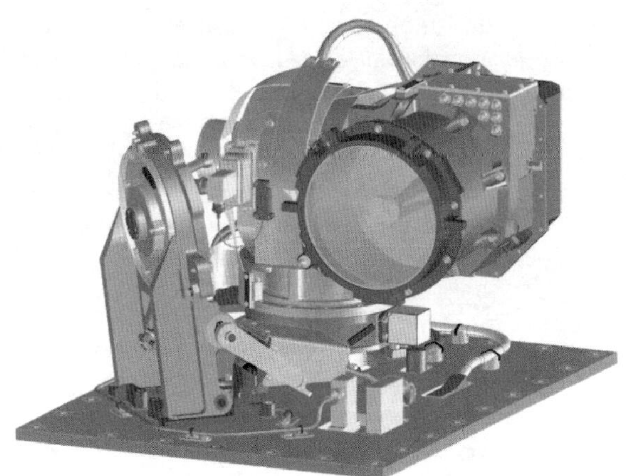

Fig. 12.1   Inertially stabilized optical module.

mounted on a two-axis gimbal via a magnetohydrodynamic inertial reference unit (MIRU). Angle-rate sensors in the MIRU detect angular disturbances, which are then rejected using voice-coil actuators for inertial stabilization of the telescope. Optical fibers couple the optical module to the modems where transmitted optical waveforms are processed. Control for each optical module and its corresponding modems are provided by a controller. Each optical module is held and protected during launch with a cover and one-time launch latch.

VI.   FLIGHT MODEMS

As stated previously, there are some differences between the technological approaches to optical communications specifically designed for near-Earth missions versus deep-space missions. This is mostly due to the vastly differing ranges and data rates for the two scenarios. One area that has been looked at for some time within NASA is the appropriate modulation, coding, and detection scheme for the two different classes of missions. Photon counting and PPM has been identified as the technique of choice for deep-space missions, while DPSK is the current preferred choice for near-Earth missions. LCRD will demonstrate both techniques.

Photon-counting PPM is highly photon-efficient, although the ultimate data rate is limited due to detector limitations and the requirement for faster electronics. LCRD leverages the PPM modem developed for NASA's LLCD [1] as a cost-effective approach to providing a PPM signal. The LLCD modem supports a variable rate downlink from 39 to 622 Mbps and variable uplink from 10 to 20 Mbps.

The PPM flight transmitter encodes data with a rate-½ serially concatenated PPM (SC-PPM) turbo code. The encoded data stream is convolutionally interleaved (to mitigate the effects of atmospheric fading) and modulated with a 16-ary PPM modulation scheme (signal is placed in exactly one of each 16 temporal slots). The maximum data rate is achieved using a 5 GHz slot clock rate; lower data rates are accomplished by combining consecutive slots, effectively lowering the clock rate, with a minimum slot rate of 311 MHz. The optical modulation is accomplished with a master-oscillator power amplifier (MOPA) architecture. A continuous-wave (CW) laser (at $\sim$1550 nm) is modulated with a Mach–Zehnder modulator, and amplified with a two-stage erbium-doped fiber amplifier (EDFA) to a 0.5 W average power level.

The PPM flight receiver is an optically preamplified direct detection receiver. After amplification and filtering, the signal is optically split to perform spatial tracking, clock recovery, and communications. The uplink communications signaling is 4-ary PPM, with a simple two-comparator demodulator performing binary hard decisions. The received uplink data stream are de-interleaved and decoded (rate-½ SC-PPM coding is applied on the uplink).

LCRD will also support DPSK, which has superior noise tolerance, can be used at extremely high data rates, and supports communications when the Sun is in the field of view. LCRD leverages an Massachusetts Institute of Technology/ Lincoln Laboratory (MIT/LL) previously designed DPSK modem [2] as a cost-effective approach to providing a DPSK signal. It can both transmit and receive data at an (uncoded) rate from 72 Mbps to 2.88 Gbps. In future relay scenarios, it could be replaced by a higher-rate DPSK modem that would support data rates beyond 10 Gbps.

The DPSK modem uses identical signaling for uplink and downlink directions. The DPSK transmitter generates a sequence of pulses at a 2.88 GHz clock rate. A bit is encoded in the phase difference between consecutive pulses. As demodulation is accomplished with a Mach–Zehnder optical interferometer, the clock rate remains fixed. The DPSK transmitter uses a MOPA architecture similar to the PPM transmitter [2]. The EDFA amplifies the optical signal to a 0.5 W average power level. Data rates below the maximum are accomplished via burst-mode operation, where the transmitter sends pulses only a fraction of the time, sending no optical power the remainder of the time. Because the EDFA is average-power-limited, the peak power during the bursts is increased, so rate reduction is accomplished in a power-efficient manner.

The DPSK receiver has an optical preamplifier stage and an optical filter, at which point the light is split between a clock recovery unit and the communications receiver. The receiver uses a delay-line interferometer followed by balanced photodetectors to compare the phases of consecutive pulses, making a hard decision on each channel bit. While coding and interleaving will be applied in the ground terminal to mitigate noise and atmospheric fading, the DPSK flight receiver does not decode nor de-interleave. The modems instead support a relay architecture where up- and downlink errors are corrected together in a decoder located at the destination ground station [3].

VII.   HIGH-SPEED ELECTRONICS

To be an optical relay demonstration, LCRD will create a relay connection between two ground stations. A significant objective of LCRD is to demonstrate advanced relay operations on the GEO spacecraft. LCRD will enable a wide variety of relay operations through the high-speed electronics (HSE) that connect the two optical terminals. In addition to real-time relay operations, the electronics will allow scenarios where one link uses DPSK signaling and the other PPM. A known challenge with optical communication through the atmosphere is the susceptibility to cloud cover. The HSE will include a significant amount of data storage in order to demonstrate store-and-forward relay services for when the uplink is available but the downlink is unavailable. The HSE will support DTN protocols [8]. To support DTN over the DPSK optical links, the HSE will implement any required decoding and de-interleaving so the payload can process and route the data (at a rate less than the maximum DPSK throughput). The link operations will be configurable to allow support for a variety of scenarios.

VIII.   GROUND SEGMENT

The LCRD Ground Segment comprises the LCRD Mission Operations Center (LMOC) and two ground stations. The LMOC will perform all scheduling, command, and control of the LCRD payload and the ground stations.

Each Earth ground station must provide three functions when communicating with one of the two optical communications terminals on the GEO spacecraft: receive the communications signal from the GEO space terminal; transmit a signal to the GEO space terminal; and transmit an uplink beacon beam so that the GEO space terminal points to the correct location on Earth.

The receiver on Earth must provide a collector large enough to capture adequate power to support the data rate; couple this light onto low-noise, efficient detectors while trying to minimize the coupled background light; and perform synchronization, demodulation, and decoding of the received waveform.

The uplink beacon, transmitted from each Earth ground station, must provide a pointing reference to establish the GEO space terminal beam pointing direction. Turbulence effects dominate the laser power required for a ground-based beacon. Turbulence spreads the beam, reducing mean irradiance at the terminal in space, and causes fluctuations in the instantaneous received power.

IX.   LCRD GROUND STATION 1

The Jet Propulsion Laboratory (JPL) will enhance its Optical Communications Telescope Laboratory (OCTL) so that it can be used as Ground Station 1 of the demonstration. In this section we describe the major modifications that will be made to the OCTL to support LCRD. These are the dome, the adaptive optics optical train, the atmospheric monitoring system, the Monitor and Control (M&C) system, and the LCRD User Service Gateway (LUSG). The OCTL is located in the San Gabriel Mountains of southern California and houses a 1 m f#75.8 coudé focus telescope [4]. The large aperture readily supports the high-data-rate DPSK and PPM downlinks from the LCRD space terminal with adequate link margin. Required to operate 24/7, in the presence of winds, and as close as 5° solar angles, the OCTL telescope shown in Fig. 12.2 will be enclosed in a temperature-controlled dome with a transparent window to allow laser beam and radar transmission. The Laser Safety System at the OCTL (LASSO) will ensure safe laser beam transmission through navigable air and near-Earth space [5].

The seven coudé mirrors will be coated with high-reflection, low-absorption coatings to reduce the amount of sunlight scattered into the receiver when pointed at the required 5° solar angle and also backscatter from the uplink laser. The estimated reflection loss from all seven mirrors is 0.4 dB.

The integrated optical system at the telescope coudé focus is shown in Fig. 12.3. A shutter controlled by a Sun sensor protects the adaptive optics system should the telescope inadvertently point closer to the Sun than specified. The downlink is collimated by an off-axis parabolic mirror incident on a fast tip/tilt mirror and dichroic beam splitter before reflecting off a deformable mirror (DM). A fraction of the beam is coupled to the wavefront sensor to measure aberrations in the downlink beam. A scoring camera monitors the quality of the corrected beam focused into a fiber coupled to the DPSK/PPM

Fig. 12.2 OCTL telescope will be modified with an optical flat to support links in the presence of more windy conditions.

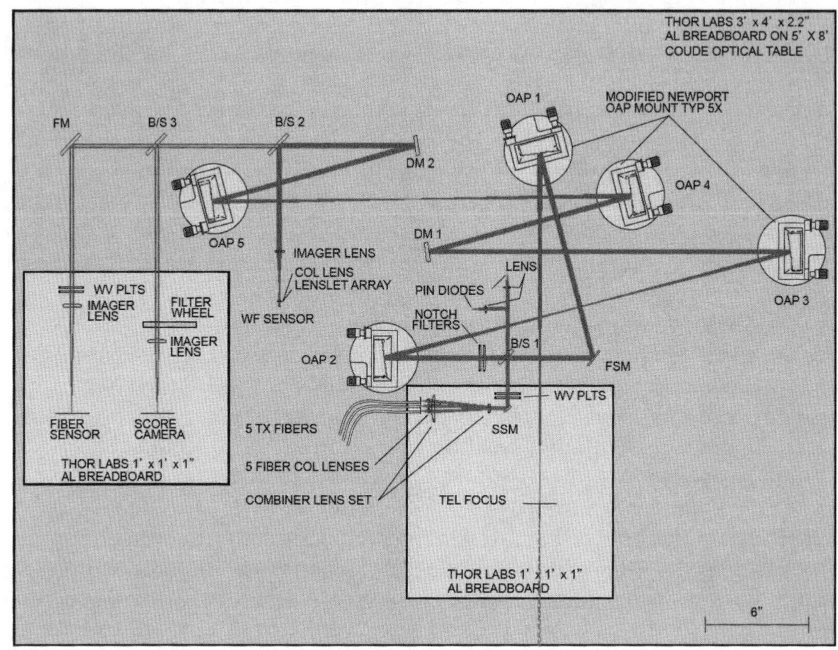

Fig. 12.3 Schematic of the integrated optical system to be located at coudé focus in OCTL.

receiver. A waveplate adjusts the polarization into the fiber to the DPSK Mach–Zehnder interferometer and a slow tip/tilt mirror ensures maximum signal input to the fiber. In the uplink system the beacon and communications beams are first reflected from the slow tip/tilt mirror to track out satellite motions, and then coupled to the telescope through a dichroic mirror.

As a prelude to an operational system, understanding the optical channel and the performance of the link under a variety of atmospheric conditions informs the definition of requirements for future operational ground stations. Figure 12.4 presents some of the atmospheric monitoring instruments that will be implemented at the OCTL. The Sun photometer measures atmospheric transmission and sky radiance, the ground scintillometer measures the boundary layer turbulence that is the major contributor to scintillation in the downlink signal, and the cloud imager measures cloud coverage and cloud optical depth. In addition, a differential image motion monitor integrated into the monitor and control software will measure the Fried coherence length r_0 using the downlink signal. The weather station measures wind speed and direction as well as relative humidity and temperature at the OCTL.

The LCRD PPM modem will support the 16-ary PPM downlink modulation format at data rates of 39, 78, 155, 311, and 622 Mbps, and the 4-ary PPM uplink at 10 and 20 Mbps with rate-½ serially concatenated coding. The modem will support transfer frame formation, convolutional interleaving, and multiplexing. The asymmetry in modulation is characteristic of deep-space optical links where the uplink is a command link and the downlink returns a large volume of science data from the deep-space probes.

The PPM modem will support real-time uplink and downlink processing in support of DTN provided by the LUSG. The LUSG will interface simulated (and potentially real) users to the LCRD optical service network, providing real-time bit stream and store-and-forward DTN services. The LUSG provides network data performance measurements, and coordinates with the M&C subsystem.

Fig. 12.4   Suite of atmospheric monitoring instruments to characterize the optical channel.

The M&C subsystem will provide intelligent control of the LCRD ground terminal. It will implement the software to provide the interface for remote control and status monitor of all of the OCTL subsystems. It will provide a gateway to the LMOC to support remote control, status reporting, and data return. The M&C subsystem will also implement a high-speed data recording system and engineering interface. The data recorder will archive all the OCTL system data for post-analysis of system performance. The engineering interface will be a temporary user interface for early evaluation of the integrated OCTL subsystem before delivery of User Simulator and LMOC connection.

The DPSK ground modem supports the same signaling structures as the DPSK flight modem, namely phase-modulated pulses at a 2.88 GHz slot rate and burst modes to vary the channel data rate between 72 Mbps and 2.88 Gbps. In addition, the ground modem must implement forward error correction coding – anticipated to be a low-density parity check (LDPC) code from the digital video broadcasting (DVB-S2) standard – and interleaving to mitigate atmospheric scintillation.

X. LCRD GROUND STATION 2

MIT Lincoln Laboratory designed and is building the Lunar Lasercom Ground Terminal (LLGT) [6] for NASA's LLCD. The LLGT, shown in Fig. 12.5, will be refurbished and enhanced to serve as Ground Station 2 for LCRD. A summary of the LLGT, as designed for LLCD, follows. The primary enhancements for LCRD will be an adaptive optics system to couple received light into single-mode fiber (to support the DPSK receiver), and further development of the single-photon detectors (to support the PPM receiver), including the development of more robust and scalable optical packaging, cabling, and readout electronics.

The LLGT is an array of four 40 cm receive reflective telescopes and four 15 cm transmit refractive telescopes. For the uplink, the optical signal (PPM for LLCD, to include DPSK for LCRD) is modulated onto four

Fig. 12.5 Lunar Lasercom ground terminal will be enhanced with adaptive optics and a DPSK modem.

separate carrier wavelengths, each very slightly detuned. Each modulated signal is amplified to a 10 W average power, and coupled to a transmit aperture via single-mode fiber. For the downlink, the receive apertures couple into few-mode multi-mode fibers connected to an array of superconducting nanowire single-photon detectors (SNSPDs) [7]. The SNSPDs must be cryogenically cooled to approximately 3 K, and it is impractical to locate them in the focal planes of the receive apertures. The multimode fiber was designed to efficiently couple the received light from the aperture to the detector over a distance of 22 m. By using multimode fiber, efficient coupling is achieved without an adaptive optics system.

For LCRD, the DPSK modem requires the received light to be coupled into single-mode fiber. For this reason, at least one of the receive apertures will use an adaptive optics system to support the DPSK receiver. The current LLGT design will continue to support PPM functionality for LCRD.

Owing to their high photon efficiency and fast reset times, the SNSPDs are a significant enabler for high-speed laser communications from deep-space terminals to Earth terminals. For this reason, LCRD will investigate updates to this detector technology. This will include efforts to make the detectors more robust, more scalable, and require reduced size, weight, and power (SWaP). The main LCRD efforts will be directed towards optical packaging and improved cabling and cryogenic readout circuitry.

XI. DEMONSTRATION OPERATIONS

Control of all activities during LCRD will take place from the LMOC. The LMOC is connected with all other segments, and communicates with the two ground stations using high-capacity connections. Connection to the space segment will be provided either through one of the ground stations, or through a lower-capacity connection to the host spacecraft's Mission Operations Center (HMOC) and then to the LCRD flight payload by RF link.

The LMOC will provide services such as planning and scheduling, control, status monitoring, and reporting and accountability.

The mission operations for the spacecraft and the optical communications demonstration are intimately intertwined. The unique nature of the demonstration is that there is a path to and from the spacecraft that is outside the usual RF connection. Commands for the GEO optical communications terminal can be sent either via the optical uplink or via the host spacecraft RF uplink. There are two paths for obtaining engineering data (health and status): optical or RF. The LMOC coordinates all optical communications activities and provides an interface to spacecraft operations.

On the telemetry side there are again two paths, though for somewhat different reasons. Data (user information or engineering telemetry) can be sent to Earth via the GEO optical communications terminal. It is possible that the GEO terminal may add/multiplex additional engineering data into the data stream. The spacecraft monitors terminal parameters like power and includes those in engineering telemetry, which is passed over the RF link. In addition to these, there are many "test points" within the GEO terminal that are sent via RF as part of the engineering telemetry.

Operations strategies for mitigation of the effects of weather and atmospheric conditions will be explored. One possibility would be to have multiple terminals within the same beam simultaneously receive the same data to guarantee getting through to at least one terminal for a reasonably high percentage of the time. On the other hand, buffering and retransmission strategies can be used to downlink the data to single geographically (and meteorologically) diverse stations in a form of temporal diversity.

The ground stations will have the capability to simulate both user spacecraft and user MOC data systems. This will allow the demonstration of high-data-rate scenarios without the requirement for high-data-rate connections external to the ground stations. The simulators will also allow multiple user and user-type scenarios. The LCRD payload itself will also include the ability to simulate user spacecraft data and multiple relay user spacecraft data systems.

The system will be continuously operating, as much as possible, over the two-year mission. The system will be configured to be demonstrating or testing a specific direct-to-Earth (DTE) scenario, a relay scenario, or be continually characterizing the optical channel and hardware. The DTE and relay scenarios will emulate different user and relay locations, orbits, or trajectories.

XII. CONCLUSION

The capacity of current and near-term RF communications technology is still increasing. However, it is limited by bandwidth allocation restrictions, power requirements, flight terminal antenna size, and weight limitations. The cost and complexity of expanding the existing Space Communications Networks to enable these higher data rates using RF solutions with large-aperture antennas is a significant undertaking. Optical communications holds the potential for high data rates with smaller systems on user spacecraft and on the ground.

LCRD will provide two years of continuous high-data-rate optical communications in an operational environment, demonstrating how optical communications can meet NASA's growing need for higher data rates or how it enables lower-power, lower-mass communications systems on user spacecraft. In addition, LCRD will serve as a developmental testbed in space. LCRD is a critical stepping stone to an eventual deep-space optical communications capability. By conducting this demonstration in GEO, we can gain experience and knowledge important to future deep-space optical communications projects without the cost of a deep-space mission. This is also a critical step towards the next-generation tracking and data relay satellite, as we believe strongly that the next-generation satellite will provide both RF and optical services. Doing this demonstration will enable initial operational capability of an optical service on the first next-generation satellite.

ACKNOWLEDGMENTS

The work described in this chapter was carried out at NASA's Goddard Space Flight Center; at Lincoln Laboratory, Massachusetts Institute of Technology; and at the Jet Propulsion Laboratory, California Institute of Technology, under a contract with NASA. It is funded by NASA's Office of the Chief Technologist and NASA's Space Communications and Navigation Program Office. This work was sponsored by NASA Goddard Space Flight Center under Air Force contract FA8721-05-C-0002. The opinions, interpretations, conclusions, and recommendations are those of the author and not necessarily endorsed by the U.S. Government.

REFERENCES

[1] Robinson, B. S., Boroson, D. M., Burianek, D. A., and Murphy, D. V. "The Lunar Laser Communications Demonstration," *Space Optical Systems and Applications (ICSOS), 2011 International Conference*, 11–13 May 2011, pp. 54–57, doi: 10.1109/ICSOS.2011.5783709. Website address: http://ieeexplore.ieee.org/stamp/stamp.jsp?tp=&arnumber=5783709&isnumber=5783648.

[2] Caplan, D. O., Rao, H., Wang, J. P., Boroson, D. M., Carney, J. J., Fletcher, A. S., Hamilton, S. A., Kochhar, R., Magliocco, R. J., Murphy, R., Norvig, M., Robinson, B. S., Schulein, R. T., and Spellmeyer, N. W. "Ultra-wide-range multi-rate DPSK laser communications," *Lasers and Electro-Optics (CLEO) and Quantum Electronics and Laser Science Conference (QELS)*, 16–21 May 2010, pp. 1–2. Website address: http://ieeexplore.ieee.org/stamp/stamp.jsp?tp=&arnumber=5500551&isnumber=5499482.

[3] Shoup, R., List, N., Fletcher, A., and Royster, T. "Using DVB-S2 over Asymmetric Heterogeneous Optical to Radio Frequency Satellite Links," *Military Communications Conference, 2010 (MILCOM 2010)*, 31 Oct. 2010–3, pp. 785–790, Nov. 2010, doi: 10.1109/MILCOM.2010.5680309. Website address: http://ieeexplore.ieee.orgsol;stampsol;stamp.jsp?tp=&arnumber= 5680309&isnumber=5679517.

[4] Wilson, K. E., Wu, J., Page, N., and Srinivasan, M. "The JPL Optical Communications Telescope Laboratory (OCTL), Test Bed For The Future Optical Deep Space Network," JPL, Telecommunications and Data Acquisition Progress Rept., 142–153, Feb. 2003.

[5] Wilson, K., Roberts, W. T., Garkanian, V., Battle, F., Leblanc, R., Hemmati, H., and Robles, P. "Plan For Safe Laser Beam Propagation From The Optical Communications Telescope Laboratory," JPL, Telecommunications and Data Acquisition Progress Rept., 142–152, Feb. 2003.

[6] Grein, M. E., Kerman, A. J., Dauler, E. A., Shatrovoy, O., Molnar, R. J., Rosenberg, D., Yoon, J., DeVoe, C. E., Murphy, D. V., Robinson, B. S., and Boroson, D. M. "Design of a Ground-Based Optical Receiver for the Lunar Laser Communications

Demonstration," *Space Optical Systems and Applications (ICSOS), 2011 International Conference*, pp. 78–82, 11–13 May 2011, doi: 10.1109/ICSOS.2011.5783715. Website address: http://ieeexplore.ieee.org/stamp/stamp.jsp?tp=&arnumber=5783715& isnumber=5783648.

[7]	Rosfjord, K., Yang, J., Dauler, E., Kerman, A., Anant, V., Voronov, B., Gol'tsman, G., and Berggren, K., "Nanowire Single-Photon Detector with an Integrated Optical Cavity and Anti-Reflection Coating," *Optics Express* Vol. 14, 2006, pp. 527–534.

[8]	Cerf, V., Burleigh, S., Hooke, A., Torgerson, L., Durst, R., Scott, K., Fall, K., and Weiss, H., "Delay-Tolerant Networking Architecture," IETF RFC 4838 informational, April 2007.

FURTHER READING

[1]	Boroson, D. M., Scozzafava, J. J., Murphy, D. V., Robinson, B. S., and Shaw, H., "The Lunar Laser Communications Demonstration (LLCD)," *Space Mission Challenges for Information Technology, 2009. SMC-IT 2009. Third IEEE International Conference*, 19–23 July 2009, pp. 23–28, doi: 10.1109/SMC-IT.2009.57. Website address: http://ieeexplore.ieee.org/stamp/stamp.jsp?tp=&arnumber= 5226852&isnumber=5226792.

[2]	Constantine, S., Elgin, L. E., Stevens, M. L., Greco, J. A., Aquino, K., Alves, D. D., and Robinson, B. S., "Design of a High-Speed Space Modem for the Lunar Laser Communications Demonstration," *Proc. SPIE* Vol. 7923, 2011, 792308.

[3]	Wang, J. P., Magliocco, R. J., Spellmeyer, N. W., Rao, H., Kochhar, R., Caplan, D. O., and Hamilton, S. A. "A Consolidated Multi-Rate Burst-Mode DPSK Transmitter Using a Single Mach–Zehnder Modulator," *Optical Fiber Communication Conference and Exposition (OFC/NFOEC), 2011 and the National Fiber Optic Engineers Conference*, 6–10 March 2011, pp. 1–3, Website address: http://ieeexplore.ieee.org/ stamp/stamp.jsp?tp=&arnumber=5875775&isnumber=5875055.

CHAPTER 13

Replacing the Central Storage System

Bernd Holzhauer* and Osvaldo L. Peinado†
DLR–GSOC, Wessling, Germany

I. INTRODUCTION

The definition of "lifetime" in the world of computers is quite different from the understanding of this term in the space industries. Indeed, hardware components of the storage attached network (SAN) system in the Columbus Control Center (Col-CC) ran out of service time much sooner than expected, and with the increasing amount of data acquired during real operations the SAN became increasingly unstable. Furthermore, a planned upgrade of application software was impractical because this required a newer version of the Linux operating system, SLES 10, but the SAN software drivers lacked the upgrade from SLES 8 to SLES 10 on the application servers.

Together with the complex system design of the old hierarchical storage management (HSM)-SAN, this led to a deadlock situation. An investigation led to the decision to design a new SAN to replace the old one, and to transfer the entire Col-CC to this new infrastructure.

The storage network in Col-CC is an infrastructure-type component, and all major subsystems are based on it, so the SAN system could not easily be switched off and replaced offline. It needed to be replaced with minimum (ideally no) interruption to Col-CC services.

Defining, planning, and testing the new SAN (called SANng) was a major undertaking. Experience with the old SAN redefined some major requirements.

A. Col-CC

Col-CC is part of the German Space Operation Center (GSOC) in Oberpfaffenhofen (near Munich), and is the European central interface to the Columbus module connected to the International Space Station (ISS). All local European User Operation Control Centers (USOC) controlling experiments in the Columbus space laboratory route their information through Col-CC. It operates 365 days a year, 24 hours a day, so there is virtually no time available in which to to replace a

---

*SAN System Engineer, Telespazio Deutschland GmbH.
†Ground Operations Manager.

TABLE 13.1   LIST OF COL-CC SUBSYSTEMS AND INSTANCES that have data (file systems) on the SAN and therefore need access to it

Subsystem	Instances
DaSS: Data Services Subsystem	Ops–Sim–Tst
MCS: Monitoring and Control Subsystem	Ops–Sim–Tst
OST: Operating Support Tools	Ops–Sim
IMS: Integrated Management Subsystem	Prime–Backup
VIDS: Video Subsystem	Single instance only
VOCS: Voice Conferencing System	Single instance only
Infra CM: Infrastructure and Configuration Management	

basic infrastructure-style system like the SAN with a central storage system. All servers in the Columbus ground segment are connected to the SAN and run on it.

If the SAN fails, the control center will go down until the SAN is restored for operations. Serious failures of the old Columbus HSM-SAN caused such shutdowns several times. This, together with the obsolescence of the installed SAN components, drove an exchange to a more practical and stable solution/system.

Col-CC is divided into several subsystems (Table 13.1). Some of these are also split into independent instances for operations (Ops), simulation (Sim), and test (Tst).

The subsystems and instances are independent from one another as much as possible. Cross access between subsystems and instances is not allowed, whereas sharing files within a subsystem and instance is mandatory.

B.   "OLD" HSM-SAN

The Columbus ground segment was designed as a complex computer system based on a SAN using an HSM approach. This was designed and implemented during Columbus mission planning. Owing to the large delay in the Columbus launch, the system became obsolete quite early after the Columbus module docked at the ISS.

The subsystems each had their own file system and were separated from one another. Data on disk were shared between servers of the same subsystem.

C.   THREE-TIERED STORAGE

The original Columbus SAN was designed as a three-tiered storage system (also called HSM-SAN). The basic idea was to have an almost unlimited SAN where the data were written to low-capacity, but fast and expensive, fiber storage. To obtain more capacity, after a while the data were moved to a cheaper Advanced

Technology Attachment (ATA) storage and subsequently to a much less expensive but large tape robotic system.

The servers in the HSM-SAN structure were operating on an EMC Clariion with low-capacity but fast fiber channel disks (Fig. 13.1). After 10 days of no access the data were moved to an ATA–Clariion (larger capacity but with less expensive drives). If data were not accessed for more than 180 days, the files where moved to a tape robotic system with almost unlimited capacity.

A database system (the HSM–DB), running on Solaris servers, managed the positions and automatic migrations of the files. Because of the separation of the HSM-SAN into subsystems and instances, 14 different file systems were in use. The vendor recommended a maximum of seven file systems per HSM cluster, leading to the installation of two Solaris HSM–DB clusters.

The HSM file system was actually a client–server system. The HSM server resided as a database in the Solaris systems and communicated to HSM client software installed on each application server system. This HSM client driver software existed as a layer between the fiber channel adapter card and the file system layer of the operating system. These drivers supported SuSE Linux Enterprise (SLES 8) and MS Windows 2000 plus Windows 2003, but there was no support for SLES 10 nor for other Windows versions. For each file access the client requested the file position via a seperate HSM network from the HSM-DB on the Solaris before loading it.

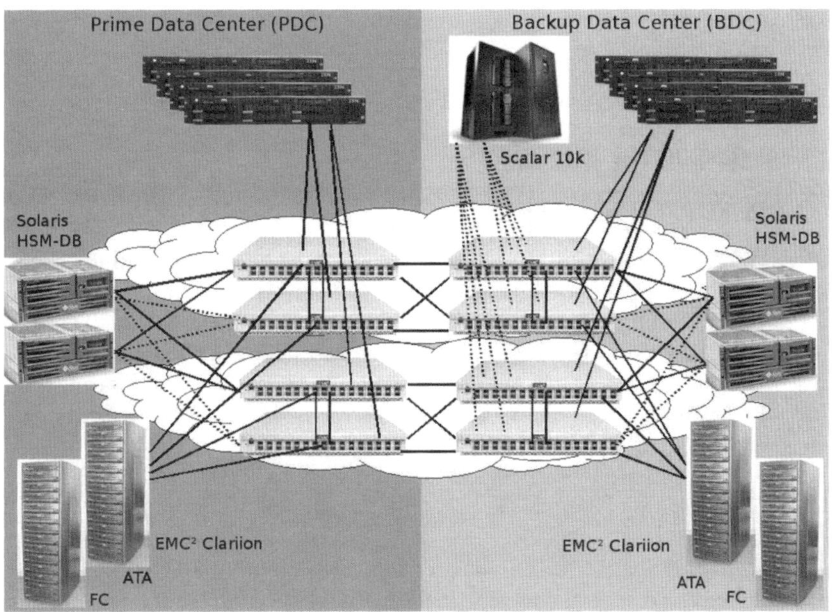

Fig. 13.1   HSM SAN overview.

The Solaris systems were set up with redundant clusters to avoid single points of failure, but during operations the HSM–DB itself showed up as a single point of failure. The database was shared between two cluster heads and was used as is by the remaining one. Before installation and during simulations of the Columbus project, the system worked fine, but as soon as the project started and the system was loaded with a constant data flow, several major problems arose:

1. As more files were stored during real operations, the database increased in size and became instable. The database itself showed itself to be a single point of failure for all file systems on this database cluster.

2. Because the HSM was designed as a client–server model, upgradability was very limited. SLES 9 was skipped during subsystem software development. The software should have been upgraded to SLES 10, but there was no HSM client–server version available that could handle SLES 8 and SLES 10 clients at the same time. An upgrade procedure worked out with the vendor of the HSM system would have caused a four- to six-week period of down time for the entire Col-CC.

3. Accessing multiple files stored only on tape caused the HSM database to crash. This then caused all subsystems on the HSM cluster to fail.

4. The Clariion storage systems reached their end of life and the service contract was not continued by the vendor.

Together, these issues led to the decision to replace this old dinosaur with a more modern and flexible system.

II.   DESIGNING THE NEW SYSTEM

A.   KISS: KEEP IT SIMPLE, STUPID

One of the major concerns was the complexity of the old system and the client–server construct (finally brought to a head by the awareness that the system was not upgradable online), so it was essential that the new system should avoid this. So, the slogan KISS—keep it simple, stupid—was adopted. A simple system is easier to maintain. Also, the components should, as much as possible, be independent of one another to make them exchangeable in small chunks if some parts become obsolete in the future.

As a result of lessons-learned and this KISS approach, new requirements were defined. For example, the new SAN should take the following into account:

1. *Update problems.* Each component of the SAN environment, including SAN attached servers, should be replaceable without affecting other items. In other words, "no special drivers should be necessary," to enable independence from different operating system versions. This leads to the next issue.

2. No *more special (vendor proprietary) software at application servers*. About 90% of the servers run on Linux (SLES 8 and SLES 10) and the rest use MS Windows. As standard protocols, NFS for Linux and CIFS for MS Windows shares should be used.

3. *Single point of failure*. The system should not have a hidden single point of failure.

However, there are some other points that also need to be fulfilled, as discussed in the next sections.

B.   FILE SHARING BETWEEN SERVERS IN A SUBSYSTEM

Some subsystems use multiple servers to store and process data, so the system must allow file sharing. A simple SAN will not do this without vendor-specific software (which should not be used). However, file sharing can be supported using standard network protocols like NFS and/or CIFS.

The deadlock situation in the old system was mainly caused by special driver software handling shared access to the SAN disks. As operating systems and software will be updated quite often in the future, there is an absolute "no-go" for special driver software from a third-party vendor. A solution was sought to provide file sharing between servers with standard operating system drivers.

Network protocols like NFS for UNIX/Linux systems and CIFS for MS Windows can handle such file sharing. A set of file servers or a network attached storage (NAS) can provide this.

C.   USE "OUT OF THE BOX" HARDWARE AND SOFTWARE

Common off-the-shelf (COTS) products should be used as much as possible. This keeps investment cost low. Also, COTS products are usually easy to replace at the end of a product lifetime. The more complex components should be supported until end of mission (set at 2018 during the planning period; now 2020) to avoid end of lifetime and other data migrations before the end of the project. If the ISS/Columbus project is extended beyond 2020, an upgrade path must exist.

D.   SYSTEM CONCEPT

With all the above in mind, a computer consultant (system house) was met at a computer exhibition. Following many meetings and discussions they understood the special space business requirements and the Col-CC specifications and concerns in particular. They also brought Hitachi Data System (HDS) into the game. Together, we developed the specific Col-CC system concept.

Hitachi offers an appliance called the Hitachi Network Attached Storage (HNAS). This resides on top of a standard SAN storage and provides, among other services, the disk space that NFS and CIFS networks share. This HNAS hardware platform is capable of emulating up to 64 virtual file servers (an enterprise virtual server, EVS) on a single hardware. The HNAS is not just a PC-style system. It specializes in file services and operates in specially designed hardware (field programmable gate arrays, FPGAs). It is very fast and (nearly) without problems on the operating system level.

The basic HNAS features are as follows:

1. It enables 24 h, 365 day operations in a cluster configuration with automatic failover.

2. It guarantees independency between subsystems and instances using the EVS concept. EVSs will automatically migrate to the remaining cluster part of HNAS in the event of an HNAS failure.

3. The split brain between HNAS heads is omitted by using a system management unit (SMU) as a quorum device.

4. Each EVS will have its own file system, and disk quotas may be used.

5. A failover (at the client site) lasts less than 30 s.

6. To keep system topology simple, as much as possible storage should be used via the EVS (i.e., used on NFS or CIFS exports).

7. For special applications, such as Oracle RAC or cluster quorum disks, direct logical units (LUNs) from storage are also available but should be used rarely.

Storage capacity was calculated from the incoming and generated data flow per year (approximately 40–50 TB) and the remaining project time (10 years). Also, Hitachi promised to support at least the Universal Storage Platform (USP) storage towers until the end of the project (2018 at time of project design).

E. FINANCIAL CONCERNS

The first system design was made with the idea of keeping all project data over 10 years online. This would have required a storage system with a 500 TB (net) online capacity, mirrored by a Prime Data Center (PDC) and Backup Data Center (BDC). The storage towers themselves would have needed to be "Enterprise Class" storage to guarantee the 365 day/24 h operation. From a cost perspective this "high end" system was too expensive, and alternatives were discussed. The most costly items in the concept were storage capacity and capacity license for online mirroring (quite expensive at Hitachi). However, with less capacity, the target of 10 year online data could not be reached. So the alternative was to (re)define a Data Retention Policy (DRP).

F. DATA RETENTION POLICY

Redefining the DRP opened the door for significant cost savings. The DRP basically allows data older than 12 months to be moved to a tape archive. System log files do not need to be archived, which means they can be deleted after 12 months.

This seems to be not far away from the old system design, but the process is completely different and the new system remains simple.

As a result of this new DRP, the total storage capacity was stripped down to 55 TB (net). Furthermore, the HNAS itself is able to replicate data from prime to backup storage. This replication is almost comparable to a mirroring function except mirroring is done without (visible) time delay whereas replication has a time scheduler as part of it. As a result, if replication is done on an hourly basis, data on the backup storage may be out of date by up to 1 h when the system fails, with the possible consequence of losing the last hour of data.

As the Hitachi USP–VM storage system used here is Enterprise Class storage, it is very unlikely this system will fail, but even a small possibility of a 1 h data loss is not acceptable for Ops instances and replication is therefore not acceptable for these data. Video, Test, and Sim instances, however, are not this critical, so they were defined to run on replicated instead of mirrored storage. This therefore means mirroring just 10 TB and replicating the remaining 45 TB of storage. It is a little in conflict with the KISS strategy, but the financial concerns force this.

G. SANng SYSTEM OVERVIEW

The system consists of two identical USP-VM storage towers, one in the PDC and one in the BDC (Fig. 13.2). The link between the PDC and BDC is an optical cable

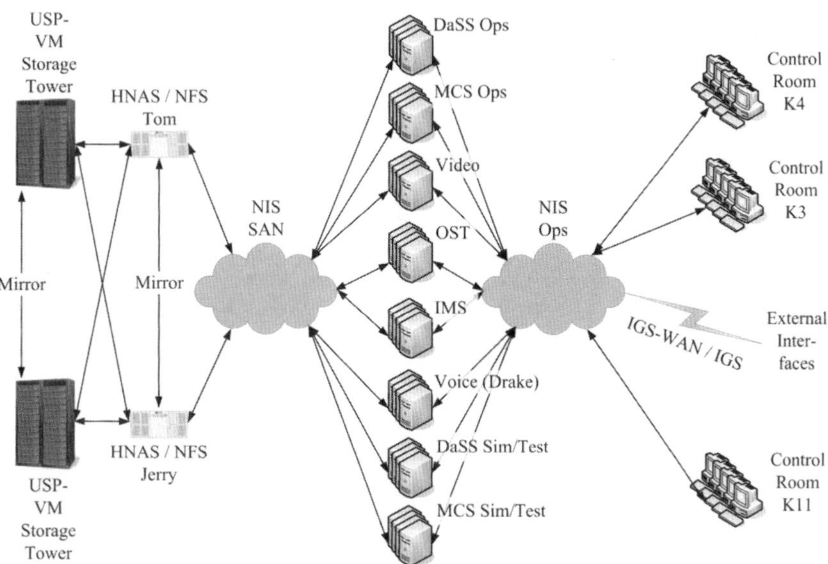

Fig. 13.2   Col-CC SANng overview.

about 400 m in length. Storage towers and HNAS systems are connected by redundant fiber channel networks (two fiber fabrics). The connection between HNAS and the NIS-SAN cloud is a 10 GB redundant link.

The TCP/IP network network infrastructure (NIS) is split into two parts. The newly installed storage network NIS SAN (replaces the old HSM-LAN) and the existing Operational (OPS) network (NIS Ops). All application servers are connected redundantly to both NIS clouds. Two clouds are used to avoid bottlenecks and cross-connections between OPS and the storage network.

H. PROOF OF CONCEPT

During the design of the new system it was not clear if the speed of data processing would be sufficient. One possible bottleneck could be the data flow through the TCP/IP network instead of using high-speed fiber-channel connections.

Before going deeper into planning and running the complete order processing procedure, a proof of concept was requested. A smaller system, a single AMS 2300 with 20 TB net capacities, was borrowed from Hitachi. This system was also equipped with a complete HNAS cluster (later named Max & Moritz) on top.

The plan was to prove the system under real operation conditions, so some Test and Sim instances were ported to the borrowed test system. After familiarizing with the usual failover scenarios and other testing scenarios, the system should also be used during real space (operating) simulations (Joint Mission Sequence Test (JMST), etc.). This should demonstrate where the total system may run into performance problems and should also verify if and how the subsystem software will react to failovers and movements of EVSs within the HNAS cluster.

I. INTERIM SAN

During the first tests on this loaned system, the old HSM–SAN became increasingly weak and failed too often. We became familiar and satisfied with the test system and found it much more stable than the old operating HSM–SAN. So, the loan system was turned into the "Interim SAN." After becoming familiar with the system, some OPS instances were transferred to this Interim SAN to reduce the stress on the old database-based HSM–SAN system.

As the Interim SAN increasingly became a productive OPS system, a backup system was purchased and installed to avoid data losses. The backup system is a single Linux server that can mount all network shares as local drives. As well as working this way for the Interim SAN, this backup system will also do so for the SANng.

In the future, the server itself will also control the tape archive. It is therefore set up as a very powerful system that can easily handle data migrations forward and backward between the two SAN systems.

III. MIGRATION

A. MIGRATING SUBSYSTEM BY SUBSYSTEM TO INTERIM SAN

For data migration from the old to the new SAN storage, a Linux server was installed and connected to both SAN systems – the old HSM–SAN and the new Interim SAN. This server had full access to all file systems on both sites. With the Linux Rsync command it was a quick and easy job to copy immediately needed data from the old system to the new one. Then, changing the mount points in the appropriate application servers from the old HSM–SAN disks to the new provided network shares, and a subsystem was operating (migrated) on the new Interim SAN.

This was a quick and dirty solution to keep Col-CC alive, but it stopped the real proof of concept in the middle because the system was now in use for real-time OPS and could not be tested seriously any more.

B. LINUX RSYNC DOES THE MAIN DATA MIGRATION JOB

Rsync, a standard UNIX/Linux command for remote file synchronization, was used to perform the migration job. This is a very handy tool for copying files from a source to a destination location. By command line parameters it can be forced to copy data only if files do not exit on the destination and to copy only newer files. Rsync can also work recursively through entire directory trees and complete file systems.

It does not matter if the initial copy takes hours, weeks, or months. Rsync can copy all files during operations, without disturbing them. Only files that are currently modified need to be copied again later. Rsync is able to find such files and synchronize them in a second or third stage. So with each run of Rsync, less data will be copied.

For the final move (migration), the application needs to be stopped. The application server must be disconnected from old and connected to new mount points, and the final Rsync job has to be executed, but these are tasks that need only a few minutes. That is all. The final subsystem move (migration) can be done within a 15–20 min Loss Of Signal (LOS) slot.

C. MOVING DURING LOS TIMES AND COORDINATION WITH OTHER SUBSYSTEM MAINTENANCE

Copying data with Rsync could be done during and without affecting operations, and more than 99% of data could be migrated without interruption of services. The final movement, that is, remounting the file shares and the final Rsync of single subsystems or instances, was coordinated within other regular subsystem maintenance or LOS slots, so the movement from the old HSM–SAN to the Interim SAN was almost invisible to the Flight Control Team (FCT), except that the total system became more stable and reliable.

D. INSTALLING THE COLUMBUS SAN

Because the proof of concept was interrupted by using the Interim SAN for operations, it was resumed later after installing the final SANng and combined with the large site acceptance test (SAT). Real-time OPS on the Interim SAN were something like a reverse proof of concept. It showed up the functionality of the system concept, and the "smaller" system never ran into performance bottlenecks. However, some tests had still not been carried out, and could not be performed on the Interim SAN. Part of the proof of concept was testing various error conditions during the running of real simulations (like JMST, European Simulation, etc.) in the control rooms.

E. EXTENSIVE TESTING AND SITE ACCEPTANCE TEST

After installation of the final SANng, a very extensive SAT was required, so the final tests were done within the SAT. The SAT took more than three weeks and included all possible failure scenarios, a complete site failover from PDC to BDC, and back. Also, traffic simulation systems were used to generate a system load up to 10 times higher than today's standard data rates in Col-CC. This was done to future-proof extensions like upgrading from standard video to high-definition video (HDVCA), which will at least triple the existing video data rates.

F. TOM & JERRY VERSUS MAX & MORITZ

All the testing, migration, and operations were carried out in parallel at the two different Hitachi HNAS clusters. So as not to get confused with the similar names of the different HNAS clusters, and knowing which is related to what job at a particular time, the SANng HNASs were named "Tom & Jerry" and the nodes from the Interim SAN were called "Max & Moritz." This caused some amusement at the beginning, but practice proved it to be a useful naming convention and the terminology was retained.

Figure 13.3 gives an overview of how the two Hitachi SANs are connected. The HNAS cluster, Max & Moritz of the Interim SAN, is placed in a single rack. The other cluster, Tom & Jerry (the SANng), is controlling the USP-VM storage, split over two different buildings (PDC and BDC) to support a more secure setup. Both SANs are connected to both clouds of NIS. The data are mainly transported via the storage Virtual Local Area Networks (VLANs), but some shares are also exported directly to the OPS Network, for example supporting user "Home" directories for the FCT. With old HSM–SAN this feature was set up as an extra NFS server cluster. The fiber channel connections of the (very few) servers supported with direct LUNs are not drawn here.

G. DATA MIGRATION FROM INTERIM SAN TO SANng

As described in Sec. III.A–C, data migration using Rsync is very easy and can mainly be done without interrupting operations. The backup server can be

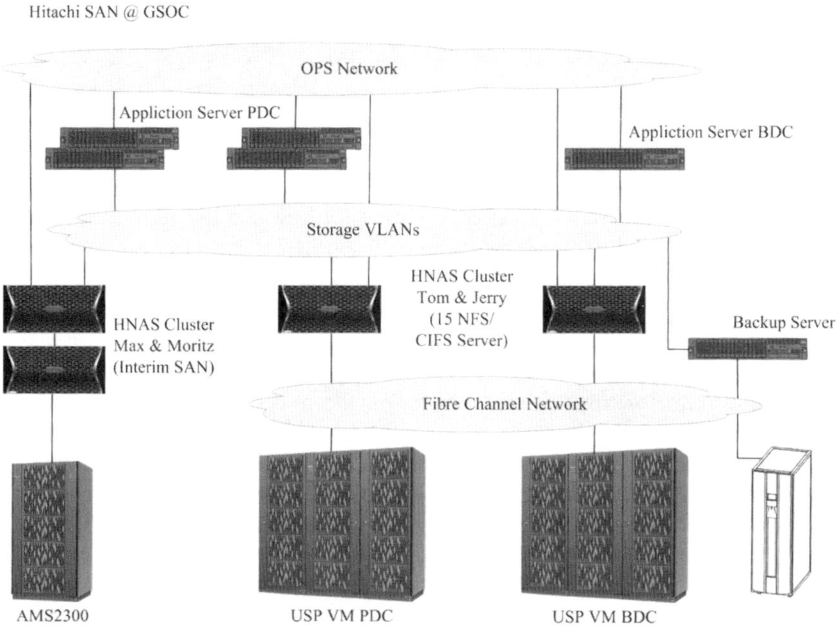

Fig. 13.3 GSOC SAN overview.

connected easily to both SAN clusters. So it is very easy to move (migrate) subsystems between Tom & Jerry (SANng) and Max & Moritz (Interim SAN). This made the total system extremely flexible and provided a good opportunity for extensive testing.

H. MIGRATION OF "HISTORICAL" DATA FROM THE OLD HSM–SAN

Migration of the old data residing on the HSM tapes caused us some extra headaches. Just doing an Rsync as with the old "online" data ended up in a mess. The HSM was a basic system and inserted tape cassette after tape cassette into the drives, which resulted in a very slow data transfer, too much delay caused by the tape robot, and also broke the tape drives themselves.

Writing a script that proved data availability on disk or on tape and invoking a semi-automated restore on the directory level instead of reading file by file sped up the process dramatically. However, even with that script, reading back all the old data from tape took about 18 months in total. Also, some old files were lost during the process because of broken AIT tapes and invalid (unusable) second copies due to more broken tapes or database errors. Most of the missing data could be recovered by getting Path-TM data on tape cassettes from NASA and merging this into our datasets.

I. MIGRATING NATIVE LUNs

During system migration, most of the old native LUNs were converted into network shares. However, some servers still need native LUNs (e.g., Microsoft Windows cluster server). These servers were connected redundantly to the old Clariions via a dual fiber link. For migration the redundancy was broken, so that one fiber link was set free, and the servers could be connected to both storage systems—the old FC network with Clarrion and the new fiber fabrics attached to USP VM—at the same time. This allowed data to be copied from old LUNs to new LUNs. By adding a second quorum disk during operations, most of the migration could be done without shutting the applications down.

With this preparation the real outage for final switching was just a few minutes, stopping for example a MSQL database and restarting them on the new disks. With the application working on the new disks the old disks could be deactivated and disconnected.

Only de-installing the old EMC Powerpath software and installing the new Hitachi HDLM (Multipath Software) – to get back to redundant data paths – required some cluster reboots. However, this could be handled without big interruptions in operations during LOS periods.

J. BACKUP

For the backup system, a very simple strategy was used. Applications and the operating system itself are standards and can be reinstalled from an installation server, so there is no need to back up those "local" data. It is only important to back up operational data. However, all operational data are written to the file systems on SAN, and these file systems are completely visible to the backup server. The backup server mounts them and all the data are backed up (like backup of PC local data) to a tape robot system. The robot library is split into two parts, a small one for backup and a larger part for the archive.

K. ARCHIVE

The tape archive is under installation and test, but not yet in use. It will operate in a manner similar to the backup scenario and will use the same infrastructure. In contrast to the backup, the archiving process will not be executed automatically. Data from the current year and the year before are defined to be available at the online storage (i.e., on disk). From time to time data will be written to the archive and deleted from online storage.

For availability (safety) reasons archive data will be written to at least two (three currently under consideration) tape copies. One tape will be kept in the robot library, and the other(s) will be taken to a safe location(s).

Data stored in the archive will not be available directly to the user. A data archive retrieval request (DART) request has to be made and the SAN team will restore the requested data for the defined period. After processing by the user, the data will be archived and removed from online storage again.

IV. CONCLUSION

In total, the migration took nearly two years, but the actual switching times were hidden within other service times and/or during LOS periods. To date, the migration from old HSM–SAN to the new SANng has been (almost) invisible to the FCT. The real visible effect for the FCT is that the SAN is now much more stable for operations. No outages have been caused by SAN in the last two years of operation.

The availability of the Interim SAN was good luck and very helpful because data could be moved very smoothly from HSM–SAN to Interim SAN and to SANng later.

No system is perfect, but ... Two years later we found servers with uptime values of 300 days and more. We have had, of course, some minor failures with the Hitachi systems, but we are proud of not having had any FCT-visible outages due to SAN failures as in the preceding years. Application servers with uptime values of more than 300 days were never seen with the old HSM–SAN, but with SANng this is very usual.

Our timeframe was as follows:

1. 2008: planning and testing different SAN concepts at vendor sites, mainly Hitachi, IBM, and NetApp;

2. December 2009: installation of Interim SAN as a test system;

3. March 2010: migration of the first subsystems to Interim SAN;

4. July 2010: installation of SANng;

5. August/September 2010: SAT;

6. September 2011: final HSM–SAN shutdown.

ACRONYMS

ATA	Advanced Technology Attachment
BDC	Backup Data Center
Col-CC	Columbus Control Center
COTS	Common off-the-shelf products
DRP	Data Retention Policy
EVS	Enterprise Virtual Sever (a virtual file server inside a HNAS)
GSOC	German Space Operation Center
FC	Fiber Channel
FCT	Flight Control Team
HDS	Hitachi Data Systems

HNAS	Hitachi NAS (appliance: virtualizes up to 64 EVS)
HSM	Hierarchical storage management
HSM-SAN	The old Col-CC SAN based on HSM technology
LOS	Loss of signal
LUN	Logical unit (see Glossary)
NAS	Network attached storage
NIS	Network infrastructure
PDC	Prime Data Center
SAN	Storage attached network
SANng	SAN next generation
USP	Universal storage platform

GLOSSARY

EVS   Enterprise virtual server: a virtualization of a file server offering network file services. From the possible file services only NFS and CIFS are used. An EVS does not need to be redundant. Redundancy is provided by the HNAS cluster itself. The EVS is migrated to the other node in case of an error.

HNAS cluster   Hitachi NAS Appliance. Tom & Jerry are the two HNAS heads for SANng. Max & Moritz are the two HNAS heads for Interim SAN.

HSM–SAN   The old SAN in Col-CC based on a hierarchical storage management; was installed in 2003 and running out of support in 2009–2010.

Interim SAN   A loaned system for proof of concept, which was actually used to keep operations up and running.

LUN   Logical unit. In a SAN environment many disks are combined to some kind of large disk pool. The pool will be partitioned into practical slices (parts). Such parts exported to a server system are typically called LUNs.

Multipath   Multipath Software combines two separate connections to the storage media and may use them in failover and/or load-sharing combination. Without this software the server will see the same disk or LUN twice and handle the disk like two independent ones.

SANng   The term "SAN next generation" was defined for the new SAN to have an easy understanding about which SAN was meant—three SANs were operated in parallel during migration.

CHAPTER 14

Ten Times More Information in Your Real-Time TM

David Evans and Ugo Moschini[†]
ESA/ESOC, Darmstadt, Germany

I. INTRODUCTION

Whereas science data are routinely compressed, housekeeping telemetry is not. This is despite the fact that many mission profiles are dominated by housekeeping telemetry (e.g., technology demonstrators) and every mission has phases dominated by it [e.g., Launch and Early Operations Phase (LEOP), recovery phases]. In previous work conducted by the Future Studies section of ESA/ESOC (European Space Operations Centre) [1–3] we argued that there would be considerable advantages in compressing housekeeping data and no increase in risk, provided it was performed correctly. The compression algorithm must have a stable performance, be lossless, and require little onboard processing power.

We showed that the Consultative Committee for Space Data Systems (CCSDS)-recommended compression algorithm for space (RICE [4]) could not compress housekeeping telemetry effectively and proposed an alternative algorithm. While implementing this algorithm we encountered two fundamental problems and solving those problems required a major rethink of our approach. This new approach resulted in the creation of a new algorithm that could not only compress stored housekeeping telemetry data but also data that was transmitted to the ground in real time. This chapter describes the problems, the new algorithm, and the many advantages of the new approach.

In Secs. II and III we recap the previous algorithm's performance, general advantages, and elaborate on how compression can be used to increase information content. In Sec. IV we describe the implementation problems with that algorithm. In Secs. V–VII the new algorithm and its test results are presented. Sections VIII–X deals with the robustness of the technique and the results of tests obtained when running the new algorithm on real space hardware. Finally, Secs. XI–XII describe a modification to the algorithm to make it self-adapting when faced with changes in spacecraft behavior or entirely new packets.

---

[*]Mission Concept Engineer, Future Studies Section.
[†]Young Graduate Trainee, Future Studies Section.

TABLE 14.1   COMPRESSION PERFORMANCE COMPARISON BY MISSION FOR BIT TRANSPOSITION-RLE

Mission name	Mission type	Compression (% of original)
Columbus	Human spaceflight	5.75
Rosetta	Interplanetary	14.06
Venus Express	Interplanetary	18.23
Proba-1	Technology demo	25.93
Herschel	Astronomy	28.00
Goce	Earth observation	38.20

II.   PREVIOUS WORK

In our three previous papers [1–3] we described an algorithm that preprocessed stored housekeeping telemetry by grouping identical packet types and reading the individual bits in a transposed manner before run length encoding (RLE) the resulting bit stream. We called it the bit transposition–RLE algorithm. It is extremely effective at compressing stored housekeeping telemetry for a wide variety of missions. The following experiment was set up. A week's worth of housekeeping data were retrieved from the ESA mission archives for different ESA missions. These were then processed to extract the CCSDS source packets, thereby effectively reconstructing the original onboard packet store. These packet stores were then compressed using the bit transposition–RLE algorithm and the size of the compressed file compared to the original packet store. The results are given in Table 14.1. Please note that all compression ratios in this chapter are given in percentage terms (size of compressed file compared to original packet store), so the smaller the percentage the better. The results proved that there is a considerable amount of information redundancy in housekeeping telemetry that could be removed using this simple compression algorithm.

III.   HOUSEKEEPING COMPRESSION ADVANTAGES

We also argued that implementing housekeeping telemetry compression would reap direct cost savings by enabling shorter or fewer ground station passes and that there were many less obvious advantages:

1.   Allowing the use of smaller, cheaper ground stations (fewer data so improved link budgets);

2.   Saving onboard power and energy (less transmission time);

3.   Relaxing operational time constraints (shorter dumps so more choice on when they are done);

4. Improving interactivity with the spacecraft (shorter reaction times possible);

5. Making packet design easier (fewer constraints on parameter and sampling rate choice);

6. Improving mission safety (can send critical packets multiple times).

In this chapter we would like to highlight a further advantage that has been cited by multiple operators: using compression to increase the information content in housekeeping telemetry. The idea would be to maintain the same level of bandwidth usage for housekeeping, but to use compression to allow the remote system to sample more parameters at higher frequencies. If compression ratios of ten are achieved it would correspond to an increase in information content of over 1000%. This is an information increase, not a telemetry rate increase. Parameters that compress well can be included in the housekeeping packets and sampled often with only a small impact on bandwidth usage. Therefore, the actual increase in telemetry rate could be far higher than 1000%.

Without compression it is necessary to carefully select all the parameters and their respective sampling rate in the housekeeping packets. The aim is to achieve a balanced compromise between bandwidth use and information content. If bandwidth is really tight it could involve removing parameters from housekeeping and designing asynchronous events to flag operationally significant changes in them. Compression fundamentally changes this trade by effectively removing all "compressible" parameters, which make up the vast majority. Hence one can simply select as many of these parameters at whatever sampling rate one wants in the housekeeping, with little bandwidth usage impact. There is no need to guess if these parameter histories might be needed one day and no need to design asynchronous triggers and events for these parameters.

This results in richer, finer information on the ground to analyze. It therefore removes the need to guess or extrapolate when trying to reconstruct onboard events. It reduces the chance of missing important information due to low sampling rates (e.g., short abnormal behavior like spikes, transients, high-frequency switching). It gives the ground more chance of discovering correlations between parameter behavior and important events or trends, which is not possible using snapshots. These are all operationally important advantages.

IV. TWO PROBLEMS

Although the bit transposition–RLE algorithm generated considerable interest two fundamental problems emerged. The first was raised by operators who pointed out that housekeeping telemetry is used as input for two processes: real-time control and offline analysis. Because the packet structures for telemetry destined for the offline analysis process and the real-time control process are interchangeable, real-time considerations would always negate the theoretical advantages gained in the offline analysis process. For example, we argued that significant savings could be made in the operations preparation phase during packet design as the need for careful selection of parameters and sampling rates could be eliminated for those parameters that compress well. However, if the same packets are used for real-time transmission then one still has to be very careful when designing the packet.

The second objection came from the onboard software area. The bit transposition–RLE algorithm requires that each packet type be processed differently; hence, all packets of a particular packet type must be read from the packet store before moving onto the next packet type. It was pointed out that packet stores are presently flat; that is, packets are stored sequentially as soon they are generated, with no indexing or hierarchy. When the time comes to dump the packet store, a pointer is set to the memory address of the last packet read and then all the packets are simply read back from this point. Any attempt to read only certain packet types from this flat store would require considerable Central Processing Unit (CPU) usage. Although one can envisage using different packet stores for each packet type, this would raise operational issues such as trying to predetermine how big each one should be to avoid overwrite. This would introduce unwanted complication into the mission operations concept.

We realized that the only solution to the operator's problem would need a compression algorithm that would intercept a packet when it was generated, convert it very rapidly into a smaller packet, and then transmit this in the real-time stream instead of the original. If this could be done then it would also solve the onboard software problem because these smaller packets could also be stored in a flat packet store and retrieved using the present simple mechanisms.

V.   THE NEW APPROACH

At first there seemed to be no way to do this. Most compression algorithms require a certain critical mass of data to be available before they can determine where information is redundant and then remove it. An algorithm that could compress data in the real-time telemetry stream would only have a single packet to work with and therefore not have that critical mass.

We began by reexamining why the bit transposition–RLE algorithm worked so well with housekeeping packets. These packets use fixed bit position mapping to determine where parameters start and finish, which means that they must allocate a fixed number of bits for each parameter. This is efficient in the sense that no parameter labeling is required, but it also means that each parameter must be allocated a sufficiently large enough bit field to cover its entire possible dynamic range. However, it is very rare that a parameter uses its entire dynamic range, so while most parameters change value from one housekeeping packet to another, the majority of their bits do not.

The bit transposition–RLE algorithm exploits this property by reading the same bit position from multiple packets in sequence. We realized this property could be exploited in another way; by performing a bit XOR of the newly generated packet with the last packet generated of that type. This would produce a new packet that contained mainly zeros at the bit level (for those bits that had not changed state). This could be then be compressed by run-length encoding those bits. By adding a new header to this compressed data and a trailer to make it an integer number of bytes, we can create a new smaller packet from the original longer one.

Inspection of the XORed packets revealed that some bits were changing state much more often than remaining in their present state. Hence, adding a simple inversion of the XOR result for that bit would produce a zero more often and improve the compression. This could be reversed in the decompression algorithm. A similar approach was taken for those bits that tended to change state at the same time. We found that XORing the two results produced a zero more often and this could also be reversed in the decompression algorithm. This was called horizontal correlation. Both inversion and horizontal correlation are applied to all the subsequent algorithms in this chapter.

We called the technique the "basic" algorithm and used it to compress a week's worth of stored housekeeping data for several ESA missions. The results are shown in Fig. 14.2. This shows that this basic algorithm was indeed capable of producing significant compression but was less well performing than the bit transposition–RLE results presented in Table 14.1.

VI.   IMPROVEMENTS ON THE BASIC ALGORITHM

The bit transposition–RLE results proved that it was possible to remove more information redundancy from the data than the basic algorithm, so we began to search for methods to improve it. We considered how a priori information could be exploited. Our first idea was to calculate the probability that each bit would change state from one packet to another based on historical data. The ground could then preload an order in which to read the bits in the XORed packet based on those values. This would have the effect of grouping bits with a high probability of being zero after the XOR and this would make the RLE compression more effective. A similar idea has already been proposed for compressing and storing housekeeping telemetry on the ground [5], so we called this algorithm "Staudinger" after the author. The results are shown in Fig. 14.2. One can see that it has very good performance, beating the basic algorithm (and even bit transposition–RLE) in all considered cases.

However, our overall aim was to produce a system that could compress the real-time telemetry stream, so it was important to test how much time it took to compress an average packet. We ran speed trials using a notebook equipped with an Intel Core i7 processor to gain relative speed information between the techniques. The results are shown in Fig. 14.3 and highlight that the Staudinger algorithm was taking over twice as long to compress a packet on average compared to the basic algorithm. This was not good news for a real-time compression system.

VII. POCKET

As well as being slower than the basic algorithm, Staudinger had other disadvantages when applied to our problem. It has a complicated ground–space interface because it requires the loading of a specific order table for each bit in every packet type. Also, in his paper, Staudinger points out that it is not stable for "real data." Staudinger proposed a recalculation of the order tables every hour to compensate, but this would be impractical for us.

We therefore started looking for a solution that would have a better performance than the basic algorithm but would be faster, stable, and have a more simple ground–space interface. The algorithm we created was christened POCKET (Probability Of Change masKEd Transformation). The idea behind POCKET is to exploit probability information available in historical data to produce a simple bit mask packet rather than a bit order table. The ground then sends this mask packet to the spacecraft and the onboard algorithm uses this in a series of bitwise operations to compress the newly generated packet with very little calculation involved. This gives the algorithm a very simple ground–space interface and makes the algorithm very fast.

The POCKET algorithm has a ground element and an onboard element, which are described in the following steps.

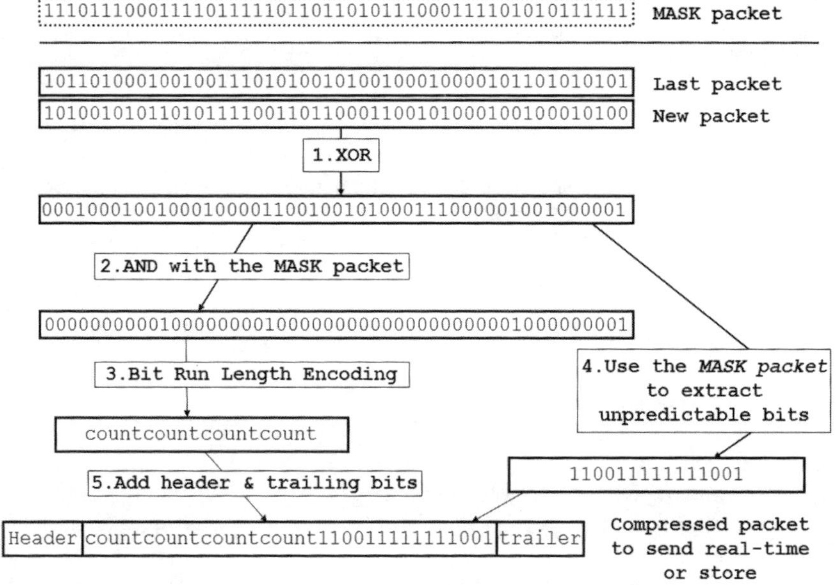

Fig. 14.1 The onboard element.

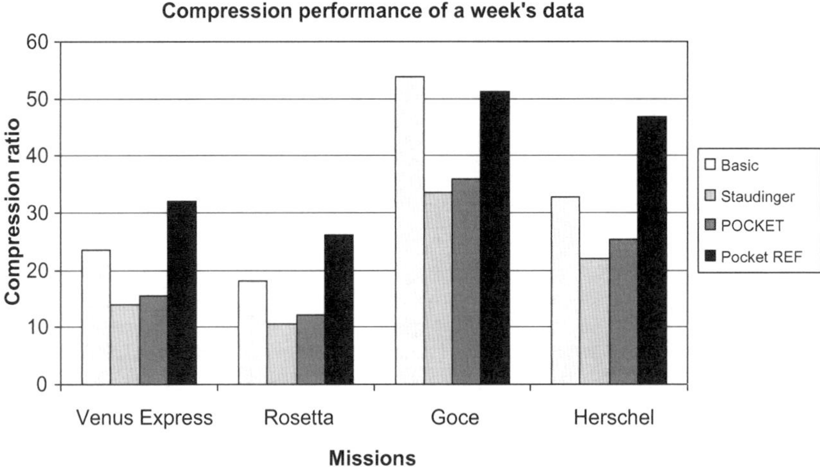

Fig. 14.2   Compression performance comparison by mission for the four methods illustrated in Secs. V, VI, VII, and VIII.

A.   THE GROUND ELEMENT

Use historical data to determine the predictability of each bit in a data packet based on the state of the bit in the last packet generated with inversion and/or horizontal correlation applied.

For each bit in the packet decide whether its state is predictable; i.e., its state can be predicted with a better probability than a certain threshold.

Use this information to create a mask packet with bits in a one state for all predictable bit positions and in a zero state for the rest.

Upload the mask packet to the spacecraft along with inversion and horizontal correlation instructions.

B.   THE ONBOARD ELEMENT

The onboard element is described in the following and in Fig. 14.1.

When a new packet is generated perform an XOR with the previous packet of that type.

Perform inversion/horizontal correlation instructions as necessary. The resulting packet is called the processed packet.

Perform a bitwise AND between the processed packet and the ground loaded mask packet. This will change all the unpredictable bits in the packet to a zero state.

Perform bit run length encoding on the resulting packet and write the results to an output buffer.

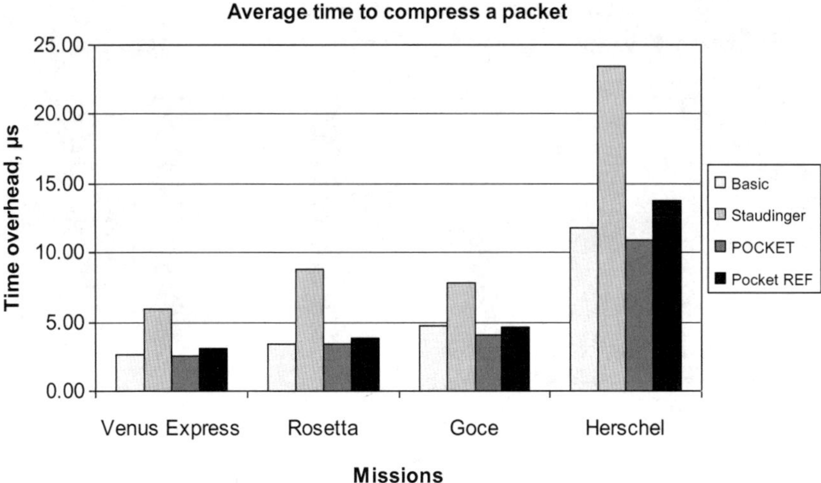

Fig. 14.3   Speed performance comparison by mission for the four methods illustrated in Secs. V, VI, VII, and VIII.

Extract the unpredictable bits from the processed packet (information in the mask packet) and append them to the output buffer in read order.

Add a header and a trailer to the output buffer to create the compressed packet.

Transmit or store the compressed packet as required.

The compression results of the POCKET algorithm are presented in Fig. 14.2. The figure shows the compression ratio percentages by mission, as size of the compressed packets compared to the original packet stores. For every mission, a week's worth of stored housekeeping data has been considered. One can see that it performs almost as well as the Staudinger algorithm, even though it is much simpler and easier to configure. Looking at Fig. 14.3 one can see that it is significantly faster than the Staudinger algorithm and even faster than the basic algorithm.

VIII.   POCKET ROBUSTNESS

In earlier papers [1–3], we argued that compression does not increase the risk of information loss if the information in a telemetry frame can be uncompressed without needing any information contained in other telemetry frames. The use of an XOR with the previous packet means that the compressed packet is effectively a delta change packet. One must work forward from a reference packet in which the absolute state was known to calculate the present absolute state. It is also possible to work backward from a reference packet in the future. If a packet is lost then all information will be lost (or delayed) until the next reference packet is transmitted. Hence, reference packets must be sent at intervals. The frequency with which one sends these reference packets depends on the probability that a packet will be lost and how much risk one is prepared to take.

One could simply use uncompressed packets as reference packets but there is also a simple way of configuring POCKET to produce reference packets. POCKET REFERENCE packets are not based on an XOR with a previous packet. Instead of using the probability of change of state between packets, the probability that each bit is in a particular state is calculated based on historical data. The ground then loads a corresponding mask, inversion packet, and horizontal correlation instructions based on these predictions. The results are given in Fig. 14.2 and show that POCKET can produce reference packets with a good level of compression. Figure 14.3 also shows that POCKET REFERENCE is much faster than Staudinger.

IX. STABILITY RESULTS

To test the robustness of the POCKET algorithm we set up the following experiment to determine its performance over time. A week's worth of housekeeping data of Venus Express was split into chunks corresponding to the seven days of the week. The first chunk was used as historical data to produce the mask packet and the inversion/correlation instructions, which were then applied to compress the packets of the remaining six days. The results are shown in Fig. 14.4. It shows that POCKET performance remains remarkably stable over time, only decreasing from 15% to 16% compression over one week without

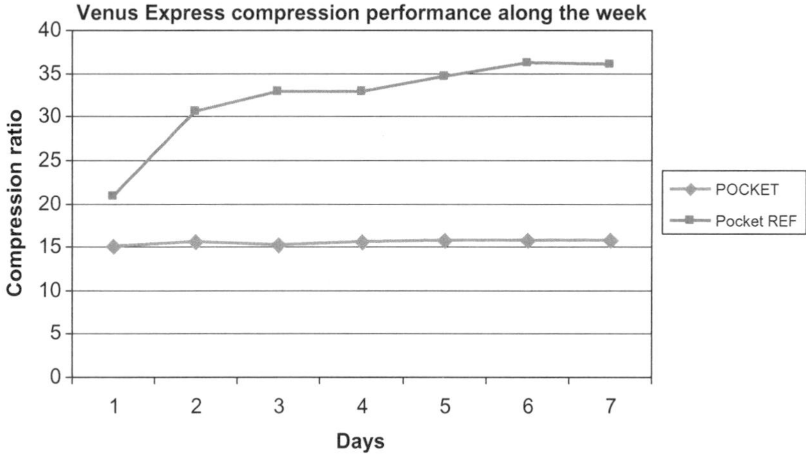

Fig. 14.4   Stability comparison between POCKET and POCKET REFERENCE.

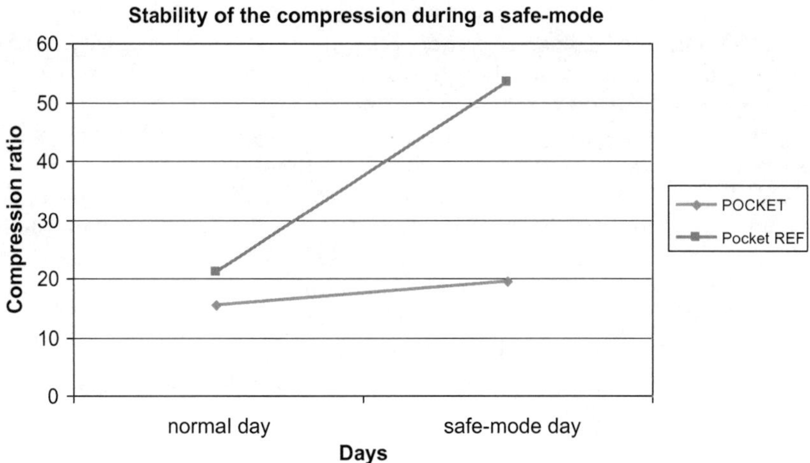

Fig. 14.5   Compression performance comparison of POCKET and POCKET REFERENCE after a safe-mode entry.

updating the mask packet. On the other hand, when a similar experiment was performed with POCKET in reference mode, significant degradation was seen (21% to 36%). As reference packets are likely to only be a small part of the data stream the impact would be limited.

We also wanted to test if the algorithm performance was stable when a major change occurred. The following experiment was performed. Mask packets and inversion/correlation instructions were created on historical telemetry, stored during a day when Venus Express behaved nominally. That configuration was then used to compress telemetry packets produced on a day when the satellite entered safe mode. The results are given in Fig. 14.5 and they show that POCKET performance remains remarkably stable even after safe mode entry, only decreasing from 16% to 19%. On the other hand, when POCKET was run in reference mode then a significant degradation was seen (21% to 52%). This shows that when a spacecraft enters safe mode many status values will change but the dynamic range of most parameters remains similar.

X.   ONBOARD SOFTWARE TESTING RESULTS

POCKET compression has been selected as the baseline compression technique for compressing the formation flying data on PROBA-3. The project is interested in increasing the information content in the housekeeping data related to the sensors and actuators used for formation flying. This is essential because this is a technology demonstration mission and so the more information one can gather during the formation flying experiments the higher value those experiments will have. The project is presently in Phase B. As part of the preparations POCKET was tested on the PROBA-3 target hardware: a LEON-2 50 MHz processor. It could easily compress the 6680 bytes/s of formation flying measurement data, with CPU usage varying between 3.5% and 5.5%. The compression performance achieved with this sort of formation flying data was between 20% and 25%.

ESA has also started to investigate the use of POCKET in other areas. An industrial study investigating its use on telecommunications satellites (which are typically real-time control only) started in March 2013 and another one targeting remote robotic operation systems is planned for later in the year.

XI. ADAPTIVE POCKET

Even though the results in Figs. 14.4 and 14.5 show that POCKET performance is remarkably stable over time, the algorithm relies on having a certain amount of historical data on the ground. Typically, we used a single day of historical data to create the configuration files used for the compression tests. We decided to investigate how we could change the algorithm so that it could rapidly adapt if the behavior of a bit in a packet changed unexpectedly, without needing to wait for ground intervention. We called this new algorithm "Adaptive POCKET."

Simply moving the historical analysis algorithms performed on the ground onto the spacecraft is impractical. There is simply too little onboard memory and processing power available to run those algorithms in near real time. After a prolonged period of testing we selected the algorithm detailed in Fig. 14.6. It

---

Adaptive Algorithm –   mask packet *maskPacket,* temporary mask packet used to build the final mask *buildNewMaskPacket, number of* packets needed before a new mask is generated *threshold,* packet counter initialized to 0 *counter,* new packet generated *newPacket.*

```
function processPackets()

    If isNewPacketGenerated Then
        counter = counter + 1;
        BitwiseOR(newPacket, buildingNewMaskPacket);
        compressedPacket = POCKET_Compress(newPacket, maskPacket);
        SendToGround(compressedPacket);

        If counter == threshold Then
            counter = 0;
            maskPacket = buildNewMaskPacket;
            SendToGround(maskPacket);
        Endif

    Endif

End processPackets
```

Fig. 14.6   The adaptive algorithm triggered when a new packet is generated.

TABLE 14.2   COMPRESSION PERFORMANCE COMPARISON BY MISSION USING ADAPTIVE POCKET COMPARED TO BASIC COMPRESSION

Mission name	Adaptive compression (% of original)	Of which mask overhead (% of original)	Basic compression (% of original)
Venus Express	28.75	12.50	23.55
Rosetta	24.38	12.50	18.09
Goce	49.79	12.50	53.77
Herschel	34.77	12.50	32.91

has the advantage that it does not need much extra memory, requiring only three packet length buffers for each packet type (the last packet generated, the current mask, and the next mask). We also believed it should be relatively fast as it only uses bitwise operations, is relatively simple, and the processing is spread (the next mask is built up gradually).

The immediate problem with this new algorithm is that it is not robust to a single packet loss. In theory the information about the contents of the new mask can be derived on the ground as it can perform exactly the same algorithm on the packets received. However, if a single packet is lost then the chain would be broken and it would be impossible to decode subsequent packets. Hence, it is sensible to introduce resets in the chain at certain intervals to mitigate this risk. As chain resets, we initially chose to simply inject the new mask packet into the downlink stream as soon as one was generated. However, it should be noted that the processes of mask generation and chain reset could be asynchronous. Of course this implied an extra overhead to transmit this information. For instance, sending the mask packet every two compressed packets means incurring an overhead of 50% in terms of compression performance. This meant that the selection of the counter threshold that triggers a new mask packet to be generated and sent became an important system parameter.

We also discovered that the threshold parameter affected compression performance in another manner. Tests showed that calculating the new mask too infrequently resulted in a loss of compression performance for the compressed packets themselves. This occurs because even bits that changed state infrequently were being masked for long periods. On the other hand, calculating the new mask too frequently meant that the overhead associated with sending the new mask information to the ground was too high. After exhaustive testing we settled on a value for the counter threshold of eight packets. The results are given in Table 14.2. The second column shows the overhead due to the uncompressed mask packets that are sent every eight packets, and the comparison with the basic algorithm presented in Sec. V is shown in the third column

XII. IMPROVING ADAPTIVE POCKET PERFORMANCE

With the mask packet being sent every eight packets it introduced an overhead of 12.5%. As Table 14.2 shows, in most cases this meant that the adaptive algorithm was usually performing worse than the basic algorithm. We therefore looked for ways to improve this. When examining the mask packets we found that unpredictable bits tended to occur next to each other. This is due to a combination of the nature of housekeeping telemetry and the encoding schemes used. In housekeeping telemetry the actual variance of parameter values is usually small compared to the whole dynamic range that the encoding allows. This means that bits representing small values tend to be much more unpredictable than those that represent large ones. As most encoding schemes arrange bits in either ascending or descending order according to their significance, it means that these unpredictable bits tend to occur next to each other.

We realized this property could be easily exploited if we performed an internal XOR on the mask packet. Each of the bits in the mask was XORed with its preceding bit. For example, a sequence such as "001111100" becomes "001000010" when it is internally XORed. The result was a packet with long runs of zeros, enabling us to reuse the POCKET algorithm, which generates run length counters to compress this mask packet as well. Using this technique the mask packets compressed to approximately 25% of their original size on average. This process significantly reduced the overhead of sending the mask packets, as shown in Table 14.3, making the overall performance results better than the basic algorithm. The second column shows the overhead due to the compressed mask packets that are sent every eight packets. Mask packets are compressed and their overhead goes down from the 12.5% shown in Table 14.2 to the values shown in Table 14.3.

XIII. ADAPTIVE POCKET RESULTS

An identical set of tests was then run using the adaptive algorithm (threshold set to eight packets, masked compressed and sent) so that a fair comparison could be

TABLE 14.3   COMPRESSION PERFORMANCE COMPARISON BY MISSION USING ADAPTIVE POCKET WITH COMPRESSION OF THE MASKS

Mission name	Adaptive compression (% of original)	Of which mask overhead (% of original)	Basic compression (% of original)
Venus Express	19.10	2.85	23.55
Rosetta	14.26	2.38	18.09
Goce	41.15	3.86	53.77
Herschel	25.71	3.44	32.91

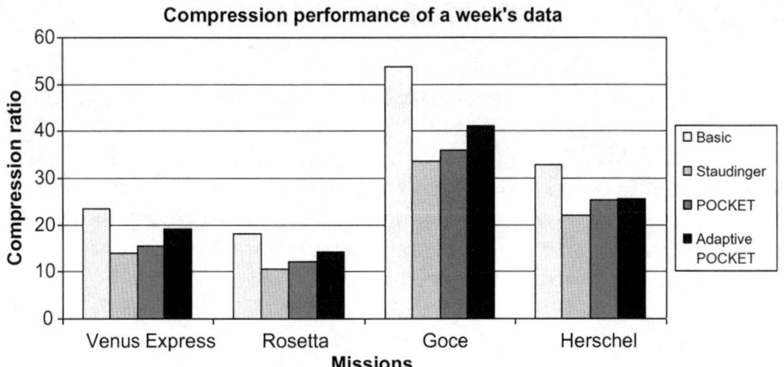

Fig. 14.7   Compression performance of the Adaptive POCKET method illustrated in Secs. XI and XII compared with three other methods.

made between it and the ground-based variant. The Adaptive POCKET algorithm results are described in Figs. 14.7–14.11.

In Fig. 14.7 we can see that for Venus Express, Rosetta, and GOCE there is a significant drop in compression performance when we compare Adaptive POCKET and ground-based POCKET. Analysis shows that the majority of the drop is caused by the overhead of having to send the extra mask packets. This implies that actual packet compression is not benefiting from the ability to rapidly adapt. In the case of Herschel there seems to be some advantage in adaption but this is canceled out by the mask packet overhead.

On the other hand, in Fig. 14.8 we have a significant advantage when using the adaptive variant if POCKET is run in REFERENCE mode. In this case the

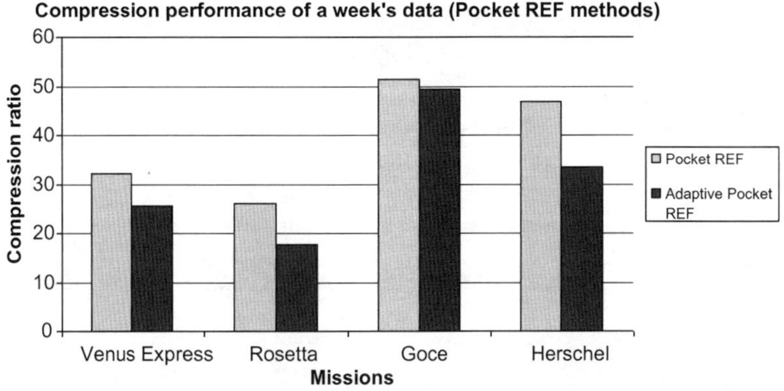

Fig. 14.8   Compression performance comparison of POCKET REFERENCE and Adaptive POCKET REFERENCE.

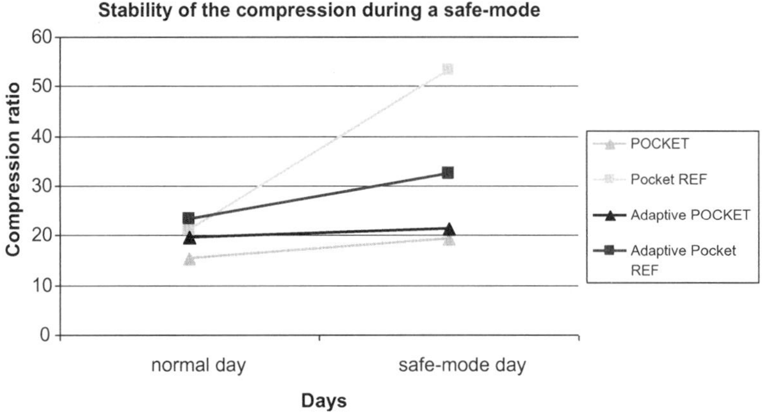

Fig. 14.9   Compression performance comparison of the methods after a safe-mode entry.

overhead of sending the new mask packets is more than compensated by the better performance obtained by the adaptive solution.

In Fig. 14.9 we see that these results are repeated. The figure shows clearly that the compression ratio of Adaptive Pocket REFERENCE remains more stable during the safe-mode period. POCKET REFERENCE alone shows a substantial performance degradation, not being able to mask the status changes in the telemetry. Although adaption brings no advantage for POCKET when this spacecraft enters safe mode, there is significant advantage if POCKET is run in REFERENCE mode.

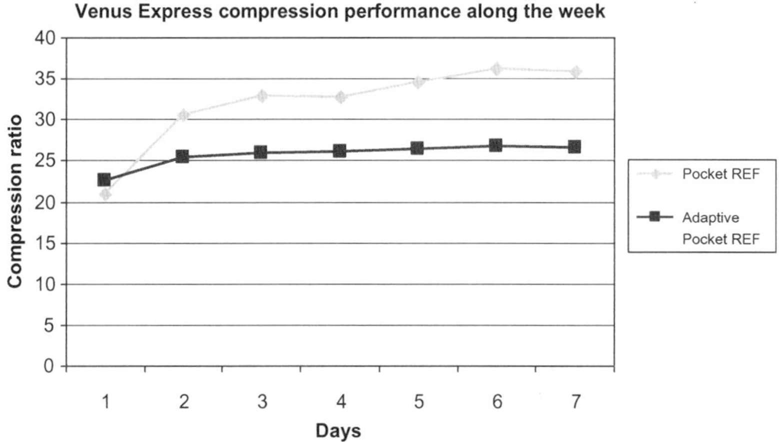

Fig. 14.10   Stability comparison between POCKET REFERENCE and Adaptive POCKET REFERENCE.

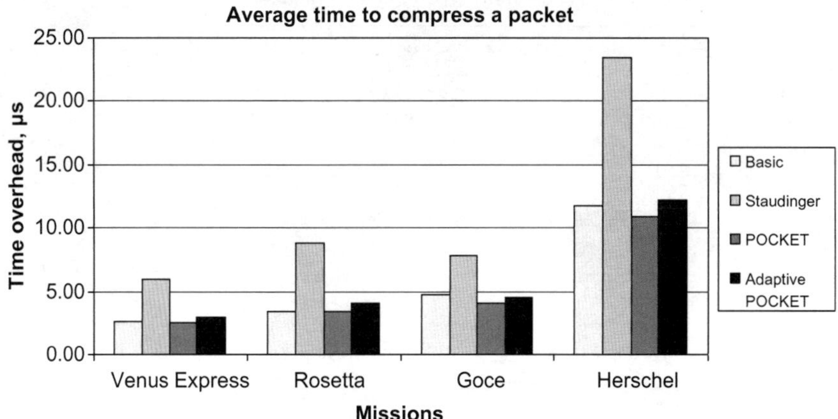

Fig. 14.11   Speed performance of the Adaptive POCKET method illustrated in Secs. XI and XII compared with three other methods.

In Fig. 14.10 we see that there is a significant improvement in the long-term stability of POCKET when the adaptive variant is run in REFERENCE mode.

We see in Fig. 14.11 that updating POCKET to be adaptive does not slow the algorithm significantly. This is because it is relatively simple and mainly uses bitwise operations that can be executed at the processor word level.

XIV.   CONCLUSION

We have presented a new algorithm for compressing spacecraft housekeeping data called POCKET. This is capable of compressing packets as they are generated. It shows very good performance when tested on real housekeeping data from various ESA missions. It is very fast because it relies mainly on bit operations that are carried out at the data word level. In all cases considered so far, this will be fast enough to compress the real-time data stream. This ability means that the numerous advantages of housekeeping compression can be applied to both the real-time and playback telemetry streams, including significantly increasing the information content of housekeeping telemetry. This can be achieved with a single algorithm and with little change to the present operation concepts or onboard telemetry storage systems. It also opens up new markets such as the control of telecommunications satellites, real-time industrial processes, robotics, and telepresence.

We have also shown how this algorithm can be updated to reduce its reliance on prior ground processing of historical data. This adaptive POCKET algorithm can rapidly adjust to changing behavior in the data without the need for ground intervention. In fact, it can begin to compress a completely new packet type efficiently after only eight packets have been generated. The performance of this adaptive variant has been found to be slightly worse in three mission cases and a similar performance was recorded in one more. It is remarkably fast and no significant slowdown is seen in the algorithm when it is implemented. It also brings significant performance and stability advantages when POCKET is based on state prediction, that is, in REFERENCE mode. This implies that adaptive POCKET should be considered if large differences in the behaviors of spacecraft parameters are expected or the scheme chosen uses a large component of POCKET REFERENCE packets.

ACKNOWLEDGMENTS

The authors would like to thank Tomas Laroche and Spacebel of Belgium for conducting testing on the LEON 2 processor.

REFERENCES

[1] Martinez-Heras, J. A., Evans, D., and Timm, R., "Housekeeping Telemetry Compression: When, how and why bother?", *International Conference on Advances in Satellite and Space Communications (SPACOMM 2009)*, July 20–25, 2009, IEEE, pp. 35–40.

[2] Evans, D., Martinez-Heras, J. A., and Timm, R., "Housekeeping Data: Can You Afford Not to Compress It?", *SpaceOps* 2010, April 26–30, 2010, AIAA-2010-2208.

[3] Evans, D., Martinez-Heras, J. A., and Timm, R., "An Idea from the World of Housekeeping Data Compression", *2nd International World Symposium on Onboard Payload Data Compression (OBPDC)*, October 28–29, 2010. ISBN: 978-90-815839-1-6, published by Congrex Holland/ESA Conference Bureau.

[4] CCSDS 120.0-G-2, *Lossless Data Compression, Green Book*, The Consultative Committee for Space Data Systems, December 2006.

[5] Staudinger, P., Hershey, J., Grabb, M., Joshi, N., Ross, F., and Nowak, T., "Lossless compression for archiving satellite telemetry data", *Aerospace Conference Proceedings 2000*, Vol. 2, IEEE, 2000, pp. 299–304.

CHAPTER 15

CNES Ground Network Renewal: Challenges to Increase Capacity and to Reduce Costs

Jean-Marc Soula
Senior Advisor, Ground Network Operations, CNES, Toulouse, France

Héléne Ruiz
CORMORAN Project Manager, CNES, Toulouse, France

Marc Palin
Ground Network System Engineer, CNES, Toulouse, France

Fabienne Vincent-Franc
IDEFIX Project Manager, CNES, Toulouse, France

Michel Recoules
CADOR Project Manager, CNES, Toulouse, France

Isabelle Hernandez-Marcellin
SLE Deployment Project Manager, CNES, Toulouse, France

I. INTRODUCTION

The Centre National d'Etudes Spatiales (CNES, The French National Space Agency) has been operating an S-band Ground Stations Network (GSN) for more than 27 years. This is a multimission asset that provides tracking, telemetry, and command (TT&C) services to all CNES satellite missions (currently 17 in-flight satellites) during their launch and early orbit phase (LEOP), their mission phase, and their end of life operations (EOL).

On occasion, the spare capacity of the network may be used to provide the same services to satellites of other space agencies in the framework of cross-support, or to launchers on a noninterference basis. Also, the provisions of support to satellites of other organizations is possible, always for short-duration operations and under special conditions, with the main rule being that CNES will not compete with commercial providers of same the services in open calls for tender. The exceptions to these rules are when the risks for a first mission must preferably be assumed by an Agency, when support is requested under an intergovernment umbrella, or when the CNES assets are the only ones capable of providing the required services (e.g., because of technical features or the geographical locations of the antennas).

In this context, the main challenges for CNES are to guarantee the availability of the required services to all users and to provide high-quality services at a reasonable cost. This must be true whether the users are CNES missions or external missions. It must also remain true during evolution of the facilities, both for refurbishment to fix obsolescence issues or when upgrades are required to establish compatibility with new satellite missions that rely on new standards or need enhanced performances.

This chapter will explain how CNES has addressed these challenges in the frame of the CORMORAN project and will describe the expectations for the associated developments. A quick overview of the CNES multimission GSN architecture and components will be provided first, followed by a description of the evolution cycle of the GSN with a definition of the criteria used by CNES to select between refurbishment, replacement, or termination of the assets, balancing each decision with the possibility to purchase services from commercial providers of station services. Application to the present situation and ongoing developments will then be described.

II. CNES GROUND STATION NETWORK

The fact that a significant number of the station sites are on French soil and that the core network is operated under the sole responsibility and control of CNES is one main strategic requirement in the definition of the GSN, with a view to guaranteeing access to the national satellites under any circumstances. The CNES network is therefore organized around central facilities located in Toulouse, France, with ground stations distributed around the world, most on French territories and some in foreign countries (Fig. 15.1).

The network central facilities assume both the role of a communication node (providing an access point to users or to external providers) and the role of Network Management Center, and comprise the following elements:

1. *The Central Scheduling Office* (Bureau Central de Planification, BCP). This is in charge of the establishment of the station utilization plan and of conflict resolution.

2. *The Network Operations Center* (NOC). This is in charge of controlling data transfers (telemetry, command, tracking, orbit, various files) and of supervising real-time operations (activation of configuration instances, execution of recovery procedures, activation of on-call maintenance services).

3. *The Orbit Computation Center* (OCC). This is in charge of orbit calculation and of the generation of predicts (station acquisition data, pointing elements, visibilities, interferences, collision risk assessment).

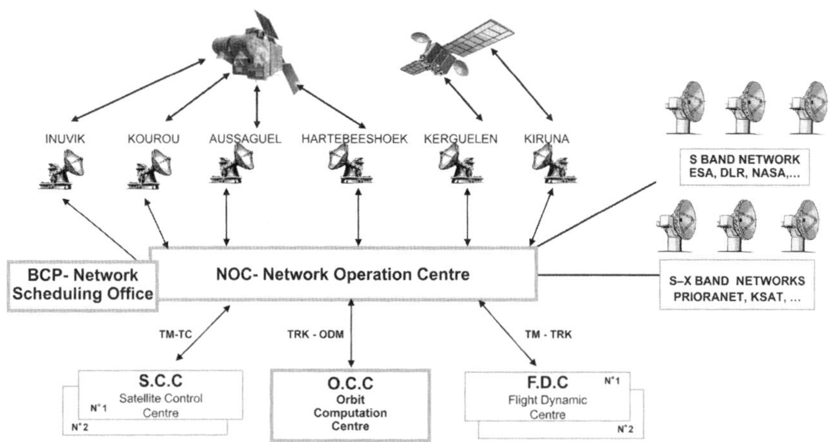

Fig. 15.1   CNES Ground Station Network architecture.

The BCP is manned only in normal working hours as the scheduling activities are executed offline and the network controllers in the NOC have the possibility to modify the station utilization plan to answer short-notice requests from users. The establishment of the station utilization plan is made four weeks in advance by means of the Outil Central de Planification (OCP) system, which is based on constraint solver software [1]. Requirements from the projects in terms of contact duration and distribution are combined with station constraints (masks, maintenance timeframes, etc.) to establish the plan that satisfies all users. In the last two weeks preceding the support, only manual modifications of the program can be made to cope with late support requests or station unavailability.

The NOC is manned 24/7, as the supervision and control of operations require the presence of a Network Controller. Operations on the data transmission systems, ICARE, and some Protocol Gateways [2] are still manual, and under nominal conditions they mainly consist in the management of support instances and in the activation of the corresponding support configurations. Only the monitoring and control (M&C) of the ground stations is driven by an automated process based on the REGATES system, by which the station activities are executed in a timely manner according to station scheduling and the pass plan files (Sequence of Events of stations activities) provided by the users.

The OCC is manned only in normal working hours, and all calculations and generation of predicts are made by the flight dynamics automated software. The presence of the OCC engineers is mainly to expertise the execution of the operations and to conduct non routine analysis or activities. For what regards the station operations, the visibility predicts are sent to the OCP and the station acquisition data to the station antenna control units, on a daily basis.

The CNES GSN provides worldwide coverage for satellites in polar, inclined, or equatorial orbits, as shown in Fig. 15.2. The ground station sites are in the following locations:

1. Aussaguel, near Toulouse, France (four S-band TT&C and two X-band receive antennas);

2. Kourou, French Guyana, South America (one S-band TT&C antenna);

3. Kerguelen, French Overseas Territory, South Indian Ocean (one S-band TT&C antenna);

4. Hartebeeshoek, South Africa [two S-band TT&C antennas through agreements with the site owner, SANSA (South African Space Agency)];

5. Kiruna, Sweden [one S-band TT&C and X-band receive antenna and several others through agreements with the site owner, SSC (Swedish Space Corporation)];

6. Inuvik, North West Territories, Canada (one S-band TT&C and X-band receive antenna and another one as back up, through agreement with the site co-owner, SSC);

7. Station Simulator in the CNES Toulouse Space Center (facility with the same equipment as in the stations, except the antennas, to perform radio-frequency (RF) compatibility tests and other test, training, and maintenance activities).

The last two stations are very new and will start operations in 2012, but the TT&C antennas in Aussaguel, Kourou, and Hartebeeshoek are getting old

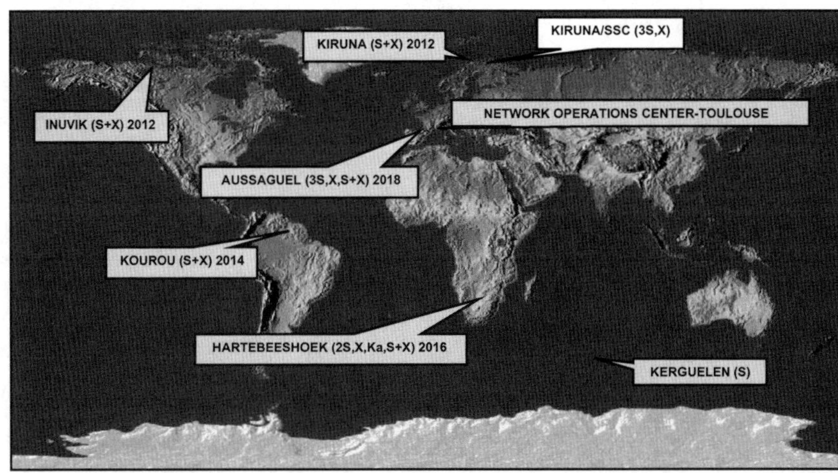

Fig. 15.2 CNES station sites.

(especially Hartebeeshoek), and maintenance and procurement of spare parts are becoming increasingly difficult.

To facilitate operations and maintenance of the ground stations, it is CNES policy to have the same equipment deployed in all stations. This is applied in all stations for processing and communication, or other back-end equipment. Only the antennas are not identical.

Routine operations are automated and driven by the REGATES M&C system [3]. This system also authorizes remote control of station operations from the NOC or locally from the station. These alternate modes on a station may run concurrent with the automated mode on the rest of the network; they are mainly used for maintenance, tests, or special operations phases for which the support instances cannot be automated and the coordination is still via voice. When conducted from the station, the operations are made by the station technicians, as no station controller team has been maintained on any station site since the automation of station operations in 2005.

To date, the loading of the network with routine operations has been on the order of 80–85% of its capacity (average over years). The remaining capacity provides security to the satellites in the case of emergency and may remain unused for some years. Also, it is used to perform LEOPs, EOLs, or other types of exceptional operations, both for CNES satellites or in the framework of cross-support to partners.

A maintenance plan is executed in the stations to guarantee the availability of the ground stations. This involves preventive, corrective, and evolution maintenance and, on average, leads to between two and three hours of station unavailability per day. Because such actions require the presence of the station technicians, maintenance activity is performed during working days and normal working hours; in some cases this means the unavailability of a station for maintenance coincides with satellite visibilities.

To cope with the risk of unavailability of the ground stations or to augment network capacity where there are peaks of activity, agreements are in place with other space agencies and with commercial providers of station services to complement the core network and to obtain external station support, usually for short-duration phases such as LEOPs, emergencies, antenna backup, and EOL operations.

III. EVOLUTION CYCLE OF THE CNES GSN

The guidelines for the evolution of ground stations are provided by an Infrastructure Roadmap in the CNES system for preparation for the future. This level of roadmap uses mission profiles and high-level requirements from the Thematic Roadmaps (Science, Earth Observation, Defense, etc.) and provides inputs to the Technical Roadmaps, which will elaborate more detailed plans, in particular in the domains of Research and Technology (R&T) and the standardization actions to be conducted. The current version of the Stations Roadmap is four years old and will be revisited in 2012 for updates and additions. It was established concurrently and consistently with other infrastructure roadmaps such as those for command and control ground segments; mission ground segments; TT&C ground-to-satellite interfaces; and payload data satellite-to-ground interfaces.

This set of roadmaps is the basis for the definition of the components of the future Space Segments of CNES, concretized in the Myriade Evolution (microsatellite platforms up to 400 kg) and the Initiative for Space Innovative Standards (ISIS; modular platforms above 400 kg) projects: it is expected that a majority of CNES future satellite missions will be based on either one of these Space System concepts and that they may share components, in particular in relation to ground architecture and ground stations.

Examples of guidelines of interest for this chapter are the utilization of the Consultative Committee for Space Data Systems (CCSDS) standards and the space link extension (SLE) for ground-to-ground exchanges, or the utilization of multiple channels in the X-band associated with bandwidth-efficient modulations for payload telemetry data dumps.

Independent of the roadmaps, the actual evolutions of the CNES GSN are triggered by either the obsolescence of some major components of the system or the need to upgrade the systems to answer requirements from decided satellite projects that cannot be covered with the existing features of the GSN assets.

Every two to four years, on average, a complete overview of the situation is therefore made by the management of the ground network, to decide the framework and the objectives of the next evolutions, and to establish the programmatic conditions for such developments. The following criteria are addressed:

1. The status of obsolescence of the systems in the network;

2. Technical requirements from future projects that cannot be satisfied with existing assets;

3. The status of network capacity with respect to evolution of the workload, as a total throughout the network, but also for each site;

4. Application of the guidelines from the Stations Roadmap;

5. Potential options to reduce the running costs of the network and to provide services to the users at a "competitive" cost.

In each development, lessons-learned from previous implementations and subsequent operations are also taken into account. This aspect will not be developed here as it has been presented in a previous *SpaceOps* conference [4].

A. OBSOLESCENCE OF THE SYSTEMS

Today, when considering that the CNES network is about 26 years old, it is obvious that most of the original systems have already been replaced once or twice, and are therefore still relatively recent. Based on experience, the lifetimes of electronic and software components will not exceed 8–12 years, after which modifications and evolutions become difficult due to the technologies becoming obsolete.

A main question mark remains regarding the antennas, which, according to studies on their aging and considering the trend of their daily duty cycle (20–35 supports per day, depending the antenna), are not expected to have lifetimes exceeding 30–35 years, assuming the execution of a strict sustaining maintenance plan.

B. TECHNICAL REQUIREMENTS

As part of new features required by projects, the following cases may justify a major evolution: the need for extended geographical coverage; the need for different frequency bands; and the need for new services or technical features. These were the main technical drivers for the recent deployment of the two new antennas in Kiruna and Inuvik, as extended coverage of the polar regions was required and an X-band telemetry reception service was needed. The polar station project was conducted with the SSC, in a partnership agreement covering both the development and utilization phases of the stations, according to which costs and capacity are shared in a coordinated way [5].

C. APPLICATION OF THE ROADMAP

In this context, the station roadmap promotes the current multimission architecture, the deployment of S + X band antennas and the utilization of commercial off-the-shelf (COTS) components and standards that are compatible with the future CNES space systems. Through the utilization of standards, it is also aimed to achieve compatibility with other space agencies.

The last project related to evolutions of the GSN – HOMERE—was conducted in 2007–2010 and was driven by the Stations Roadmap. It had federated several specific ground stations into the multimission network, centralizing their scheduling (the OCP system) and communications management. This project was also the occasion of the initial deployment of the SLE in the CNES metwork.

D. EVOLUTION OF NETWORK CAPACITY

Ideally, the network capacity should remain as close as possible to the cumulated total requirements of the supported missions. This being said, it is obvious that some margins must exist so that special operations, quick reconfigurations, or emergency situations may be addressed efficiently. For example, in the network availability scheme, it is said that in the case of failure of a station to support one pass, the same or another station will be made available to provide a replacement pass, within the next revolution for a LEO satellite. To satisfy this type of requirement, the CNES network relies on its own internal margins and on the spare capacity of the external networks in support.

Over the years, it has become clear that spare capacity is desirable in each site, but most of the time it is not homogeneously distributed throughout the network. To make sure of the situation in the future, simulations are required to evaluate the global margins and the margins per station, with evolution of the network mission model and distribution of the workload. CNES makes use of the same scheduling software as that based on the OCP to evaluate loading feasibility and distribution.

For CNES, such simulations are conducted as often as necessary to analyze the changes in the mission models. Such changes may relate to schedule changes of new missions, to changing requirements, or to some events with in-flight satellites (lifetime extension, failure, planned end of life, etc.). The conclusions of these loading simulations may be quite different depending the analyzed periods. For instance, the following findings may occur:

1. The capacity of the stations and network is sufficient with reasonable margins.

2. Capacity is lacking in the network, on a short- or long-term basis, globally or on a site.

3. There is excess capacity in the network, on a short- or long-term basis, globally or on a site.

Of course, the potential consequences and decisions may be very different from case to case.

Before any significant investment is made on the network, it is the rule in CNES to verify for the long term that the subsequent fluctuations in workload will not justify the opposite decisions for the very assets on which investments are to be made. In such cases, loading for the global network and for each station is the criterion, as it will justify the sustainability of the asset. This analysis also allows making sure that the return on investment will be satisfied long in advance of any period of uncertainty.

E.　RUNNING COSTS AND SERVICE USAGE COSTS

The running cost of a GSN may be assimilated to a flat rate, whatever the actual loading of the stations, as long as the architecture is stable and the operating modes remain unchanged. When purchasing station services, the price for such services is proportional to the quantities. The drawback is that the cost of station services just increases if the satellite missions require more passes. The advantage is that if your requirements diminish, the costs will decrease and remain adapted to your needs.

In this context, a comparison between the two options, to own an asset or to purchase services, if made at the level of a single project, generally concludes on the side of purchasing services, as the procurement of totally new assets is too expensive. The same comparison may reach the opposite conclusion in the situation of an existing network with capacity margins and a very high workload. This is the case for the CNES network, in spite of the number of ground stations, as the current workload with 17 satellites and some more to come justifies the choice of a CNES-owned network.

Nevertheless, the situation may change with time and events, and, of course, it must be analyzed with each new project with respect to the available capacity of the network:

1. A new satellite mission for which requirements may be accommodated within the available capacity of the network may very well be at no extra cost for CNES; this has been the situation in the last ten years, with new satellites replacing, in an almost consistent schedule, those that were de-orbited.

2. By contrast, an increase in mission requirements beyond the existing capacity may result in the need for additional stations, either to be procured if the total requirements are stable in the long term, or in the need to purchase external services if the need is temporary. In either case, the running cost of the network increases according to its capacity. This is the case CNES anticipated when making the decision regarding the new polar stations in 2008.

3. Finally, a drop in requirements from projects may lead to a situation where capacity is unused and running costs are no longer justified. In this case, capacity must be reduced and some stations must be placed in dormant mode (survival maintenance only and reduced or no operations) or finally phased out. This is the situation CNES has faced with its Ku-band stations in Aussaguel and Kourou, which were eventually phased out recently after several years with minimal maintenance.

The close relationship between network capacity and cost for rendered services was translated into the definition of the "network hourly rate," according to which CNES projects are charged for their share of the utilization of the multimission network. This hourly rate (HR) is computed every year using the following expression:

$$HR = E/C$$

where E represents all expenses made to run the network: external costs for maintenance and operations, manpower and share of the technical means used for the network activities (computers and networks, communication lines, quality supports, etc.). It must be pointed out here that the cost of all communication lines between the NOC and the stations is accounted for in the calculation of E. C represents the network capacity.

The network capacity that may actually be used by the satellite missions takes into account structural unavailability and limitations. It results from the addition of the individual multimission station capacities weighted with a network limitation factor, as follows:

$$C = (C1 + C2 + \cdots C6) \times t$$

where t is the limitation factor of the NOC, which can only support four simultaneous passes with one network controller shift (the situtation today), whereas the multimission network is fitted with six main antennas, and each station has an average daily capacity of:

$$Ci = \mu * p * 24\,h$$

where μ is the remaining availability of stations not in testing or maintenance activities (currently 90%), and p is the remaining availability of the stations considering the lost time between passes that cannot be used due to too short duration slots or low probability of a satellite pass at that time (currently 80%).

This calculation is typical of the CNES mission profile, which comprises mainly LEO satellites. However, for the sake of simplicity, the same rate is used when, on occasion, the network is used for medium-Earth orbit (MEO) or Geostationary Earth Orbit (GEO) satellites.

The predicted network hourly rate is a perfect indicator that can be used to measure the efficiency of the investments made by CNES with its GSN evolutions and to anticipate the effect of such evolutions on the total costs of the CNES projects.

IV.   RATIONALE FOR THE CORMORAN PROJECT AND ASSIGNED OBJECTIVES

CORMORAN is a French acronym that stands for "COnsolidation et Renouvellement des MOyens Réseau et des ANtennes" (consolidation and renewal of the network and the antennas). This project was the result of a new iteration on the evolution cycle of the network conducted in 2009–2010. The rationale for the evolutions was this time a combination of several factors, as discussed in this section.

A.   SATELLITE PROJECTS' TECHNICAL REQUIREMENTS

Considering the projects in the CNES mission model for the upcoming 10–15 years and also the trends coming out of the CNES roadmaps and the Phase 0 studies, it was confirmed that the major evolution required for the CNES GSN is the availability of X-band data reception in low- or mid-latitude sites. As the recommendations of the CNES Stations Roadmap in this domain had been recently applied in the specifications of the polar stations, the CORMORAN project was tasked to align to these recommendations and specifications. Among other characteristics, this includes the following:

1.   Large dishes, enabling very high data rates;

2.   The possibility of overhead passes with minimum altitude of satellites around 400 km;

3.   Enhanced phase noise features to enable the use of present and future bandwidth-efficient modulations;

4. Polarization diversity in tracking and telemetry;

5. Homogeneous architecture and the same choice of processing equipment for the receive chain in all stations.

On the TT&C side, no critical change was detected in the requirements from the projects.

Two studies were conducted to consolidate the overview of the users' requirements and assess the needs regarding the following:

1. To maintain the capability to provide S-band TT&C supports to geostationary satellites;

2. To track and receive S-band telemetry from the European launchers (Ariane, Soyouz, and Vega).

Both studies had a positive conclusion:

1. There is an obligation to continue the backup services to existing geostationary satellites and, in addition, the geostationary arc is still part of the options studied in CNES Phase 0 projects. Moreover, the link budgets are similar to those required to support MEO satellites (currently Galileo LEOPs) or, under some restrictions, the high elliptical orbit (HEO) satellites.

2. Multimission antenna in Kourou will be used to support all launches from French Guyana, and the one in Aussaguel should continue to support the Soyouz launches for the Galileo series; however, due to the many different standards used by the launchers, it was decided to limit the specifications to the minimum set of features to support such missions (2 GHz couplers to connect a hosted telemetry processing unit, tracking in frequency diversity, real-time pointing of the antenna).

B. EVOLUTION OF THE WORKLOAD

It has been identified that the project requirements will progressively (and in less than seven years) bring the loading of the network to a level that is three to four times higher than the level of 2009, as reported with the simulation shown in Fig. 15.3.

An extension of the network capacity had to be considered. However, it was not clear initially if some stations would be impacted more than others, so loading simulations were conducted. It was concluded that not only polar sites were concerned, but also equatorial and mid-latitude stations. Nevertheless, no new need for additional sites was evidenced; only the need for additional X-band assets on existing sites was established.

As the impact was global over the existing network and it is not reasonable to imagine a solution with multiple CNES antennas on each site, it became clear that, at least in some periods, the utilization of backup antennas on the same sites or the purchase of external supports could be required, more than

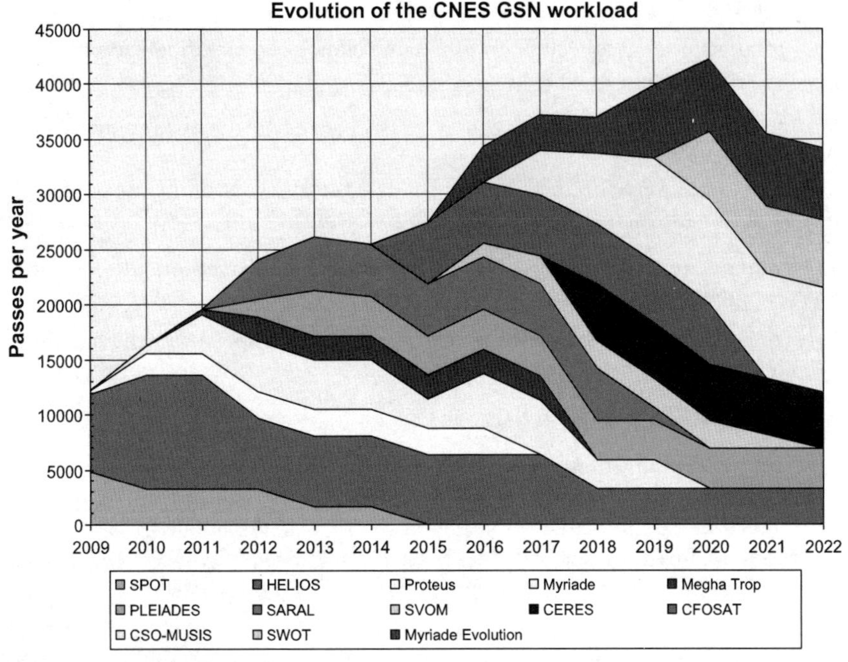

Fig. 15.3   Evolution of the CNES Ground Station Network workload.

today, to find viable solutions. It was verified that the existing agreements for procurement of such services could cover the needs but alternate offers should be evaluated, with the objective of a wider range of solutions and prices.

To minimize the utilization of such solutions, as their cost may be significant, the need to augment the intrinsic capacity of the assets was recognized and was noted as a special objective of the CORMORAN project.

C.   MAJOR OBSOLESCENCE ISSUES

As the requirements from the projects clearly extend the need for the S-band ground stations beyond 2020 and the existing stations will be more than 35 years old at that time, it became clear that with the new satellite project requirements there was an opportunity to make a decision regarding the replacement of the old antennas for which obsolescence issues had been identified. This concerns the multimission antennas in Aussaguel, Kourou, and Hartebeeshoek, as previously indicated, but also one of the two existing X-band stations, the one in Toulouse used for the Parasol satellite, which does not meet the requirements for future projects (small dish) and in addition would need significant investment to guarantee its sustainability. The objective of replacement of these old antennas was assigned to the CORMORAN project, including the constraint of continuity of service to in-flight satellites.

Being less than 20 years old, Kerguelen antenna and other backup antennas on various sites are out of the scope of the CORMORAN project and will be considered in future iteration of the evolution cycle, as it has already been identified that the project requirements on those antennas also extend beyond 2020.

The provider company of the telemetry command and ranging (TCR) processors in the CNES multimission stations, Zodiac Data System, had also announced the termination of the manufacture of their product line, based on the Enertec 3801, and that software evolutions and maintenance would also be terminated, in 2012 and after 2013, respectively (depending the evolution of their stock of spare parts). This is confirmed now. Although the equipment in place covers most of the requirements for present and upcoming satellites, enhanced specifications were engaged with a view to replacing these units in the CNES stations under the CORMORAN project [e.g., Gaussian minimum shift keying (GMSK) for downlink telemetry; quadrature phase shift keying (QPSK) uplink with Doppler compensation; code tanging as per ECSS E50-02].

D.   COST REDUCTIONS

After integration of the new polar stations into the network, a significant increase in running costs was expected, mainly related to expenditures to operate the new stations rather than related to the CNES manpower to drive the operations and maintenance. In the case of the CORMORAN project, the objective assigned to the team was clearly to maintain the global running cost at a level close to that resulting from the previous evolutions. A limited increase was expected as a result of the following combined effects:

1. On the increase side, the augmented bandwidth on communication lines and the deployment of additional systems in the network, in particular for multimission X-band services;

2. On the decrease side, modernization and automation of assets and their operations.

From a combination of the increases in capacity and running costs, it is expected that the network hourly rate would remain stable.

V.   COMPONENTS OF THE CORMORAN PROJECT

Based on the required evolutions of the Network and on the objectives assigned to the project, the following components as shown in Fig. 15.4 were identified for CORMORAN, to be developed in a loosely dependent schedule:

1. Replacement of the existing multimission TT&C antennas in Kourou, Hartebeeshoek, and Aussaguel, in that order, with new TT&C and data reception

antennas on the same sites. The satellite project requirements imply the availability of a mid-latitude capacity for data reception from 2014, and the availability of a complete upgraded set of stations in 2017. The order of development of the new antennas was chosen to also take into account the obsolescence issues and the availability of backup antennas in Hartebeeshoek and Aussaguel.

2. Replacement of the telemetry command and ranging (TCR) processors in all of the CNES multimission stations. The main driver is the obsolescence of the equipment; the detailed schedule of such replacement is not strictly related to project needs. It was decided to start the deployment with the polar stations and to then deploy the new antennas. Some flexibility was left in the deployment schedule for Kerguelen, the backup antennas on various sites, and the test facilities. The project is free to adapt this schedule according to its own needs, for instance to validate the new system in the station simulator facility, or to take any opportunity for deployment, according to the schedule changes, with other components of the project and the manpower resources then becoming available.

3. Development of a multimission ingestion and file distribution system – Idefix – devoted to the X-band data, to be deployed in the stations and in the NOC, as shown in Fig. 15.5.

This new system should first be installed in the polar stations, to replace the current monomission solution provided by SSC, and then in the new antennas

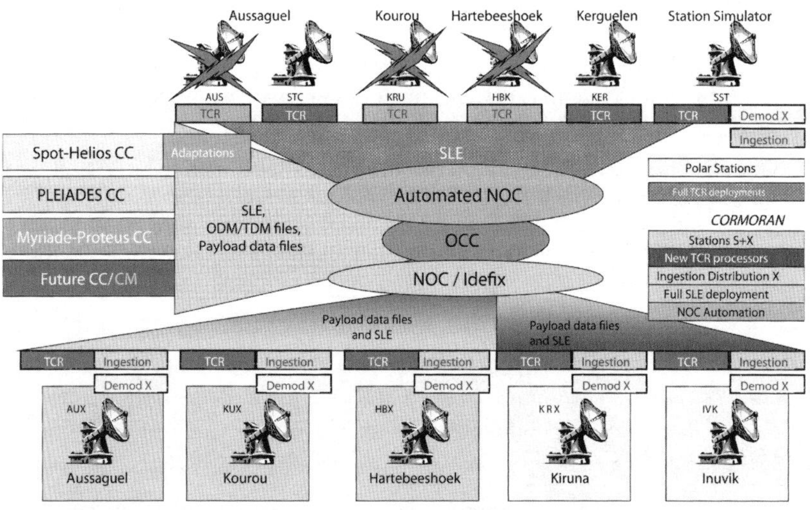

Fig. 15.4   Components of the CORMORAN project.

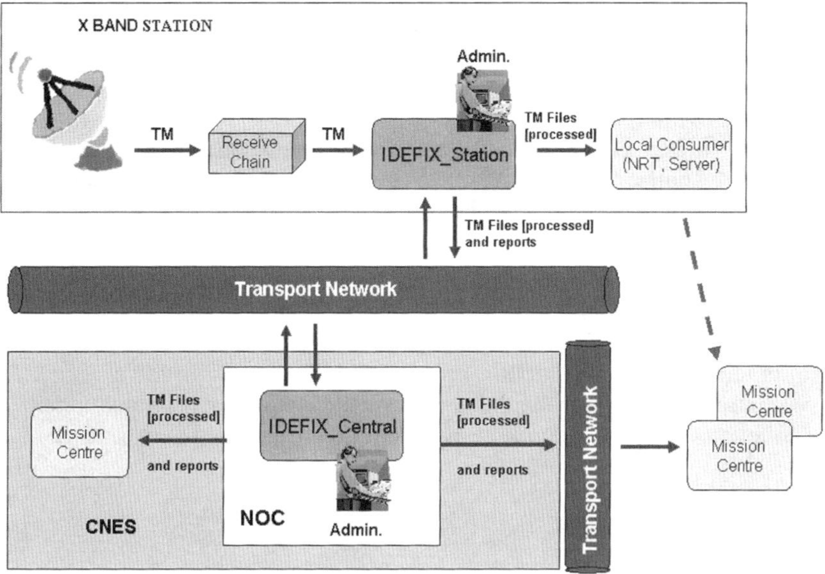

Fig. 15.5   Architecture of the Idefix system.

in a consistent deployment schedule. Among the features of interest in the Idefix system are the following:

(a) Processing of single or multiple channels from the satellite to ground interface, and parallel processing as required.

(b) Processing of frames or packets and creation of files with data sorted per virtual channel (VC) or application process identifier (APID).

(c) Local optimization and elimination, before transmission from the station to the users of the useless data (trash, filler data, duplicated segments, not supported channels, etc., according to configuration options).

(d) Global optimization and elimination, before transmission from the station to the users of the data already received by other stations (according to configuration options).

(e) Reporting and statistics to users.

(f) Temporary storage in the station or in the NOC.

(g) File transfer to a server that may be located in the station (e.g., for near real time application), in the CNES premises in Toulouse or at the user's premises.

(h) Management of security and of priorities in data transfers.

4. Extension of the current automation in the NOC to the management of all communications and data distribution systems. This development is based on the Regates M&C system that already drives the automated operations of the multimission stations. All instances of network services may be executed in parallel under the single control of the Network Operations Automated Control System (CADOR). The system will control all systems and activities in the stations and the NOC.

5. Full deployment of the SLE over the network and of the CCSDS interfaces for exchanges of orbit, pointing, and tracking data between the OCC and users of the network. This evolution had been initiated with the implementation of the CCSDS provider capacity in the frame of the Homere project and will continue under the CORMORAN project with the retrofit of the eldest CNES control centers. Depending the expected lifetime of the supported missions, the solutions in this domain range from the upgrade of the data acquisition software in the control centers to the implementation of protocol gateways between the control centers and the network facilities (NOC or OCC).

VI.   CONTRIBUTION OF EACH COMPONENT TO THE OBJECTIVES

This section provides an overview, for each component, on the choices and evaluations made by the CORMORAN project to comply with both objectives to master the running costs and to increase the intrinsic capacity of the network. Only those choices having a significant impact are mentioned here (if we were going into detail, many more could have been mentioned).

A.   NEW ANTENNAS

1. COST REDUCTIONS

The main factors for a cost reduction in the development of antennas are as follows:

1. The reuse of existing sites, such as infrastructure and site services already in place.

2. Planning for the development of the three antennas, spread over six years, such that there is no peak requirement on budgets or on project team manpower.

3. The decision to have fully identical antennas on the three sites so that there are recurrent costs for the second and third antennas and easier duplication of M&C software from site to site.

4. Integration of the antenna into the existing network architecture, implying the reuse of existing interfaces for communications and M&C interfaces, so

there is no further development, system security is already covered, validation is reduced, operator training is reduced, and so on.

5. The utilization of COTS for the majority of the subsystems: recurring costs already with initial implementations.

Some of the above choices are also expected to have positive effects regarding the running costs of the stations:

1. Identical antennas and systems will reduce maintenance and operation costs.

2. The utilization of COTS will reduce maintenance costs, among the other usual advantages expected from the utilization of COTS.

Finally, some proposals made and features specified by the CORMORAN project should have additional effects on the running costs:

1. The replacement of one antenna in Aussaguel and one antenna in Toulouse with just one new antenna in Aussaguel, thereby reducing maintenance costs.

2. The proposal to phase out the old antenna on each site very soon after the qualification of the new antenna (about a six-month overlap for the first site, less for the next ones), requiring no duplication of maintenance over a long period.

2. CAPACITY INCREASE

Some specifications by the CORMORAN project should have a positive effect on the capacity of the new antennas:

1. As the stations are designed to simultaneously support S-band TT&C and X-band data reception operations, this feature will allow requirements that were previously supported in distinct passes with different antennas to be addressed in just one pass. This provides an increase in capacity, as it allows two requirements to be addressed in the same instance of station service, which in the case of CNES is a real advantage, as many of the satellites to be supported require both services from the network.

2. Points a) and b) are expected to increase station availability out of maintenance activities (μ in the calculation of capacity, Sec. III) and station intrinsic capacity.
   (a) The selection of modern antennas: obviously, these will require less time for normal maintenance and shorter downtimes for repair compared to the existing antennas, including in the case of mechanical parts failures.
   (b) The specification of many measurement points in the station architecture so as to connect test and measures equipment and to enable automated maintenance activities: automation of maintenance is critical with respect to the definition of station capacity, as it enables the

performance of preventive maintenance out of normal working hours and provides enough flexibility to avoid conflicts with operational support to satellites.

3. The time to change configurations and to prepare for the next support is critical in a multimission ground station. The following aspects will improve network performance in this regard and should increase station availability between passes (p in the calculation of capacity, Sec. III) and station intrinsic capacity.

   (a) Fast M&C interfaces of the new equipment: these are expected to provide faster response times because of the technologies used to support them (e.g., Ethernet versus IEEE).

   (b) The equipment is supposed to have a more efficient interface in terms of integration and relevance of the information in the M&C parameters: this should improve response times and security of the operations. One special feature is the capacity of the equipment to memorize a complete support configuration, which may be called back for each satellite with one command instead of individually setting up hundreds of parameters to establish the next support configuration.

B. TCR PROCESSORS

Being station equipment, the TCR processors will bring the same advantages as mentioned above, in particular concerning costs (benefits of the implementation of COTS) and capacity [increase of station availability between passes (see also comments on the deployment of SLE in Sec. VI.E)].

C. IDEFIX

1. COST REDUCTIONS

As the X-band data reception and distribution is a new service added to the legacy TT&C services of the network, the implementation of an additional system means some costs that were not supported in the previous definition of the CNES GSN. However, two characteristics must be mentioned that clearly aim to reduce costs for this new service:

1. Idefix was designed as a multimission system. The objective was to offer a service that may provide support to all satellites without multiplicity of hardware in the station. In many cases, the ingestion and storage equipment in the station was designed as monomission equipment, resulting in many racks of equipment and their associated maintenance costs or ad hoc operations requirements.

2. Other features of Idefix that are expected to reduce the network running costs are the local and global optimization software that will eliminate any data that

do not need to be transferred. This is intended to reduce requirements for communication line bandwidth, on individual mission passes or accumulated over parallel/consecutive mission passes and data transfers, and therefore the associated line costs. Along with the start of operations on new satellites, CNES has a plan to progressively increase the bandwidth from the current 1–2 Mbps, to 30–45 Mbps (polar stations) in the next five years; the rate of this increase will be reduced and mastered using the optimization features of Idefix.

It is difficult to quantify the savings on the communication lines as this will depend on the final strategy and schedule for upgrades of the line bandwidths, but it is estimated that it is worth the investment, in the order of 1 M€ for Idefix, because of the high cost and number of communication lines in the network.

2. CAPACITY INCREASE

Rather than an objective of increasing the capacity, the concern of the COR-MORAN project with the Idefix component was not to reduce the capacity by combining the constraints of the real-time passes and of the data transfers after the passes. The requirements on processing times are very high on the system as satellite passes may be separated by as little as five minutes. Accordingly, the Idefix system has been designed in a multimission approach and with a capacity for the parallel processes of real-time acquisition, near-real-time ingestion, and post-pass data transfer. That way, each satellite pass will be supported independent of the previous and next ones. Of course, the communication line capacity has to provide a sufficient bandwidth to deliver the data with the latency required by the projects.

Based on these specifications of the Idefix system, it is anticipated that the scheduling of the data reception passes will just consider limited pre-pass activities (e.g., antenna pointing elements loading, equipment configuration setup, antenna pointing), pass timeframe from acquisition to loss of signal, and very short post-pass activities (e.g., antenna into standby safe position). No extra time will be added for post-pass data transfers.

D. NOC AUTOMATION

1. COST REDUCTIONS

Regarding the automation of the NOC systems, one has to remember that, today, network operations are conducted by NOC controllers 24/7. The controller on shift is required to support up to four station passes in parallel, controlling the systems in the NOC, monitoring the station automated operations, and fixing any problem. The team is composed of eight controllers, in compliance with French laws for workers on shifts. The increase in network loading would justify having a second controller on shift from 2014, as the number of simultaneous passes will very often reach five or six. Therefore, the automation of the NOC, as it permits continuing network operations with one shift, might correspond to savings on the order of €500,000–600,000 per year. The expense of about 1 M€ on CADOR is justified as the return on investment may be very short.

Of course, automation is also expected to bring reliability and security of operations. This is the case with CADOR as the M&C software will address and fix a number of degraded or failure cases (switchover to backup equipment, configuration changes, reaction to events, etc.). At this stage, it is not planned to rely only on the system. The decision was just to convert the positions of controllers into positions of supervisors of operations in charge of managing major failures or short-notice requirements. At a later stage, unmanned operations could be considered but this will certainly have prerequisites such as CADOR application software validated as fully robust to failures and the improvement of service management interfaces with the users, in particular in the domain of short-notice operations. Additional cost reductions could then be considered.

2. CAPACITY INCREASE

At the same time, automation will cancel the network limitation factor (t in the calculation of capacity, Sec. III), which is effective today in the station scheduling procedures: only four passes are accepted in parallel, although there are more stations (six antennas, not counting the backup antennas). When this limitation is removed this will lead to a major gain in overall network capacity.

E. SLE FULL DEPLOYMENT

1. COST REDUCTIONS

The previous evolution project conducted with the network–Homere–set the grounds for the implementation of the new CCSDS standards for ground-to-ground data exchanges. The objective of Homere was to prepare for the future, as it had been decided that such standards would be compulsory for all new projects requiring services from the network, the first one being Pleiades. However, the retrofits to SLE of the interfaces with the Proteus and Myriade satellite series had been achieved. The objectives were cost reduction in the long term, but also reactivity and efficiency in the implementation of cross-support with partners.

The CORMORAN project will now take into account the retrofit of all remaining control centers. The various solutions that will be implemented may certainly increase the running costs of the control centers wherever new components are to be added (e.g., protocol adaptations or gateways). However, to minimize the procurement and running costs of the additional systems, these developments will be based on existing gateways that will be adapted for purpose. Such cases are limited in number and in duration, because of the remaining lifetimes of the missions (SPOT and HELIOS control centers).

For other control centers and flight dynamics facilities, the retrofits consisted in the implementation of software already used and qualified with other projects. The risks and costs are then minimized. This applies mainly to the orbit and tracking data interfaces to the satellites in the current Proteus and Myriade families.

2. CAPACITY INCREASE

The deployment of SLE throughout the network should be as effective on the increase in capacity as on cost reduction. Today, the TCR processors in the stations still have a dual boot for selecting between the legacy protocol and the SLE protocol; as a consequence, the station setup between passes takes two to three minutes more than it should when consecutive satellite supports do not use the same ground-to-ground protocol. The decision to only operate SLE over the network on the provider side will be beneficial both for the replacement of the TCR equipment (because only one interface is specified) and for the station setup times. This will increase station availability between passes (p in the calculation of capacity, Sec. III) and also station intrinsic capacity.

VII. EXPECTATIONS FROM ONGOING DEVELOPMENTS

As previously mentioned in this chapter, among the objectives assigned to the CORMORAN project are increasing the capacity of the network and limiting the increase in running costs. The indicator to evaluate the success of the project in fulfilling these objectives will be the network hourly rate.

As the project was initiated in 2010–2011, only predictions are available at this stage. Such predictions are based on the following main assumptions on the planning:

1. Full Space Link Extension (SLE) deployment in the first quarter of 2013.

2. Delivery of Idefix and implementation in the polar stations in the second half of 2013. Subsequently, Idefix will be implemented in the new $S + X$ band antennas along with their own development schedule.

3. Delivery of the new Telemetry, Command and Ranging (TCR) equipment in the first quarter of 2013 and progressive implementation in all sites, according to their own schedule for the new $S + X$ antennas and according to needs or resource availability for the other stations. The global targeted deadline for complete implementation is the end of 2015, excepted to occur with the last CORMORAN antenna.

4. Delivery of CADOR and the start of fully automated operations in the NOC, as from the first quarter of 2014.

5. Delivery of the new S + X band antenna in Kourou in the last quarter of 2013
    and start of operations six months later.

6. Delivery of the new S + X band antenna in Hartebeeshoek in the third quarter
    of 2015 and start of operations six months later.

7. Delivery of the new S + X band antenna in Aussaguel in the third quarter of
    2017 and start of operations six months later.

Considering this planning and the plans elaborated by the CORMORAN project,
relative to the capacity of the GSN, a prediction of the evolution of capacity has
been established as shown in Fig. 15.6.

This illustrates the anticipated efficiency of the measures proposed by
the project. However, it must be understood that the actual capacity will need
to be measured throughout the development of the CORMORAN components, to verify the predictions and to initiate any corrective action that may be
required.

Moreover, as the prediction gets close to the mission model in Sec. IV but does
not exceed the figures, this is already leading the CNES network team to consolidate the following actions:

1. Perform network loading simulations, as required with the project schedule
    changes (launch postponements, lifetime extensions, failures) to evaluate
    the resulting model and identify peaks or drops in activities.

2. Extend existing agreements for support from external stations, either as
    backup on the same site or as supplements from other sites.

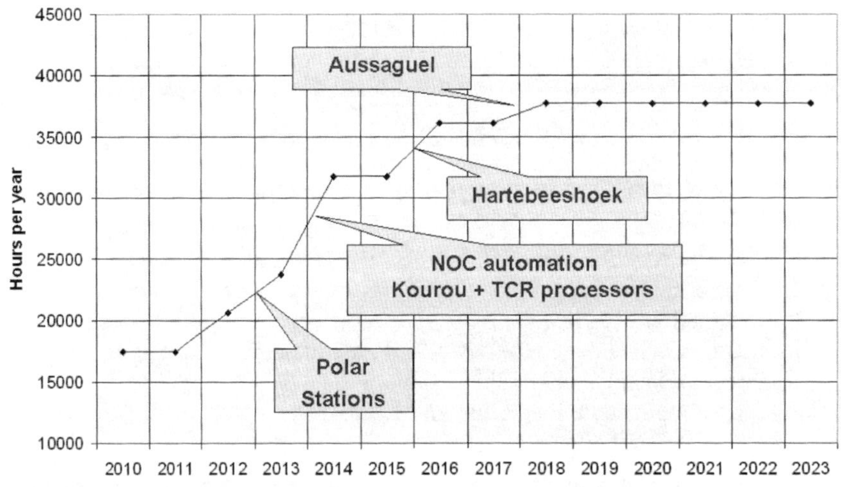

Fig. 15.6   Predicted evolution of CNES network capacity.

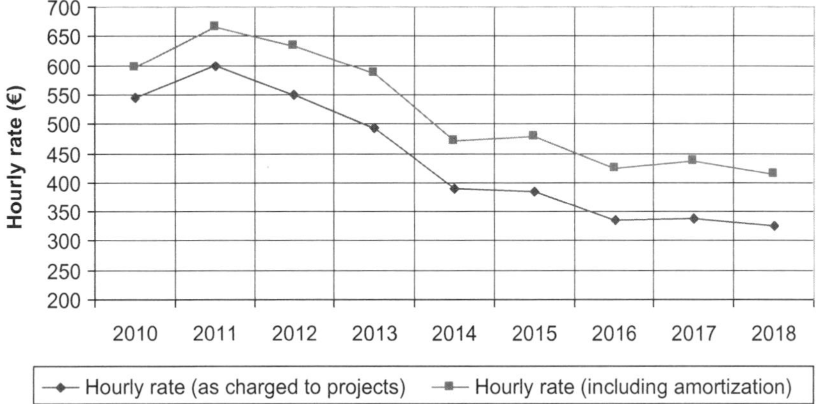

Fig. 15.7   Predicted evolution of the CNES network hourly rate.

3. Establish new agreements with other providers of network services to diversify the options and increase the chances to find solutions at short notice.

4. Consider new iterations and studies on the subsequent evolution of the CNES network to identify the course of appropriate actions to be taken (unlikely before 2014).

A model has also been established for the evolution of the running costs of the network. As a conservative and pessimistic assumption, it was modeled that, after the increase due to the addition of the polar stations, the running costs of the network would remain stable or within $\pm 10\%$ of that level. It is expected that the network expenses will remain stable and the manpower to operate the systems will increase slightly in the initial periods of operation of the new systems, in particular for those temporarily being operated in parallel with the system they are to replace. As a result of this model, the predicted evolution of the network hourly rate is foreseen to follow an evolution as shown in Fig. 15.7.

VIII.   CONCLUSION

The analysis conducted in the frame of the CORMORAN project goes beyond the usual criteria of the management of a development project, essentially based on the performance–schedule–cost triplet, and takes into account, in the project objectives, parameters related to the objectives of the operators of the ground assets, in this case the capacity and running costs of the stations. One advantage of the approach taken for this project is that these objectives are shared between the engineering team developing the systems and the operations team who will provide services while making use of them. The effects of any choice in the design or features of the new system may be evaluated through comparison with the replaced systems or the previous operational concepts. The indicator selected – the network hourly rate – is calculated every year based on the results of the previous year, so it will be easy to measure the actual performance of the CORMORAN project.

ACRONYMS

APID	Application Process Identifier
BCP	Bureau Central de Planification (Central Scheduling Office)
CADOR	Contrôle Automatique Des Opérations Réseau (part of the CORMORAN Project, for the Network Operations Automated Control System in the NOC)
CCSDS	Consultative Committee for Space Data Systems
CNES	Centre National d'Etudes Spatiales (The French National Space Agency).
COTS	Commercial off-the-shelf
DLR	Deutsche Luft und Raumfahrt
ECSS	European Cooperation for Space Standardization
EOL	End of life
ESA	European Space Agency
FDC	Flight Dynamics Center
GMSK	Gaussian minimum shift keying
GN	Ground network
GSN	Ground Station Network
HEO	High elliptical orbit
HOMERE	Harmonisation Operationnelle du Management et des échanges au sein du REseau (a precursor project to CORMORAN to initiate the SLE deployment)
ICARE	Infrastructure de Communications et Applications pour le REseau (the communications management system in the CNES GSN)
IDEFIX	Ingestion et de Diffusion Externe des FIchiers reçus en bande-X (part of the CORMORAN Project, for the definition of the ingestion and distribution of payload telemetry files received in X-band
IEEE	Institute of Electrical and Electronics Engineering
ISIS	Initiative for Space Innovative Standards
KSAT	Kongsberg Satellite Services
LEO	Low Earth orbit
LEOP	Launch and early orbit phase
MEO	Medium Earth orbit
NASA	National Aeronautics and Space Administration
NOC	Network Operations Center
OCC	Orbit Computation Center

OCP	Outil Central de Planification (Scheduling System in the CNES GSN)
QPSK	Quadrature phase shift keying
REGATES	Renouvellement de Gestion, Automatisation et Télé-Exploitation des Stations (Monitoring and Control System of the CNES GSN)
SANSA	South African Space Agency
SCC	Satellite Control Center
SLE	Space link extension
SSC	Swedish Space Corporation
TCR	Telemetry, command, and ranging
TT&C	Tracking, telemetry, and command
VC	Virtual channel

GLOSSARY

CCSDS   An organization of space agencies that produces space data standards mainly for flight and ground systems and their interface to space systems.

CORMORAN   A project to replace and enhance the performances and capacity of the CNES Ground Stations Network.

Hourly Rate   The unit price for the utilization of ground-station services

ACKNOWLEDGMENTS

The work described in this chapter was performed at the Toulouse Space Center of CNES. The authors gratefully acknowledge the collaboration of the operations and development teams involved with the components of the CORMORAN project.

REFERENCES

[1]   Servant, D., Hernandez-Marcellin, I., and Charlot, P. (CNES), "The New CNES Station Scheduling System: Both an Operational and Simulation Tool," *Proceedings of Space Ops 2012*.

[2]   Rubio, J.-C., and Roquebert, J.-M. (CNES), "Icare – Migration of CNES Ground Station Communications to Internet Protocol," *Proceedings of Space Ops 2004*.

[3]   Landrodie, P., Houpert, L., Anesi, J., Goudy, B., Suarez, G., and Racaud, T. (Cap Gemini), "Regates – A New Generation of Automated M&C System," *Proceedings of Space Ops 2006*.

[4]   Soula, J.-M., de Beaumont, O., and Palin, M (CNES), "The CNES Ground Networks: Lessons Learned and Future Plans," *Proceedings of Space Ops 2008*.

[5]   Stern, M., Diedrich, E., and Soula, J.-M. (CNES), *Proceedings of Space Ops 2010*.

CHAPTER 16

Operations for Parallel Satellite Support

Marcin Gnat* and Peter Willburger†
DLR, Oberpfaffenhofen, Germany

I. INTRODUCTION

The future of space exploration is becoming increasingly interesting, as well as challenging, as new fields of operation open up and many already known aspects develop new facets, all under a regime of permanent budget pressure and cuts. The job of engineers is to provide solutions to these challenges, allowing ongoing development. In this chapter we focus on three aspects of space operations, trying to collect experiences and find solutions to consolidate them into one universal coherent operations concept.

Our first area of interest is the increasing problem of space debris. Even if all the rules for debris avoidance [1, 2] are enforced, this would only keep present numbers from increasing exponentially. The topic of active debris removal is therefore being discussed more intensively. Among the several realistic concepts based on current technological levels [3, 4] there are some concrete proposals for targeted robotic missions. These discuss the possibility of catching objects (old or defective satellites) and perform de-orbiting in the case of low Earth orbit (LEO) debris [5–7, 23] or removal to the graveyard orbit in the case of geostationary Earth orbit (GEO) debris [6, 7]. In some cases, for example where a defective satellite is very valuable and there is the possibility of repair, the concept of such robotic missions is being extended to on-orbit servicing. Highly sophisticated robotic fixtures (arms, docking mechanisms [5, 8, 24–25]) would be used to perform servicing tasks such as inspection, catching, berthing, docking, fueling, spare part replacement, and so on. All these tasks require high precision and autonomy of spacecraft operation while simultaneously imposing similarly strict requirements on communication with the servicing satellite and ground operations.

*Ground Data System Manager, Communication and Ground Stations; marcin.gnat@dlr.de.
†Ground Data System Manager, Communication and Ground Stations; peter.willburger@dlr.de.

In preparation for such on-orbit servicing, several demonstration robotic missions have already been performed or are now being planned [8]. Some such demonstrations include two spacecraft, one acting as a servicer and another as a client. This requires a specific approach for ground operations, especially in the case where the distance between the spacecraft changes dynamically from tenths of kilometers at distant range to practically zero when berthed or docked. Another example of a multispacecraft environment is formation flight, for which there is complexity in terms of the varying numbers of spacecraft, distances [millions of kilometers (LISA) down to hundreds of meters (TanDEM)], and aspects of communication, among other factors. There are additional requirements in terms of ground track accuracy, effective resolution of observation instruments, how it is importance that these spacecraft fly over the same place on Earth, and payload data availability, which in turn impose tight conditions for absolute and relative position, attitude, number of ground contacts, and link capacity.

The second area of interest is the close formation flight or principally the dual/multiple simultaneous spacecraft support. When discussing link capacity, we actually touch on the third area of interest. High-data-rate links for space to ground and also within the ground segment are becoming more common than before. On the one hand this follows general developments (e.g., in relation to the Internet or mobile networks), but on the other hand there are a lot of specific constraints that require special attention in this particular environment. High bandwidth is required by all applications using video cameras (in human spaceflight and for robotic missions for inspection and telepresence) or for the data produced by observation payloads [including high-resolution optical or synthetic aperture radar (SAR) images and other high-fidelity sensor data]. The amount of data is increasing, so link capacity and quality also have to increase in order to enable full onboard memory dumps during very short contact periods. There are an increasing number of scenarios where this high-bandwidth payload information is required to be available as soon as possible, so the requirement for high bandwidth also extends to the ground network. However, it is not just pure link bandwidth that is of importance; questions about redundancy (e.g., is it still economically acceptable to pay for two or more expensive high-bandwidth links throughout the mission lifetime?), archiving and storage, monitoring of components, and the capabilities of software and hardware also need considering.

Many publications have already considered the topic of robotics in space, limited mostly to robotic arms or the space segment in general, whereas communications or issues related to the ground segment are rarely touched upon. There are some interesting works, however, on signal delay in space (depending on the distance of the spacecraft to the ground station) [9]. Here we aim to discuss a further consideration – both technically and operationally – the requirements of ground segment (Ground Station – Link – Control Center) operations for conducting low-delay, low-jitter teleoperation robotic missions with two spacecraft.

II. TECHNICAL SOLUTIONS

Within this section we present technical solutions and areas of possible future technical improvement for a typical telerobotic mission at DLR. The focus is placed on the communication solution, and other elements are covered only briefly.

A. LOW-DELAY COMMUNICATION CONCEPT FOR ROBOTIC MISSIONS

A new communication concept is being prepared within the generic framework for robotic missions at the German Space Operations Center (GSOC). At first, the communication concept of the ROKVISS mission [5, 10, 11] was seen as a potentially simple reuse option. However, as the analysis of the requirements progressed, it was decided to take a different approach. The ROKVISS-solution remains as a fallback position, nevertheless.

The main requirements of the ground communication segment can be summarized as follows:

- Uplink with 256 kbps;

- Downlink with real-time terrestrial transfer (or online) of a few Mbps;

- Very low delay in the commanding loop (round-trip between sending a command and receiving feedback from the robot in space) below 100 ms;

- Very low jitter (few milliseconds);

- Multiple-source simultaneous commanding possible (simultaneous operations of robotic payload and housekeeping).

The ROKVISS-like solution covers most of these requirements; however, there is a question of implementation costs and general plausibility, especially in relation to remote ground stations. The Weilheim Ground Station is still viable, but for other stations there are many issues to be clarified (Is the modem solution applicable on distant heterogeneous networks? Is it allowed to install third-party equipment at the partner site?).

Our new concept is based on three derivative requirements: enable low-delay and low-jitter connection over larger distances; enable easy integration into _existing infrastructure and systems; allow simultaneous commanding. These targets can be achieved (among other things) by separation of the transport layer from the application layer, and integration with Space Link Extension (SLE) services [12].

The transport layer (and all other layers below) is to some extent dependent on the chosen technical solution for the link. Up to now, the Transmission Control Protocol (TCP) has typically been used. However, due to its inherent mechanism for retransmission, this protocol is the source of jitter. This does not lead us to reject the TCP, as such, for space applications over terrestrial communication; however, for the specific robotic use we are considering, it seemed interesting to look at the User Datagram Protocol (UDP). UDP is widely used in many streaming applications (see http://tools.ietf.org/html/rfc768 and http://en.wikipedia.org/wiki/User_Datagram_Protocol), especially where the completeness of information is not crucial (video, voice). We now need to consider this issue in more detail: to what extent can we allow losing some data packets in exchange for lower delay and jitter? This question needs to be considered in relation to specific circumstances; on the one hand, UDP does not guarantee the delivery and sequence, but on the other hand we use fewer links over the "open Internet." Typically, when ordering a link between control room and the ground station we have specific requirements for the telecom provider, including guaranteed bandwidth, availability, and quality of service in general. We can therefore assume that the delivery is guaranteed by the link (provider) itself, whereas the UDP layer guarantees us low (at least to some extent) latency and jitter.

Moreover, there are several solutions that assist in the use of UDP, such as the Real-time Transport Protocol (RTP) (see http://en.wikipedia.org/wiki/Real-time_Transport_Protocol and http://tools.ietf.org/html/rfc3550). There are also some devices on the market, which, placed at the ends of the link, take care of timely and sequenced delivery while keeping jitter very low (see http://www.avtec.com/index.php/ioplex-ip-access-gateway.html).

Our requirement for easy integration is covered mostly by the use of SLE [or, in future, Cross Support Transfer Service (CSTS) and compatible protocols] at the application layer. SLE is now used widely, and is being integrated in an increasing number of agencies and other space-related institutions and companies.

So far, with current implementations, there is no real approach for using SLE services for robotic mission such as the Deutsche Orbital Servicing Mission (DEOS). The SLE as of today is more focused on safe provision of the command or telemetry, and is based on TCP/IP. Using SLE, one cannot guarantee low jitter, although some tweaking of TCP/IP parameters (such as Window Size) can improve the situation, especially in broad heterogeneous networks. Accordingly, one of the tasks we are going to approach is to check the feasibility of a transport-layer separated SLE protocol, or at least attempt to achieve some similar functionality (in the future, of course, it would be desirable to have full protocol support). In the upcoming months we plan to perform experiments with SLE-like support on top of UDP.

Another topic related to the SLE is the question "Which services can be used for robotic support?" At present, the Forward Command Link Transfer Unit (FCLTU), Return All Frames (RAF), and Return Channel Frames (RCF) services are used. The FCLTU service could be used principally (see also the issue of multiple sources, discussed later on in this section) for commanding, but there is potentially an issue when using the RCF (or RAF) service for robotic feedback information. Data is to be delivered to the robotic operator (or better to say the joysticks' force-feedback servo motors) in the same manner as commands, possibly with low delay and no jitter. When using RCF we have the problem of choosing the right virtual channel for the data, and also the issue of how to guarantee that each frame transported within a channel contains the information we need. All this needs to happen at a constant rate (low jitter). Furthermore, the principal data link capacity could have an influence on the entire performance when the virtual channel carries larger amounts of data. This could lead to inhomogeneous transmission and again result in high jitter.

There are two possible solutions with which we could deal with the feedback path of the robotic data. The first is based on a new SLE service, the Return Frame Secondary Header (RFSH). The Consultative Committee for Space Data Systems (CCSDS) specifies that each telemetry (TM) frame can contain a so-called secondary header section with ancillary data (up to 80 bytes) [13]. This capacity would be enough to support the force feedback information. This would guarantee that feedback information is practically carried by each single frame coming down in the TM stream (see Acknowledgments). Also, use of the SLE service would lead to very early separation of this information and thus asynchronous delivery over the critical path to the joystick. As the required data bandwidth for such information is fairly low it could be envisaged to obtain a separate physical link for the RFSH service (thus removing the risk of being degraded by other data delivered over standard RCF or RAF services on the main data link).

Figure 16.1 presents an example of such a configuration, with CLTUs on the command path being merged within SLE User as a single coherent data stream is forwarded to the ground station and radiated from there. On the way back, the telemetry is split within SLE Provider, allowing all parties to receive only the telemetry they are interested in, allowing separate connections for sensitive data (such as a separate UDP link for RFSH).

Figure 16.2 presents an example of the underlying transport layer, which principally allows both housekeeping and payload operations to talk to all ground stations, choosing the link and transport layer (USP or TCP) depending on current needs.

Another possibility that we visualize for our use requires more effort (Fig. 16.3). This is based on the use of Space Packets [14] and their respective Forward Space Packet (FSP) [15] and Return Space Packet (RSP) services. In this case one does not even really need to define the RFSH service; simply a

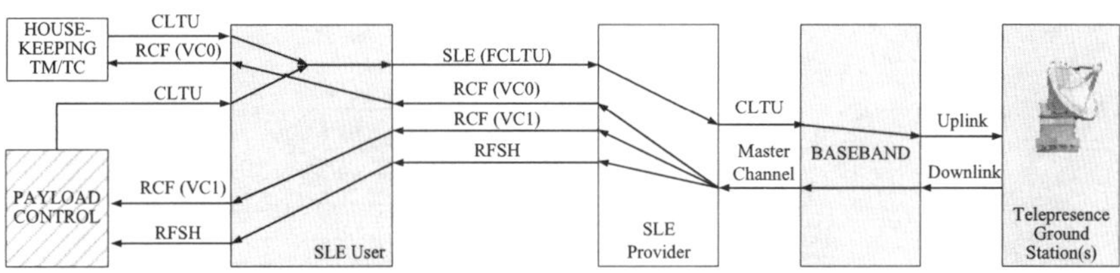

Fig. 16.1   Application layer (SLE).

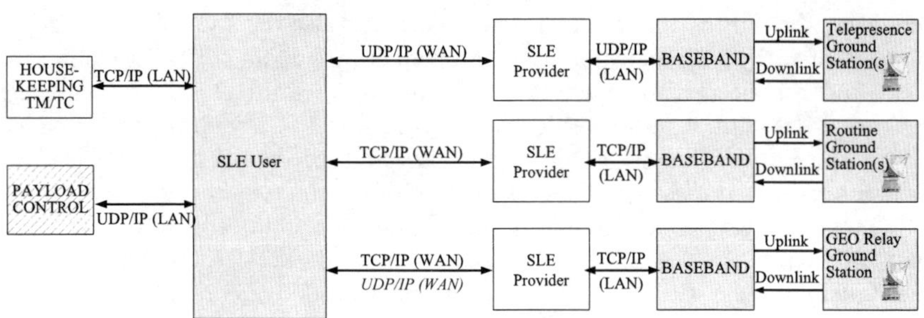

Fig. 16.2   Transport layer.

specific RSP instance could cover the robotic feedback data. The actual advantage is provided on the forward side (FSP). It allows easier multisource commanding such that different parties may command their respective applications onboard. It also closes the loop for the Communications Operations Procedure (COP-1) [16] already at the ground station, thus relieving the link to the control center. As one can see, most of the software load is placed on the SLE Provider (merging and splitting the data), and the SLE User plays the role of the protocol gateway.

Where the effort for all these implementations appears to be too high at the beginning, we can imagine some specific combination of the abovementioned scenarios, where RFSH for time-critical robotic feedback and all other TM information is provided within RCF or RAF and in the forward direction with FSP for commanding. Many current ground systems [telemetry and telecommand (TM&TC) processors, robotic control] already inherently work with space packets. Accordingly, we do not anticipate any paradigmatic change, although some basic interface implementations will need to be performed.

Finally, we come to the issue of parallel commanding. As already mentioned for the use of the FSP service and space packets, in principle one can rearrange the solution for the SLE service provider. However, this does not resolve the issue. We have to assume that multiple FSP services will supply the SLE provider with asynchronous packets. The SLE service provider will need to use a

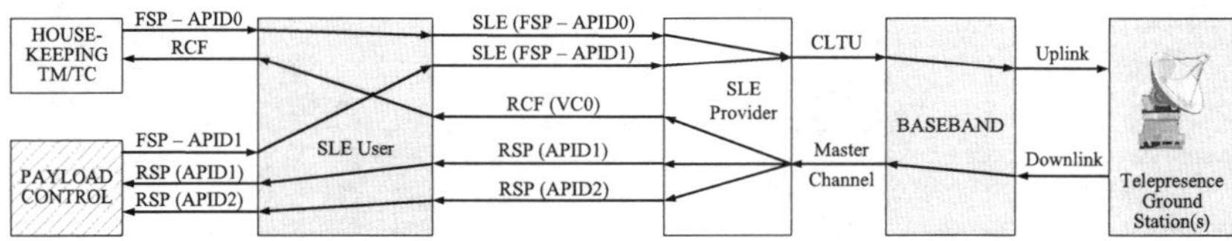

Fig. 16.3   Application layer (SLE) realized with space packets.

complicated algorithm to sort the incoming space packets, and prioritize the robotic payload commands to avoid delay while keeping the command rate constant to avoid jitter. When working on the CLTU level, one can imagine just interleaving the CLTU from robotic control and the common housekeeping CLTU in the manner whereby "every second has to be robotic." This makes the algorithm easier to implement and keeps timings constant. Also, in such case it is possible to decide where to perform the multiplexing (merging): at the SLE service user or at the SLE provider. The first makes it simpler for implementation (one is not dependent on the ground station, which principally needs to receive just one FCLTU stream). The second provides for asynchronous commanding over a large distance, possibly over a separate communication line. In this case, however, a more fundamental change in the SLE service provider will be needed. The merging algorithm stays the same for both cases.

As we have already discussed, in most paths between ground station and control center we could still have multiple issues when dealing with the timing behavior of the equipment. We plan to perform several tests on basebands to find the optimal settings and check their feasibility. Where not enough performance is achieved, it is thinkable to develop a customized logical part of the baseband [Telecommand Unit (TCU) and Telemetry Unit (TMU)] integrated with the SLE provider. This would need to generate all the information required for the uplink and only feed it as a bit stream into the modulation unit with Phase Change Modulation (PCM), analogue to the ROKVISS, and receive a bit stream out of the modulator, decode and synchronize frames and feed it into the SLE provider for the return direction. This could form a high-performance teleoperation station, whereas stations with only an SLE provider could support all the functionality, but with higher delay or without fulfilling all jitter requirements. For the operations team such a setup would provide much simplification (only one configuration and teleoperations depending only on the choice of physical link and the ground station). This concept offers a plug-in solution for ground segment communication.

B. DUAL-UPLINK ANTENNA SYSTEM

To control multiple spacecraft flying in close formation, a possible solution that the DLR has implemented at its ground station in Weilheim is a dual-uplink antenna system. Since 2010 the DLR has been operating TerraSAR-X and TanDEM-X in close formation, with a distance of around 200 m between the two spacecraft and an orbit of 514 km. Until recently, different antennas had to be used to control the spacecraft in parallel. The altitude, distance between spacecraft, and ground antenna diameter provide the opportunity to use one of the Weilheim ground antennas' 3 dB beams for transmission of telecommands, reception of telemetry, and tracking of both spacecraft.

More detailed information on the operation of such an antenna is presented in [17]. That paper describes the installation, configuration, and operation of the ground antenna system, as well as both spacecraft, when only one ground antenna is active. An example describes the results in terms of signal processing and Monitor & Control (M&C) system design. Operations of one and two satellites are compared and the analysis of the advantages and disadvantages will help for future work.

This solution can also be used for telerobotics missions where two independent spacecraft (or in general two communication paths that fit into a ground antenna's 3 dB beam) are used. Where the separation of the two spacecraft is very small, or even for a docked configuration, such an antenna setup can bring several advantages (saving resources, reduced cost, and easier tracking).

Another dual use of a single dish is of the dual-feed type. This is rather common, and the principle is to use the same dish and antenna mount for two (or more) different feeds in different frequency bands (e.g., S- and X-bands). Although it is not as novel as the dual-uplink system just discussed, it can be a source of several operational issues and such a specific resource must have special attention when mission planning or scheduling.

C. OTHER ELEMENTS OF THE GROUND SYSTEM

Other relevant elements of the ground segment include Flight Dynamics services (orbit and attitude analysis, production of orbit and attitude change commands, predictions for the ground stations, event files), Mission Operations (performing the actual monitoring and control of the spacecraft), Mission Planning, Payload Operations (control of the payload, analysis, evaluation, and further processing of the payload data), and Ground Facilities or Network Operations (including Ground Data System, Scheduling, Ground Station Operations, and Network Monitoring).

Owing to budget constraints, the automation of ground operations is becoming a key requirement. At a minimum, routine tasks should be executed in an automatic or semi-automatic manner (using workflows or scripts). Automation brings another advantage in that it allows for monitoring and control of more spacecraft with fewer resources. However, it should be noted that automation does not come at zero cost; it needs to be developed, adapted to the existing system, and maintained. A trade-off analysis should be performed depending on individual constraints.

Today, ground stations are usually remotely controlled from Network Control Centers (NCCs). Routine operations are generally performed from the NCC, whereas launch and early orbit phases (LEOPs) might be supported directly from the ground station. This extends the need for complete M&C and fault reaction, but also routine remote actions performed from an NCC.

All this requires a large effort on the part of software development for automation, monitoring, and control. Such systems need to be scalable, redundant, and reliable, and must be fully integrated into the existing environment.

Another aspect in terms of cross-support is the availability of a standardized interface for monitoring information, as a minimum.

The scheduling system is one of the central elements of the ground system. Its complexity ranges from very simple implementations to very sophisticated ones. This is an expanding area, and its several aspects will also be covered by the CCSDS in the near future [18].

Scheduling needs a good interface with the Mission Planning System (MPS) for the exchange of all the information required to plan specific contacts with the spacecraft. Information flow is required not just in one direction (a schedule request based on general project constraints), but also in the form of feedback to the project or even explicitly to the MPS. Already scheduled contacts are used to plan further actions regarding the project (such as offline data transfers or triggering of data processing). Also, information about the current status of ground stations is transferred, allowing for better long-term planning (i.e., information about antenna maintenance).

When talking about the interface to the MPS, we touch on one of the main issues in today's scheduling systems: different data formats. There is some effort (as from the CCSDS) to standardize the interface and file format for scheduling information. Currently, however, the scheduling system needs to support a number of different formats and ways of exchanging information. Much schedule information is exchanged in simple text form via e-mail. This is rather error-prone, and hard to verify. Standardized and primarily automated processing systems should be pursued, leaving special tasks (such as ad hoc scheduling, emergencies, and conflict resolution) in the hands of a human (scheduling officer).

Specific examples of scheduling issues that can appear during parallel support became obvious during the TanDEM mission some time ago. The mission was performed by two satellites flying in close formation [19], operating the same type of instrument and contacting the Earth in the same frequency bands (S-band for housekeeping and X-band for payload). As a consequence of the close proximity of the satellites it was discovered that it is possible to perform parallel contact with only one antenna equipped with a dual-feed [17]. As long as the contact requests for the two satellites are consolidated there is no issue. However, this is usually the case only for housekeeping data. Payload data contacts are planned much more dynamically and the requests are often received from different sources. As it happened, a housekeeping contact was already planned in the S-band for one of the satellites in this close formation when another request for a data dump in the X-band on the same antenna, but for the other spacecraft, was received. Unfortunately, at that moment, no one was aware of this specific situation. The X-band request arrived significantly later and the scheduling software was not prepared. The outcome was that the first request (S-band housekeeping) was deleted and overwritten by the X-band request. Such constellations of requestor–antenna–spacecraft have now been implemented into procedures, and we are working on changes in the scheduling software to support such cases.

III.   EXAMPLE MISSION SCENARIOS

Several simple communication scenarios that we need to consider when working on the new communication concept are presented in Figs. 16.4–16.6. These do not cover all the possible permutations, and one could expand them to cover more than two spacecraft (swarm), multiple control centers, or a complicated delay-tolerant network (DTN) for a deep-space planetary mission (such as the Mars Rover).

If we wanted to go into more detail, we could divide the communication elements into groups, which in turn could be used to create different permutations of the communication scenario:

1. TM processing facilities (spacecraft control centers, user data center, payload control center);

2. Commanding facilities (spacecraft control center, payload control center);

3. Ground data connections/data links;

4. Ground stations and antennas (single with S- or X-band, dual feed with S- and X-band, dual uplink with two S-Band uplink frequencies, etc.);

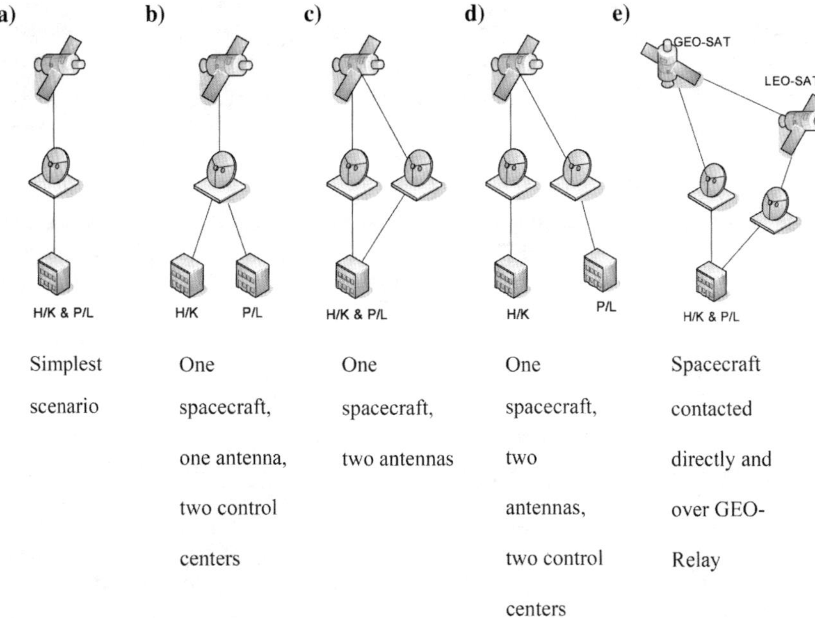

Fig. 16.4   Examples of single spacecraft scenarios.

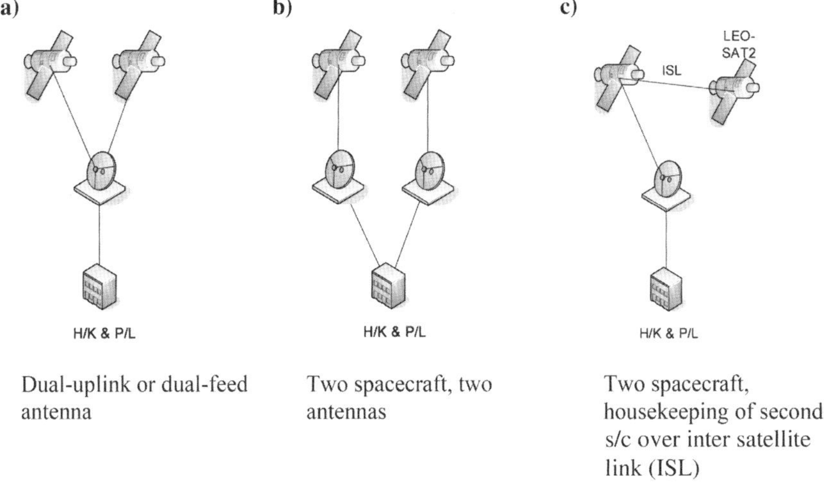

Fig. 16.5   Examples of two spacecraft scenarios.

5. Space links (different frequency bands like S-, X-, Ka-bands, direct link, link over GEO relay, ISL, laser communication);

6. Spacecraft (single, multiple, swarm, constellation, formation, client–servicer).

As soon the scenario(s) are chosen, each of these elements needs to be analyzed in terms of its capabilities and especially the interfaces between elements (i.e., ground station–data link–control center).

In the following section a few general mission profiles are presented in an example that we have used in our analysis.

The TanDEM mission consists of two sibling satellites (TerraSAR-X and TanDEM-X) flying in a close formation (as little as 200m from one another). The satellites were launched separately, about two years apart. The satellites either perform Earth imaging, each of its own accord using SAR, or in chorus, with both SAR instruments synchronized to allow for the generation of Digital Elevation Model (DEM) image, which, in principle, gives us a 3-D map of the Earth's surface [19].

The mission generates large amounts of data, which needs to be dumped in a timely manner. Also, the formation flight requires high-precision position information for both spacecraft and frequent command possibilities. These factors impose very specific requirements on the ground station network (redundant stations and connections), flight dynamics services, and mission planning [19].

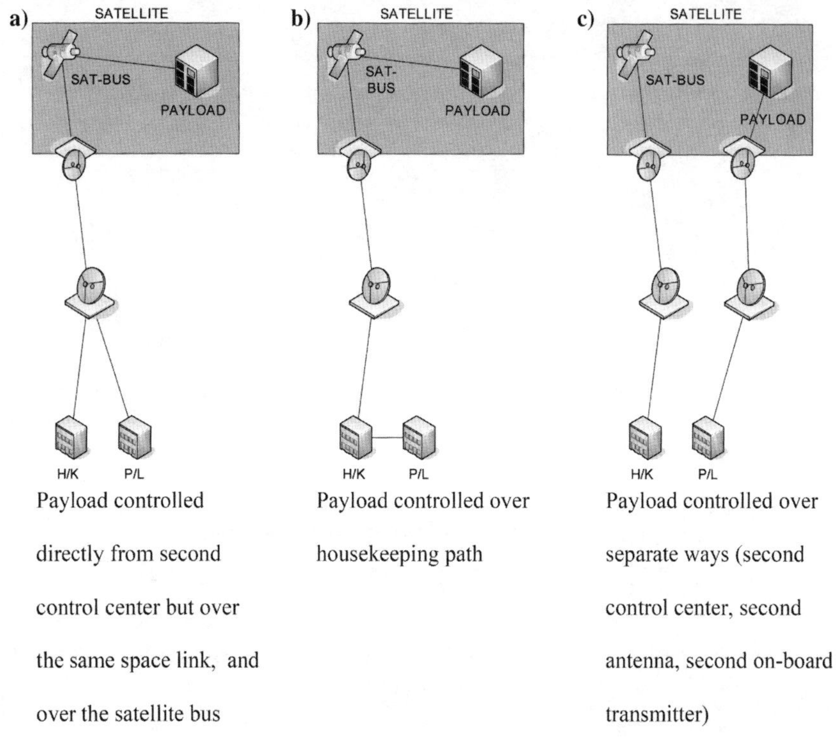

Payload controlled directly from second control center but over the same space link, and over the satellite bus	Payload controlled over housekeeping path	Payload controlled over separate ways (second control center, second antenna, second on-board transmitter)

Fig. 16.6   Examples of payload control.

The PRISMA mission also consists of two satellites. These were started in a combined (docked) configuration and after the commissioning phase were separated. The main satellite has a propulsion system, and the secondary (or target) satellite is solely controllable with respect to attitude (reaction wheels). The target satellite is commanded and monitored over the ISL, with the main satellite as a relay. The main satellite's payload includes cameras (for video-based navigation and inspection) and GPS (for positioning and navigation). The main mission objectives are imaging, guidance, far and close navigation, as well as rendezvous operations (the mission supports several other experiments, which are not relevant to this discussion). Because the main satellite is used as a relay for the target as well as for video and other sensors, the downlink capacity is in the range of 1 Mbps (megabits per second). Because the scientific payload control is on the ground, the scientific data archive is fed directly from the real-time data stream. This imposes a specific requirement for ground communication (RAF service usage [20]).

The ROKVISS mission focused principally on payload operation as the robotic arm was placed on the International Space Station (ISS). It was controlled over a separate space link, which was exclusively used during contact times. The space link was realized in the S-band with a 256 kbps uplink data rate and 4 Mbps downlink. On the ground, a joystick was used to perform real-time operation of the distant (ISS) robotic arm. Feedback information for the operator on the ground was delivered as force-feedback directly to the joystick (giving a haptic response) and a video signal from the stereoscopic cameras. The main requirements could be briefly summarized as high-bandwidth video transmission and low-delay and low-jitter round-trip signal propagation for the joystick-to-robotic arm operation [5, 10, 11].

The project DEOS is being prepared right now. This mission will be a demonstrator mission consisting of two satellites and delivering on-orbit servicing capabilities. The spacecraft will be launched in a docked configuration. After commissioning and some experiments, the satellites will separate and the main part of the mission will begin. The main objectives are to demonstrate various relative navigation, rendezvous, and docking scenarios, as well as catching the client satellite (which should act as an uncooperative spacecraft) with the robotic arm, followed by different possible servicing operations (refueling, module exchange, etc.). At the end of the mission, controlled de-orbiting in a docked configuration will be performed. The mission presents several challenges, not only for the space segment, but also for the ground operations and infrastructure, as already presented in detail in Sec. II.

There are also several interplanetary robotic missions where communication plays an important role (e.g. Mars Rover). However, these missions inherently include large signal delays and are addressed with techniques such as DTN, so we will not deliberate on these in this chapter.

IV.   CHALLENGES FOR OPERATIONS

In this section we discuss several issues or challenges that we consider important for close formation flight and telerobotics missions. We do not cover all aspects of ground operations, and focus solely on ground communication and infrastructure.

During formation flight missions, one of the issues is radio-frequency interference (RFI) between the satellites, or one satellite influencing the ground contact of another. It is possible for two spacecraft to have the same or very similar space link frequencies (which is especially true in the X-band and for payload data, as they typically use almost the entire available bandwidth). The results are massive data drops or even complete lost passes. It is crucial to perform coordination in the early phase [International Telecommunication Union (ITU; see http://www.itu.int/) or Space Frequency Coordination Group (SFCG; see https://www.sfcgonline.org/)] to account for RFIs and perform planning to avoid interference. Also, shortly before the mission begins and also during the mission, this coordination must continue. Proper localization (and assignment to the spacecraft) of ground stations can help in avoiding these issues.

Crucial for any mission, but especially for robotics and close formation, is Flight Dynamics support and products. Maintaining high availability of these services and products is one of the main tasks of the Ground Data System. This can be approached from different sides, so the frequent availability of tracking data helps increment the precision of orbit estimation. Timely distribution of visibility and orbit information to all interested parties (mission, ground stations) helps in the better planning of activities. Often, in the case of critical maneuvers, the immediate availability of specific attitude or orbit information becomes a major element of mission operations. It is important to support the dissemination of these data with controlled data transfers, backups, and frequent tests of data distribution paths.

Missions with multiple spacecraft and very specific payloads that need to be controlled directly by their own engineering teams lead to a very specific environment for scheduling services. What was previously not an easy task becomes even more complicated, and one has to ask the question "How far can this scheduling be handled strictly by the operator?" For simple missions (like the case in Fig. 16.4a) the scheduling process is based on the "first in, first out" principle and is serialized. Mostly intermission conflicts and some priority regulations are the cause of intense work. With the missions under discussion here, there are additional complications, such as the booking of the same antenna resource for two spacecraft (dual-uplink or dual-feed antennas), parallel requests from two different sources, and so on. As long as the number of missions and cross-supports remains lower than just a few per day, with some effort and training these situations can be handled with the old tools. When this frequency increases, however, support from a more sophisticated scheduling system is required. Such a system could be realized on the platform of Web services, for example, allowing easy user access and extensibility.

The mission operations team is also one of the topics that require closer investigation. The cost of personnel in long-lasting missions becomes a main driver of cost. For close formation and robotics mission the team needs to be expanded for crucial operations such as maneuvers and teleoperation. Afterwards, there is often a need for observation of the spacecraft for a longer period of time, easily leading to 24/7 operations and a full three-shift setup. There is therefore a strong push for automation of specific routine operations, more Failure Detection, Isolation and Recovery (FDIR) mechanisms onboard, and better monitoring of all elements (e.g., with direct notification of on-call personnel in case of problems).

When planning robotic mission, one has to analyze the availability of contacts to the spacecraft. Typically, a spacecraft in LEO has a maximum of approximately 10 minutes of visibility with a ground station. Assuming a small buffer for locking onto the spacecraft and synchronizing, what is left for real operation is something in the range of 6–8 minutes. It is possible to plan contacts in such a way that they become virtually one long contact period (composed of consecutive single contacts). This, however, causes operational issues, as the handovers need to be planned carefully and there will still be gaps in commanding. Such contacts in series cannot be planned for every orbit, as the ground stations are not evenly placed geographically. Also, issues with respect to the ground link (availability of an equal-quality link to all stations) can arise. As a solution, a link over a GEO relay satellite is proposed. This gives homogeneous contact times to the robotic spacecraft in the range of 30–45 minutes. However, this comes with the cost of longer space signal flight time (for a typical "ground to GEO to ground" stretch, one must take into account a round-trip time of approximately 500 ms). This makes real-time operation with joystick feedback more difficult and places very strict requirements on the terrestrial link (which then needs to compensate for the extra time used via the GEO relay). An example of such an experiment is presented in [21].

Ground communication lines become one of the most important parts of the ground segment for real-time control of telerobotics missions. The capacity of the link needs to be high enough to avoid backlog and allow low-delay commanding and feedback, simultaneously with real-time video signals. Where in previous missions one needed a solution in the range of 64 kbps (ISDN) for normal housekeeping and routine operations, suddenly, for such a robotic mission, the link capacity needs to be a minimum of a few megabytes per second. This drives the cost, particularly if we plan to use multiple ground stations, as discussed above. Also, some technical issues can arise. When crossing national borders we are probably confronted with a change of telecom provider and therefore the technology in use. This can again lead to longer delay times or sudden jitter. Extensive testing of ground communication before the actual mission start is thus required. Because of several complicated elements in the system (multiple receivers, several commanding sources), end-to-end testing is strongly recommended.

One additional item regarding signal delay should also be noted. A slowly increasing signal delay remains unnoticed for a relatively long time, causing a sudden interruption either in communication protocols (SLE) or within TM&TC processors. Unpredictable data drops will result. Therefore, the operations team should request some sort of current delay monitoring. If it is displayed in real time or otherwise communicated to the team, it is of secondary importance. It is essential that the operations team be able to prepare for the worst conditions by altering their plan, changing experiments, or even completely postponing an action until line conditions improve.

As an example for the abovementioned issues, unexpected problems occurred when using Virtual Private Network (VPN) connections over the Internet with data rates of 1 Mbps [22]. Something that does not seem challenging for most home Internet users became very difficult. The mission required 1 Mbps of real-time downlink from the spacecraft directly to the TM/TC processing software. As it turned out, the connection over the Internet was unreliable and the actual bandwidth considerably below the requested 1 Mbps. This led to serious data drops and thus to problems feeding the archive with scientific data. The solution was to implement the SLE service in the online complete mode (this at least guaranteed no data drops); however, it had the result that the total delay of the "real-time" data was measured in minutes, and the backlog was considerable. In the final stage, however, a separate connection [over the Multi-Protocol Label Switching (MPLS) network; see http://en.wikipedia.org/wiki/MPLS) was implemented. This was an important lesson: connection lines are cost drivers, and because of the availability and "daily applicability" of the Internet, there is real temptation to use it instead. One can do so, but there is a need to perform enough tests in advance and to be prepared for some circumstances where the VPN over the Internet is not a viable solution.

V. CONCLUSION

In our chapter we have tried to collate different topics regarding close formation flight and general concerns with multiple spacecraft, high-data-rate connections, and real-time applications, showing the analysis one needs to perform in preparation for such missions. Special focus has been made on an early concept of a communication setup for future robotics, multiple satellite, and multiple control entity missions.

First plausibility tests for this new communication concept and work on prototypes have been started, and within a year the first results are expected. This will be followed by step-by-step implementation. The ultimate target is to create a communication system that will allow flexible and easy usage in a cross-support environment (interagency support) and will fulfill all payload requirements.

ACRONYMS

CCSDS   Consultative Committee for Space Data Systems
CSTS   Cross support transfer services
COP-1   Communications operation procedure
DEOS   Deutsche Orbital Servicing Mission
DTN   Delay-tolerant network
FCLTU   Forward command link transfer unit
FD   Flight dynamics
FDIR   Failure detection, isolation, and recovery
FSP   Forward space packet
GEO   Geostationary Earth erbit
GSOC   German Space Operations Center
H/K   Housekeeping
IP   Internet protocol

ISL	Inter-satellite link
ISDN	Integrated services digital network
ITU	International Telecommunication Union
LEO	Low Earth orbit
LEOP	Launch and early orbit phase
MPLS	Multipacket level switching
MPS	Mission planning system
NCC	Network Control Center
OOS	On-orbit servicing
P/L	Payload
RAF	Return all frames
RCF	Return channel frames
RFI	Radio-frequency interference
RFSH	Return frame secondary header
RTP	Real-time transport protocol
SLE	Space link extension
SFCG	Space Frequency Coordination Group
TC	Telecommand
TCP	Transmission control protocol
TM	Telemetry
UDP	User datagram protocol
VPN	Virtual private network

GLOSSARY

Antenna feed   This refers to the components of an antenna that feed radiowaves to the rest of the antenna structure, or in receiving antennas collect the incoming radiowaves, convert them to electric currents, and transmit them to the receiver.

Baseband   This is an adjective that describes signals and systems whose range of frequencies is measured from close to 0 Hz to a cutoff frequency. In space communication the baseband frequency is also the name for the intermediate frequency between the modulator and the upconverter (typically 70 MHz). Colloquially, the device containing the modulator as well as telecommand and telemetry units is called the baseband unit, or just baseband.

Force-feedback   Force-feedback (or haptics) is a tactile feedback technology that takes advantage of the sense of touch by applying forces, vibrations, or motions to the user.

Housekeeping   This is the colloquial name for the operations or data (telecommands and telemetry) that serve the base operation of the satellite bus (power, thermal, attitude, etc.).

Jitter   Jitter is the undesired deviation from true periodicity of an assumed periodic signal.

Round-trip time   The round-trip time delay time (RTD) or round-trip time (RTT) is the length of time it takes for a signal to be sent plus the length of time it takes for an acknowledgment of that signal to be received. More specificly, in the case of teleoperations, this is the time from the command being sent (moved joystick) to the moment when the force-feedback information is being received.

S-band   The S-band spans the frequency range 2–4 GHz (IEEE).

Teleoperation   Teleoperation means "doing work at a distance"; in our specific case we are referring to the distant robotic arm operations performed remotely with a joystick.

ACKNOWLEDGMENTS

The authors thank Stefan Funk (Astrium GmbH) for his idea of interleaving CLTUs and using the frame secondary header for feedback data on the Space Link.

REFERENCES

[1] "UN Space Debris Mitigation Guidelines," UN Office for Outer Space Affairs, 2010.

[2] Taylor, E. A., and Davey, J. R., "Implementation of Debris Mitigation Using International Organization for Standardization (ISO) Standards," *Proceedings of the Institution of Mechanical Engineers: G*, Vol. 221, No. 8, 2007, pp. 987–996.

[3] Campbell, J. W. (Colonel, USAER), "Using Lasers in Space: Laser Orbital Debris Removal and Asteroid Deflection," Occasional Paper No. 20, Center for Strategy and Technology Air War College, Montgomery, AL, Dec. 2000.

[4] Kaplan, M. H., "Space Debris Realities and Removal," *Improving Space Operations Workshop Spacecraft Collision Avoidance and Co-location*, Goddard Space Flight Center, Greenbelt, Maryland, 2010.

[5] Landzettel, K., Albu-Schäffer, A., Preusche, C., Reintsema, D., Rebele, B., and Hirzinger, G., "Robotic On-Orbit Servicing – DLR's Experience and Perspective," *Proceedings of the 2006 IEEE/RSJ International Conference on Intelligent Robots and Systems*, Bejing, China, 2006, pp. 4587–4594.

[6] Sellmaier, F., Boge, T., Spurmann, J., Gully, S., Rupp, T., and Huber, F., "On-Orbit Servicing Missions: Challenges and Solutions for Spacecraft Operations," *AIAA SpaceOps*. Huntsville, Alabama, USA, 2010.

[7] Kaiser, C., Bellido, E., and Hofmann, P., "Space Debris Mitigation Using On-Orbit Servicing Solutions," Prague, Czech Republic, 2010.

[8] Reintsema, D., Landzettel, K., and Hirzinger, G., "DLR's Advanced Telerobotic Concepts and Experiments for On-Orbit Servicing," Advances in Telerobotics, edited by Ferre, M., Buss, M., Aracil, R., Melchiorri, C., and Balaguer, C. (Eds.), *Springer Tracts in Advanced Robotics*, Vol. 31, Springer, Berlin/Heidelberg, 2007, pp. 323–345.

[9] Stoll, E., Letschnik, J., Walter, U., Preusche, C., and Hirzinger, G., "Concept of an Algorithm to Determine the Signal Delay Time for Telepresence Space Applications."

[10] Landzettel, K., Albu-Schäffer, A., Brunner, B., Beyer, A., Gruber, R., Krämer, E., Preusche, C., Reintsema, D., Schott, J., Steinmetz, B.-M., Sedlmayr, H.-J., and Hirzinger, G., "ROKVISS, Verification of Advanced Light Weight Robotic Joints and Tele-Presence Concepts for Future Space Missions."

[11] Preusche, C., Reintsema, D., Landzettel, K., and Hirzinger, G., "Robotics Component Verification – ISS ROKVISS – Preliminary Results for Telepresence," *Proceedings of the 2006 IEEE/RSJ International Conference on Intelligent Robots and Systems*, Beijing, China, 2006, pp. 4595–4601.

[12] CCSDS 910.4-B-2, "Cross Support Reference Model – Part 1: Space Link Extension Services," Blue Book, Issue 2, Oct. 2005.

[13] CCSDS 132.0-B-1, "TM Space Data Link Protocol," Blue Book, Issue 1, Sept. 2003.

[14] CCSDS 133.0-B-1, "Space Packet Protocol," Blue Book, Issue 1, Sept. 2003.

[15] CCSDS 912.3-B-2, "Space Link Extension – Forward Space Packet Service Specification," Blue Book, Issue 2, July 2010.

[16] CCSDS 232.1-B-2, "Communications Operation Procedure-1," Blue Book, Issue 2, Sept. 2010.

[17] Dikanskis, D., Wiedemann, K., and Preuß, M., "Dual Operation of TerraSAR-X and TanDEM-X with one Ground Antenna," *AIAA SpaceOps*, Stockholm, Sweden, 2012.

[18] CCSDS 910.11-B-1, "Space Communication Cross Support – Service Management Service Specification," Blue Book, Issue 1, Aug. 2009.

[19] Hofmann, H., and Kahle, R., "The TanDEM-X Mission Operations Segment: Close Formation Flight: Preparation and First Experiences," *SpaceOps*, Huntsville, Alabama, USA, 2010.

[20] CCSDS 911.1-B-3, "Space Link Extension – Return All Frames Service Specification," Blue Book, Issue 3, Jan. 2010.

[21] Stoll, E., Letschnik, J., Walter, U., Artigas, J., Kremer, P., Preusche, C., and Hirzinger, G., "On-Orbit Servicing, Exploration and Manipulation Capabilities of Robots in Space", *IEEE Robotics & Automation Magazine*, Dec. 2009, pp, 29–33.

[22] Furtuna, C., Kruse, W., and Garcia, C., "SLE Experience Over Unreliable Data Links," *SpaceOps*, Stockholm, Sweden, 2012.

[23] Eberle, S., Ohndorf, A., and Faller, R., "On-Orbit Servicing Mission Operations at GSOC," *AIAA SpaceOps*, Huntsville, Alabama, USA, 2010.

[24] Landzettel, K., Brunner, B., Deutrich, K., Hirzinger, G., Schreiber, G., and Steinmetz, B.-M., "DLR's Experiments on the ETS VII Space Robot Mission," *Proceedings of the 9th International Conference on Advanced Robotics (ICAR)*, Tokyo, 25–27 Oct. 1999.

[25] Yoshida, K., "Achievements in Space Robotics, Expanding the Horizons of Service and Exploration," *IEEE Robotics & Automation Magazine*, IEEE, Dec. 2009, pp. 20–28.

CHAPTER 17

Location Independent Mission Operations: Systems Engineering Approach to Mobile Device Data Dissemination

Edward Birrane* and **Robert Berardino†**
Johns Hopkins University Applied Physics Laboratory, Laurel, Maryland

I. INTRODUCTION

The role of the mission operator involves the command, diagnosis, and correction of in-flight space assets. Based upon the mission characteristics and severity of fault, operators must accomplish these tasks with varying degrees of timeliness and often within constrained budgets. When responding to mission events, operators may require coordination with subject-matter experts not normally present in a Mission Operations Center (MOC), including flight hardware/software engineers, test teams, instrument subcontractors, and scientists. It is a common occurrence to require a variety of support staff to be physically present in MOCs during critical mission events and maneuvers, and there is also precedent for engineering support to be called to the MOC at any time, day or night, to assist in responding to unexpected events. Critical mission events, expected or otherwise, will always require infusions of expertise. However, advances in mobile computing, broadband infrastructure, and software frameworks may reduce the logistical cost associated with this support. Specifically, removing the requirement for support activities to only occur in the MOC reduces costs (experts do not need to be idle in the MOC when not needed), reduces logistical challenges associated with synchronizing availability, and reduces delays in getting expert dispositions on unexpected events. We term this set of capabilities Location-Independent Mission Operations (LIMO).

The challenge associated with implementing LIMO capabilities stems from the evolution of supporting technical means. Simply adding laptops and mobile devices behind Virtual Private Network (VPN) connections in support of legacy data and control flows will not effectively solve the location-independent issue. A systems engineering approach is necessary to both identify new data and control flows *and* demonstrate how these flows provide benefits to adopting missions.

---

*Senior Staff, JHU/APL, Space Department.
†Senior Staff, JHU/APL Space Department.

We propose that the continued evolution of tablet computers and broadband access provides a new technical means making feasible the command and control of space assets with appropriate data rates and security models. We further propose that these evolving technical means require changes to existing operational concepts and infrastructure to be used effectively. Efforts to incorporate mobile devices must account for their unique characteristics in the areas of input methodology, screen size, battery life, and security models. Server-side infrastructure must be built to perform data conditioning, user registration, preferences processing, and validation. User applications must undergo human–computer interface (HCI) analysis to determine how to maintain a consistent "look and feel" across a variety of platforms to reduce training and support costs. Ultimately, this analysis may change the number and nature of applications supporting a mission. We predict that the flexibility and reduced operation and maintenance cost associated with a distributed data model will recoup initial systems engineering and architecture investments.

The remainder of this chapter is organized as follows. Section II details the motivation of our approach, both in the use of new technical means and in the need to reevaluate flows in the system. Section III presents an initial systems engineering analysis of a LIMO capability, including the characteristics, operational concepts, and metrics used to evaluate technical means. Section IV provides a candidate architecture conformant to our system-level concepts. Section V discusses our experiences with a reference implementation of LIMO capabilities for operational missions. Section VI summarizes an implementation of this technology during the launch and commissioning of the NASA Radiation Belt Storm Probes (RBSP). We summarize our work in Sec. VII.

II. MOTIVATION

In this section, we consider smart devices in the broader context of computing machinery to understand their unique role in bringing computational benefits to operational systems. That laptop/ultrabook computer represents the latest achievement in a tradition of general-purpose computer miniaturization that has previously seen the migration of user computation from mainframes to servers to desktops. Each new technical achievement in this area makes the same HCI models of the mid-1980s [1, 2] available in different cost/mobility profiles. To date, reducing the cost and administrative complexity of these systems has increased reuse and decreased the cost of ground systems. Standardized applications for complex visualization, data sharing, automation, and communication reduce operator error and have increased the ability of operators and systems engineers to respond to problems more quickly and with more accuracy. Furthermore, the reuse profile for ground systems in this architecture is very high; applications are not hard-coded to a particular server, desktop, or laptop computer. However, this legacy HCI hinders the continued miniaturization of the general-purpose computer; keyboards must be of a minimum size for typing, screens must be of sufficient size to accommodate standard windowing views; and mice/trackpads must be of usable size.

Hardware capabilities continue to mature, enabling increased battery density, low-power-consumption components, touch/gesture screens, and overall reductions in cost. Hardware advancements have created a new category of computing device that, by necessity, deviates from the legacy HCI model. These devices are referred to as "tablets" or "smart phones" and contain gigahertz CPUs, gigabytes of RAM, individual graphics processors, and gigabytes of persistent storage. The continued maturation of wireless technologies connects these devices at broadband link speeds. Collectively, we refer to them as "personal mobile computing devices" (PMCDs). PMCDs offer a significant, new user capability: personalized, mobile data consumption. Because they introduce changes to the HCI, their incorporation into existing systems requires analysis.

Typical sizes of PMCDs preclude built-in keyboards, mice, or multiwindowed screen displays. Touchscreen inputs impose minimum sizes to interactive widgets based on average finger sizes [3]. Gesture controls impose a physical manipulation metaphor that changes the familiar button-click view model [4]. Beyond changes to the interface, PMCDs function differently from desktop and laptop computers. Their smaller size limits effective battery life and secondary storage. Their slim designs and lack of active cooling present thermal considerations when performing sustained computation. Therefore, although PMCDs support powerful processors, the loading of these processors is dedicated to low-power operation, monitoring of thermal environment, and a priority towards system-level scheduling and touchscreen servicing versus user application number crunching. The relative strengths and weaknesses of currently offered PMCDs are listed in Table 17.1.

As ground system architects seek to reduce cost and increase capability there is a natural desire to evaluate new technologies. Because a significant portion of ground operations relies on communication with appropriate resources at appropriate times, the unique benefit of increased mobility has increased interest in PMCDs for operational use. However, beyond the desire to incorporate socially popular devices into operations centers, we have seen no systematic analysis of the unique functional benefits of these devices, what architectural changes must be made to MOCs to realize those benefits, and whether the benefits justify the associated cost and risk. To date, many commercial vendor ground systems incorporate PMCDs as if they were any other remote thin client, with few architectural changes to support their unique properties. There is acknowledgement that alternate graphical user interfaces are necessary, but the quality and consistency of these interfaces vary widely.

We term this the "bolt-on" approach, and although it is a low-cost way of incorporating socially popular devices into operations centers, it is not an efficient way of achieving the operational goal of leveraging data mobility. As PMCDs grow in popularity, increase in computation capability, and decrease in cost, their role in mission operations will increase. However, architectures must

TABLE 17.1 APPLICATIONS MUST ACCOUNT FOR BOTH STRENGTHS AND WEAKNESSES OF PMCDS

Functional area	PMCD strengths	PMCD weaknesses
Processing	Multicore, GHz processors and Gb of RAM	Embedded processors with reduced instruction sets; high cost of polling/decoding complex touchscreen gestures; thermal issues via lack of active cooling
Power	Dense, small batteries last for days compared to hours for laptop devices	Battery life drops dramatically under moderate, persistent computational load
Storage	Very fast, solid-state internal storage typically supporting additional secure-digital (SD) cards	Less than 80 Gb total storage on most devices, even using largest SD cards; not all devices support external storage media
Networking	Wireless/3G/4G network availability	Typically no wired network access
Input devices	Touchscreen, some support for Bluetooth keyboards and mice	Not all devices support keyboard input; keyboards and mice not typically available for use when mobile
Display sizes	High-resolution, lightweight, rugged displays	Very small size (3 – 10 in.)

adapt to the relative strengths and weaknesses of this emerging technology means. The inappropriate focus on the popularity of PMCDs, rather than analysis of how they evolve operational concepts, motivates our work. We seek to describe the unique characteristics of this new category of computing device and perform a systems-level assessment of how to appropriately incorporate it into ground-system architectures.

III. SYSTEMS ANALYSIS

This section summarizes our systems analysis associated with the characteristics of PMCDs in MOCs, how such characteristics benefit mission operations, the operation concepts to achieve these benefits, and what metrics justify the expense of architecting PMCDs into operational systems.

A. CHARACTERISTICS

From Table 17.2 we decompose the concept of "mobility" into three unique characteristics governing the development of applications for a PMCD: screen size, alternate input methods, and processor/storage. We add the fourth characteristic of security to the list as a special consideration for passing operational data outside of a dedicated flight network. Although we discuss security only from the point of view of MOC networks, we recognize there is a very large body of research regarding the securing of data and services for PMCDs for any application [5, 6]. The manner in which each of these characteristics informs the application architecture is as listed in the following subsections.

1. SCREEN SIZE

In support of lower power consumption and higher mobility, screen sizes for PMCDs are significantly smaller than those of wall monitors, desktop displays, and laptops. They cannot therefore support the same level of information density in their displays. This constraint requires developers to reduce the "footprint" of information on the device, either by using pictures/graphics in lieu of words, or by displaying smaller data sets. In either case, the associated changes to the user experience (UX) levy requirements on the architecture and protocols used to generate data for the device, especially when the device will contain fused or otherwise summarized information.

TABLE 17.2   PMCDS PROVIDE THREE KEY BENEFITS TO MISSION OPERATIONS SUPPORT

Benefit	Associated characteristic			
	Screen	Input	CPU/ storage	Security
Decreased operator error through natural input methods and focused visualization	•	•		
Faster anomaly resolution with coordination amongst geographically distributed experts				•
Reduced training cost via consistent interfaces and configurations across a set of interfacing devices	•		•	•

2. ALTERNATE INPUT

PMCDs struggle to support general-purpose inputs. Keyboards and mice are typically too bulky to physically couple with the device, mechanical parts wear excessively on mobile devices, and relying on wireless peripherals requires carrying additional equipment and dedicating physical workspaces to their use. These issues are counter to a hand-held mobility model. On-screen keyboards take up valuable screen space and result in slow, error-prone inputs. PMCD developers exploit gesture-enabled touchscreen devices to provide interactive controls to manipulate data and control in their applications.

A reliance on constrained, interactive controls over free-form text input aligns precisely with the existing practice of building syntax checkers, command compilers, validators, and other tools to prevent user error in the MOC. In this regard, the evolution of error-correcting input methods provides a compelling alternative to the more complex task of verifying free-formed user data.

3. PROCESSOR/STORAGE

Hardware resources in PMCDs are not used in the same way as on traditional computers. Storage is orders of magnitude smaller and processing is constrained by thermal considerations, battery density, and secondary loads from the touchscreen. Applications running on PMCDs lack the working memory, battery life, and persistent storage to perform sustained, complex computation. Where moderate computation occurs for extended periods of time, even state-of-the-art devices experience thermal issues.

Applications built for MOCs vary from simple data-conditioning tasks to complex visualization and calculation tasks. Some applications may be ported directly to a PMCD after changes to the user interface, but others must be completely re-architected around these limitations. For example, many mobile application developers support a thin-client architecture where server-side infrastructure performs complex computation and sends processed results to the mobile device for visualization. Understanding how MOC utilities and visualizations map to this or other models is a key activity when incorporating PMCDs into the ground segment architecture.

4. SECURITY

The literature is densely populated with wireless security models operating at every level of the networking stack. Implementations of secure ciphersuites can be ported to PMCDs, especially on institutionally administered devices. However, the implementation of security protocols, and the security of data at rest, remains inconsistently implemented across various vendors. Part of this stems from constraints in the device hardware and part from the level of security in the implementing operating system and application software.

Adopting PMCDs for use in flight environments requires that ground segment architectures adopt policies to standardize the security model across all supported devices, including implementation of security at the individual application layer if not otherwise provided by a particular operating system/hardware combination.

B. BENEFITS

Based on these characteristics, and experience building applications for MOCs, several potential benefits arise from the incorporation of PMCDs into the ground architecture. Of interest, none of these benefits is unique to PMCDs, but their mobility, intuitive interfaces, and focus on personal data consumption provide novel ways to achieve long-standing application goals. This section describes three benefits identified through our analysis (Table 17.2).

1. DECREASED OPERATOR ERROR

Traditional MOC displays are dominated by large screens that provide multiple dimensions of information, certainly more than an individual operator can digest in real time. This is a time-saving convenience when screens are shared/projected for view by multiple operators at once. Similarly, tools for command construction benefit from the availability of nearby, multiple computer systems to visualize flight rule/constraint documentation, perform simulation runs, and validate command sequences. Conversely, the smaller screen size of the PMCD is meant for individual viewing of information, and often in a more graphica form.

Smaller data sets must thus be constructed using filtering, fusion, and other aggregation methods to present targeted information in a smaller footprint. This reduces the cognitive load of the operator by focusing only on those data associated with a particular function at a particular time. Furthermore, the growing practice of providing graphical widgets, preselected options, and other means to select amongst a finite set of options removes free-form text entry as an input choice. When operators can focus on simplified views of their data and interact with it in natural ways by selecting amongst preconfigured options, opportunities for error decrease.

2. FASTER ANOMALY RESOLUTION

Robust security models, combined with increasing wireless speeds, enable near real-time (NRT) data interaction with devices outside the MOC enclave. Authentication ensures that data provided, and controls received, are in accordance with access control policies configured on the system. Integrity ensures exchanged data is unmodified/uncorrupted. Confidentiality prevents unauthorized users from accessing these data while in transit. The ability to provide these levels of security allows multiple, geographically separated experts to coordinate in support of mission operations events with NRT access to telemetry and other spacecraft state.

Anomaly-resolution activities are therefore no longer time-constrained by delays in getting experts into the MOC, delays in releasing information to remote experts, and delays in applying expert-recommended configurations and controls. This ultimately allows the anomaly resolution process to begin sooner, and with more focused input.

3. REDUCED TRAINING

The use of thin-client architectures on PMCDs, coupled with the enforcement of roles and responsibilities from a security layer, provides a consistent visualization and processing context to users. This experience can be migrated to any PMCD, laptop, or desktop supporting the thin-client application. By storing user preferences and configuration information on server-side resources, thin clients across multiple devices do not need to be reconfigured. Preserving look and feel reduces the number of systems on which an operator must be trained. Enforcement of roles and responsibilities at the application layer further reduces training to just those areas of the system where a user has permissions.

Because PMCDs replace unnatural motions (point-click, double-click, right-click, drag-drop) with natural motions (swiping, pinching, multipoint scrolling), the learning curve for applications is further reduced.

C. OPERATIONAL CONCEPTS

We translate the benefits of PMCDs, derived from their characteristics, into a set of three discrete operational concepts: location-independent mission operation, HCI, and data filtering. These concepts capture the benefits of PMCDs as a series of tangible activities based upon our experience testing PMCDs in mission operations contexts at the Johns Hopkins University Applied Physics Laboratory (JHU/APL).

To provide context for these Concepts of operation (CONOPS), we contrive an operational scenario where a faulty heater is causing thermal imbalances on a deep-space asset. Mission operations are monitoring the heater and its effect on spacecraft autonomy in preparation for an instrument observation.

1. LOCATION-INDEPENDENT MISSION OPERATION

A mission operations PMCD begins the day as a wireless node on a personal home network (802.11n) communicating over the public Internet. The operator checks heater state for the previous 8 h and notes some period of anomalous current draw through the night. The PMCD transitions to a cellular network (3G, 4G, or 4G Long Term Evolution (LTE)) during the operator's commute to work. Upon arriving at work, the PMCD transitions off the carrier's network and onto either an

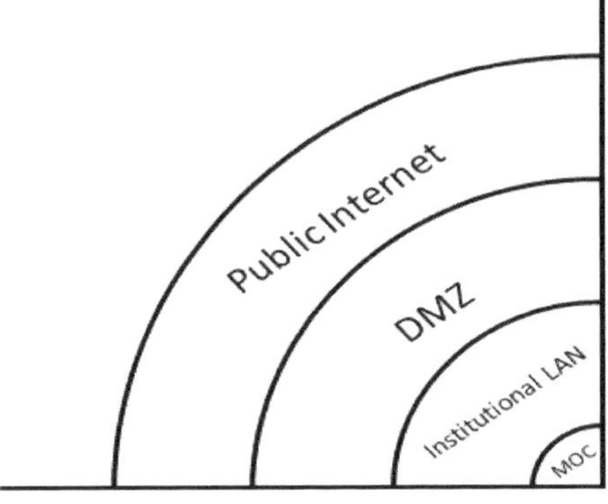

Fig. 17.1   LIMO services operate across a variety of subnetworks.

institutional demilitarized zone (DMZ) wireless network or other institutional LAN subnet. During morning meetings, feedback on the PMCD alerts the operator to another heater anomaly. Curious to observe the event, the operator excuses herself and migrates to the MOC to discuss the event with other operators. Coming out of the MOC, the operator locates the hardware engineer associated with the heater and they spend time in her office reviewing telemetry and associated spacecraft state on the PMCD. The hardware engineer obtains the heater manufacturer log in a Web version of the thin client from their offices in a different part of the country, and a teleconference takes place to discuss the NRT observations of the heater in situ.

Regardless of where the user physically travels throughout the day, a LIMO solution provides functionally equivalent application access regardless of the traversal through various subnetworks (Fig. 17.1). Tracking radio hand-offs, authentication, and policies for data-at-rest are technical implementation hurdles. In this operational concept, public radio hand-offs are automatic and institutional access is granted by prompting the user to authenticate to the institutional subnetwork on first data access. Authentication keys are maintained on the device itself as part of configuring the device. A default, and conservative, security policy requires that MOC data either not be serialized to flash memory, or be strong-encrypted before persistence. In either case, institutionally approved thin-client applications conform to this policy.

2.   HUMAN–COMPUTER INTERFACE

Two popular operating systems for PMCDs, Google's Android and Apple's iOS [7], provide software development kits (SDKs) enabling developers to decode touch-screen inputs including gestures. Further, best practices regarding the use of these input devices are available to maximize the use of these SDKs [8]. When designing a thin-client application for a PMCD, a software engineer uses these SDKs and best practices to develop graphical metaphors associated with data manipulation.

For example, the data visualization of the heater telemetry in our scenario is a series of line plots over time. The operator interacts with these data in a variety of ways uniquely enabled by the touchscreen interface. Users may use single-touch selection of a data point, pinch-based zooming, and two-finger horizontal and

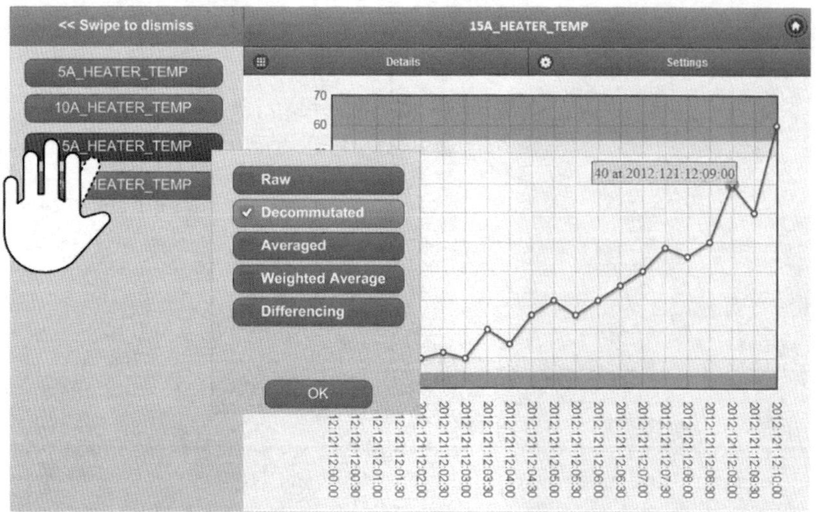

Fig. 17.2   PMCD interfaces use natural gestures to choose among preselected data.

vertical scrolling to adjust axis scaling. Data sets may be selected for inclusion in the graphing area by swiping. The type of data (raw, decommutated, averaged, weighted averaging, differencing) is selected by touching the data source icon for 1 s and then selecting type from a pop-up submenu of icons. A conceptualized view of this interface is shown in Fig. 17.2.

The user performs all critical visualization functions without the use of text-based input. The selection of data sets is as provided from a predefined list based on the user's role and functional areas providing the data. The application prevents incompatible data sets from displaying on the same graph and constrains the number of data sets that can be concurrently visualized. The process of data selection and visualization is not error-prone, occurs rapidly and naturally with a minimum of training, and works the same way across a variety of data-set visualizations. In fact, the operator is able to perform visualization on the PMCD faster through the gesture interface than they would otherwise be able to using a mouse and keyboard interface at a laptop.

3.   DATA OFFLOAD/FILTER

The application software engineer allows users of the data visualization application to personalize their display settings by supporting customized navigation screens (e.g., a "My Favorites" page), remembering recent searches for data points to plot, default plot characteristics such as colors and units, and preconfigured plots of commonly visualized data sets. These settings are saved on the local device, and optionally on a user-preference setting at the application server.

In our scenario, as the operator continues to monitor the state of the heater and collects evidence for anomaly resolution, the PMCD is exchanged for another with a fresh battery from a surplus of such devices kept in the MOC. For example, an iPad may be exchanged for an Android tablet. The operator takes two, providing one to the hardware engineer and takes the other one home at the end of the day. Neither of the two "new" PMCDs requires configuration, as they receive their configuration from the PMCD server; the views and preconfigurations of the heater telemetry data sets were saved on the original PMCD, migrated to the server side, and downlinked to the new PMCDs without significant user effort. An example architecture supporting these CONOPS is given in Fig. 17.3.

As the hardware engineer continues to evaluate past telemetry values, more complex data conditioning activities are requested, including three-day moving averages of the 1 Hz data, histograms, and number of threshold crossings. Computing these across larger data sets on the mobile device is impractical from a time and battery-life perspective, so they are off-loaded to the server, as part of the application software, and results are displayed to the user when the computation is complete.

D.   METRICS

In the set of operational concepts outlined above, there are three values that we can measure to quantify the benefit of the PMCD approach to mission operations: 1) the time it takes to identify data related to a problem, 2) the amount of time it takes to request and receive relevant data, and 3) the amount of time before subject-matter experts can review data. These metrics are captured in Table 17.3.

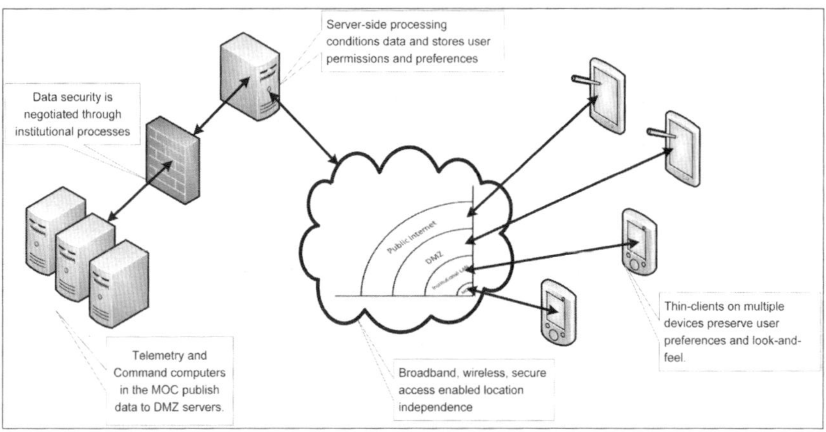

Fig. 17.3   A multitiered architecture reduces reliance on a single PMCD.

TABLE 17.3   THE BENEFIT OF PMCDS IN THE OPERATIONAL ENVIRONMENT CAN BE MEASURED WITH THREE METRICS

Metric	Description	Value in optimizing
Problem identification time (PIT)	The lag associated with collecting information from the spacecraft and identifying anomalies in that information set Critical faults and large anomalies are captured automatically through threshold crossings Subtle anomalies, such as those captured from trending, may require manual detection	Subtle anomalies caught early may resolve with preventative, rather than corrective action. More rapid anomaly identification may reduce the change of failure cascades in flight systems
NRT deviance	The age of the data measured as the lag between when the data was received in the MOC and when it was available for consumption by experts	Up-to-date data are required for health monitoring and anomaly resolution
Data mean time to expert (DMTE)	The time between problem identification and evaluation of data by a subject-matter expert	Shorter DMTE implies faster initiation of the anomaly-resolution process

IV.   DATA DELIVERY ARCHITECTURE

A.   ARCHITECTURAL GOALS

Based on our systems analysis of the benefit of PMCDs to mission operations, we have constructed a reference data-delivery architecture and implementation at JHU/APL. The goals of this reference architecture are provided in Table 17.4.

B.   RELATED PRODUCTS AND STANDARDS

The Object Management Group (OMG) published a Data Distribution Specification for Real-Time Systems (DDS) in 2004 to define an interoperability protocol for DDS. Its purpose and scope were to ensure that applications based on different vendors' implementations of DDS could interoperate. Commercial vendors like Real-Time Innovations (RTI) and PrismTech have DDS products that are implemented in accordance with the OMG DDS specification. This specification works to address the need for a common networking middleware and publish/ subscribe model for data access in large distributed systems. The types of large-scale systems that utilize a DDS implementation include: air traffic control, railway traffic management, radar processors, naval combat management systems, and UAV control. These systems could also benefit from extending data in a location-independent manner.

Commercial T&C vendors offer products that extend their primary product in an attempt to extend an operation center's reach across network boundaries. Harris Corporation, L-3 Telemetry West, and Integral Systems, Inc., all have T&C products that support secondary bolt-on offerings to extend their T&C product into some form of data distribution mechanism. They do this by providing some type of API to allow implementers to retrieve data from proprietary data storage and transmit that data back to the client application. Depending on the vendor, and the surface area of their API in relation to the product's functionality, implementers can build solutions as narrow as a plug-in to a vendor's proprietary architecture or as expansive as developing a standalone high-fidelity thin-client desktop application [9–11]. None of these offerings directly addresses data consumption on mobile or personal computing devices. Regardless of extensibility provided through bolt-on or premium services, all COTS vendors attempt to drive system integrators into a "better together" value proposition that keeps the COTS vendor's products in the mission critical system data flow and leaving customers even further surrounded in a walled garden.

At the time we began our initial research, OMG had just released a Web-enabled DDS Request for Proposal (RFP). That RFP is still being matured by OMG [12]. DDS in its current implementation did not have implementations

TABLE 17.4   IMPLEMENTATION AND SUPPORT CONSTRAINTS DRIVE ARCHITECTURAL GOALS

Goal	Rationale
Highly performing data distribution	LIMO data flows should achieve near real-time data processing The round-trip time includes a MOC to DMZ server data caching layer and DMZ server to PMCD time When not communicating real-time data from a pass, the delay is effectively from the DMZ server to the PMCD
Commercial off-the-shelf (COTS) vendor agnostic	As COTS products, telemetry and command (T&C) products, and other services may vary among supported missions and implementing institutions, a vendor agnostic approach is required
Wide PMCD platform penetration	Because of the multitude of existing devices, the proliferation of new devices and ever dynamic nature of the PMCD marketplace, the architecture must provide a way to make NRT data available to as many PMCD platforms as possible

available for iOS or Android and would have required substantial upfront engineering to demonstrate our concepts. Furthermore, the required upfront engineering would have only supported two PMCD platforms, and our overarching goal was to maximize the number of supported PMCD platforms.

To minimize long-term software development costs by maximizing code reuse, we posit the most cost-effective approach is to design as many components of LIMO as possible in a COTS agnostic way, reducing or eliminating the need to purchase bolt-on or premium services.

C. PROPOSED DATA FLOW

Based on our architectural goals, the state of current vendor products, and the maturation of associated protocols/standards, we propose the data flow shown in Fig. 17.4. In this figure we group together three primary responsibilities: data retrieval/rendering, data conditioning/storage, and data extraction.

Data retrieval covers the applications resident on the PMCDs along with the mechanism those applications use to access data from the data cache/data store. Client applications running on various PMCDs render retrieved data in ways that are appropriate for the device.

The data storage layer transforms extracted data into formats necessary for easy search, query, merge, and other conditioning operations performed by DMZ servers. Data loading populates these transformed data into the datastores serving as the cache for future retrieval by PMDCs. In our proposed architecture we provide a one-way mechanism out of the MOC during the data storage phase; data, once transformed and loaded for delivery to PMDCs, are not flowed back to the MOC systems based on current security policies. These data stores expose their data via a Web Service application programming interface (API). This technique provides the widest PMCD platform penetration, as it is supported by all devices in our use study.

The interface between the MOC T&C computers and the DMZ servers is enabled through institutionally controlled firewalls using well-administered

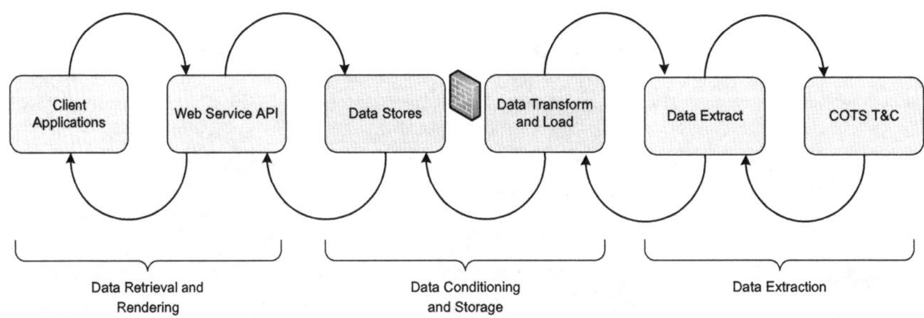

Fig. 17.4 Data extraction, storage, and retrieval are the primitives of any LIMO solution.

security protocols. The query mechanisms are similar to those used by other MOC applications that query data from the underlying data archive and real-time data streams.

D. SOFTWARE ARCHITECTURE

Our software architecture conforms to the architectural goals established from our systems analysis. When implementing the various tiers in the LIMO solution we began with an analysis of the low-level networking stacks, Web architectures, and COTS T&C core software integration necessary to generate and publish NRT telemetry. This included a survey of available security frameworks, development kits, and JHU/APL networking policies. We chose the Representational State Transfer (REST) [13] software model, which requires architecture to support the following:

1. Client-server system operation;

2. Stateless operation (there should be no need for the service to keep users' sessions);

3. Cache support (the network infrastructure should support cache at different levels);

4. Uniform accessibility (each resource must have a unique address and a valid point of access);

5. Layering to support scalability.

These constraints do not dictate what kind of technology to use; they define how data is transferred between components. A RESTful solution can be implemented in any networking architecture available using existing networking infrastructure and protocols [14].

E. PROTOTYPE IMPLEMENTATIONS

Over the course of two years, we implemented a technical demonstration and an operational prototype for NASA's Solar Terrestrial Relations Observatory (STEREO) and the Radiation Belt Storm Probes (RBSP), respectively. We built working prototypes of thin-client applications for the iPad, iPod Touch, and the Android OS.

F. INITIAL PROTOTYPES

Two working prototypes for PC Web browsers were also made to support thin clients on existing desktop and laptop infrastructure. The first used HTML, CSS, and JavaScript, and applied Asynchronous JavaScript and XML (AJAX) techniques. The second used Adobe Flex, an open source framework for building Web applications. With the exception of the Flex prototype that used XLM, all other client implementations exchanged resources as JavaScript Object Notation (JSON) objects. The client implementation subteam had no difficulties finding capable JSON libraries to use across the various platforms (iOS, Android, etc.).

In our earlier technical demonstration, it became apparent that supporting the various native platforms was costly and required many resources from different technical skills. For our more recent operational demonstrations with RBSP, we focused our attention towards our overarching principle of wide PMCD platform adoption. We looked into approaches for keeping the development costs of client implementations down and reduce the total number of lines of source code that must be maintained. We shifted our client implementation strategy to focus on standards-based responsive Web design. Responsive Web design is a term used to describe Web applications that adapt to the media that renders them [15]. We took advantage of an existing web application framework, jQuery Mobile, to rapidly prototype a client implementation that could be used on a variety of PMCDs.

G. MISSION OPERATION INTEGRATION

To achieve our research objective of demonstrating a NRT data distribution platform, the operational concept had to integrate the various layers of our back-end system and properly interface with our client prototypes, both on the JHU/APL private network and outside the JHU/APL network on the public Internet. This required a Web server with a valid Secure Sockets Layer (SSL) certificate to exchange authenticated Hypertext Transfer Protocol Secure (HTTPS) requests and responses. Further, it required the Web service API implementation to access data from the database and process the response back to the necessary clients.

We utilized Apache and Apache Tomcat for the Web server and Web application server, respectively. Apache has been the world's most popular Web server since the mid-1990s and Apache Tomcat is an open-source software implementation of the Java Servlet and JavaServer Pages technologies. To maximize the development time allocated on this research project, we used Jersey, a pure Java RESTful framework. Jersey is the reference implementation of Sun's Java API for RESTful Web Services, also known as JAX-RS. The JAX-RS project was started by the Java Community Process (JCP) with the goal of creating an API for Java RESTful Web services.

V. LESSONS-LEARNED

Given the short duration of the technical demonstration, there was only a small portion of time dedicated to the various client application HCIs. There is a wave of multiform factored devices hitting the consumer electronics market, many with rich touch/gestural interfaces, but our focus was to validate the concept of NRT, secure data distribution. The resultant platform and Web service API was demonstrated for multiple operational missions, using their mission data via various mobile devices and two Web applications.

It was possible to query, decommutate, and transfer multiple telemetry point values to a data-caching layer. During an active pass with the spacecraft, telemetry can be decommutated by the COTS T&C product, processed by the data ingest layer, and inserted into the database in the data cache in millisecond times. When accessed by a Web service client, the telemetry data stored in a MySQL database in the data cache layer are queried out by the classes that implement the Web service handlers, again in millisecond times. By keeping all the various processing and translation times down to milliseconds, the end-to-end time from spacecraft to LIMO users has remained in the subsecond to second time frame.

Users of all the various client prototypes supplied username and password credentials before logging into their applications/browsers. The username and password were passed in the HTTP authentication request header (Base64 encoded) of each call made to the Web service. Each call is an HTTPS call, and Apache Tomcat was configured to validate the username and password against a table in the MySQL database. Requiring logon credentials, communicating requests over HTTPS, and keeping a data cache for telemetry data, instead of reaching back into the MOC for data, further mitigated the potential security risk to mission-critical data. Not supporting a mechanism to push data or effectuate a change in the MOC also helped mitigate the risks in providing Web service access to the Internet. Finally, when users exit/suspend their applications, no data are written to the device's persistent memory.

VI.   LAUNCH AND COMMISSIONING OF RBSP

The operational prototype of our system successfully streamed data for the NASA RBSP mission, which launched on 30 August 2012 at 0405 hrs from NASA's Kennedy Space Center and Cape Canaveral Air Force Station. During this launch, and subsequent commissioning activity, Mission Operations deployed our infrastructure and associated PMCDs for both data visualization and event notification. Specifically, mobile devices were used to support location-independent mission operations and to offload/filter volumes of data for more rapid situational awareness during this portion of the mission. Although the MOC remained fully staffed with critical support personnel, a larger technical team wanted to observe and evaluate a variety of telemetry associated with the launch of the spacecraft. Given the very early morning launch, and the fact that the launch was delayed by environmental issues, the ability to view data from home was appreciated. Figure 17.5 lists a sample screen capture of operational telemetry (Sun offset angle) as it appears on a client mobile device.

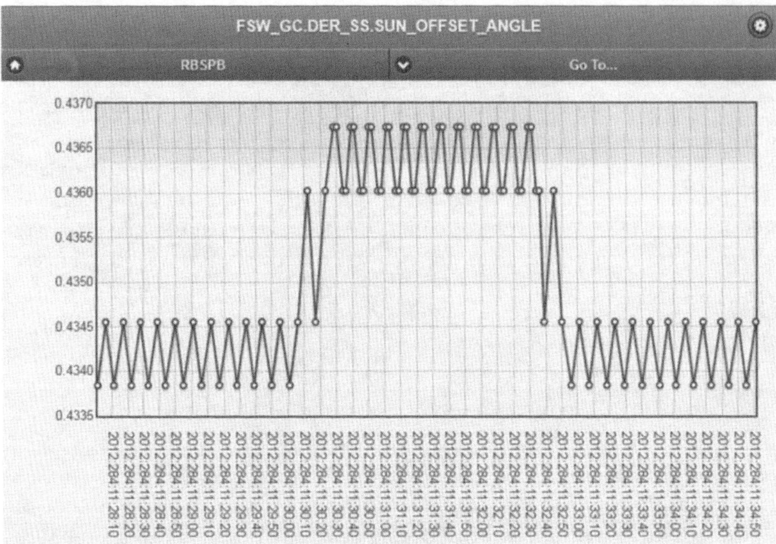

Fig. 17.5   Telemetry from each spacecraft is visualized using custom widgets on mobile devices.

The experiences related to the use of this software back-end and mobile-device front-end in the context of an actual mission were invaluable to understanding the overall benefit, and impact, of such an infrastructure on a ground segment. In supporting the RBSP launch we were able to both validate assumptions and performance metrics as well as implement new features and functions necessary to make the tools more useful to our engineering staff. As expected, the data delivery latency was very small through our data distribution tiers, with no noticeable lag or delay in the visualization of the information. This is a particularly important measure necessary to demonstrate the concept of using mobile devices to keep data close to experts who, otherwise, are unable to be in the MOC. By operating without significant delays, the visualized data are considered "fresh" and therefore usable. To ensure that the community of "location-independent" users could envision differing operational concepts, we expanded the prototype to handle multiple types of data. Whereas the development prototypes focused solely on the visualization of spacecraft telemetry, the operational prototype also supported ground telemetry, event messages, and a notification service.

Ground telemetry support includes the visualization of contact schedule information, as calculated from contact scheduling applications incorporated into the RBSP ground software. This information was particularly appreciated by off-site users during launch windows; during one launch delay we knew that the launch had been delayed before its televised announcement by viewing changes to the contact schedules on an iPhone.

In addition to ground software application information, the RBSP T&C core product produces a rich set of event messages of interest to operators and engineers. As is the case with spacecraft data, these event messages may be viewed on PMCD displays, as illustrated in Fig. 17.6.

However, because these messages are naturally event-driven, a push-notify ability has been added to the prototype whereby any user may select notifications based on any of the information available to the prototype. The configuration screen for our notification interface is shown in Fig. 17.7. The notification service is a completely dynamic, user-configurable filtering system with full access from mobile devices. The sent notifications include hyperlinks to the appropriate data trending display. For example, configuring the notifications for higher Sun offset angle in Fig. 17.7 will produce a message that hyperlinks to the Sun offset angle chart from Fig. 17.5.

Notification rules are published from the PMCD to the data server used to collect telemetry through the operations firewall. A notification service running on that server then evaluates rule conditions. We found the data server to be a natural place in the processing chain to evaluate rule conditions as this same server houses the database storing conditioned telemetry as extracted from the MOC. We configure the embedded hyperlinks in the generated

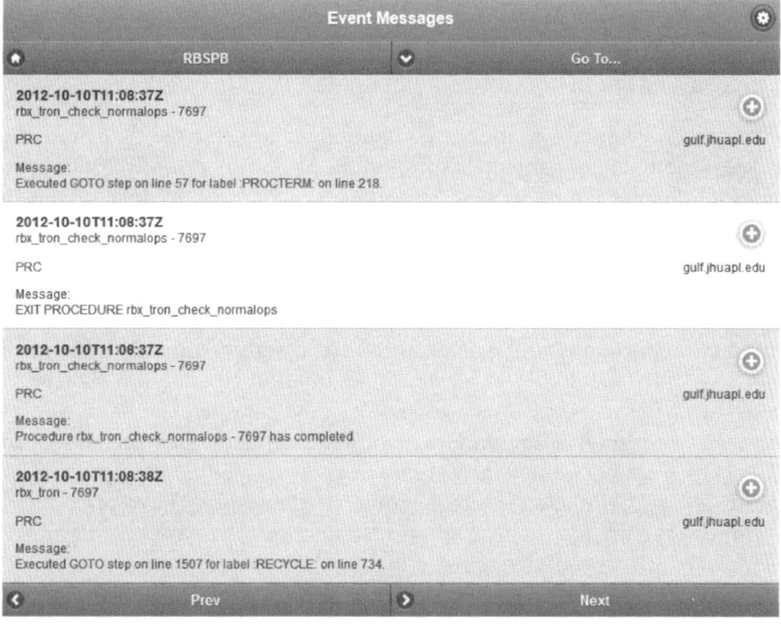

Fig. 17.6 Event notifications from ground applications are also supported in our system.

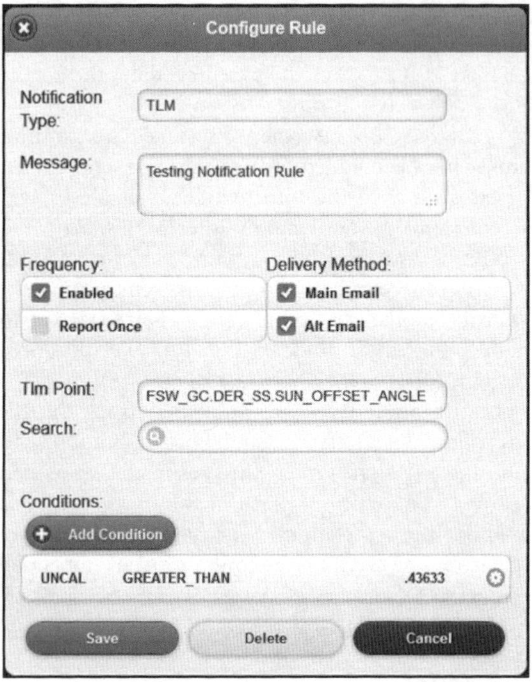

Fig. 17.7 Notifications may be sent to users based on telemetry-based conditions.

notification e-mails to launch the appropriate visualization plotting the last 50 points of related information. Throughout the launch and commissioning of RBSP the system has worked efficiently and with no noticeable latency or lag. In addition, we have found that, as expected, offloading the complex rule analysis and alerting computation to the server preserves battery life and CPU cycles of the PMCD.

VII. CONCLUSION

The continued evolution of mobile computing devices, for the first time in two decades, is changing fundamental aspects of the HCI. Concurrently, ground architectures for new missions review emerging technical means to evaluate their ability to reduce cost and increase effectiveness of applications for mission operations. The personal, mobile computing device presents the ability to provide focused, secure, intuitive data access to mission operators regardless of their geographic context. However, the benefits of these devices are most effectively realized when applications and architectures are designed around their inherent weaknesses (processing power, screen size, input methods, and battery life). Architectures that support multiple devices based on open standards show the most utility and lower adoption cost than dedicated solutions tied to a specific PMCD API. Reference implementations of these capabilities using operational missions at JHU/APL successfully demonstrate the viability of secure NRT data distribution. We predict that the appropriate incorporation of these devices into ground architectures will disseminate flight data to experts as necessary, enable new opportunities for collaboration, and remove logistic burdens and costs associated with continuous support of missions. Just as PMCDs are replacing laptop computers for mobile data consumption of entertainment media, as ground architectures evolve to support them correctly, we see them becoming an indispensible tool for future mission operations.

ACRONYMS

API   Application programming interface
COTS   Commercial-off-the-shelf
DMTE   Data mean time to expert
HCI   Human–computer interface
LIMO   Location-Independent Mission Operations
MOC   Missions Operation Center
NRT   Near real time
PMCD   Personal mobile computing device
REST   Representational state transfer
T&C   Telemetry and command
UX   User experience
VPN   Virtual private network

REFERENCES

[1] Rutkowski, C., "An Introduction to the Human Applications Standard Computer Interface, Part 1: Theory and Principles," *Byte*, Vol. 7, No. 10, 1982, pp. 291–310.

[2] Kent, L. N., Weldon, L. J., and Shneiderman, B., "Cognitive Layouts of Windows and Multiple Screens for User Interfaces. *International Journal of Man–Machine Studies*, Vol. 25, No. 2, 1986, pp. 229–248.

[3] Benko, H., Wilson, A. D., and Baudisch, P., "Precise Selection Techniques for Multi-touch Screens," *Proceedings of the SIGCHI Conference on Human Factors in Computing Systems (CHI '06)*, edited by Grinter, R., Rodden, T., Aoki, P., Cutrell, E., Jeffries, R., and Olson, G., ACM, New York, 2006, pp. 1263–1272.

[4] Hurtienne, J., Stößel, C., Sturm, C., Maus, A., Rötting, M., Langdon, P., and Clarkson, J., "Physical Gestures for Abstract Concepts: Inclusive Design with Primary Metaphors," *Interacting with Computers*, Vol. 22, No. 6, 2010, pp. 475–484.

[5] Shabtai, A., Fledel, Y., Kanonov, U., Elovici, Y., Dolev, S., and Glezer, C., "Google Android: A Comprehensive Security Assessment," *IEEE Security and Privacy*, Vol. 8, No. 2, 2010, pp. 35–44.

[6] Posegga, J., and Schreckling, D., *Next Generation Mobile Application Security*, Vieweg + Teubner Verlag, Wiesbaden, Germany, 2011, pp. 181–199.

[7] "Mobile/Tablet Operating System Market Share," http://netmarketshare.com/mobile-market-share [last accessed 24 April 2012].

[8] Wroblewski, L., "Touch Gesture Reference Guide," http://www.lukew.com/ff/entry.asp?1071 [last accessed 24 April 2012].

[9] Harris Corporation, "Network Enabled Operations," http://download.harris.com/app/public_download.asp?fid=2093 [last accessed 8 April 2012].

[10]   Integral System, "Webic," http://www.integ.com/ProductBriefPDF/ Webic%20Datasheet.pdf [last accessed 8 April 2012].

[11]   "L-3 Telemetry West InControl," http://www2.l-3com.com/tw/pdf/datasheets/ ML_InControl_Brochure_2010.pdf [last accessed 8 April 2012].

[12]   Pardo-Castellote, G., "Web-Enabled DDS: Accessing Real Time DDS Data from Web-Enabled Clients," http://www.omg.org/news/meetings/GOV-WS/pr/ rte-pres/DDS_WebEnabled_RTEW09.pdf [last accessed 18 April 2012].

[13]   Fielding, R., "Architectural Styles and the Design of Network-based Software Architecture," Ph.D. Dissertation, Univ. of California, Irvine, 2000.

[14]   Sandoval, J., "RESTful Java Web Services." PACKT Publishing, Birmingham, United Kingdom, 2009.

[15]   Marcotte, E., *Responsive Web Design*, A Book Apart, New York, 2011.

CHAPTER 18

Automating Mid- and Long-Range Scheduling for NASA's Deep Space Network

Mark D. Johnston,[*] **Daniel Tran,**[†] **Belinda Arroyo,**[‡]
Sugi Sorensen[§] **and Peter Tay**[¶]
Jet Propulsion Laboratory, California Institute of Technology, Pasadena, California

Butch Carruth,[**] **Adam Coffman**[††] **and Mike Wallace**[‡‡]
Innovative Productivity Solutions, Inc., Bulverde, Texas

I. INTRODUCTION

The scheduling of NASA's Deep Space Network (DSN) presents a variety of challenges. As a unique and mission-critical capability serving an increasingly large user community, it is essential that the network be efficiently scheduled and utilized. At the same time, DSN users represent an enormous range of types of scheduling requirements and constraints that vary substantially depending on the type of mission, on time as the spacecraft enter different mission phases, and seasonally as visibility intervals overlap in different ways. Against a backdrop of increasing budget pressure, the DSN has undertaken a project to replace its existing aging scheduling software with a new and unified tool suite intended to improve all aspects of the DSN planning and asset allocation and scheduling processes.

In this chapter we first give a general overview of the DSN and the nature of its scheduling problem, followed by a brief description of the scheduling process and software systems (Sec. II). We then describe the Service Scheduling Software (SSS, or S^3) system, covering some key design elements as well as operational lessons learned from its initial deployment phase (Sec. III). This is followed by a description of the ongoing extension of S^3 to encompass long-range planning and forecasting functionality (Sec. IV). Finally, in our conclusions, we summarize progress to date and plans for future work (Sec. V).

---

[*]Principal Scientist, Planning and Execution Systems Section.
[†]Member of Technical Staff, Artificial Intelligence Group, Planning and Execution Systems Section.
[‡]Deep Space Network Planning and Scheduling Manager.
[§]Systems Engineer, Deep Space Network Planning and Scheduling Section.
[¶]Mission Operations Engineer II.
[**]President/Architect, IPS Inc.
[††]Senior Programmer, IPS Inc.
[‡‡]Programmer, IPS Inc.

II. DSN SCHEDULING OVERVIEW

NASA's DSN [1–3] is composed of a set of large (34-m- and 70-m-diam.) antennas and the associated equipment required to communicate with spacecraft, from those in high Earth orbit to the most distant man-made objects. These antennas are situated at three Deep Space Communications Complexes (DSCC), as listed in Table 18.1. The complexes are spaced roughly equally in longitude, to provide round-the-clock coverage for missions anywhere in space. Although capabilities vary from one antenna to another, and one complex to another, overall the DSN provides a range of S-, X-, and Ka-band up- and downlink services to all of NASA's missions, as well as a number of international partner missions. These services include support for spacecraft telemetry, command, and tracking, as well as radio science, radio astronomy, very long baseline interferometry (VLBI), radar, and calibration.

Currently the DSN supports 37 spacecraft or service users, counting all those with regular requirements for scheduled time on any antenna. The mission users span a wide range of distance and orbit type: high Earth orbit, lunar orbit, solar orbit, probes at Mercury, Venus, Mars, and Saturn (and en route to Jupiter and Pluto/Charon), and the asteroids, out to the two Voyager spacecraft in interstellar space. Ground-based users conduct radio science and radio astronomy using the antennas, including coordinated programs with international partners. Other activities that must be scheduled include routine and special maintenance, calibration, engineering, and test activities. The collected set of DSN users imposes a very wide range of usage requirements on the network due to differing designs and operating modes. Some users require occasional contacts of only a few hours per week, but this ranges up to continuous coverage during certain mission phases, such as post-launch and during critical mission events. At the present time, a typical week includes about 500 scheduled activities on the antennas of the three DSN complexes.

TABLE 18.1 DSN COMMUNICATIONS COMPLEXES AND SOME OF THEIR CHARACTERISTICS

Complex	GDSCC	CDSCC	MDSCC
Location	Goldstone, California, United States	Canberra, Australia	Madrid, Spain
Longitude	117° W	149° E	4° W
Latitude	35° N	35° S	40° N
Antennas	1–70 m 5–34 m	1–70 m 2–34 m	1–70 m 3–34 m
Capabilities	S, X, Ka	S, X Ka downlink only	S, X Ka downlink only

A. PHASES OF THE DSN SCHEDULING PROCESS

The DSN scheduling process consists of three phases, which do not have sharply defined boundaries. In the following we describe these phases as they exist today, and later in this chapter we discuss plans for how they may change in the future.

1. LONG-RANGE PLANNING AND FORECASTING

In today's system, long-range planning is based on user-provided high-level requirements, specified in the form of a spreadsheet that is interpreted by analysts and entered into a database at Jet Propulsion Laboratory (JPL). The forecast software uses a statistical allocation method [4, 5] to estimate when these requirements translate into DSN loading over various time frames. Long-range planning has several major purposes:

1. *Studies and analyses*: periods of particular interest or concern are examined to determine where there is likely contention among missions, for example around launches or critical mission events (maneuvers, planetary orbit insertion, or landings), or when construction of a new DSN antenna is under investigation.

2. *Downtime analysis*: identifying periods of time when necessary antenna or other maintenance can be scheduled, attempting to minimize the impact on missions.

3. *Future mission analysis*: in the proposal phase, missions can request analysis of their proposed DSN coverage as part of assessing and costing proposals for new missions.

The time range for long-range planning is generally six months or more into the future, sometimes as much as years.

2. MID-RANGE SCHEDULING

The mid-range scheduling phase is when detailed user requirements are specified, integrated, negotiated, and all tracking activities are finalized in the schedule. Starting at roughly four to five months before execution, users specify their detailed scheduling requirements on a rolling weekly basis. These requirements include tracking time and services required; constraining time intervals and relationships (e.g., minimum and maximum gaps); visibility constraints; and flexibilities.

Further discussion of the nature of these requirements and flexibilities is included in Sec. III.C; more detail is provided in [6] and [7].

Once the deadline passes and all requirements are in, the full set is integrated into an initial schedule in which conflicts are reduced by taking advantage of whatever flexibilities have been specified. There follows an optimization step where an experienced DSN scheduler interactively edits the schedule and further reduces conflicts by taking advantage of unspecified flexibilities and making further adjustments. At the conclusion of this phase, the schedule usually contains fewer than 30 conflicting sets of activities. It is then released to the scheduling user community who negotiate to reduce conflicts and further optimize coverage for their missions. This phase generally lasts seven to eight working days, after which the schedule is conflict-free or has only waived conflicts for specific reasons. This is considered the "negotiated schedule" that missions use to plan their integrated ground and spacecraft activities, including the development of onboard command loads based in part on the DSN schedule.

Following this point, changes to the schedule may still occur, but new conflicts may not be introduced. There is a continuing low level of no-impact changes and negotiated changes that occur all the way down to real time.

3. NEAR REAL-TIME SCHEDULING

The (near) real-time phase of DSN scheduling starts roughly eight weeks from execution and includes the period through execution of all the scheduled activities. Late changes may occur for various reasons (sometimes also impacting the mid-range phase):

1. Users may have additional information or late changes to requirements for a variety of reasons.

2. DSN assets (antennas, equipment) may experience unexpected downtimes that require adjustments to the schedule to accommodate.

3. Spacecraft emergencies may occur that require extra tracking or changes to existing scheduled activities.

For many missions that are sequenced well in advance, late changes cannot be accommodated readily.

B. STATUS OF DSN SCHEDULING SOFTWARE

The DSN scheduling software systems represent a collection built over many years and interfaced in a very heterogeneous manner [8–10]. At the present time, the different stages of the scheduling process are supported by different tools and databases, as described in Table 18.2.

The DSN has undertaken an overall unification and simplification of the scheduling software systems [6, 7, 11–13], of which the first increment has been operational since June 2011. This is called the SSS (S^3) and has initially been applied only to the mid-range process. S^3 is described in more detail in the following Section III.

In mid-2012, DSN deployed the first major software update to S^3, as well as initiated the development of replacement software for long-range planning and

TABLE 18.2 DSN SOFTWARE TOOLS THAT SUPPORT THE DIFFERENT PHASES OF THE DSN SCHEDULING PROCESS

Process phase	Software tools (software/ database)	Characteristic activities
Long-range	TIGRAS (RAP version) + MADB database	Identify and resolve periods of contention Plan for extended downtime Assess proposed missions Assess long-range asset options
Mid-range	S^3 Webapp/database	Schedule normal science operations Schedule preplanned spacecraft activities (maneuvers, unique science opportunities) Generate negotiated schedules for spacecraft sequencing Schedule network maintenance
Near real-time	TIGRAS (SPS version) + Service Preparation System (SPS) database	Predict generation for execution Reschedule due to unplanned resource unavailability Respond to spacecraft emergencies Activate preplanned launch contingencies

forecasting. In the following sections we provide an overview of S^3, including lessons learned from its initial deployment and how these have been addressed. We then discuss the current development effort for the long-range planning software, and how it leverages the mid-range software design to provide an improved set of long-range planning functionalities.

III. DSN SERVICE SCHEDULING SOFTWARE

The DSN S^3 project has had a long gestation period, but implementation of the mid-range stage started in December 2008. The system was developed in iterations, which were made available to users as prototypes so as to get early feedback. The initial operational deployment was carried out in June 2011, but the system has been in use for requirements entry and integration since December 2010, and for negotiation since April 2011.

The mid-range scheduling process was described briefly in Sect. II.A and has several unique characteristics worth elaborating on as the basis for several key architecture and design decisions about S^3.

A. MID-RANGE SCHEDULING PROCESS IMPLICATIONS ON S^3

Unlike other NASA network scheduling processes, DSN does not have a centralized scheduling authority that develops and publishes the schedule. Instead, the schedule is developed collaboratively via a negotiation process, on a rolling weekly basis. About 20 individual schedulers (not all of whom are full time), representing 37 DSN users, provide scheduling requirements to start the process each week. They then participate in negotiation where these requirements come into contention for oversubscribed assets. This negotiation process consists of a series of proposals and counterproposals (typically up to around 40) that eventually converge on a mutually agreed schedule with essentially no conflicts.

Before S^3, requirements were provided via e-mail, spreadsheets, or word processor documents, and integrated manually by a team dedicated to this purpose. Negotiation took place via e-mail or face-to-face meetings or teleconferences. There was no generally accessible schedule database for this phase of the process.

S^3 provides support for all the key elements of the mid-range process, based on a modern Web application and integrated, accessible database [14] (see Fig. 18.1). Users can directly enter their own scheduling requirements [6] and verify their correctness before the submission deadline. The database in which requirements are stored is logically divided into "master" and "workspace" areas. There is a single master schedule representing mission-approved requirements and DSN activities (tracks). Each user can create an arbitrary number of workspace schedules, initially either empty or based on the contents of the master, within which they can conduct studies and "what if" investigations, or keep a baseline for comparison with the master. These workspaces are by default private to the individual user, but can be shared as readable or read–write to any number of other users. Shared workspaces can be viewed and updated in real time; although there can only be one writer at a time, any number of other users can view a workspace and see it automatically update as changes are made. These aspects of the Web application architecture and database design support the collaborative and shared development nature of the DSN schedule.

In addition, S^3 offers specialized features to facilitate collaboration, including an integrated wiki for annotated discussion of negotiation proposals, integrated chat, notifications of various events, and a propose/concur/reject/counter workflow manager to support change proposals. Details on the design and use of the S^3 collaboration features are provided elsewhere [14].

Underlying the Web application and database is a scheduling automation component, the DSN Scheduling Engine [13] (DSE). The DSE provides a range of functions based on the semantics of the DSN scheduling domain, including

1. Expanding scheduling requirements into tracking or other activities;

2. Checking for and identifying conflicts in the schedule, i.e., situations that violate any DSN scheduling rules;

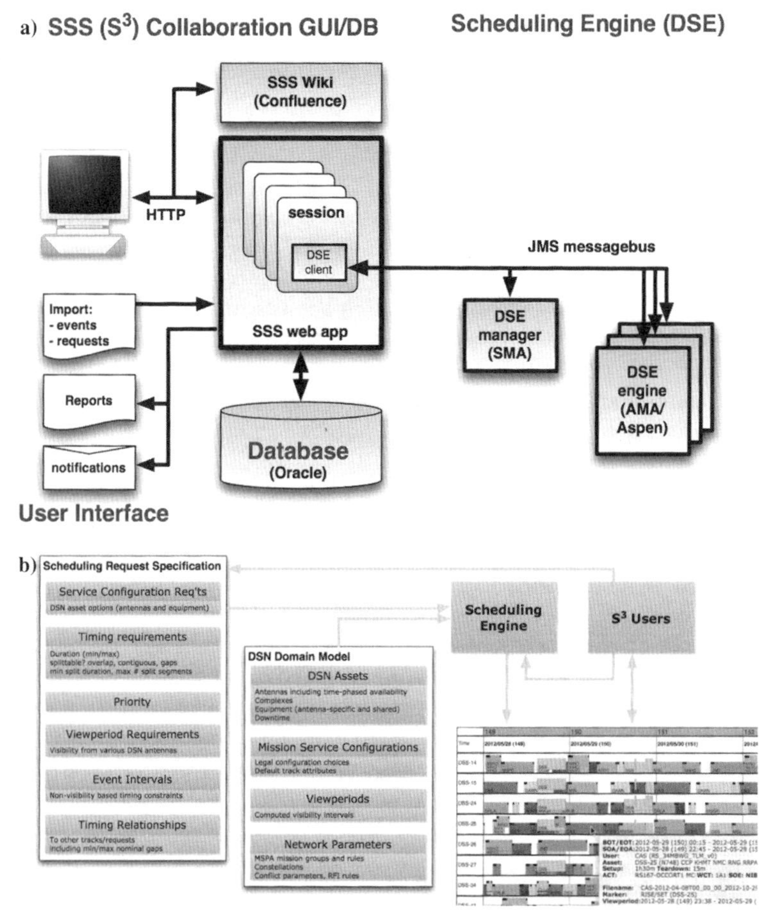

Fig. 18.1   (a) Block diagram of S³ software elements and how the user interacts with the system. (b) Major elements of the S³ request specification and domain model, also showing the S³ HTML5 canvas graphical user interface.

3. Checking for and identifying requirement violations in the schedule, i.e., situations where activities in the schedule do not meet the user's specified requirements and constraints;

4. Deconfliction algorithms that attempt to reduce conflicts or violations while preserving satisfied requirements.

The DSE is based on a distributed session-oriented infrastructure running the ASPEN planning system [15] with a DSN domain adaptation layer.

In the following we focus on some specific design elements of S^3 and the DSE, as the basis both for "lessons-learned" as well as the extension to long-range planning functionality.

B. S^3 KEY DESIGN ELEMENTS

Figure 18.1 shows a pair of block diagrams illustrating the main elements of the S^3 design. Users interact with the system as indicated on the left, via a Web browser that communicates securely with the Web application running on a server at JPL. A wiki is used to record general information about scheduling process status, as well as details relevant to negotiations in progress and to individual user scheduling requirements. File imports and exports are directly via the browser to/from the Web application. Notifications of various events (proposal status changes, shared workspaces, new conflicts) are provided via an in-application mechanism, but at least twice a day an e-mail notice is sent as a reminder of current negotiation status.

The central database is Oracle, which stores schedule and requirement data for both the baseline master schedule as well as a large number of user workspaces. The scheduling engine (DSE) runs in a distributed manner on separate hosts, and communicates via a Java Messaging System messagebus with the Web application. The DSE Schedule Manager Application (SMA) mediates requests for new scheduling sessions on behalf of Webapp users, and each user session is dedicated to an Aspen instance via the Aspen Manager Application (AMA). Although there is only one instance of the S^3 Web app running at a time, there are about 100 AMA/Aspen instances available, to eliminate any delays when users require the services of the scheduling engine.

C. DSN SCHEDULING REQUESTS

One of the key innovations at the heart of S^3 is the notion of request-driven scheduling, that is, the schedule is determined not by specific track allocations (which are an output product of the process), but by scheduling requests, requirements, and constraints that represent a service-oriented approach to scheduling. The intent is to move towards a more abstract basis for scheduling, where users are allocated services that can be more flexibly provided by the network. Scheduling requests as implemented in S^3 include the following types of information. Note that much of this is optional, either by defaulting or as only relevant if some dependent choices require elaboration.

DSN users represent their needs to the S^3 software system as scheduling requests. The main elements of a scheduling request are as follows:

1. *Service specification.* S^3, via the DSE, provides an abstraction level on top of DSN asset specifications that may be referenced by users much more simply than specifying all the possible options. At the physical level, the

spacecraft onboard electronics (frequency band, data rates, encoding), radiated power, distance, along with the DSN antennas, receivers and transmitters, and other equipment, determine what space and ground configurations are feasible. The abstraction level provided in S^3 is called a "service alias" such that a single service alias encapsulates a wide range of options, preferences, and associated information that is required to schedule the network. For example:

(a) Some users need the added sensitivity of more than one antenna at a time and so must be scheduled as antenna arrays using two or more antennas at once [16] (as many as four at a time).

(b) For navigation data, there are special ranging scenarios that alternate the received signal between the spacecraft and a nearby quasar, over a baseline that extends over multiple DSN complexes.

(c) For Mars missions, there is a capability for a single antenna to communicate with several spacecraft at once (called Multiple Spacecraft Per Aperture, or MSPA); although more than one at a time may be sending data to Earth, only one at a time may be receiving an uplink.

A more detailed description of service alias functionality is provided below.

2. *Timing constraints.* Users need a certain amount of communications contact time to download data and upload new command loads, and for obtaining navigation data. How this time is to be allocated is subject to many options, including whether it must be all in one interval or can be spread over several, and whether and how it is related to external events and to spacecraft visibility. Among the factors that can be specified in a schedule request are the following:

(a) *Reducible*: whether and how much the requested time can be reduced; e.g., to resolve conflicts;

(b) *Extendable*: whether and how much the request time can be extended, should the option exist;

(c) *Splittable*: whether the time must be provided in one unbroken track, or can be split into two or more separate tracks;

(d) *Split duration*: if splittable, the minimum, maximum, and preferred durations of the split segments, and the maximum number of split segments;

(e) *Split segment overlap*: if the split segments must overlap each other, the minimum, maximum, and preferred duration of the overlaps;

(f) *Split segment gaps*: if the split segments must be separated, the minimum, maximum, and preferred duration of the gaps;

(g) *View periods*: periods of visibility of a spacecraft from a ground station, possibly constrained to special limits (rise/set, other elevation limits), and possibly padded at the boundaries;

(h) *Events*: general time intervals that constrain when tracks may be allocated, such as day of the week, time of day (for accommodating shift schedules, daylight, and so on); orbit/trajectory events (occultations, maneuvers, surface object direct view to Earth, etc.); different event intervals may be combined (with optional inversion) and applied to a request.

3. *Track relationships.* In some cases, contacts need to be sufficiently separated that onboard data collection has time to accumulate data but not overfill onboard storage. In other cases, there are command loss timers that are triggered if the time interval between contacts is too long, placing the spacecraft into safe mode. During critical periods, it may be required to have continuous communications from more than one antenna at once, so some passes are scheduled as backups for others.

4. *Priority.* The DSN currently has a priority scheme ranging from 1 to 7, with 7 being nominal tracking and 1 a spacecraft emergency. Priority is relatively infrequently used, but it does have the effect that the scheduling engine will try to avoid conflicts with higher-priority activities if possible. Depending on their degree of flexibility, missions trade off and compromise to meet their own requirements, while attempting to accommodate the requirements of others. As noted above, one of the key goals of S^3 is to facilitate this process of collaborative scheduling.

5. *Preferences.* Most preferences are incorporated in the service alias and timing requirements described above, but some are directly representable in the scheduling request. For example, users may choose to schedule early, centered, or late with respect to the view period or event timing interval.

One characteristic of DSN scheduling is that, for most users, it is common to have repeated patterns of requests over extended time intervals. Frequently, these intervals correspond to explicit phases of the mission (cruise, approach, fly-by, orbital operations). These patterns can be quite involved, because they interleave communication and navigation requirements. S^3 provides for repeated requests, analogous to repeated or recurrent meetings in calendaring systems, so as to minimize the repetitive entry of detailed request information.

D. S^3 SERVICE CONFIGURATIONS

One of the challenges of modeling the DSN scheduling domain is the wide range of options available for making use of the network. As previously described, one of the primary attributes of a scheduling request is the specification of the DSN services that are needed, which must be transformed into a set of specific resource reservations to satisfy the request. It has been a key element of the S^3 design that users can specify their needs at a more general and abstract level, and that the system will translate that into the details, ensuring the right antennas and equipment are scheduled. This has the obvious advantage that there is flexibility in the implementation of a request that can be used by the DSN systems, for example, to optimize the schedule or to reschedule on short notice in case assets go down. At the same time, the scheduling system needs to handle a very detailed specification of requested tracking time, down to the selection of individual antennas and equipment types to be reserved. A design to accommodate this spectrum of possibilities has been developed and implemented in the DSE, and is illustrated in Fig. 18.2. In the figure, red highlights the information related to a single-track choice (left) and blue that related to a two-antenna array choice (right). More complex aliases are used to represent up to four station arrays, specialized ranging tracks Delta Differential One-way Ranging (DDOR), separate uplink and downlink options for multiple spacecraft tracked all at once, and maintenance activities that affect an entire complex or the entire DSN at once.

Each DSN service user or mission must define one or more service configurations, which are referred to by a name or "alias." Each configuration specifies the following information:

1. One or more choices for how antennas and equipment can be allocated to meet the user's DSN requirements.

2. For each choice, which sets of antenna and equipment are acceptable.

3. For each antenna/equipment combination, what are the default values for associated tracking parameters:
    (a) Setup and teardown time before and after the track;
    (b) 16-character activity description for the track;

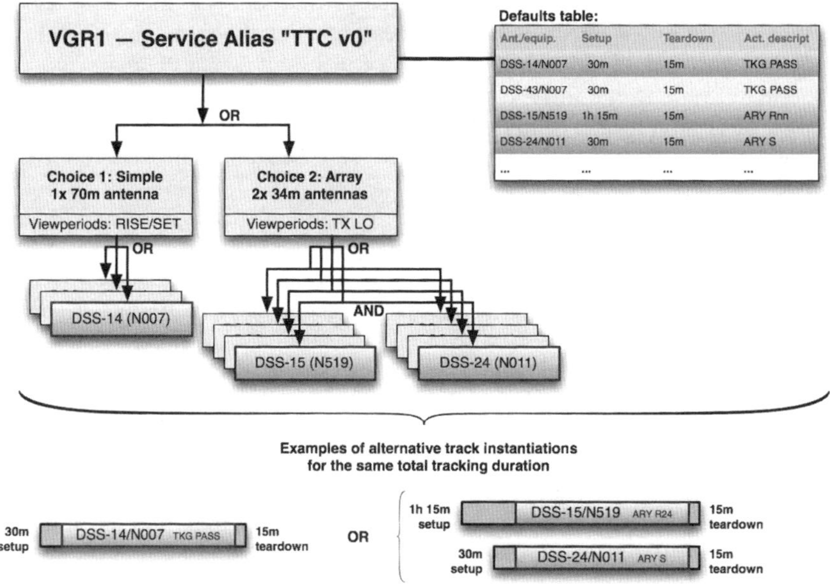

Fig. 18.2 Structure of a service alias representing a choice between single antenna and multiple antenna (array) implementations of the same tracking duration.

    (c) Standardized work category used to identify the kind of activity;
    (d) If applicable, a code for a specific sequence of events that define all steps that occur during the track.

A "choice" within an alias represents a high-level configuration option. For example, some missions may require either a single 70 m antenna, or two or more arrayed 34 m antennas. Each of these possibilities corresponds to very different antenna selections, while still satisfying the requirements of the overall service specification. Within a choice, all acceptable sets of antennas/equipment combinations must be specified, in preference order if applicable. Antenna/equipment combinations within a single antenna choice are in the form of a single list, while those in array choices contain multiple such lists. The same antenna may play different roles within these options, for example as a reference or slave antenna depending on how the equipment is to be configured.

Depending on the nature of the activity, different times must be scheduled for the activity setup (before tracking starts) and teardown (after it completes). Typical setup times are 30–90 min, while teardown times are usually shorter. The alias definition specifies the default (minimum) setup and teardown time for each antenna/equipment option. In special circumstances these times may be lengthened, but may not be shortened without violating DSN operational rules (and causing a setup or teardown conflict).

Once aliases are defined and validated, their usage in S^3 is straightforward. Whenever a user creates a scheduling requirement, a service alias must be specified, simply by name. The selected alias then determines all the remaining DSN asset requirements and options, while the remainder of the requirement goes on to specify parameters such as timing, duration, and relationships to other tracks. By separating the definition of aliases from their usage, it becomes easier to validate them to ensure that any selection is a legal DSN configuration for that service user.

Most DSN service users will define at least several aliases corresponding to their commonly used scheduling configurations. For example, one alias might specify downlink-only configurations, while another might be used for both downlink and uplink (the latter requires the allocation of transmitters as well as receivers and decoders).

In addition to specifying which service alias applies to a given requirement, S^3 provides a capability for overriding the definition of that alias in any requirement in which it is used. An alias override can only restrict the full set of choices allowed by the alias, not add additional ones. As a result, validating the alias is sufficient to ensure that only legal configurations can be generated by the scheduling system. Examples of possible alias overrides include the following:

1.   Single antenna versus arrayed configuration;
2.   One or more DSN complexes (Goldstone, Canberra, or Madrid);

3. Specific antenna subnet (70 m, 34 m, etc.);

4. Single specific antenna and equipment combination.

As well as filtering the set of antenna and equipment choices, users can also override the default values associated with any choice. For example, a particular requirement might need an extended setup time, or customized activity description string that differs from the default. These can be specified using alias overrides.

In addition to antenna and equipment options, certain other attributes of any corresponding activities are also specified by the alias. These include the following:

1. Which kind of view period must be used for scheduling, i.e., geometrical rise and set versus higher elevation transmitter limits;

2. Whether the activity is downlink or uplink only, which is used when scheduling MSPA activities as described in the next section;

3. Special activity description suffixes that must be included to indicate certain types of activities;

4. Effective date and time range, which allows for phasing alias definitions in or out of service as ground or spacecraft configurations change.

Service alias definitions are currently captured in XML files that specify all properties of the alias. A key design feature of the service alias concept in S^3 is that the same XML files are used by the DSE as by the domain-specific model of DSN activities and assets, and in the S^3 Web application graphical user interface (GUI) as the set of all legally selectable choices. Any changes to assets, aliases, or other mission parameters are immediately reflected in the DSE as well as the GUI, without code changes.

E. LESSONS-LEARNED FROM INITIAL S^3 DEPLOYMENT

The initial deployment of S^3 to operations occurred in June 2011. Since that time, a number of areas have been identified as most in need of revision and updating. In this section we highlight the most important of these and what has been done leading up to the first major update to the software.

1. PERFORMANCE

Performance has proven to be the single largest issue. A typical week of the DSN schedule consists of approximately 500 tracking activities. All versions of all requirements and activities are retained in the database. In addition, all past versions of conflicts and requirement violations are also retained. The access patterns for this volume of data, with its complex linkages, proved to be significantly slower than originally anticipated. In addition, the graphical scheduling components of the user interface were implemented in Adobe Flash, which requires a Web browser plug-in as well as specialized data pathways back to the host Web server. As a result, the system was insufficiently responsive when working on populated schedules.

Following an analysis and incremental change to the database storage approach, certain tables were denormalized, leading to a significant speedup in database queries for retrieving and displaying data. After these changes were made, the largest contributor to slow performance was determined to be the Adobe Flash component. Adobe Flash is not intrinsically slow, but for large volumes of graphical data (thousands of DSN tracking activities), it requires a great deal of Flash software design and optimization to provide a responsive GUI. This would have required the investment of substantially more resources than available. The decision to use Adobe Flash was made at a time several years ago when there were few alternatives. However, since then, the evolution of Web standards has made tremendous strides, and the development of HTML5 and its adoption as a standard by the major Web browsers has enabled several alternative approaches, including the HTML5 canvas and Scalable Vector Graphics (SVG) features.

Following a feasibility investigation in August 2011, it was determined that the HTML5 canvas feature would be sufficiently performant and flexible to support all the scheduling graphics needs of S^3, so the replacement of Flash by the HTML5 canvas was initiated. Among the benefits of this approach are the following:

1. *Simpler architecture.* No plug-in is required, and the canvas is managed by javascript that co-resides with other elements of the Webapp GUI, thus simplifying integration.

2. *No vendor lock-in.* The HTML5 canvas element is supported by all major browser platforms.

3. *Performance improvements driven by browser vendors.* All of the browser platforms have made large javascript performance improvements, which can be leveraged by a canvas GUI scheduling component.

As a rough measure, the HTML5 canvas component (Fig. 18.3) turned out to provide about ten times faster performance when rendering schedules in the Web GUI. As a representative point, a one-week schedule containing about 500 activities can be displayed in ~2 s, while an 8-week schedule with 4000 activities takes less than 8 s. These times can be slower when running over a low-capacity network connection, or on a computer where the browser is competing for resources with other applications. (In Fig. 18.3, mousing over a track brings up a transient window with detailed information about the activity (lower right). In this view, different missions are color-coded, and setup/teardown is indicated by the black segments at the top left and right of each activity. Each timeline represents one of the DSN antennas.)

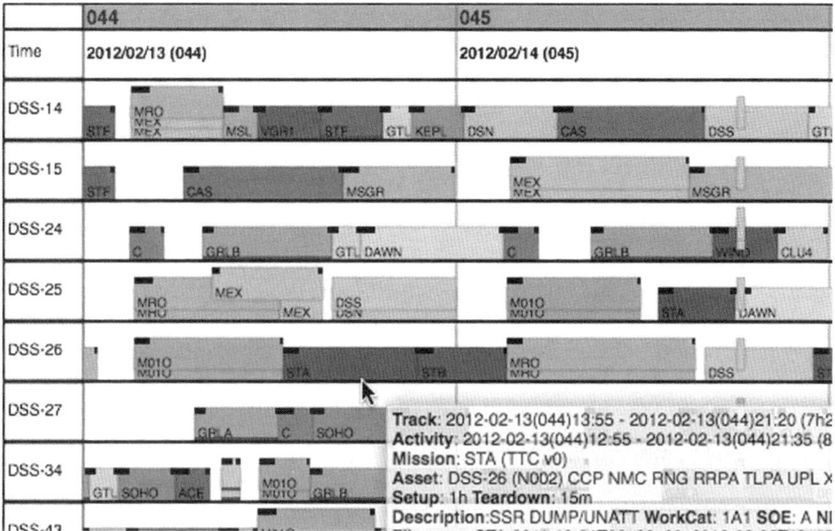

Fig. 18.3   Example of an HTML5 canvas view of a portion of the DSN schedule.

2.   TRACEABILITY

Although S^3 retains the history of every track and requirement in the database, there were no features initially included to report this information to users. From a usability perspective, users wanted to know not just what the current proposed or approved schedule contained, but how it got that way. In response, several elements were added into the latest update to make it possible for users to see this information:

1.   All of the actions taken in a workspace (shared or private), with timestamp, details about the activity, and who performed it;

2.   History of previous versions of an activity in a workspace or the master schedule;

3.   History of all activities derived from a specified requirement, in a workspace or the master.

These history features make it possible to tell in detail how activities got to their final states in the schedule, arbitrarily long after the fact.

3.   UNDO

As is the case for many Web-based database applications, changes to the S^3 schedule were committed to the database when made, without a capability to "back up" in time and revert recent changes. As well as traceability, this is a usability concern that has been addressed to some degree in the current S^3 release.

The use case of particular concern for "undo" was schedule editing, where users could potentially make a succession of track changes while trying to resolve a conflict, but then need to back up should the approach taken lead to a dead end. To address this, the HTML5 canvas revision of the schedule editor GUI provides for a client-side "change list" that is only committed to the database when specified by the user (or in certain other circumstances described in the following). Items in this change list are constantly displayed in the GUI and can be undone in reverse order, just as in a conventional desktop application such as a word processor. The format of the change list makes it easy to spot exactly what has changed, and multiple attribute changes to the same track are merged into a single item in the change list.

The trade-off with a client-side change list is that work can be lost if the browser crashes, network connectivity is lost, or the user switches context and forgets to commit their pending changes. To mitigate this, S^3 has implemented an "auto-save" feature that commits changes after 20 items are put into the change list, or after 30 min of unsaved changes. In addition, the user is prevented from refreshing the page or switching to another S^3 context without confirming that the pending changes will be lost.

There is a second use case for "undo," which has not yet been addressed, that of reverting back past explicit user "save" operations. S^3 already provides a "save point" function that allows the user to make a snapshot of the state of a workspace, then revert back to that state at any time in the future. A feature under consideration is a way to create automatic save points before each user-initiated database operation, thus allowing the option to return to any previous state with minimal effort.

F. EXTENDED SCHEDULING REQUEST TYPES

The initial deployment of S^3 has focussed on the most frequently encountered types of scheduling request types, which directly affect how DSN antenna allocations are to be constructed. Direct requests and requirements specify such attributes as

1. Tracking duration, and duration flexibility;

2. Whether activities can be split and, if so, whether the split segments must be overlapping, contiguous, or separated by gaps;

3. Which antennas and equipment combinations may be used to satisfy the requirement;

4. Timing linkages among activities;

5. Constraints on when activities can be scheduled based on occurrence of specified events.

Since the initial deployment of S^3, work has been ongoing on a second category of scheduling requirement, which indirectly affects allocations in a nonlocal manner. By this we mean that an extended time period and multiple activities may have to be examined to determine whether some preferred condition is satisfied. These conditions can have a varying degree of preference, ranging from very high to quite weak. It can also be the case that there is a trade-off between satisfying these types of requirements and the direct requirements noted above. Examples of these indirect requirements included the following:

1. Three out of every ten tracking passes must be scheduled at the Canberra complex (i.e., in the southern hemisphere).

2. There must be scheduled 6 h of uplink per day and 12 h of downlink per day, regardless of how this is divided among different antennas, measured midnight-to-midnight UTC.

3. There must be at least 24 h of tracking time scheduled per week, added up over four related missions.

4. Downlink tracks of sufficient duration must be scheduled to ensure that onboard recorder capacity is not exceeded.

We have denoted these types of scheduling requests as timeline constraints or preferences, as they are best assessed by considering the overall timeline of activities (or subset of activities) for a DSN service user over some time period. Table 18.3 includes a more detailed list of major timeline requirement types and their parameters.

Because these requests have a varying degree of preference, and therefore need to be accessible to the judgment of scheduling users, we have pursued their incorporation into S^3 in two phases:

1. Integrated with the scheduling system GUI for visualization along with the schedule itself;

2. Incorporated into the DSE algorithm set, for invocation as strategies or heuristic repair and rescheduling options that can be included or not into the normal scheduling process.

Integration with the S^3 GUI has built upon the newly deployed S^3 HTML5 canvas-based GUI (Fig. 18.3), which has enabled the rapid extension of the GUI to additional visualization elements. We provide examples of the visualization of each of the major categories of timeline requirements in the following.

The Total Time timeline requirement applies to about 25% of the DSN user set, but over a wide range of timescales, from a full week on, down to a fraction of a single day. An example for the GRAIL A/B mission (two spacecraft in lunar orbit) is shown in Fig. 18.4a.

The Tracking Gaps timeline requirement applies to about a third of the DSN user set. In some cases, the gaps of concern are only for certain activity types,

TABLE 18.3   TIMELINE REQUIREMENT TYPES, WITH EXAMPLES AND PARAMETERS

Request type	Examples	Parameters
Total time	8 h of tracking per day 6 h of uplink tracking each midnight to midnight UTC 24 h of specific activity types per week summed over four different but related spacecraft	Mission(s) Service aliases Time frame (1 day, 1 week, etc.) Min/max tracking times with yellow/red limits
Tracking gaps	6–12 h gap between tracks, measured mid-point to mid-point Gaps no greater than 8 h measured EOT to BOT	Mission Service aliases Min track gap Max track gap Yellow limits Measured by (BOT–BOT, EOT–EOT, mid track to mid track)
DSN complex distribution	3 of 10 tracks per week must be scheduled at Canberra DSN complex At least one track per week must be scheduled at each DSN complex	Mission Duration List of (complex, count)
Recorder	Do not exceed onboard recorder volume capacity limit	Mission Track overhead duration Recorder collection rate (X units/s) Yellow/red recorder max capacity Recorder downlink rates (antenna, downlink rate X units/s) Initialization rule

BOT: beginning of track; EOT: End-of-Track.

as illustrated in Fig. 18.4b where gaps are only significant between adjacent ranging passes.

About 20% of users have DSN Complex Distribution requirements, but this varies depending on the phase of the mission. These requirements are typically driven by navigation considerations, where it is important to have ranging data from widely separated baselines so as to reduce ephemeris errors. Examples are shown in Fig. 18.4a–c, where satisfaction or violation of the distribution requirement is clearly visible. In Fig. 18.4a, there is a gap constraint and a minimum tracking time constraint in a 24h UTC day (both of which are violated and shown as a dark color in the Fig. 18.4, but in red on the application GUI);

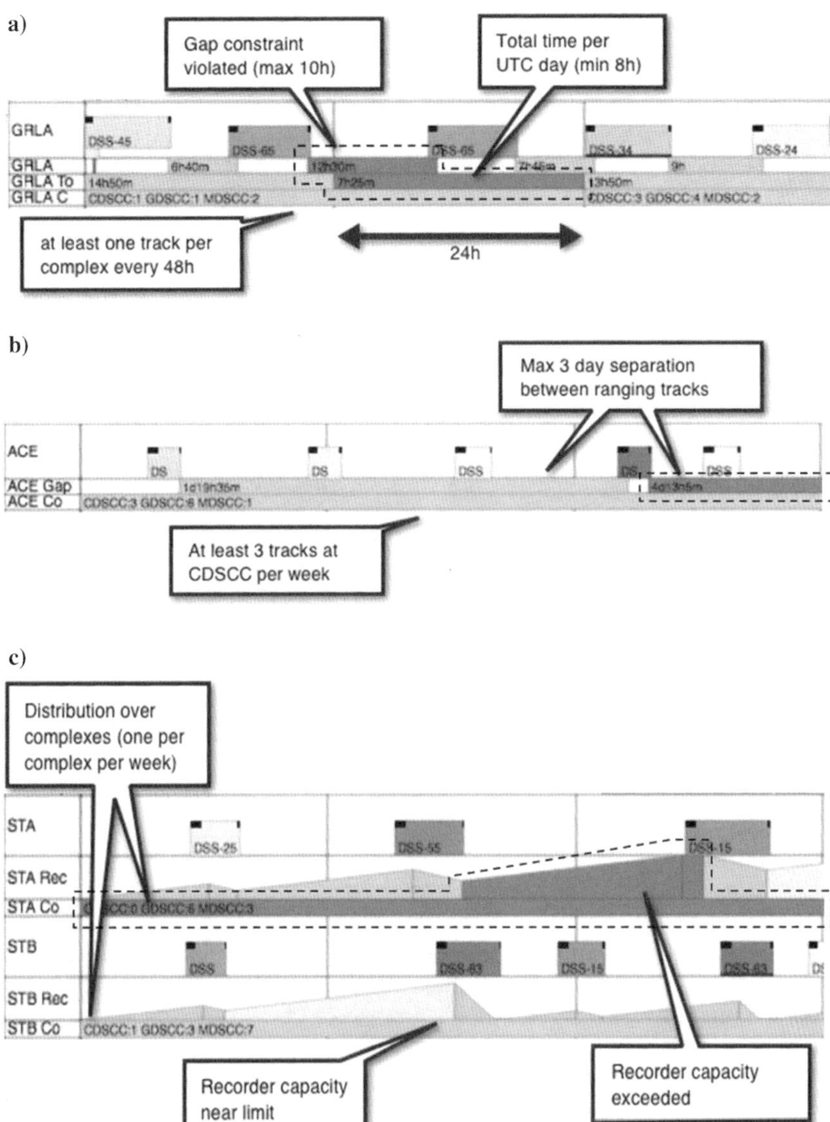

Fig. 18.4 Timeline constraints for three representative spacecraft, depicted in the S^3 scheduling HTML5 GUI: a) Example of multiple timeline requirements applied to a single spacecraft, here GRAIL A, one of a pair of lunar orbiters; b) Example of a gap constraint between ranging passes only, i.e., ignoring the intervening tracking passes; c) Example of a recorder timeline constraint applied to the STEREO A/B mission pair.

there is also a requirement to track on all three DSN complexes within a 48h period (satisfied). In Fig. 18.4b, the second maximum gap requirement has been violated and the resulting interval again shows the dark shade representing red (violation) on the GUI.

Although most missions have onboard recorders, only a handful can potentially be modeled simply enough to include in the early stages of DSN scheduling. For those missions with uniform data collections rates and well-defined downlink rules, the Recorder timeline requirement can provide early visibility into recorder capacity and how it is affected by specific scheduling choices. An example is shown in Fig. 18.4c for the STEREO A/B spacecraft. Here, the figure shows violation of the constraint in an interval where the accumulated data would exceed the recorder capacity. Note that the recorder volume drops more quickly when a 70 m contact (e.g., DSS-63) is scheduled, due to the higher downlink data rate. The STEREO spacecraft also have a requirement to schedule at least one track per week at each complex, here satisfied only for STEREO B.

By providing a highly visual view of these timeline constraints and preferences, users who are working on schedule changes to resolve conflicts can immediately see whether their proposed changes would introduce any violations. Presently, many scheduling users have custom scripts that they use to evaluate proposals from other users, but by providing for common models and visibility, feedback can be provided much more rapidly. This feedback has the potential to reduce the overall negotiation process effort and duration.

IV.   LONG-RANGE PLANNING AND FORECASTING

A.   SIMILARITIES AND DIFFERENCES BETWEEN DSN LONG-RANGE AND MID-RANGE PROCESSES

By necessity, there are many similarities between the mid- and long-range planning and scheduling functions. Underlying both is the set of current and future DSN assets, including antennas and equipment, some coming into service and others being decommissioned. Both are based on DSN usage requirements from a varying mission set with a wide range of time-dependent tracking and navigation needs. Both are charged with arriving at an ultimately feasible allocation of DSN resources by balancing user needs and resolving periods of resource contention.

However, long-range planning has some significant differences from mid-range planning:

1. Long-range planning has to deal with numerous and sometimes intrinsic sources of uncertainty, including
   (a) Unpredictable spacecraft locations for some missions and trajectory types, leading to uncertainties in visibility times from the different DSN antennas;
   (b) Unknown science targets beyond some time horizon in the future;

(c) Uncertainties in the mission set, due to funding changes, launch date changes, or mission extensions.

2. Optimization criteria and scenarios differ from those of mid-range planning, where the main objectives are to minimize conflicts in the schedule and violations of user requirements. For long-range planning a variety of other objectives may come into play, including

(a) Identifying the best times to schedule extended downtime for preventive maintenance, minimizing the impact on active missions;

(b) Identifying the best times to schedule special flexible but resource-intensive operations, such as reference frame calibration activities;

(c) Maximizing the satisfaction of requirements where, due to contention, not all requirements can be satisfied across the entire DSN user base.

3. In addition, long-range planning needs to provide information to mission planners about where contention with critical events may occur, so that this can be taken into account as early as possible in each mission's planning process. In many cases this needs to be provided during the mission proposal phase when, for both feasibility and costing, it is necessary to map out DSN allocation needs to some preliminary level of accuracy. Such proposal studies also impose a requirement for protection of proprietary or competition-sensitive information, whereas the mid-range process for DSN allows general access to scheduling requirements and to the schedule itself.

4. Finally, long-range planning needs to support the specification of a more abstract type of requirement with less detail than would be acceptable in mid-range planning. This serves two purposes: it represents at a coarse level some of the uncertainty in requirements, and it makes it easier to specify "what if" alternate scenarios.

B. LEVERAGED DEVELOPMENT OF LONG-RANGE PLANNING AND FORECASTING TOOLS FOR DSN

Building first on the similarities noted above, the first phase of development of the Local Analysis and Prediction System (LAPS) tools will make direct use of a number of capabilities already deployed operationally in the mid-range S^3 software, including

1. A model of DSN asset availability including antennas and equipment, with time-varying availability for new construction or new types of equipment, and out-of-service dates for retired assets;

2. A model of DSN user and mission types, including

(a) Ground- and space-based users, schedulable on non-interference basis or not,

(b) Multi-spacecraft constellations,

    (c) Multiple spacecraft per antenna (MSPA) groupings and their special scheduling rules;

3.   A service alias model, which defines what asset sets are allowable and preferable for a user, depending on the service desired (described in more detail in Sec. III.D);

4.   A view period model, specifying legal visibility intervals of various types, calculated by the Service Preparation System and imported in a form optimized for scheduling;

5.   A scheduling requirement model, allowing (but not requiring) allocation needs to be specified to the same level of detail as mid-range requirements (Sec. III.C), should such detail be both available and necessary for the type of study to be undertaken;

6.   DSN Scheduling Engine algorithms used in the mid-range process, which would allow for fully detailed "what if" generation of hypothetical mid-range schedule periods in those cases where sufficient detail is available to warrant this level of analysis.

Compare Fig. 18.5 and Fig. 18.1b; the existing S^3 data model and algorithms provide a basis that can be readily extended as indicated for long-range planning.

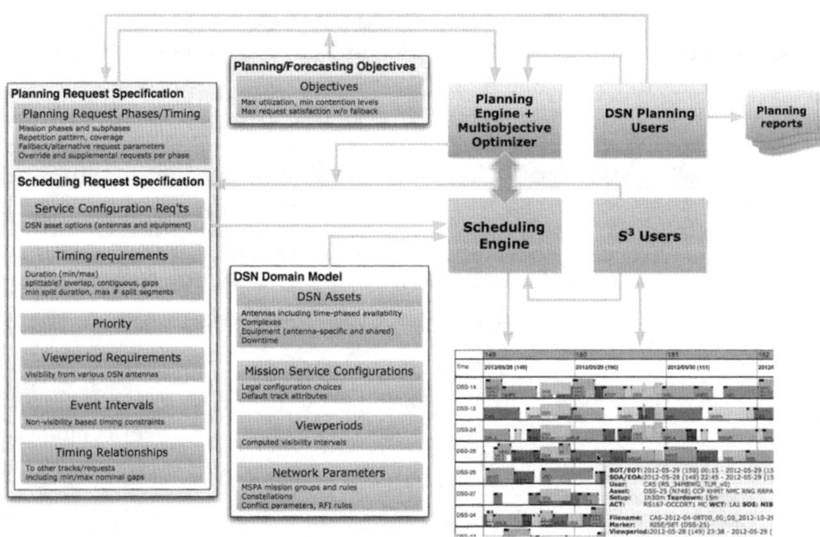

Fig. 18.5 Extension of the S^3 data model to support long-range planning, forecasting, and downtime analysis.

Reuse of the S^3 software base in these areas provides a large degree of leverage in the development of LAPS, but several other areas are also being addressed with additional capabilities:

1. Planning request representation to allow for more abstract and high-level specification of allocation needs than the scheduling requirement model allows (e.g., "3 × 8 h tracks/week on 34 m Beam Wave Guide (BWG) for the 6 months of interplanetary cruise"); at the same time, planning requests will be convertible automatically into mid-range scheduling requests to minimize duplicate data entry and speed up the mid-range process.

2. Capability to define and run planning scenarios in an automated way, such as

   (a) To assess a range of options for downtime placement;
   (b) To evaluate nominal and fallback requirement options for resource contention periods;
   (c) To quantify the impact of a mission's alternative launch dates on projected resource loading.

3. Multi-objective optimization mechanism to automatically generate a portfolio of candidate plans/schedules optimizing the trade-offs among multiple quantitative objectives.

The incorporation of multi-objective optimization into LAPS offers a new way to optimize DSN resource allocations, taking into account that there is no single objective that captures all the disparate goals and objectives that are important. Multi-objective optimization has been used in a wide variety of problem domains, including scheduling for science missions [17–20] and generating some requirements inputs to the DSN mid-range process [21].

The initial phase of LAPS development will encompass the modeling and optimization noted above. The second phase will extend the user interface elements of the software to allow end-users, such as mission planners and schedulers, to directly enter their own planning requirements and conduct "what if" analyses using a baseline DSN asset and mission model. It will also include an extended report generation mechanism to generate a wider variety of tabular and graphical output formats.

V. CONCLUSION

In this chapter we have described the DSN scheduling process and software, including the initial operational deployment of the S^3 system, and its ongoing extension to support long-range planning and forecasting. S^3 represents a new approach to scheduling the DSN, embodying a request-driven approach to scheduling along with a collaborative peer-to-peer negotiation environment using modern Web application and database technology. Future work is expected to address a number of areas, including the following:

1. *Extension to real-time scheduling.* As described above, this third phase of the DSN scheduling process covers the period from execution to some number of weeks in the future. Extending S^3 to support this phase involves some challenging technical problems of integration with existing systems and support for contingency scheduling (e.g., launch slips, unplanned asset downtime). At the same time, bringing the information model of S^3 into the real-time domain will allow for decision making considering options that are not now accessible.

2. *Cross-network scheduling.* NASA has recommended [22] integrating access to the capabilities provided by its three major networks: DSN, the Space Network (SN), and the Near Earth Network (NEN). For those users requiring services from two or all three of these networks, such integration would be a source of significantly improved efficiency and cost savings. S^3 has the potential to serve as a common scheduling platform in this regard. It is interesting to note that nowhere on the S^3 scheduling request editor main user interface is there any indication that the user is working with the DSN; this is apparent only when drilling down into the detailed visibility and event intervals, and service alias definitions.

ACKNOWLEDGMENTS

The research described in this chapter was carried out at the Jet Propulsion Laboratory, California Institute of Technology, under a contract with the National Aeronautics and Space Administration. We gratefully acknowledge the support of the DSN scheduling community over the course of this work.

REFERENCES

[1] "NASA's Deep Space Network," http://deepspace.jpl.nasa.gov/dsn/, 2010 [last accessed 15 May 2013].

[2] Imbriale, W. A. *Large Antennas of the Deep Space Network*, Wiley, New York, 2003.

[3] Berner, J. B., and Statman, J. "Increasing the Cost-Efficiency of the DSN," *SpaceOps 2008*, spaceops.org, Heidelberg, 2008.

[4] Lacey, N., and Morris, D. G. "JPL RAPSO Long-Range Forecasting," *12th AAS/AIAA Space Flight Mechanics Meeting*, Vol. AAS 02-223, AAS/AIAA, 2002.

[5] Morris, D., and Burke, E. "The JPL Resource Allocation Planning and Scheduling Office (RAPSO) Process," *12th AAS/AIAA Space Flight Mechanics Meeting*, Vol. AAS 02-219, AAS/AIAA, 2002.

[6] Johnston, M. D., Tran, D., Arroyo, B., Call, J., and Mercado, M. "Request-Driven Schedule Automation for the Deep Space Network," *SpaceOps 2010*, Paper No. AIAA-2010-2265, AIAA, 2010.

[7] Johnston, M. D., and Tran, D. "Automated Scheduling for NASA's Deep Space Network," *6th International Workshop on Planning and Scheduling in Space (IWPSS)*, 2011.

[8] Smith, J., and Wang, Y. "E-Scheduling the Deep Space Network," *Proceedings of the Aerospace Conference 2001*, Vol. 7, IEEE, 2001.

[9] Werntz, D., Loyola, S., and Zendejas, S. "FASTER – A Tool for DSN Forecasting and Scheduling," *Proceedings of the 9th AIAA Computing in Aerospace Conference*, AIAA, 1993, pp. 230–235.

[10] Loyola, S. J. *PC4CAST: A Tool for DSN Load Forecasting and Capacity Planning*, Jet Propulsion Laboratory, Pasadena, 1993, pp. 170–184.

[11] Johnston, M. D., and Clement, B. J. "Automating Deep Space Network Scheduling and Conflict Resolution," *International Conference on Autonomous Agents and Multiagent Systems (AAMAS)*, ACM, 2006.

[12] Clement, B. J., and Johnston, M. D. "Design of a Deep Space Network Scheduling Application," *International Workshop on Planning and Scheduling for Space (IWPSS-06)*, Space Telescope Science Institute, Baltimore, 2006.

[13] Johnston, M. D., Tran, D., Arroyo, B., and Page, C. "Request-Driven Scheduling for NASA's Deep Space Network," *International Workshop on Planning and Scheduling for Space (IWPSS)*, 2009.

[14] Carruth, J., Johnston, M. D., Coffman, A., Wallace, M., Arroyo, B., and Malhotra, S. "A Collaborative Scheduling Environment for NASA's Deep Space Network," *SpaceOps 2010*, AAS/AIAA, 2010.

[15] Chien, S., Rabideau, G., Knight, R., Sherwood, R., Engelhardt, B., Mutz, D., Estlin, T., Smith, B., Fisher, F., Barrett, T., Stebbins, G., and Tran, D. "ASPEN – Automating Space Mission Operations using Automated Planning and Scheduling," *SpaceOps 2000*, AAS/AIAA, 2000.

[16] Luong, I., and Bliss, D. "DSN Antenna Arraying; Its Past, Its Contributions to Achieving the Spitzer Space Science Mission Objectives, and Its Future Promise," *SpaceOps 2008*, AAS/AIAA, 2008.

[17] Johnston, M. D. "Multi-Objective Scheduling for NASA's Deep Space Network Array," *International Workshop on Planning and Scheduling for Space (IWPSS-06)*, Space Telescope Science Institute, Baltimore, 2006.

[18] Johnston, M. D. "An Evolutionary Algorithm Approach to Multi-Objective Scheduling of Space Network Communications," *International Journal of Intelligent Automation and Soft Computing*, Vol. 14, 2008, pp. 367–376.

[19] Johnston, M. D., and Giuliano, M. "MUSE: The Multi-User Scheduling Environment for Multi-Objective Scheduling of Space Science Missions," *IJCAI Workshop on Space Applications of AI*, International Joint Conference on Artificial Intelligence (IJCAA), 2009.

[20] Giuliano, M., and Johnston, M. D. "Multi-Objective Evolutionary Algorithms for Scheduling the James Webb Space Telescope," *International Conference on Automated Planning and Scheduling (ICAPS)*, ICAPS conference, 2008.

[21] Johnston, M. D., and Giuliano, M. "Multi-Objective Scheduling for the Cluster II Constellation," *6th International Workshop on Planning and Scheduling in Space (IWPSS)*, 2011.

[22] SCAWG, "NASA Space Communications and Navigation Architecture Recommendations, Final Report," Space Communications Architecture Working Group (SCAWG), NASA, 2006.

CHAPTER 19

Intrinsic Interoperability of Services: A Dream or a Key Objective for Mission Operation Systems

Mehran Sarkarati[*], **Mario Merri**[†] **and Mariella Spada**[‡]
European Space Agency, ESA/ESOC, Darmstadt, Germany

Sam Cooper[§]
SciSys UK Ltd, Bristol, United Kingdom

I. INTRODUCTION

A. INTEGRATION VERSUS INTEROPERABILITY

Interoperability and integration are closely related concepts in general software engineering, and are occasionally used, by mistake, as exchangeable terms. They refer, however, to two very distinct software design concepts, which can be considered the exact opposites. The ultimate goal of interoperability, the so-called intrinsic interoperability of software components, is eliminating the need for integration.

Each of these two design concepts has significantly influenced the IT landscape over the last two decades. Integration has been the focus of the Enterprise Application Integration (EAI) era, while interoperability concept has been declared one of the strategic goals of Service Oriented Architectures (SOA). The EAI era has been marked by the emergence of a set of enterprise integration patterns [1] and a multitude of both open source and commercial integration middleware products.

The Enterprise Service Bus (ESB) compound pattern [2] has complemented the EA integration patterns with supplementary aspects in support of SOA. The term ESB has become an ambiguous IT buzzword in recent years, because software product vendors have used it to brand their products. It is therefore helpful for further discussions in this chapter to regard the ESB purely as a compound integration pattern, as shown in Fig. 19.1.

---

[*]Ground Data Systems Manager, HSO-GDA, CCSDS SM&C Working Group Member; mehran.sarkarati@esa.int.
[†]Head of Mission Data Systems, HSO-GD, CCSDS SM&C Working Group Chair; mario.merri@esa.int.
[‡]Head of Applications and Special Projects Data Systems, HSO-GDA, mariella.spada@esa.int.
[§]CCSDS SM&C Working Group Member, sam.cooper@scisys.co.uk.

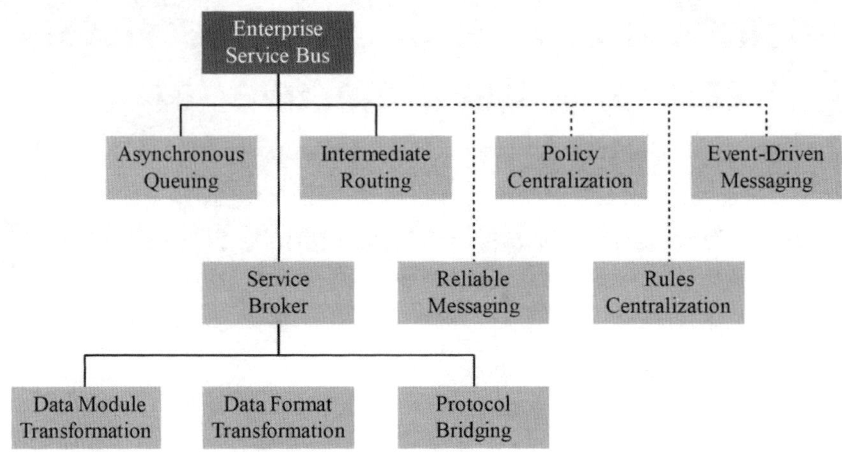

Fig. 19.1   The Enterprise Service Bus compound pattern [2].

Apache Camel, Apache ServiceMix, Microsoft BizTalk Server, IBM Web-Sphere Message Broker, IBM WebSphere ESB, TIBCO BusinessWorks, BEA System's AquaLogic Service Bus, Oracle Service Bus, SoftwareAG WebMethods Integration Server, Talend ESB, Progress Software's Sonic ESB, Fuse ESB, Mule ESB and many other middleware products are implementations of these patterns complemented with a set of out-of-the-box adapters for specific software technologies and products. Figure 19.2 illustrates the ESB from a middleware platform perspective.

Although the ESB concept, design pattern, and software products are usually associated with SOA, many of the abovementioned middleware platforms are often used for solving the traditional integration problem of overcoming disparity between interacting non-service-oriented software applications. The philosophy behind all EAI patterns has been to allow disparity at different architectural levels among the interacting software applications (e.g., implementation technology, communication, data model, and data format). The main objective of the EAI middleware platforms is to reduce the resulting integration effort.

In today's mission operations world, integration is the primary means of interaction between mission operation software applications, both within the boundaries of a space organization and in cross-organization cooperation scenarios. This is typically achieved by agreeing on an interface control document (ICD) and the communication protocols to be used between any two interacting software applications and transforming to and from the agreed ICD within the boundaries of each application, with or without integration middleware.

In contrast to the EAI paradigm, one of the strategic goals of service-oriented computing is to provide at the time of design for abstraction from the implementation and communication technology of units of service-oriented computing logic – the so-called services – hence avoiding, as much as possible, disparity at run time when service providers and consumers engage in interaction scenarios. This goal is addressed by adopting certain service design principles, such as service abstraction, discoverability, composability, autonomy, loose coupling, and statelessness [3]. Applying a set of domain-specific design standards consistently to all services within a domain service inventory and agreeing on a reference service model can also significantly increase the interoperability of services. This design principle is referred to as the "standardized service contract" [3] and will be elaborated on in the next sections of this chapter.

B. SOURCES OF DISPARITY AMONG INTEGRATING SOFTWARE APPLICATIONS

When discussing integration and interoperability it is worth analyzing the sources of incompatibilities between two interacting software applications. The main sources of disparity are illustrated in Fig. 19.3 and can be summarized as follows:

1. Incompatible software implementation technologies; e.g., a telemetry (TM) provider application is implemented in C programming language,

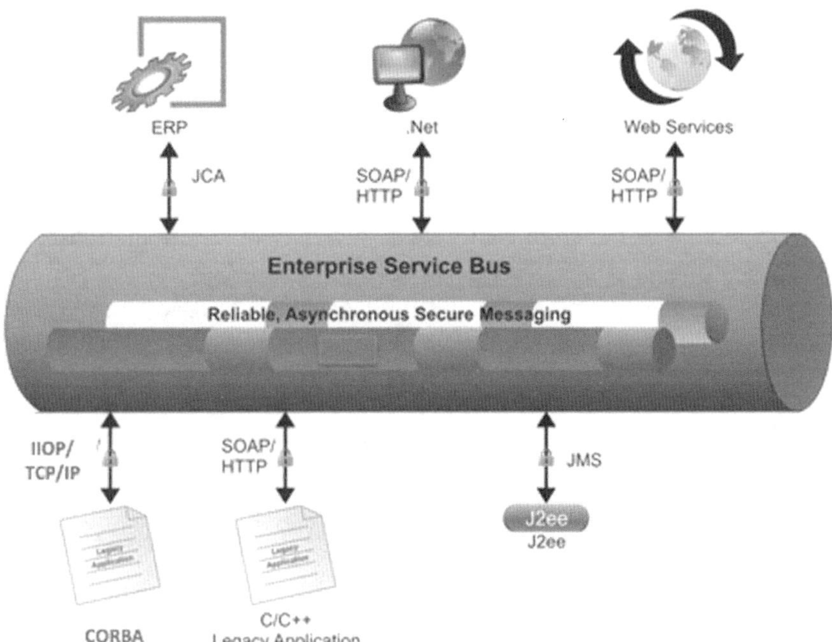

Fig. 19.2   Enterprise Service Bus as integration middleware.

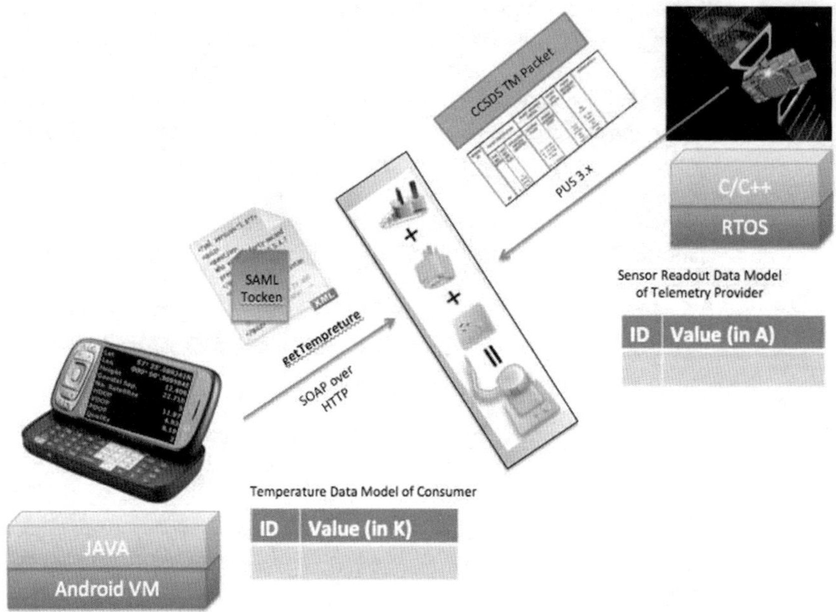

Fig. 19.3   Sources of disparity in interacting software applications.

running on a real-time operating system onboard a spacecraft, while the TM consumer is a mobile application, developed in Java running in an Android environment.

2. Incompatible data models used by each application; e.g., the specification of TM parameters can be different between a TM data provider and consumer.

3. Incompatible data format and communication protocols; e.g., the TM data provider publishes data in Consultative Committee for Space Data Systems (CCSDS) packet TM format while the consumer, being a mobile device, can only understand Extensible Markup Language (XML) encoded data received via the Hypertext Transfer Protocol (HTTP).

4. Incompatible syntax and semantics of the exposed capabilities of interacting applications; e.g., the exposed interface of the TM data provider for retrieving TM data provides raw TM data in the Système International d'Unités (SI) system, while the consumer expects calibrated engineering data in imperial units.

5. Nonfunctional requirements such as quality of service (QoS) and security policies of the exposed interfaces; e.g., the TM data consumer application attempts to authenticate itself using a Security Assertion Markup Language (SAML) token while the TM providing system expects encrypted username

and password. To give a QoS example, satellite visibility in a particular location may be limited to few minutes, while the consumer is supposed to send requests at any given time.

To achieve intrinsic interoperability (i.e., avoid the need for custom integration), interacting applications (both service consumer and service provider) must comply with the following:

1. A set of technical industry standards for abstracting from software implementation and communication technologies to allow a generic middleware to establish, at run time, communication between them.

2. A set of domain-specific design standards to achieve a common service model that ensures that the content of the communication is interpreted the same way by all communicating parties.

Adherence to the first point does not mean that all interacting parties must use the same software implementation and communication technology. It requires instead that the capabilities of each application be exposed through an implementation and communication technology agnostic service contract. The Web services specifications framework, the so-called WS-* specifications (WSDL, WS-Security, WS-Addressing, WS-Reliability, WS-Policy, WS-Transaction), the Service Component Architecture (SCA) specifications [4] and the CCSDS Message Abstraction Layer (MAL) specification [5] are examples of such agnostic service specification frameworks.

The implementations of these specifications rely on a run-time environment (middleware) that can map the abstract service specifications dynamically to the concrete implementation and communication technologies on the service provider and consumer side. This mapping from the abstract service contract to the concrete implementation and communication technology is called service binding.

Despite the importance of the first point as a prerequisite for achieving interoperability, it is important to acknowledge that no service abstraction framework and corresponding middleware can alone provide intrinsic interoperability between service providers and consumers, as they do not imply any assumptions on the content and quality of the communication. To fully appreciate the difference, consider the example of a USB keyboard for a PC. The USB standard harmonizes the physical interface between the keyboard and the computer as well as the electronic communication between the two devices for successful exchange of messages. However, it makes no assumptions on the content of the exchanged messages and how they should be interpreted. This becomes apparent when plugging a PC keyboard into a Mac computer and pressing the Windows key. The interpretation of the received messages from the keyboard and the associated capabilities are very different in Windows and Mac computers. This is where standardizing the semantics of the service contracts for each domain service

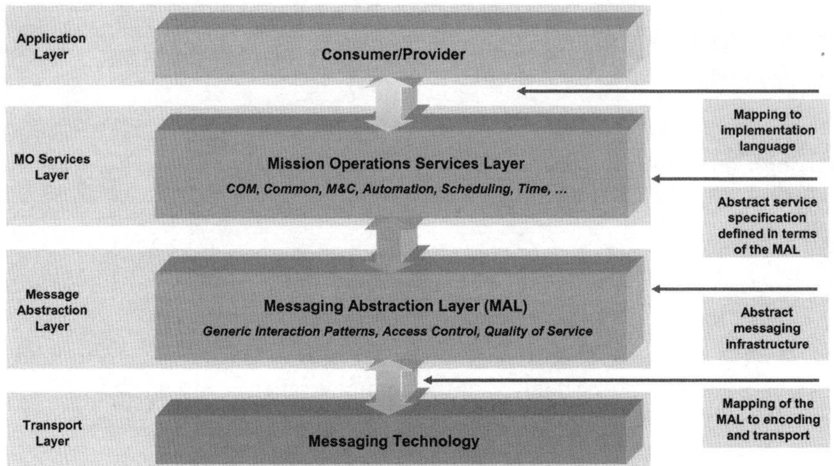

Fig. 19.4   The layered architecture of the CCSDS Mission Operations framework.

and the role of a reference service model come into the picture, as we will explain in Sec. III.

C.   THE CCSDS MISSION OPERATIONS FRAMEWORK BACKGROUND [6]

The Spacecraft Monitoring & Control (SM&C) Working Group of the CCSDS, which sees the active participation of 10 space agencies, has been working since 2003 on the definition of a service-oriented architecture for space-mission operations. The ambitious goal of the Working Group is to define a set of standardized, interoperable mission operations services, which will allow the rapid and efficient construction of cooperating space systems (Ground Segment, but also part of the Space Segment).

For this purpose the Working Group has defined a mission operations layered service framework, shown in Fig. 19.4, which allows these services to be specified in an implementation and communication agnostic manner. The core of the mission operations service framework is its MAL, which ensures interoperability between mission operations services deployed on different framework implementations.

The mission operations services are defined in compliance with a reference service model using an abstract service description language specified by the MAL (similar to WSDL). For each concrete software implementation and communication technology the abstract service contracts must be bound to that particular technology. The MAL, in turn, layer provides standardized interfaces in the form of Application Programming Interfaces (API) towards both upper and lower layers.

II. THE CANONICAL MESSAGE FORMAT AND THE ESB ROLE IN ACHIEVING INTEROPERABILITY

The scenario shown in Fig. 19.5 depicts a point-to-point interaction between the service provider and service consumer. In this scenario the integration layer is responsible for bridging between different service provider and consumer bindings at run time. As the number of service providers and consumers grows, the bridging can become a challenge and would require upfront agreements between any two interacting parties, hence a barrier to out-of-the-box interoperability (i.e., no plug-and-play). The hub-and-spoke pattern, also shown in Fig. 19.5, solves this issue by introducing a canonical message format, which is used by an ESB to bridge from a given service binding to any other binding, hence allowing n-to-n communication among all plugged service providers and consumers.

It is again important to appreciate that the utilization of a canonical message format by the ESB does not mean that all services must adopt the same message format and communication protocol. It means, instead, that by implementing once a single transformation from a given binding to the canonical message format, the ESB can establish communication to any other already bridged binding. The advantages of the hub-and-spoke concept can best be explained with an example of natural languages. Consider the example of a medical research association with fifty member states. The objective of the association is to make the work results of each researcher available, as soon as possible, to other researchers in all member states. If all researchers published their work results in a common language, e.g., English, the association would need no translators. This would be similar to requiring all interacting applications to use the same software implementation and communication technology, which is often not a realistic

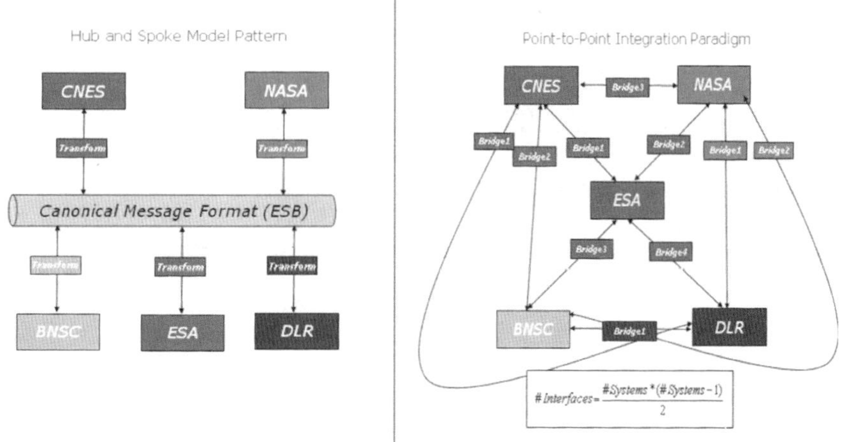

$$\#Interfaces = \frac{\#Systems * (\#Systems - 1)}{2}$$

Fig. 19.5   The hub-and-spoke pattern vs. point-to-point integration.

scenario. In our example many renowned researches may publish their work in their native language. The association could hire a large number of translators for translating from any of the fifty member state languages to any other language. In this scenario the association would need 49 translators per language, hence 1225 translators in total. Moreover, each time a new member state joins the association, many new translators must be hired (as many as the previous number of member states). This would be an example of a point-to-point communication pattern. A more appropriate scenario, however, would be to hire only 50 translators, who would translate the research results from any member state language to a common language, e.g., English, for internal use in the translation department, allowing the other 49 translators to further translate it from English to their own native languages. In this hub-and-spoke scenario, when a new member state joins the organization, only a single new translator must be hired and its researchers would have automatic access to the work results of all other researchers linked to the association, and vice versa.

III. THE ROLE OF DOMAIN-SPECIFIC DESIGN STANDARDS AND REFERENCE SERVICE MODELS IN ACHIEVING INTEROPERABILITY

In Service-Oriented Architectures, the capabilities of each service are exposed through so-called service contracts. This is typically done by specifying the

Operation Identifier	List	
Interaction Pattern	REQUEST	
Pattern Sequence	Message	Body Type
IN	REQUEST	List Operation Request
OUT	RESPONSE	List Operation Response

Structure Name	List Operation Request	
Extends	Composite	
Field	Type	Comments
Folder Name	String	Directory name to list the files

Structure Name	File	
Extends	Composite	
Field	Type	Comments
Name	String	Name of the file
Last updated	Time	Last updated time of the file

Structure Name	List Operation Response	
Extends	Generic Operation Response Type	
Field	Type	Comments
Files	File	Array of type file

Fig. 19.6   The abstract representation of CCSDS MAL concepts in table format [14].

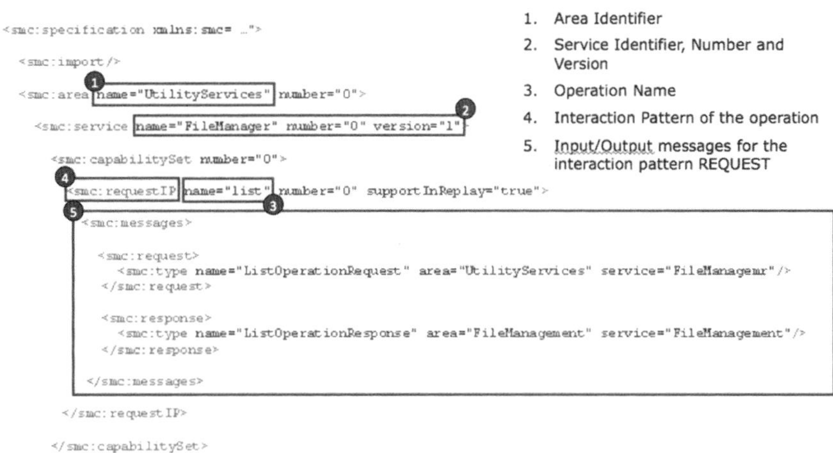

Fig. 19.7   The XML representation of the CCSDS MAL concepts [14].

exposed capabilities of each service as a list of operations and specifying the order and structure of the exchanged messages for each service capability (operation) in an abstract manner. Figure 19.6 shows an example of an abstract service contract of an imaginary file management service, specified in the CCSDS MAL service description language. The service provides an operation called "List" for retrieving the list of all files in a directory. The same abstract service contract can also be represented in a machine-readable XML format, as shown in Fig. 19.7, which is used by the run-time middleware (the MAL layer in this case) to establish a concrete binding to technology platforms of the service provider and service consumer.

As illustrated in these figures, formalized service contracts define the functional semantics of the exposed service capabilities. As is the case with legal contracts in real life, service contracts also establish the grounds for reaching a common understanding between the service provider and consumer of the modalities of transaction, the duties of each party, and their expectations of the other side. Again, as with legal contracts in real life, the clarity and quality of agreements formed by a service contract can impact significantly the smoothness of a transaction. To achieve intrinsic interoperability between software services at run time, it is often not enough to specify the functional semantics of the service interface. The service contract must also cover agreements on a number of nonfunctional aspects related to communications and interaction management, including addressing; security; QoS; communication reliability; taxonomy, description, and discovery; and transaction management. Otherwise, supplementary point-to-point agreements must be reached between the service providers and consumers before the interaction can succeed. Questions of the type "Where do I find your service?", "Do I need to authenticate myself before consuming the service?", "What kind of authentication do you expect?", "Do I have to use a secure communication line and encrypt all messages?", "Make sure the account number is encrypted when you are sending the request to query operation," "How many consumers can your service support at the same time?", "Please config. your service to send the answer to my request to this endpoint, which is different than the one I use for calling your service," "The integrity of this operation is essential, so make sure you use a reliable communication so that all messages are received at each end," would be unavoidable if the service contracts were limited to specification of functional service capabilities.

The so-called WS-* specifications (e.g., WSDL [7], WS-Security [8], WS-Addressing [9], WS-Reliability [10], WS-Policy [11] and WS-Transaction [5], as well as others [12]) provide a set of standards for addressing each of the above mentioned architectural concerns for Web services. These specifications define a set of standardized attributes that can be used at design time to extend the technical service contracts with nonfunctional requirements of the service. At run time, the required information (e.g., the replyTo endpoint address or the credentials of the service consumer) is exchanged between the service provider and consumer, using metadata embedded in the message headers.

The SCA Policy Framework specification [4] is based on a similar concept. However, it is important to realize that while these standards provide a generic framework for specifying metadata related to nonfunctional requirements of a service contract, they do not provide a reference service model for a particular domain. In other words, they do not specify which of the many possible nonfunctional aspects must be specified for all services of a given domain, e.g., for mission operations services. As a result, again, adherence to these generic industry standards alone does not guarantee interoperability of services. For instance, the WS-Security standard provides the means for specifying many different kinds of security requirements in a service contract. It does not, however, make any assumptions about which of these aspects shall or shall not be specified for all services of a particular domain. If a service provider would like to specify as part of its service contract that a certain authentication mechanism is required for one of its exposed operations, it could use the corresponding WS-Security policies to do so. The standard does not, however, require that an authentication mechanism be specified for all operations in all service contracts nor put any preference on a particular authentication mechanism (e.g., plain or encrypted username/password vs. use of certificates). The same applies for specifying whether whole or portions of a service operation message shall be encrypted. The WS-Security standard provides the means for specifying such requirements in a standardized manner, but it does not require the presence of such policy assertions in service contracts.

This is exactly where the standardized service contract design principle [3] steps in. Domain-specific design standards specify how and to what extent the generic industry standards shall be adopted in service contracts of all services within a domain service inventory. The result of application of this design

Design-Time

Area Identifier	Service Identifier	Area Number	Service Number	Service Version
Interaction Pattern	Operation Identifier	Operation Number	Support in replay	Capability Set

Run-Time

Field	Value
URI From	
Authentication Id	
URI To	
Timestamp	Message generation timestamp
QoSlevel	The QoS level of the message
Priority	The QoS priority of the message
Domain	Domain of the message
Network Zone	Network zone of the message
Session	Session of the message
Session Name	Session Identifier for Simulation and Replay sessions
Interaction Type	
Interaction Stage	
Transaction Id	
Service Area	Area Identifier
Service	Service Identifier
Operation	Operation Identifier
Service version	Service Version number
Is Error Message	FALSE

Fig. 19.8   The design and runtime MAL metadata [14].

principle is a Reference Service Model that specifies which exact architectural concerns must be addressed by all services of a domain inventory; i.e., which metadata must be present in all service contracts. The Reference Service Model must consider both design-time and run-time architectural concerns to ensure interoperability of compliant services. The abstract service specification template and the related run-time MAL header of the CCSDS mission operations service framework shown in Fig. 19.8 define such a Reference Service Model for mission operations services.

IV.   CONCLUSION

Achieving true intrinsic interoperability of services at run time (real plug-and-play) is a challenging objective for any SOA. The traditional notion of application integration is replaced in service-oriented computing with the concept of service orchestration, which aims at composing new service-oriented solutions from the repository of existing services and extending them at run time. This requires that all services be specified in an implementation and communication agnostic manner. A set of cross-domain industry standards such as WS-* and SCA specifications exists today for this purpose. However, the adoption of these standards alone cannot provide the envisaged intrinsic interoperability of services, as they do not specify a Reference Service Model for any particular application domain and do not address the semantics of the services of any domain.

The CCSDS mission operations MAL specification provides an abstract service specification template for design time and a related message header model for exchange of corresponding metadata at run time. Together, they form a Reference Service Model for the mission operation services domain. This Reference Service Model is focused on the immediate needs of typical space mission operation scenarios. Dynamic discovery and plug-and-play of services at run time is not the prime concern of real-world space mission operations scenarios, because many agreements at organizational level are usually required before software services can engage in interactions. The MAL Reference Service Model captures a more focused set of metadata related to the non-functional requirements of mission operations services. These metadata include information related to service taxonomy, service addressing, QoS and transaction management. The security aspects are deliberately limited to authentication and authorization, for which the Reference Service Model metadata only include an identifier, delegating the rest to the implementation of the run-time environment.

The current specification of the mission operations service framework does not use a canonical message format for bridging between different technology bindings at run time. Hence, dedicated technology bridges are required for any incompatible point-to-point interaction of services. Given the scenarios of today's mission operations, this approach is sufficient for the primary objective of enabling cross-organization interoperability of mission operations services. It could, however, be a point for improvement of the mission operations services framework for increasing run-time interoperability in the future. Abstraction from the implementation and communication technology, which is currently provided by the CCSDS mission operations SOA framework and its Reference Service Model, can significantly reduce the effort required for integration of mission operations services.

The Mission Operations Monitor and Control Services Specifications [13], currently under agency review, are the first set of mission operation service specifications that build upon the mission operations SOA framework. These specifications standardize the semantics of the services involved in spacecraft monitoring and control by providing an implementation and communication agnostic definition of the involved service in terms of MAL. The specification of each service assigns the service taxonomy and defines the set of operations it offers, the order and semantics of the exchanged messages, as well as the expected behavior of the service in case of error, paving the path towards implementation of better and more interoperable mission operations systems.

ACRONYMS

API	Application programming interface
CCSDS	Consultative Committee for Space Data Systems
EAI	Enterprise application integration
ESB	Enterprise service bus
HTTP	Hypertext Transfer Protocol
ICD	Interface control document

IT	Information technology
MAL	Massage abstraction layer
MDA	Model-driven architecture
MO	Mission operations
PIM	Platform Independent Model
QoS	Quality of service
SAML	Security Assertion Markup Language
SCA	Service component architecture
SI	Système International d'Unitées
SLA	Service level agreement
SOA	Service oriented architecture
SOAP	SOAP is not an acronym but a name. (Multiple abbreviations exist for SOAP including Simple Object Access Protocol and Service Oriented Architecture Protocol; none, however, reflects the current state of SOAP as the basis for many WS-* standards specification.)
TM	Telemetry
WS	Web service
WSDL	Web Services Description Language
XML	Extensible Markup Language

REFERENCES

[1]   Hohpe, G., and Woolf, B., "Enterprise Integration Patterns: Designing, Building, and Deploying Messaging Solutions," Addison-Wesley Professional, ISBN-13: 978-321200686.

[2]   Erl, T., "SOA Design Patterns," Prentice Hall, ISBN-13: 978-0136135166.

[3]   Erl, T., "Principles of Service Design," Prentice Hall, ISBN-13: 978-0132344821.

[4]   OASIS, "SCA Specifications," http://www.oasis-opencsa.org/sca [last accessed 15 May 2012].

[5]   OASIS, "Web Services Transaction Specification," http://www.oasis-open.org/committees/tc_home.php?wg_abbrev=ws-tx [last accessed 15 May 2012].

[6]   Merri, M., and CCSDS SM & CWG, "Cheaper, Faster and Better Missions with the CCSDS SM&C Mission Operations Framework," *SpaceOps 2009*.

[7]   W3C, "Web Services Description Language Specification," http://www.w3.org/TR/wsdl [last accessed 15 May 2012].

[8]   OASIS, "Web Services Security Specification," https://www.oasis-open.org/committees/tc_home.php?wg_abbrev=wss [last accessed 15 May 2012].

[9]   W3C, "Web Services Addressing Specification," http://www.w3.org/Submission/ws-addressing/ [last accessed 15 May 2012].

[10] OASIS, "Web Services Reliable Messaging Specification," http://www.oasis-open.org/committees/tc_home.php?wg_abbrev=wsrm [last accessed 15 May 2012].

[11] W3C, "Web Services Policy Framework Specification," http://www.w3.org/TR/ws-policy/ [last accessed 15 May 2012].

[12]   Bustamante, M. L., "Making Sense of All These Crazy Web Service Standards," http://www.infoq.com/articles/ws-standards-wcf-bustamante [last accessed 7 Jan. 2013].

[13]   CCSDS SM&C WG, "Mission Operations Monitor and Control Services," CCSDS 521.1-R-2.

[14]   CCSDS, "Mission Operation Massage Abstraction Layer," Blue Book, CCSDS-521.0-B-1.

CHAPTER 20

Delay-Tolerant Networking Engineering Network: Constructing a Cross-Agency Supported Internetworking Testbed

Edward Birrane* and Kristine Collins[†]
Applied Physics Laboratory, Johns Hopkins, University Laurel, Maryland

Keith Scott, Ph.D[‡]
The Mitre Corporation, McLean, Virginia

I. INTRODUCTION

The operational concept (CONOPs) of the delay-tolerant networking (DTN) architecture is to connect assets separated by interplanetary distances with automation and functionality equivalent to those in the terrestrial Internet [1]. Critical to this concept is the ability to support packetized, multipath, multihop data exchange between mobile assets separated by large distances incurring significant signal propagation delays. The operational environment of these networks presents unique challenges to the construction of reliable data communications protocols and, therefore, to the testbeds used for their validation.

The NASA DTN Space Readiness Project performs those protocol engineering, reference implementation, and tool construction tasks necessary to implement enabling technical means for constructing a Solar System Internet (SSI). The verification and validation (V&V) activities for this project require a testbed able to address three unique properties of the SSI: 1) disruptive link characteristics, 2) overlay network operation, and 3) multiple administrative boundaries. To accomplish this goal, the project constructed a DTN Engineering Network (DEN) to serve as both a simulation and emulation testbed. The DEN may be configured to simulate necessary signal propagation delays and link disruptions. As a distributed, heterogeneous set of machines independently managed by a variety of NASA centers, it represents a ground-based analog of the space internetworking model. The portions of the DEN that have been constructed and used for testing to date have validated software implementations and illuminated the challenges of a decentralized administrative layer.

---

*Senior Staff, Space Department, Guest Member.
[†]Associate Staff, Space Department.
[‡]Principal Engineer.

This chapter describes a systems analysis and design of a comprehensive SSI testbed and presents the construction, test execution, and anticipated evolution of the DEN. This includes the unique testing requirements of space internetworking, including the ability to operate across agency boundaries using heterogeneous hardware. The chapter concludes with a discussion of lessons-learned and how to extend the DEN configuration to include hardware managed by other space agencies, more flight-like hardware, and, potentially, incorporating flight assets such as the International Space Station (ISS). Ultimately, the DEN provides a unique mechanism to raise the Technology Readiness Level (TRL) [2] of both software and CONOPs so that missions may more deterministically adopt a space internetworking model.

The remainder of the chapter is organized as follows. Section II discusses the motivating factors behind a customized test framework for DTN protocols and tool implementations. Section III details the system-level design and capabilities that must exist for such a testbed. Section IV provides an architectural overview of the reference implementation of the DEN used for early DTN validation. Section V outlines experiences and lessons-learned from constructing and running early tests in this reference environment. Section VI discusses the future of the testbed as it migrates from a simulation to an emulation environment. Section VII summarizes our work.

II. MOTIVATION

Creating component- or subsystem-level testbeds to test individual implementations of networking protocols for predefined sets of functionality is neither new nor revolutionary. Additionally, there is precedent for system-level testbed models that exercise capabilities to-scale for particular applications. However, there is much less precedent for testbed models that both validate individual components, provide to-scale, system-level simulation and emulation, and operate generally outside of the assumption set and problem space of a particular mission. The construction of such general-purpose, to-scale testbeds remains an active area of research [3, 4]. The motivation for developing new testbed architectures within the NASA DTN program stemmed from a system analysis of the testing activity required and the levels of fidelity necessary to advance the TRL of the protocols being tested. The unique characteristics of a DTN are not well represented by terrestrial, to-scale testbeds. There are additional, unique challenges associated with cross-linked space services amongst differing administrative domains such as those encountered in a space internetwork.

III. SYSTEMS APPROACH

The goal of creating an SSI testbed is to produce a controlled environment that emulates the characteristics of the protocol operational environment. Such a controlled environment must address the challenges inherent with constructing and administrating a DTN.

A. DTN CHARACTERISTICS

DTN protocols in general, and the SSI in particular, must operate in environments that challenge the progression of data through the network in a variety of ways. Some of these challenges stem from the physics of the operating environment, such as occultation and extreme distances imposing significant signal propagation delays. Some challenges stem from a lack of resources on individual platforms that limit data rate, transceiver duty cycling, and a lack of contacts due to the sparse population of the network physical footprint. Furthermore, as DTNs may be constructed from multiple, otherwise independent networks, contacts may be challenged due to administrative policy. Any testbed that would substantively increase the TRL of a DTN protocol must either emulate, or simulate with appropriate fidelity, each of these challenges. We summarize the salient aspects of each of these categories of challenges, and the requirements of a testbed to address them, in Table 20.1.

B. TESTBED FIDELITY

Testbeds must provide control over the range of tunable characteristics, outlined in Table 20.1, associated with the protocols, applications, and CONOPs being tested. However, the level of fidelity associated with the tuning of these characteristics varies based on the economics of construction and the goals of individual test cases. We propose three levels of fidelity differentiating the tuning and application of testbed characteristics: effect simulation, algorithm approximation, and emulation.

1. EFFECT SIMULATION

Effect simulation refers to the practice of creating an offline model of a particular effect, either from separate analysis, user configuration, or random process. The impact of the effect on the network is exactly as prescribed from the offline model and is not calculated during the simulation. Examples of this practice include configuring lookup tables of bit error rate over look angle, or signal propagation delays over transmission distance. The value of this model is that fundamental characteristics of the effect can be changed offline with no subsequent change in the testbed. The weakness of this approach is that the fidelity of the effect is limited because it is not calculated continuously throughout the model.

2. ALGORITHM APPROXIMATION

The term algorithm approximation refers to the practice of configuring core aspects of the operational environment and then using in situ algorithms to

TABLE 20.1   CHALLENGES TO DATA PROGRESSION IN THE SSI

Challenge	Testbed responsibilities	Examples
Environmental impairments	Physical impairments relating to the implementations of the link layer	Bit error rates Attenuation and interference
	Physical impairments incurred by the geometries of the operational environment	Signal propagation delays Look angles and occultation
Network configuration	Physical configuration of the network	Number of nodes and their location Node mobility models
	Configuration of node assets related to messaging	Pointing (directional transmission) Processing power and storage capacity Transmitter power and configured rate Constraints imposed by the rest of spacecraft operation (e.g., science data gathering)
Administrative controls	Service negotiation, including selective asset access	Bandwidth or priority limitations Radio/storage limitations by role
	Security configurations	Key management Per-link security configuration
	Differing configurations	Role-based changes to network topology Emergency mode changes (safe modes, critical commanding, new pointing for science observation, etc.)

compute an approximate impact of the effect on the network. Because the impact of the effect is calculated in real time, the impact may change as any variable in the algorithm changes. An example of algorithm approximation is to real-time calculate the signal propagation delay as the distance between two objects divided by the speed of light. As the object distances change during the test run, the signal propagation delay changes continuously.

3. EMULATION

Emulation refers to the use of operational hardware or software in a testbed. The benefit of using operational hardware is that there is no need for offline simulation or algorithm development. Presumably the hardware performance in the testbed is indicative of the hardware performance in the operational system. Emulation as it applies to software implies the use of the same code in the testbed as in the operational system, to include the same configuration settings. Additionally, operational software *must* be tested on hardware that closely models the hardware in the operational system. The fundamental challenge associated with emulation is the cost and complexity of incorporating an operational device. Even when such a device is incorporated into the testbed, it must be fed information that is commensurate with the data it would receive in the operational environment. Typically, emulation assets are brought into a testbed as part of preflight testing and/or early operations testing. Occasionally, dedicated in situ flight payloads may also participate in testbeds as part of flight tests. While costly, there is precedence for including flight assets into testbeds [5].

To explain our use of these terms, we consider the common SSI concept of implementing the signal propagation delay and occultation encountered between an Earth and Mars orbiter. Table 20.2 shows how this simple task would be implemented with various levels of fidelity.

TABLE 20.2   TUNABLE TESTBED CHARACTERISTICS MAY BE IMPLEMENTED WITH VARIOUS LEVELS OF FIDELITY

Fidelity Level	Approach
Effect simulation	Offline, the orbits of the two landers would be generated (through an analysis tool or by hand) and fed into the simulation as a set of delays and outages over time. The simulation would refer to this "table" based on the current simulation time.
Algorithm approximation	The orbits, pointings, and other geometries of the system would be specified in a set of algorithmic inputs. The algorithm would implement orbit propagation and other models, with some imperfect-but-appropriate fidelity. Delays and occultation may then be calculated as necessary based on time, tracked geometries, and state.
Operational emulation	Flight hardware at Mars, such as on a hosted payload, participates in the testbed concept and provides actual delays and occultation effects associated with Mars communication. Even if the flight hardware does not exactly match associated orbits, the level of fidelity and emulation is as close as can be without implementing the system.

C.   REQUIREMENTS OF AN SSI TESTBED

Ultimately, a successful SSI testbed must emulate, or simulate, each of the core space internetworking challenges; this implies that not only does the testbed provide mechanisms to measure the response of the system in the presence of these challenges, but that the testbed further provides the mechanisms necessary to control their magnitude and duration. Impacts from these challenge areas must be consistent during the course of a particular test phase and must further be repeatable as tests are re-run.

The mapping of SSI challenge areas to envisioned levels of fidelity is given in Table 20.3. From this table, we see that the ultimate end state of the testbed is to support sufficient hardware and software emulation to provide a very high-fidelity set of test results as new utilities and CONOPs are tested.

The environmental challenges of link disruption and delay must be modeled in such a way that software down to the link layer reacts to these changes as if they were occurring as part of the environment. Initially, simulating the effects of link disruption by precalculating allowed periods of transmission (such as by configuring contact outages even when there is full connectivity in the testbed) provides a mechanism to test upper layers of the software. Environment simulation of these effects involves working with operating system primitives to inject delays and disruptions into the network outside the software being tested. Finally, the incorporation of flight assets into the testbed model provides the ability to work with actual delays and disruptions across operational radio-frequency links.

The SSI concept implements an overlay network using multiple constituent networks. In this model, multiple space assets coordinate their application-layer support to provide a homogeneous messaging service atop otherwise hetero-geneous link services. Therefore, coordination of networking activities is a completely separate activity from coordinating the individual link services of a platform. The naming, addressing, routing configuration, and local state of nodes as they participate in the overlay must be part of a testbed feature set.

TABLE 20.3   THE GOAL OF AN SSI TESTBED IS TO SUPPORT TESTING WITH OPERATIONAL HARDWARE AND SOFTWARE

Challenges	Effect simulation	Algorithm approximation	Operational emulation
Environment impairments	X	X	X
Network configurations		X	X
Administrative controls			X

Furthermore, the impacts of network configurations on the system are broad and complex, preventing a mechanism to simply simulate their effect. Rather, the configuration is specified in the testbed, and the testbed must implement simulated algorithms or operational software to apply that configuration to the nodes in the network.

Administrative control of the network is difficult to capture. Questions of how a distributed network administration role affects data flow, the mean time delays between correcting network error, or negotiating key management policies and space asset access are driven by individual institutional policies not easily summarized as table lookups or encoded algorithms. The most straightforward way to gain experience with this model is to build the testbed such that it crosses administrative boundaries from the beginning. With such a model, connections amongst testbed assets are then constrained by operational administrator and other institutional policies.

IV. REFERENCE ARCHITECTURE

This section describes a reference architecture and associated implementation of the DTN DEN, developed under the auspices of the NASA Space DTN Readiness Project. The purpose of the DEN is to provide the initial SSI testbed, starting with ground assets and evolving to full capability over time. We include here the architectural layout of the DEN and the way in which it simulated and emulates crucial parts of the SSI testbed concept.

A. OVERVIEW

The DEN is a closed, multicenter network used to prove out new technologies, test interoperability, analyze performance of applications and the network in general, and experiment with different mission operations scenarios and procedures. The DEN is implemented as a star-configuration of virtual private networks (VPNs) connecting machines at the Johns Hopkins University Applied Physics Lab, the University of Colorado at Boulder, Goddard Space Flight Center, Glenn Research Center, the Jet Propulsion Laboratory, Johnson Space Center, and Marshall Space Flight Center. The DEN provides implementations of the protocols comprising DTN capabilities: the Bundle Protocol (BP) [6] and the Licklider Transmission Protocol (LTP) [7]. It supports Contact Graph Routing [8, 9], and the Bundle Security Protocol (BSP) [10]. Because both BP and LTP can use a transport layer such as Transmission Control Protocol (TCP) or User Datagram Protocol (UDP) for communication between DEN nodes, separation of the various locations by multiple Internet Protocol (IP) hops is not an issue. Therefore, the establishment of DEN connectivity is possible over the terrestrial Internet.

The DEN provides two services. First, it provides a standing, secure method for pairs of centers to test particular capabilities over the network without having to first instantiate those capabilities outside their institutional firewalls. The DEN also enables large-scale, multicenter testing and experimentation where different centers can bring unique hardware and capabilities to the shared test environment. For example, we are working to make "flight-like" hardware available to allow more realistic measurements of bundle throughput both with and without security. In this way, Goddard Space Flight Center can make available hardware with computing and storage capabilities representative of Earth-observing satellites, the Jet Propulsion Laboratory can place representative deep-space hardware on the DEN, and they or another center can provide a "ground station/ground network" analogue. Multicenter experimentation also helps uncover coordination issues that may not be readily apparent with single-center experimentation. Requiring personnel from different locations and different time zones to coordinate the operations and scheduling needed to perform tests with scheduled connectivity is providing insights into the unique network monitoring and control capabilities needed to manage a network with time-varying connectivity and store-and-forward operation. The current architecture of the DEN is illustrated in Fig. 20.1.

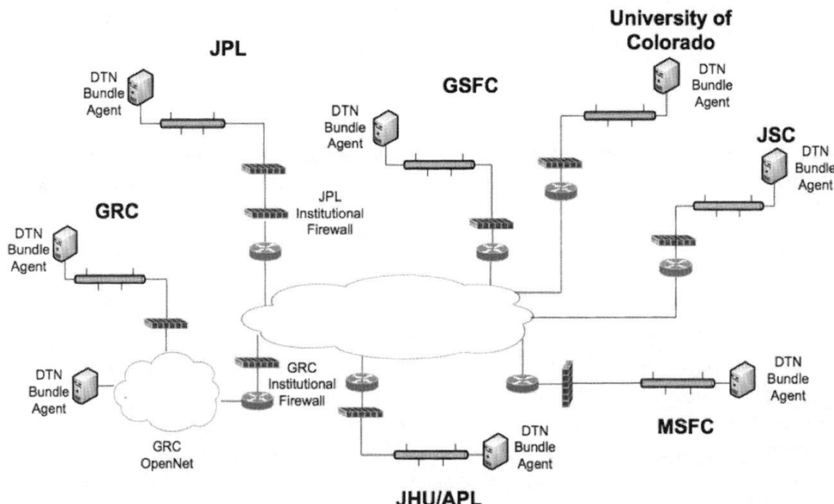

Fig. 20.1   The DEN is a secure network comprising nodes at multiple participating organizations.

B. CONFIGURATION ITEMS

The hardware and software comprising the DEN support configurations in each of the three identified challenge areas.

1. ENVIRONMENTAL IMPAIRMENTS

Currently, environmental impairments are configured on the DEN using operating system primitives for changing the transmit and receive buffers associated with the networking interface. Simulation and approximation of these effects are accomplished with standard primitives that control the effective data rate, link delays, and link disruptions.

Data rate may be selected by either inserting a rate buffer at the operating system layer or configuring the DTN networking software to pay attention to data rates and not transmit beyond the calculated maximum capacity of a link. Link delays are also simulated with buffers where "transmitted" data are held in a signal propagation delay buffer, either on the transmitting node or receiving node, until it is ready to be "received." Finally, link disruptions may be configured as percentage of bit disruptions in the communicated data stream, with sufficient link disruption causing a loss of link.

Further, the DEN supports the ability to form a ground-truth backplane comprising physical links amongst ground assets in the network. This backplane provides high-availability, high-bandwidth information exchange between nodes. While the use of such information as part of the experiment would be counter to the exercise of delay-tolerant protocols, it does provide an excellent visualization and recording tool for the experimenter to collect coordinated networking states for debugging and performance measurement purposes.

2. NETWORKING CONFIGURATION

Network configuration is accomplished by synchronizing the setup and configuration files associated with the various hard and software systems comprising the testbed. We separately consider the configuration of the operating system, the configuration of the physical networking layer, and the configuration of the DTN support software.

The only important piece of operating system information that must be synchronized amongst assets in the DEN is the time. Significant clock drift is both counter to the operational realities of space systems, which invest quite heavily in this regard, and prevents other aspects of the DEN support system (such as propagation delay queues) from operating. Configuration of the physical networking layer ensures that the physical components of the network are able to communicate outside of their respective organizational boundaries through a VPN. This includes ensuring that there are no subnetwork collisions and that ports are enabled as appropriate.

TABLE 20.4   THE NETWORK CONFIGURATION OF THE DEN SPECIFIES PROTOCOL, ROUTING, AND ADDRESSING INFRASTRUCTURE

Configuration Item	Function
Protocol exchange	When using the DTN communication protocols (LTP, BP) there are several configurations that must be synchronized throughout the network so that the protocols interoperate correctly. LTP configurations include messaging sizing, retransmission windows, and other timing-related configurations. BP includes synchronization on naming conventions, and interoperability and support for the extension blocks used as extensibility mechanism within the protocol.
Routing information	The DTN suite uses a graph-based routing mechanism called contact-graph routing (CGR). In this scheme, transmission opportunities are decomposed into contacts and the transmission ranges associated with those contacts. These contacts and ranges must match between any pair of transmitters and receivers in the network, and must further match (or at least not run counter to) the configuration of the underlying physical network.
Naming and addressing	Each node in the overlay network is assigned a name within the overlay network called an end-point identifier (EID). This identifier may be associated with unicast, multicast, or broadcast and it is used to communicate to other nodes at the overlay. However, this overlay exists over one, or more, physical networks. The mapping between the DTN EID and the underlying networking identifier (such as an IP address or a Media Access Control (MAC) address) must be configured and consistent at each node in the network.

The DTN support software provides several configuration mechanisms that must be synchronized to make the overlay network function, as listed in Table 20.4.

3.   ADMINISTRATIVE CONFIGURATION

The administrative configuration of the network is composed of two major activities: the coordinated administration of machines on the network and the management of security information. Coordinated machine administration ensures that the machines representing networking nodes are enabled, appropriately configured, and will have the appropriate resources, including disk space, to run the experiment. If there are problems with a particular machine, policies are in place to communicate this difficulty. On the DEN, individual institutions manage their own set of machines. Although there is a coordination function to ensure that the various administrators communicate with one another, no single entity has administrative privileges across the network.

The physical security of the network is provided by the boundary defense mechanisms at participating institutions. The security of the network layer comes from the establishment of a VPN and its associated hardware configurations. Finally, a manually negotiated key management process exists to share keys amongst various nodes in the network, allowing the security mechanisms that operate at the overlay to function.

V. RESULTS

The DEN, as the evolving reference implementation of the SSI testbed, has already been used for testing of DTN-related software and operational concepts. This section outlines the software that has been tested on the DEN, demonstrates a typical configuration of the network, and discusses lessons learned from these efforts.

A. MAJOR FUNCTIONAL TESTING

Although the DEN has hosted a variety of individual tests during its construction, there have been five major functional tests that have tested not only a software subsystem but the overall testbed infrastructure (BP interoperability, video streaming, and file synchronization).

1. INITIAL CONNECTIVITY AND INTEROPERABILITY

The first set of tests conducted across the multicenter DEN network used a combination of the NASA reference implementation of DTN protocols for spaceflight software (ION) and the Delay-Tolerant Networking Research Group (DTNRG) provided DTN2 reference implementation to verify both IP and BP-layer connectivity. This ensured that the various centers' VPN connections were functioning properly and that the firewalls were configured to allow BP traffic. Several issues were uncovered by even this simple test, including differences in the TCP convergence layer implementations and issues related to interoperability of naming schemes. Both issues were addressed by later versions of the ION and DTN2 implementations.

2. BUNDLE AUTHENTICATION BLOCK (BAB) SECURITY TESTING

This test validated that various nodes in the network, especially those that operated across administrative boundaries, were able to authenticate using the BAB from the BSP suite. The BAB is one of the fundamental security mechanisms in the DTN architecture and provides hop-by-hop authentication of traffic. The test demonstrated that a user without the correct BAB authentication key was unable to inject traffic across the emulated space link. This ensures that "rogue" users, even if they have access to the ground network, cannot mount a denial-of-service attack against the space link by sending "junk" traffic to a spacecraft. As with the initial connectivity tests, several interoperability issues between the ION and DTN2 implementations were uncovered and fixed during the course of the testing.

3. VIDEO STREAMING

The DEN has also been used to demonstrate the concept of delay-tolerant video streaming over BP. While the concept of streaming video over a high-delay, disrupted link seems counter-intuitive, the concept provides mission operations a valuable service: the ability to view a real-time stream tolerating dropouts and to go back and evaluate a more complete version of the stream at a later time. The ability to "rewind" a video stream allows operators to view retransmitted frames that would otherwise have been lost in a less tolerant network. A data repository at the video stream receiver reconstructs streams from frames based on their construction timestamp, and the transmitting video source may offload all reliability concerns to the underlying network. The DEN testing demonstrated the viability of this approach using the DTN protocols and configurations. Notably, the delay-tolerant nature of the transaction allows the "library" of streaming video to continue to accumulate well after the initial video streaming occurred.

4. LONG-TERM FILE SYNCHRONIZATION

This test demonstrated the ability to perform Rsync-style file synchronization across multiple nodes in the DEN with simulated link disruptions. The test was run for multiple months during which periodic underlying directory and file changes, and the performance of both the synchronization utility and the ION suite were reviewed. This test demonstrated the manner in which multiple administrators interact to ensure that DEN machines performed as necessary for long-term tests.

5. DELAY-TOLERANT NETWORK MANAGEMENT TESTING

A significant capability provided by the DEN is to test the interoperability of emerging

```
------------------
DTNMP DATA REPORT
------------------
Sent to  : ipn:1.1
Rpt. Size: 88
Timestamp: 1347429287
Num Mids : 5
Value(s)
-------------------------------------
Data Name: BP NODE ID
MID      : 0x11092a030102017a010101
Value    : 1
Data Name: BP_BUNDLE_VERSION
MID      : 0x11092a030102017a010102
Value    : 6
Data Name: CUR_RES_CNT_0
MID      : 0x11092a030102017a010103
Value    : 0
Data Name: CUR_RES_CNT_1
MID      : 0x11092a030102017a010104
Value    : 0
Data Name: CUR_RES_CNT_2
MID      : 0x11092a030102017a010105
Value    : 0
==================
STATISTICS:
MIDs total 55 bytes
Data total: 33 bytes
Efficiency: 37.50%
-------------------
```

Fig. 20.2 Name/value pairs result in low message efficiency.

```
------------------
DTNMP DATA REPORT
------------------
Sent to  : ipn:1.1
Rpt. Size: 36
Timestamp: 1347429262
Num Mids : 1
Value(s)
-----------------------------------
Data Name: BP NODE ID
MID      : 0x11092a030102017a010101
Value    : 1
Data Name: BP_BUNDLE_VERSION
MID      : 0x11092a030102017a010102
Value    : 6
Data Name: CUR_RES_CNT_0
MID      : 0x11092a030102017a010103
Value    : 0
Data Name: CUR_RES_CNT_1
MID      : 0x11092a030102017a010104
Value    : 0
Data Name: CUR_RES_CNT_2
MID      : 0x11092a030102017a010105
Value    : 0
=================
STATISTICS:
MIDs total 3 bytes
Data total: 33 bytes
Efficiency: 91.67%
------------------
```

Fig. 20.3 Preconfiguring report definitions significantly reduces message costs.

standards. One such emerging standard is that proposed for the autonomous management of DTNs. The Delay-Tolerant Network Management Protocol (DTNMP) [11] offers a mechanism to perform fault protection, performance monitoring, configuration, and administrative control of network nodes even when they are partitioned from human operators due to disruptions and delays. A reference implementation of the performance monitoring portion of the DTNMP has been deployed to the The Johns Hopkins University Applied Physics Laboratory (JHU/APL) nodes of the DEN and used to validate the performance, stability, and feasibility of the protocol. This implementation uses the NASA ION DTN software to exchange network management information between an agent located on one DEN machine and a manager located on another DEN machine.

A significant metric generated by these tests is the savings associated with bundled data distribution versus sending name/value pairs of data. In Fig. 20.2 a network management bundle containing performance data for five data points is sent by a DTNMP agent. From this figure, we calculate a small set of statistics associated with the amount of overhead information that must be used to identify the encapsulated data. In this case, 55 bytes of overhead is used to label 33 bytes of actual performance data, so only 37.5% of the message payload is used for novel data exchange. However, when customized reports are used, as described in the DTNMP specification, a single identifier may be used to label the ordered collection of data values. When custom reports are enabled within the system, the same 33 bytes of performance information may be identified using only 5 bytes of identifier, making the resultant bundle payload 91.67% efficient. This is shown in Fig. 20.3.

The DTNMP reference implementation currently supported on the JHU/APL DEN nodes supports approximately 100 separate performance data values. The use of this testbed has been instrumental in capturing the performance characteristics and operational concepts necessary to validate management behavior. Based on the success of these tests, DTNMP agents and managers are anticipated to migrate to other NASA center nodes on the DEN to expand the test to include management across administrative boundaries. Additionally, work is beginning to implement DTNMP reference implementations in the DTN2 DTN software codebase, at which time interoperability testing may be run on these NASA DEN machines.

B.   GENERALIZED TEST SETUP

Nodes in the DEN run the ION software and communicate over the terrestrial Internet using a VPN. Certain NASA centers house nodes running on flight hardware and other centers/organizations run different Linux distributions natively and on virtual machines.

The network graph illustrated in Fig. 20.4 shows how, at the overlay level, each center is physically connected to at least one other center, but also requires bundle forwarding to at least one other center. Contact graph routing is able to test the routing of bundles between each center, and can also include disruption of links to show how bundles are routed when a direct connection is no longer present, or how bundles are stored when there is not even an indirect route to be used. Each node is configured with a global routing configuration file that specifies the contact times for each node on the network; an example of such a file is given in Fig. 20.5. The global configuration file sets up the routing table for the network and is essential for when there are multihops from node to node, or for when delays and disruptions are introduced into the system. Along with a global configuration file, each node must implement a unique configuration that includes the means for communicating with the other nodes. This file includes turning on or off various features, such as security, and provides the means for the nodes to be able to communicate with one another.

Testing involves either leaving each link to run continually without delay/ disruption, or introducing link impairments. For example, the topology specified in Figs. 20.4 and 20.5 was in support of the file synchronization test. During this test each center ran DTN protocols for 48 h and sent messages to each respective center every 15 min. At the end of the test, the message traffic was analyzed and graphed to see the bandwidth of messages within the system. A common version of the ION suite (in the case of this test, version 2.4.1 was used) ran at each node. Once each center had properly configured their node, each node was started at 0000 hrs EST by a script running on the machines. During testing, the point of contact at each center monitored the node to watch for any node failures. In the event of a failure, the point of contact restarted the node as appropriate. The initial connectivity test allowed for the restarting of nodes, and a failure presented a good opportunity to simulate the event of a delay within the network. Each center configures their own nodes, which emulates organizations managing the configuration of their own node on the network.

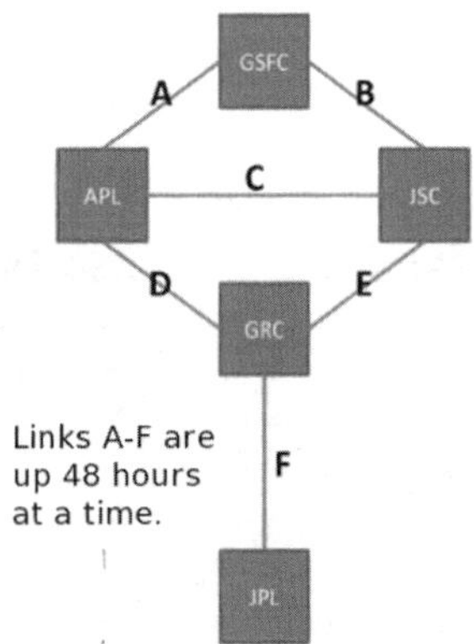

Fig. 20.4   A typical configuration of the DEN involves multiple organizations.

```
# Global Contact Plan

#JPL  ; 21216
#APL  ; 21403
#GSFC ; 21302
#GRC  ; 21111
#JSC  ; 21610

a_range +1 +604800 21216 21111 1
a_range +1 +604800 21111 21610 1
a_range +1 +604800 21610 21302 1
a_range +1 +604800 21302 21403 1
a_range +1 +604800 21403 21111 1
a_range +1 +604800 21610 21403 1

a_contact +1 +604800 21216 21111 100000
a_contact +1 +604800 21111 21216 100000
a_contact +1 +604800 21111 21610 100000
a_contact +1 +604800 21610 21111 100000
a_contact +1 +604800 21610 21302 100000
a_contact +1 +604800 21302 21612 100000
a_contact +1 +604800 21302 21403 100000
a_contact +1 +604800 21403 21302 100000
a_contact +1 +604800 21403 21111 100000
a_contact +1 +604800 21111 21403 100000
a_contact +1 +604800 21403 21610 100000
a_contact +1 +604800 21610 21403 100000
```

Fig. 20.5　Contact plans specify overlay contacts over time.

C.　LESSONS-LEARNED

The initial set of DEN tests provided multiple insights into the greater challenges that will be faced during deployment of DTN on systems, specifically in the realms of network administration and competing administrative domains. The greatest challenges, and lessons-learned, are in the non-trivial task of resource management (both human and machine).

For the initial testing, each link was left to run continually without delays or disruptions; however, once delays were introduced a new problem surfaced in the form of timing synchronization. To properly implement contact periods designed during a test, network time must be precise and shared amongst all nodes in the network. However, different organizations, especially those supporting their own local time servers, encountered enough drift and jitter to provide problems in network synchronization. Some nodes at come organizations were not connected to any time server and experienced drift over weeks and months of testing. To solve this problem, a Network Time Protocol (NTP) server must be housed at one of the DEN nodes, and this must be used to synchronize time in order to start the nodes at a certain time.

A logistics hurdle in assembling all components of the DEN involved assembling points of contact (POC) from each center that would be participating in various tests and scheduling their time concurrently to discuss the setup and administration of the network. This included juggling very busy schedules and time zone differences. As more nodes are added to the DEN, and the levels of tests increase, knowing how to manage the personnel running the tests is an important task that can very easily be overlooked. Once all of the personnel resources were assembled, it became apparent that there needed to be a full-time administrator of the DEN at one center who would then coordinate the efforts of all others. The DEN coordinator then became the global network administrator. This eased the challenge of having each center attempt to create global network configuration files; instead, the DEN administrator would send out the files for each center to load onto their respective machines. The concept of truly distributed network administration, absent a coordinator role, is not recommended based on our experience.

Fault detection and recovery across administrative boundaries provided a different set of challenges. When nodes within a particular organization experienced faults, especially at waypoint nodes, other centers typically discovered the fault first. However, lacking administrative permissions within a different administrative domain, the fault-detecting centers could not perform diagnostic analysis. During these outage times, the network available waited for administrators from organizations housing troubled nodes to complete their anomaly resolution processes. This proved to be a bottleneck to testing because of time zone differences and loading of the administration staff. In our experience, having an automatic notification go to a point of contacts e-mail list would help to prevent the lag in the periodic, manual checking of nodes' health.

VI.   ROADMAP

NASA is currently designing experiments to support DTN development in FY12 through FY16. The experiments in this timeframe will focus on evaluating the performance of the protocol stack (apps/BP/LTP) in configurations representative of current missions (e.g., communication with the ISS, single-hop communication with spacecraft) as well as near-future missions such as two-hop relay communication. These tests will evaluate whether current routing mechanisms can scale to accommodate realistic contact lists and rates of bundle forwarding when using flight-like hardware. The large-scale tests will also be used to experiment with and further refine the DTN management mechanisms under development.

NASA will also use the DEN to prototype and demonstrate capabilities being considered for experiments and/or use on the ISS. These will include a prototype "border gateway router" for ISS to allow a single point of contact with the ground and compression of the custody acknowledgements that form the basis for DTN reliability.

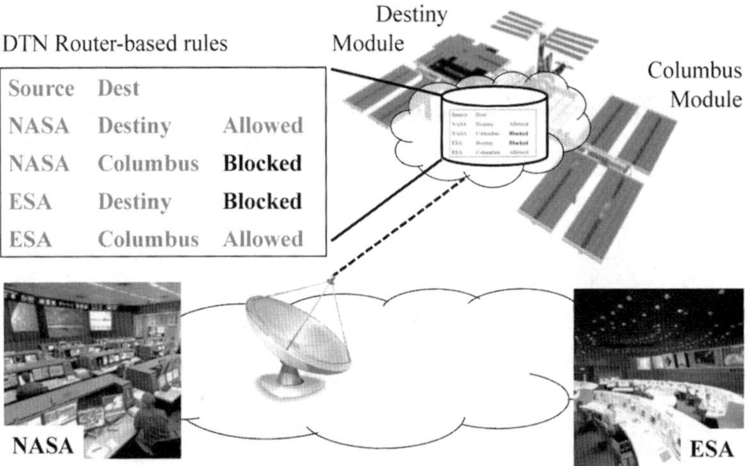

Fig. 20.6   Future test scenarios will incorporate fine-grained security and flight assets, such as the ISS.

In preparation for more extensive international testing and increased use of net-worked communications, the team is also experimenting with router-based security control measures such as access control/firewall capabilities. These will ensure that policy rules can be imposed to limit the forwarding of certain traffic in the network.

Figure 20.6 shows an example of a set of router-based firewall rules to prevent unintended commanding of other agencies' assets. In the figure, bundles with source endpoints within NASA can only be routed to the Destiny module onboard ISS, and bundles with source endpoints in ESA can only be routed to the Columbus module. Such a capability is complementary to the BAB security mechanisms. The BAB keeps unauthenticated traffic out of the network, while router-based security will keep authenticated traffic from inadvertently being routed contrary to policy. This capability will help assure operators that net-worked communications can be implemented without necessarily having to check the content of every uplinked bundle. Obviously, allowing particular for-warding operations under such a policy requires that the consequences of the allowed bundles do not impact overall operation.

VII. CONCLUSION

The construction of a to-scale testbed for the SSI involves the simulation or emulation of a variety of operational characteristics, including individual link-layer effects, overlay network configurations, and administrative boundary conditions. Constructing a cohesive, controlled, deterministic testbed that addresses each of these areas with the necessary fidelity is a nontrivial endeavor that has been under-taken by NASA. To date, the DEN represents the initial implementation of such a testbed, and it has been used to provide concepts relating to file synchronization and video streaming over DTN protocols. The DEN has also provided interoper-ability testing for portions of the BP security suite. NASA will continue to evolve the DEN into the fully envisioned SSI testbed by incorporating additional oper-ational hardware and software, including flight assets such as the ISS.

We have described the system-level objectives of the SSI testbed, described the mechanisms for the construction, test execution, and anticipated evolution of the DEN, and listed several lessons learned from this initial construction and test phase. Ultimately, the DEN provides a unique capability for validating DTN concepts in the space environment. Increasing the TRL of flight software by incor-porating flight-like hardware in the current DEN (and flight assets in future evolutions of the DEN) provides adopting missions with the assurances necessary to baseline space internetworking models.

ACRONYMS

BAB	Bundle authentication block
BP	Bundle Protocol

BSP	Bundle Security Protocol
CCSDS	Consultative Committee for Space Data Systems
CGR	Contact graph routing
CONOPS	Concepts of operation
DEN	DTN Engineering Network
DTN	Delay-tolerant networking
EID	End-point identifier
ION	Interplanetary overlay network
ISS	International Space Station
LTP	Licklider Transmission Protocol
NTP	Network Time Protocol
SSI	Solar system Internet
TRL	Technology readiness level
V&V	Verification and validation
VPN	Virtual private network

ACKNOWLEDGMENTS

The authors would like to thank the management and technical members of the NASA Space DTN Project for their hard work and dedication in the construction, maintenance, and evolution of the DEN.

REFERENCES

[1] Cerf, V., Burleigh, S., Hooke, A., Torgerson, L., Durst, R., Scott, K., Fall, K., and Weiss, H., "Delay-Tolerant Networking Architecture," RFC4838, IETF, April 2007, http://tools.ietf.org/html/rfc4838 [last accessed 29 April 2013].

[2] Mankins, J. C., "Technology Readiness Levels," White Paper, Advanced Concepts Office, Office of Space Access and Technology, NASA, 1995, http://www.hq.nasa.gov/office/codeq/trl/trl.pdf [last accessed 39 April 2012].

[3] Chun, B., Culler, D., Roscoe, T., Bavier, A., Peterson, L., Wawrzoniak, M., and Bowman, M. "PlanetLab: An Overlay Testbed for Broad-Coverage Services," *SIGCOMM Computer Communication Review*, Vol. 33, No. 3, 2003, pp. 3–12.

[4] GENI, "Global Environment for Network Innovations," http://www.geni.net [last accessed 29 April 2012].

[5] Wyatt, J., Burleigh, S., Jones, R., Torgerson, L., and Wissler, S., "Disruption Tolerant Networking Flight Validation Experiment on NASA's EPOXI Mission," *First International Conference on Advances in Satellite and Space Communications*, 2009, (SPACOMM 2009), pp. 187, 196, 20–25 July 2009, Colmar, France.

[6] Burleigh, S., and Scott, K., "Bundle Protocol Specification," Nov. 2007, http://tools.ietf.org/html/rfc5050. [last accessed 29 April 2013].

[7] Ramadas, M., Burleigh, S., and Farrell, S., "Licklider Transmission Protocol–Specification," RFC5326, Sept. 2008, http://tools.ietf.org/html/rfc5326 [last accessed 29 April 2013].

[8]   Burleigh, S., "Contact Graph Routing: draft-burleigh-dtnrg-cgr-01," July 2010, http://tools.ietf.org/html/draft-burleigh-dtnrg-cgr-01. [last accessed 29 April 2013].

[9]   Segui, J., Jennings, E., and Burleigh, S., "Enhancing Contact Graph Routing for Delay Tolerant Space Networking," *Global Telecommunications Conference.* (GLOBECOM 2011), IEEE, pp. 1,6, 5–9 Dec. 2011, Houston, Texas.

[10] Symington, S., Farrell, S., Weiss, H., and Lovell, P., "Bundle Security Protocol Specification," RFC6257, May 2011, http://tools.ietf.org/html/rfc6257 [last accessed 29 April 2013].

[11] Birrane, E., and Ramachandran, V., "Configuration and Monitoring Across Delay-Tolerant Networks: Robust Management of Space Internetworks," *63rd International Astronautical Congress Proceedings*, International Astronautical Federation, Naples, Italy, 2012.

CHAPTER 21

CCSDS Mission Operations Services for Mission Planning

Roger S. Thompson[*]
SCISYS UK Ltd, Chippenham, United Kingdom

I. INTRODUCTION

The Consultative Committee for Space Data Systems (CCSDS) is an international standards organization affiliated to the International Organisation for Standardisation (ISO). Its Spacecraft Monitoring & Control Working Group is developing a set of standardized Mission Operations (MO) services that enable interoperable information exchange between collaborating agencies or organizations involved in the operations of space missions. The approach uses service-oriented concepts and focuses on meaningful (semantic level) end-to-end information exchange between software applications supporting MO functions. These applications may be distributed between organizations and also between a range of space- and ground-based systems. The resultant MO services will support both live information exchange and open access to operations history. Figure 21.1 presents this CCSDS MO service layer, which sits between MO applications and the technologies used to integrate them, supporting meaningful information Exchange between applications.

The focus of the Working Group to date has been on the definition of an extensible framework for the definition of such services that is independent of the technology used to deploy the services. This allows for the evolution of implementation technology during the long lifetime of many space systems and also for a diversity of transport protocols that may be required to support communication in different environments.

The CCSDS Mission Operations Services Concept [1] identifies a range of application-level services, including several that are relevant to the Mission Planning function: planning request; scheduling; and navigation.

To date, only the Monitoring & Control service has been developed. As part of the CCSDS Technical Meetings held in Darmstadt, Germany, in April 2012, a call for interest was issued to members of the Mission Planning community to initiate the process of service standardization relevant to Space Mission

---

[*]Technology Manager, Space Division.

Planning. A dedicated session on the topic was attended by 58 specialists, mostly from Europe. A similar session was held later in the year in Cleveland, Ohio, attended by 42 specialists, predominantly from North America. Providing sufficient support is obtained from member agencies, the formal process of standardization within CCSDS will be initiated in 2013.

This chapter provides background on the CCSDS Mission Operations Services and the

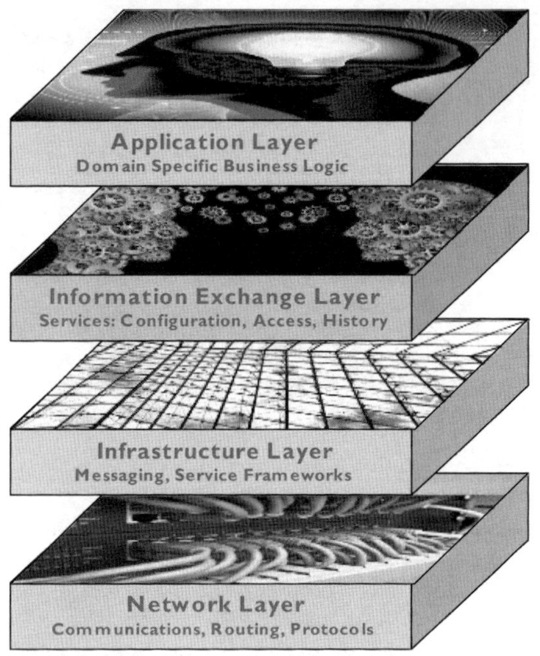

Fig. 21.1   The CCSDS MO service layer.

potential scope and benefits of MO Mission Planning services, before proposing an initial concept for such services, incorporating initial feedback obtained during the sessions in Darmstadt and Cleveland. It also summarizes the results of a short case study considering the Mission Planning interfaces for the European Space Agency (ESA) Venus Express mission.

II.   OVERVIEW OF CCSDS MISSION OPERATIONS SERVICE CONCEPT

MO functions are increasingly distributed more widely than just in a central Mission Control Center (Fig. 21.2). There may be separate Payload Operations Centers, Payload Data Processing Centers, as well as principal investigator (PI) teams and end-users. Indeed, the distribution of functions exposes potential MO services at interoperable boundaries between organizations/entities and systems, including the space segment itself. The spacecraft and payload manufacturers may play a continuing role from initial development into MO, and the increasing capability of onboard computers allows the migration of intelligence from ground- to space-based systems. This distribution of functions often crosses organizational boundaries due to the collaborative nature of space missions and requires interoperability between agencies. It can also highlight the boundaries between functions and systems within an organization where intraoperability between major system components is desirable to enable reuse and rapid integration of mission systems.

The CCSDS MO Services Concept introduced previously seeks to establish an extensible set of standard MO services to support inter- and intraoperability between applications at organizational, functional, and system boundaries.

Standards already developed include a Reference Model [2], a message abstraction layer (MAL) [3], and a Common Object Model (COM) [4]. Application-level MO services are defined in terms of the MAL and COM for specific types of MO information exchange. This layered framework for service specification is illustrated in Fig. 21.3. MO services support semantic-level inter-action between MO applications. These are defined in terms of the MAL and can be deployed over different messaging technologies. To allow deployed appli-cations to interoperate, a common binding must be used.

The MO services themselves support meaningful message exchange between applications, independent of programming language or underlying message encoding and transport. An extensible set of MO services can be defined, each based on a shared model for a particular class of information exchanged, together with the set of operations that the service consumer can invoke.

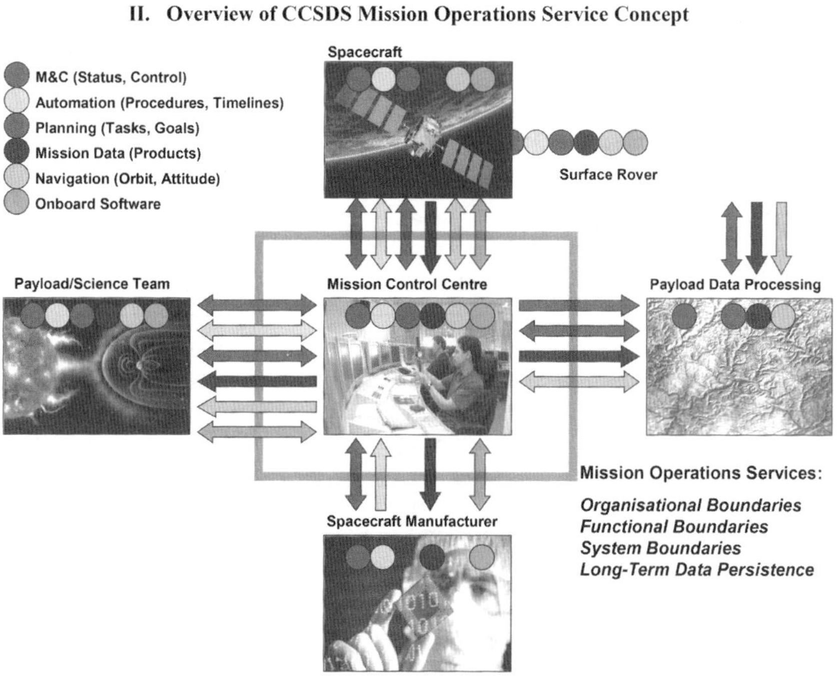

II. Overview of CCSDS Mission Operations Service Concept

Fig. 21.2   Distributable MO functions.

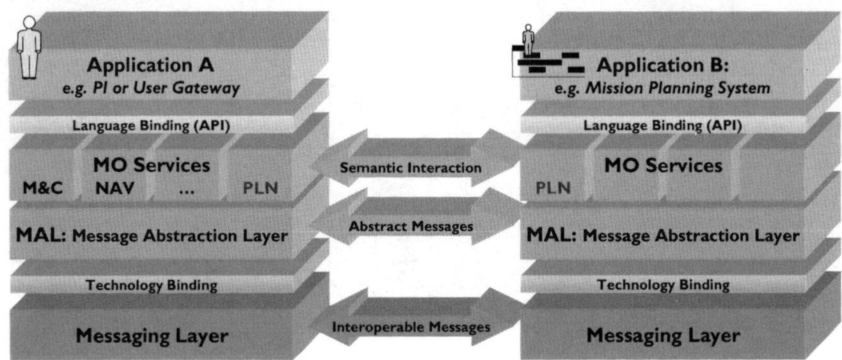

Fig. 21.3   Layered MO service framework.

The COM provides a generic template for an MO service and the object classes it defines simplify the specification of individual MO services and ensure a harmonized approach across multiple services.

The MAL defines an abstract message structure and a set of standard interaction patterns for message exchange, including both request–response and publish–subscribe patterns. The MAL isolates services from deployment technology and may be "bound" to multiple message transport and encoding technologies, including both terrestrial technologies and space communications protocols. Two types of MAL binding exist:

1. *Language bindings* define how to express the application programming interface (API) for a service in a particular programming language. This defines a transformation rather than a specific service API and therefore defines the API for all services specified in terms of the MAL. Communicating applications can be implemented in different languages and use different language bindings, but still interoperate, as the underlying communication is defined in terms of the MAL.

2. *Technology bindings* define how the MAL messages and interaction patterns are implemented for a specific messaging technology. A common technology binding must be used to enable interoperability between applications, but which technology is used in deployment is transparent to the application layer and can be specific to deployment requirements. Bridging between technologies is also possible at the MAL layer. Standardization of technology bindings allows for interoperability between independently developed systems, but private bindings can also be developed for intraoperability between applications within a single system context. All MO services can be migrated to a different deployment technology through the definition or adoption of an alternative MAL technology binding.

III. POTENTIAL SCOPE AND BENEFITS OF MO MISSION PLANNING SERVICES

A. BENEFITS OF STANDARDIZING MO MISSION PLANNING SERVICES

The specification of standardized interoperable interfaces between PIs, users, operating agencies, and spacecraft would in itself bring a number of benefits. Each organization would be able to develop or integrate its own multimission systems, which could then be rapidly integrated via the common service layer with other systems as required to support a particular mission. This does not preclude the reuse of legacy systems, but only one adaptation layer is required to support all missions, rather than many mission-specific bespoke interfaces.

The development of a standard will require the alignment and agreement of common concepts across multiple missions and mission classes. This does not imply a "one size fits all" approach – separate standards or optional elements within standards may be required to support different concepts – but where commonality exists, alignment of concepts and terminology will simplify integration. Standards would need to recognize different approaches to mission planning, including

1. Task-, constraint-, or goal-based algorithms;

2. Time-, position-, or event-based plans and schedules;

3. Discrete, multipart, and repetitive operations.

Many current mission planning interfaces are supported by exchanging files in an agreed format; sometimes, many file formats are used within a single mission system, with different formats being used to exchange similar data across different interfaces. Simple agreement on common, standard file formats would bring benefits in itself, but it has limitations:

1. It assumes file transfer as the exchange protocol.

2. It only defines the static aspects of information exchange (the data structure) and not the dynamic interaction protocol required to initiate an exchange or provide feedback on status. This is currently often achieved through ad hoc operational mechanisms, including phone calls and person-to-person e-mail exchanges.

MO Planning Services would define both the static information exchanged and the dynamic interaction protocol for those exchanges, as a set of standard operations that include the provision of status feedback to the initiator. Multiple messaging technologies can be supported based on the same service specification: file exchange can be supported, but so can other technologies, such as e-mail, enterprise service buses or packet telemetry/telecommand (TM/TC) links. It is also an enabler for the recording of service history.

Despite these benefits, the question has been raised as to whether there is a net cost benefit. The cost of developing standards has to be set against cost savings in deployment and operations of mission systems, which are difficult to predict. However, an advantage of developing MO Planning Services is that the cost of standards development is minimized by building on the existing MO service framework.

B. SCOPE OF MISSION PLANNING SERVICES

Figure 21.4 illustrates the anticipated context of MO Mission Planning services as envisaged by the MO Services Concept [1]. This is an initial perspective that is subject to revision in the course of any future standardization program. Mission planning is loosely coupled to other functions via a set of standard services. These services support both planning inputs and outputs and can also be used to integrate distributed mission planning functions.

The scope of MO services is to standardize the exposed interfaces between applications and not to standardize the applications themselves. The objective is to enable integration of a range of different Mission Planning solutions and algorithms into the wider mission operations system. This allows for reuse across missions, evolutionary change, and innovation.

The Mission Planning function is shown surrounded by a set of services that support loosely coupled interaction with other systems.

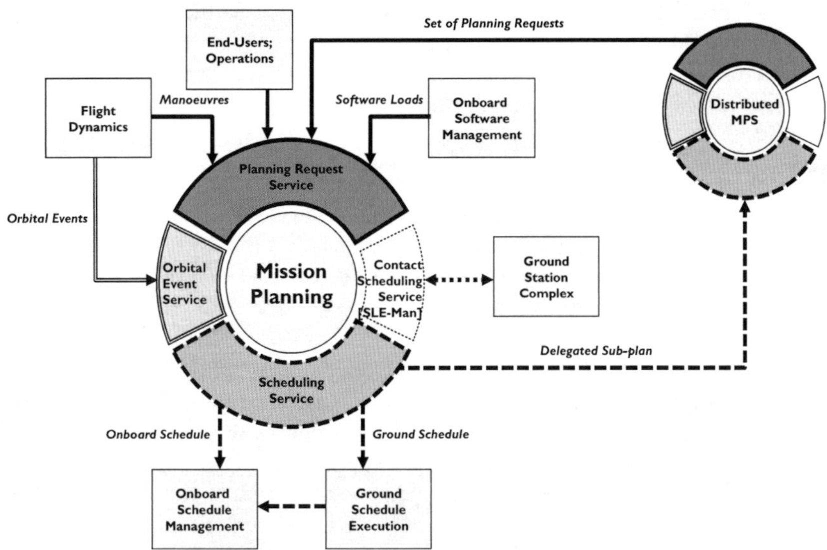

Fig. 21.4   Potential MO mission planning services scope.

Mission Planning System interfaces include the submission of planning requests from end-users and members of the operations team that correspond to operations to be performed or goals to be achieved (shown as a Planning Request service as bold lines); the receipt of orbit vectors or predicted events that may be needed to identify when such operations should be performed (Orbital Event service as double lines); interaction with network service providers to identify the availability of ground station and relay satellite resources and to negotiate allocation of these to support mission operations (Contact Scheduling service as dotted lines); and the provision of plans or schedules to delegated operations execution functions located within the ground segment and/or onboard the spacecraft (Scheduling Service as dashed lines).

The Orbital Event service is anticipated as an MO Navigation service, and contact scheduling is already addressed by the CCSDS Cross-Support Services Service Management standard. MO Mission Planning standardization is therefore expected to focus on Planning Request (input) and Scheduling (output) services.

In the case where the Mission Planning function is itself distributed, then either the Planning Request or Scheduling services can be used to delegate responsibility for a subplan/schedule to another system.

IV. IDENTIFICATION OF POTENTIAL MO MISSION PLANNING SERVICES

This section outlines the approach taken during the workshop held at the CCSDS Technical Meeting in Darmstadt, April 2012, in response to the Call for Interest in MO Mission Planning services. The focus of this meeting was to engage the Mission Planning community in the proposed standardization process and identify candidate services for standardization. Following an introduction to the MO Services Concept by the Working Group, several short position papers were presented by Mission Planning specialists before a discussion forum was held structured around the following questions:

1. Which mission planning scenarios should be considered?

2. What are the communicating entities and functions involved in mission planning?

3. Which candidate services can be identified, and what should be explicitly excluded?

4. Has work already been done that can be used as a basis for standardization in this area?

These issues are discussed below, together with a summary of what needs to be done to define an MO service.

A.   MISSION PLANNING SCENARIOS

The distribution of mission planning functions and users for a given mission is dependent on a number of factors. First, there is the class of mission:

1.   Earth observation missions, which can be both systematic (little end-user interaction) and request-based (driven by end-user interaction);

2.   Observatory missions, which can also be systematic or request-based;

3.   Science/exploration missions, which can have complex constraints because of the competing resource requirements of different instruments;

4.   Communications and navigation missions;

5.   Surface landers and rovers, for which increasing levels of autonomy lead to onboard planning capability;

6.   Manned space missions.

Second, there is the structure of the ground-based MO organization, which includes the following:

1.   Single Mission Control Center;

2.   Distributed operational responsibility (e.g., Payload and Platform);

3.   Separate Payload Data Segment.

Third, there is the degree of onboard autonomy:

1.   Limited autonomy, in which real-time operations are driven from the ground (typical for communications satellites);

2.   Onboard schedule providing mission timeline capability;

3.   Onboard position-based scheduling, which is becoming increasingly common for Earth observation missions;

4.   Onboard autonomy, including goal-based commanding or autonomous replanning (e.g., on rovers).

It is clear that there are commonalities across multiple mission classes and an indication that it would be possible to derive a smaller set of mission criteria that impact mission planning. No full analysis has yet been performed, but potential criteria include the following:

1.   Whether the mission is systematic or request-driven;

2.   Whether tasking is by time, position, or event;

3.   Whether the spacecraft requires pointing to satisfy individual requests;

4.   Whether planning responsibility is distributed;

5.   Whether the spacecraft can autonomously modify the plan/schedule.

B. COMMUNICATING ENTITIES AND FUNCTIONS

A generalized view of the functions involved in Mission Planning and their inter-actions is given in Fig. 21.5. This is consistent with the view previously given in Fig. 21.4 and the line style of interactions indicates potential services. Two instances of the Planning function are shown to illustrate the scope for distributed Mission Planning. Multiple functions can generate tasking requests as input into planning, and planning can distribute plans/schedules to multiple ground-based or onboard execution environments. Predicted events are required by the plan-ning process and interaction is also required with ground-station scheduling.

These functions may be distributed over a number of distinct entities (organ-izations and systems) within a given space mission system. There is no fixed set of such entities, but typical examples include the user community/PIs, the Science/Payload Operations Center, the Payload Processing Center, the Spacecraft Oper-ations Center, Flight Dynamics/Navigation (usually part of SOC), the Ground Tracking Network, unmanned spacecraft, surface landers/rovers, and manned space vehicles.

Figure 21.6 illustrates the potential deployment of each of the functions shown in Fig. 21.5 to the entities listed above. It is where interactions between functions are exposed across one or more boundaries between these entities that there is jus-tification for standardization within CCSDS as a potentially interoperable inter-face between agencies. Functions can be distributed over various organizations and systems that together form a space mission system. The arrows in Fig. 21.6 indicate the interactions in a typical current deployment, but the potential distri-bution of functions indicated by the circles shows that all the functional interfaces shown in Fig. 21.5 can be exposed to the boundaries between these entities and are therefore candidates for standardization as MO services.

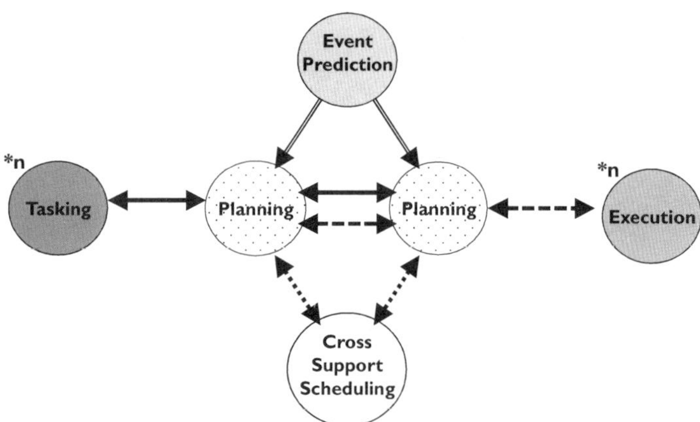

Fig. 21.5   Functions involved in (distributed) mission planning.

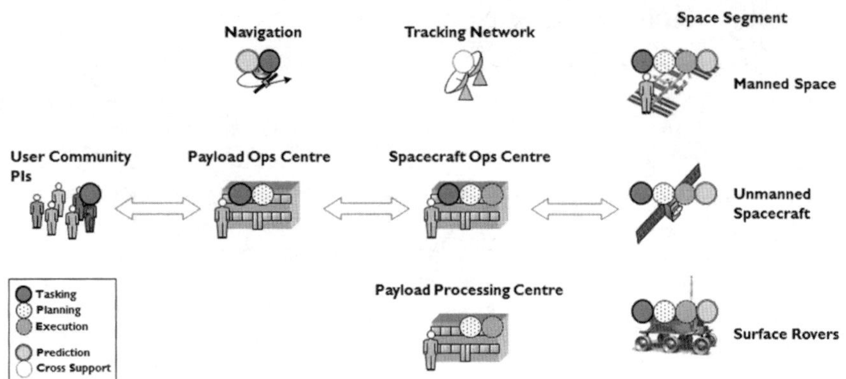

Fig. 21.6   Entities and functions involved in mission planning.

C.   POTENTIAL SERVICES AND EXCLUSIONS

Before the discussion, the Working Group had identified two candidate MO Mission Planning services, corresponding to the interactions shown in Fig. 21.5 between Tasking and Planning, and between Planning and Execution: Tasking/ Planning Request Service and Plan/Schedule Distribution Service (including execution feedback).

These correspond to operational interfaces used routinely to support MO. The feedback received from the planning community is that they are seen as candidates for service standardization and that there would be a clear benefit to service-level standardization that would also address the interaction protocol rather than just limiting standardization to a data exchange format. Several participants expressed the opinion that standardization of the Scheduling service would be easier than standardization of the Planning Request service, because of the diversity in the content of tasking requests between missions. It was also proposed that a simple approach should be adopted initially for Planning Requests: standardizing the interaction between the user community and Mission Planning system, rather than the full content of the tasking requests themselves. This may then be refined as greater commonality between requests is identified.

In addition to these dynamic information exchanges, several participants identified a need to standardize the exchange of static planning configuration data. This is used by the Mission Planning function itself as part of the planning process and typically comprises planning constraints, rules, and resource usage profiles. Although this may be considered "private" to the Mission Planning function, in practice the nature of the planning constraints often needs to be understood by several actors in the system, including spacecraft and payload manufacturers; PIs and users; and Distributed Mission Planning systems (e.g., Payload and Spacecraft Operations Centers).

A generally expressed concern is that the nature of this information may be specific to the Mission Planning algorithms and tools used. Standardization may therefore be difficult to achieve, and the interactions supported are both infrequent and offline. The consensus was that this area should not be considered initially for standardization, but that emphasis should be given to the operational interfaces already identified.

Standardization of the provision of predicted orbital events is considered to be a candidate for MO Navigation services, rather than specifically associated with Mission Planning, but it was suggested that the mission planning community should also be involved in the standardization process.

It is accepted that the process by which space link service providers and space missions exchange information needed to arrange spacecraft contact periods is excluded from the scope of MO services, because it is already addressed by the CCSDS Communications Cross Support Service Management Standard [5].

D. MO SERVICE DEFINITION

Specification of application-level MO services requires the following:

1. Definition of the associated information model (service objects);

2. Definition of the service operations, each mapped to a MAL interaction pattern;

3. Provision of the service configuration (object definitions) for service deployment.

The following do not require specification as they are provided by the MO service framework:

1. Message encoding/binding to the messaging technology (providing the required binding is already available);

2. Specific definition of the API (standard language transformations apply to all MO services for supported languages and can be used to autogenerate the API);

3. Definition of service discovery, login, authentication, and so on, as these are covered by the MO Common Services Specification;

4. Specification of a dedicated service history model as this can be derived from the MO COM.

A key aspect of service definition will be to define the information model associated with these interfaces. The first challenge in standardization will be to agree common terminology that can be used to describe the service information model and operations.

Representation of plans or schedules is not limited to simple time-tagged lists of commands, but increasingly will need to represent operations tagged by other initiation criteria, including position, predicted events, state conditions, or user interventions. Other aspects may need to be reflected in the model, such as applicable configuration, allocation of operational responsibility and priority. Similarly, the information model for planning requests may need to address discrete, serial, and repetitive operations.

In addition to information structures, service definitions also address the pattern of information exchange that is required between collaborating systems or components. This includes the dynamic update of service data, for example to provide feedback on the execution status of schedules, or to reflect changes to a plan or schedule made by a collaborating system. Service operations can be defined to support the following:

1. Injection of planning requests, either individually or as a batch;

2. Provision of feedback on the status of a planning request;

3. Publishing a new or updated plan or schedule;

4. Modification of the plan/schedule, either discrete changes or in a batch;

5. Provision of feedback on the status of the plan/schedule, either live or as a batch offline update, and for both planning changes and execution status.

E. EXISTING STANDARDS AND OTHER INPUTS TO STANDARDIZATION

Before development of any MO Mission Planning Services takes place, a full survey of existing standards and other potential inputs to the standardization process will need to be performed to ensure that there is no unnecessary duplication of standards and that full advantage is taken of any existing body of work. A number of potential sources were identified by participants in the workshop. These include

1. The OGC OpenGIS Sensor Planning Service [6];

2. The European Ground Segment Technology Harmonization Programme;

3. The Planning Domain Definition Language (PDDL) [7].

PDDL has been under development in the planning academic community for some years and offers a potential solution for the exchange of planning configuration data.

V. CASE STUDY

As a case study, the Mission Planning infrastructure the ESA Venus Express mission was considered. Although Venus Express is considered an example of a mission with relatively streamlined interfaces, it was found that there were a total of 48 different planning products being generated by all parties to the mission. These may be grouped into 28 functional interfaces that use a total of

36 different containers, mostly file based, and at least five different transport mechanisms, including FTP, HTTP, and e-mail. The exchange mechanisms themselves, even if described in documentation, are mostly ad hoc and require manual intervention by users. When the full set of interfaces is considered, it is apparent that there is room for consolidation and standardization of some aspects of information exchange.

More detailed analysis showed that Mission Planning exchanges could all be classified as one of the following three categories of data exchange: timelines (or schedules); notifications; or configuration data.

Similarly, all interfaces followed one of a limited number of patterns of information exchange, which could be mapped to the MAL interaction patterns.

This leads to the conclusion that planning services could be standardized, although the complexity and variety in approach of configuration data is likely to make the standardization of meaningful content itself difficult to achieve for this category. Consequently, standardization should initially concentrate on a common representation and exchange of timelines and notifications.

VI. CONCLUSION

Development of standard Mission Planning Services based on the MO service framework developed by the CCSDS Spacecraft Monitoring & Control Working Group would promote software reuse and simplify the integration of distributed mission planning and wider mission operations systems. A call for interest in the topic resulted in significant support at a workshop held during the CCSDS technical meeting held in Darmstadt, Germany, in April 2012. This did not result in universal agreement on the scope of potential MO Mission Planning services, but recommended an initial focus on operational interfaces including planning request and scheduling services, rather than attempting to address the more complex problem of exchanging planning constraints and other configuration data. An important aspect of service standardization for the operational planning interfaces is to address the interaction protocols used to exchange information as well as the information content of those interactions. This is expected to deliver operational benefits in the short term as it avoids manually intensive exchanges via ad hoc mechanisms including telephone and e-mail. For the information content itself it is recommended that this is initially kept simple, but extensible to allow incremental development of the standards to support different mission classes and deployment scenarios. The first step in standardization of MO Mission Planning services will be harmonization of the terminology to be used in the specification of standards, to ensure a common understanding between different organizations.

Following the workshop, a draft concept paper was produced on MO Mission Planning Services, of which this chapter is effectively a summary. The Mission Planning workshop itself was repeated at the subsequent CCSDS technical meeting in Cleveland, Ohio, in October 2012. It is anticipated that the process of establishing a new CCSDS Working Group to define MO Mission Planning Services will now be initiated in 2013.

ACRONYMS

API	Application programming interface
CCSDS	Consultative Committee for Space Data Systems
COM	[CCSDS MO] Common Object Model
GIS	Geographic information system
ISO	International Organization for Standardization
M&C	Monitoring and control
MAL	[CCSDS MO] Message abstraction layer
MO[S]	[CCSDS] Mission operations [services]
NAV	[CCSDS MO] Navigation services
OGC	Open Geospatial Consortium
PDDL	Planning Domain Definition Language
PI	Principal investigator
PLN	[CCSDS MO] Mission planning services
SLE	[CCSDS] Space Link Extension Standards
SOC	Spacecraft (or Science) Operations Center
TM/TC	Telemetry/telecommand

REFERENCES

[1] "Mission Operations Services Concept, Informational Report (Green Book)," CCSDS 520.0-G-3, Dec. 2010, http://public.ccsds.org/publications/archive/520x0g3.pdf [last accessed 18 April 2013].

[2] "Mission Operations Reference Model, Recommended Practice (Magenta Book)," CCSDS 520.1-M-1, July 2010, http://public.ccsds.org/publications/archive/520x1m1.pdf [last accessed 18 April 2013].

[3] "Mission Operations Message Abstraction Layer, Recommended Standard (Blue Book)," CCSDS 521.0-B-2, March 2013, http://public.ccsds.org/publications/archive/521x0b2e1.pdf.

[4] "Mission Operations Common Object Model, Draft Recommended Standard (Red Book)," CCSDS 522.0-R-2, Sept. 2011, http://public.ccsds.org/sites/cwe/rids/Lists/CCSDS%205220R2/Attachments/522x0r2.pdf [last accessed 18 April 2013].

[5] "Space Communication Cross Support–Service Management–Service Specification, Recommended Standard (Blue Book)," CCSDS 910.11-B-1, Aug. 2009, http://public.ccsds.org/publications/archive/910x11b1ec2.pdf [last accessed 18 April 2013].

[6] "OpenGIS Sensor Planning Service, EO Satellite Tasking Extension," OGC, 10-135, March 2011, http://www.opengeospatial.org/standards/sps [last accessed 18 April 2013].

[7] Kovacs, D. L. "BNF Definition of PDDL3.1: Completely Corrected, Without Comments," Unpublished manuscript from the IPC-2011 Web site, 2011, http://www.plg.inf.uc3m.es/ipc2011-deterministic/Resources?action=AttachFile&do=view&target=kovacs-pddl-3.1-2011.pdf [last accessed 18 April 2013].

CHAPTER 22

EDRS Precursor Systems at GSOC: Relevant Heritage and New Developments

Ralph Ballweg* and Frank Wallrapp†
DLR/GSOC, Oberpfaffenhofen, Germany

I. INTRODUCTION

With the increasing amount of data being generated by Earth observation satellites, the traditional strategy of dumping the data during a ground station pass has reached its limits and is not feasible for future programs such as the European Global Monitoring for Environment and Security programme (GMES). It is estimated that this constellation will generate approximately 4 TB per day [1]. Furthermore, the need for timelier access to data by users is growing.

The characteristics of the ground station scenario are as follows:

1. Contacts with a spacecraft are only possible while direct visibility exists with the antenna. Unless the ground station is located in a polar region the number of such contacts are limited to 4–5 per day with an average time span of 10–12 minutes. This severely restricts the amount of data that can be downloaded and also the time commanding the spacecraft.

2. Having the ground station in polar regions does increase the number of available satellite passes, but has the disadvantage that data are received at remote locations, and then have to be transferred to the end-user.

The alternative is to place a relay satellite into geostationary orbit (GEO) that is capable of transferring data at high rates, in real time or near real time, to ground stations with more favourable infrastructure, or indeed directly to the user. Part of this data transfer could be performed using laser technology to achieve the desired speed. This chapter describes the technology and concept the German Space Operations Center (GSOC) has been involved with in the operations of laser communication terminals (LCTs) that have been specifically designed for this purpose.

---

*Project manager TSX-LCT, project manager TDP-1.
†Group Lead Geostationary satellites, project manager EDRS

II.   DATA RELAY CONCEPT [2]

A schematic showing the concept of a geostationary data relay for Earth observation data is presented in Fig. 22.1.

The components of the system are described in the following sections.

A.   LOW-EARTH-ORBITING SATELLITE (LEO)

The LEO is typically an Earth observation satellite carrying one or more instruments that generate user data that have to be transferred to ground. In the conventional approach, data are transmitted to ground via ground stations, for example in the X-band (not shown in Fig. 22.1). In the case of a relay satellite the LEO transfers the data to the GEO satellite (link "U_1"). Consequently the LEO has to be equipped with a high-data rate LEO to GEO communication device. To be able to transfer large amounts of data, at least two technologies might be used:

1.   Optical communication with a laser terminal (laser communication terminal, LCT). Key figures and operational aspects are described briefly in Sec. III.

2.   Radio-frequency (RF) communication in the Ka-band. This technology and related operations are described briefly in Sec. IV.

Note that, for the sake of clarity, only one LEO is shown. In fact, various LEOs may use the relay satellite either in parallel or one after the other depending on the technical implementation.

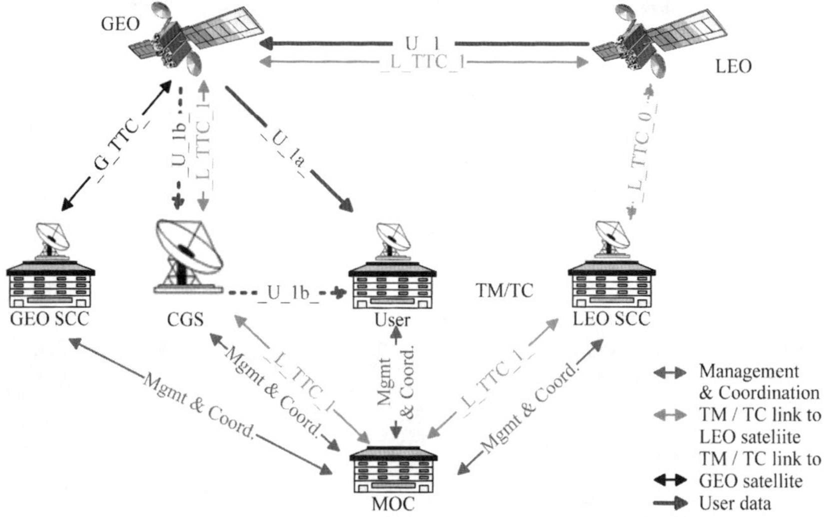

Fig. 22.1   Concept of a GEO relay system for Earth observation data.

B.   LEO SATELLITE CONTROL CENTER (LEO SCC)

The LEO SCC operates the LEO satellite. It is responsible for housekeeping and payload operations. Among these tasks is the correct pointing and activation of the LEO to GEO communication device. For correct pointing of the communication device towards the GEO satellite, the GEO orbit has to be known to the LEO SCC.

C.   GEOSTATIONARY RELAY SATELLITE

The GEO satellite receives the user data from the LEO satellite and relays it to ground. For this purpose it needs a receiver that is compatible with the terminal of the LEO. The same technologies are therefore considered, namely LCT or Ka-band. As for the LEO, these devices are usually steerable and have to be pointed towards the LEO, depending on the technology.

To complete the relay function, the GEO satellite needs a high-data-rate terminal to send data to the ground. The technology that can be used for this purpose is not dependent on the LEO to GEO link. To receive a comparable data rate the Ka-band is used. The space-to-ground beam for GEO satellites may cover a very large portion of the Earth, which should allow various ground stations spread over large distances to receive the data in parallel (link U_1a and U_1b).

D.   GEO SCC

The GEO SCC operates the GEO satellite. It is responsible for housekeeping and payload operations. Among these tasks is the correct pointing and activation of the receiver. To correctly point the receiver in the direction of the LEO, the LEO orbit has to be known to the GEO SCC.

E.   USER CENTER

In the user center the data generated by the user platform aboard the LEO is processed. The data may be received at the user center directly with the user ground station (link U_1a) or from a central ground station from which it is transferred via conventional telecommunications infrastructure to the user (link U_1b).

F.   CENTRAL GROUND STATION (CGS)

The CGS receives and stores all data from all users. This enables checking of the correct execution of the data relay and storage of all data. Also, if a user does not have his own ground station, the user data may be transferred from the CGS to the user using conventional telecommunications infrastructure.

G. MISSION OPERATIONS CENTER (MOC)

The MOC is the core component in the system. It interfaces with all other components and coordinates them. Its main purpose is to receive all the link requests from the different users and generate a link session timeline taking all known constraints into account. It also monitors and controls all involved infrastructure.

III. KEY TECHNOLOGIES

The two key technologies under consideration for the relay tasks are Ka-band RF and optical/laser. The final design of the European Data Relay Satellite (EDRS) program led to the combination of LCT for the transfer from LEO satellites to the relay satellite and Ka-band technology for the downlink of data to the ground.

A. LCT

The LCT being used on TerraSAR was developed by TESAT Spacecom of Germany with funding by the German Space Agency (DLR). It was designed with the goal of a high-rate data transfer from space to space and space to ground. Data rates of 5.625 Gbps have been successfully demonstrated between the NFIRE and TerraSAR-X satellites (the distance between these satellites was 1000–5000 km). The key parameters for the LCTs that have been verified in orbit are shown in Table 22.1 [3].

B. KA-BAND TECHNOLOGY

Currently, there are several commercial satellite missions worldwide that provide high-rate communication services at Ka-band frequencies (18–40 GHz) to various

TABLE 22.1   KEY FIGURES FOR THE LCTS

Mass	35 kg
Power	120 W
Dimension	$0.5 \times 0.5 \times 0.6 \, \mathrm{m}^3$
Telescope diameter	125 mm
Maximum optical transmit power	0.7 W
Bit error rate	$<10^{-9}$
Link distance	1000–5100 km
Link duration	<8 min
Data rate	5.625 Gbps

ground-based users. Putting higher frequencies into service gives several advantages. For example, the Ka-band provides up to 600% link advantage over the X-band. This advantage could be translated either into high-data-rate communication, longer-distance communication, or smaller and therefore much more cost-effective ground stations. Additionally, the smaller antenna beam of the Ka-band ground station considerably reduces RF interference with other systems. So, although the Ka-band technology is an inevitable part of the modern ground station complex and could be used for future data-relay satellites, a lot of effort is required to design, install, and operate such a system.

IV.   LCT OPERATIONS DEVELOPMENT AT GSOC

A.   OPERATIONS TERRASAR-X

GSOC became involved in LCT operations for the first time with the program TerraSAR-X, which hosted an LCT as a secondary payload. A second LCT is flying onboard the U.S. satellite NIFIRE, which is operated by the company Orbital Sciences Corporation (Orbital). NFIRE was launched in April 2007, with TerraSAR following in June of the same year. GSOC is the operator of the TerraSAR-X and also commands the LCT. The TerraSAR LCT is designed for two types of contact, a satellite-to-ground link and intersatellite links (ISLs). As the objective in this case is test and evaluation of LCT operations, responsibilities are divided between GSOC as the satellite operator and the LCT manufacturer TESAT. The first LCT tests on TerraSAR were space-to-ground links (SGLs) performed with ground terminals located on DLR property in Oberpfaffenhofen near Munich and on the island of Tenerife. The first ISLs were executed beginning in January 2007.

Essential to all LCT operations is the planning cycle, which is an iterative process. Starting with the different orbit information, one party (in the case of TerraSAR-NFIRE ISLs this was Orbital) calculates the link options and makes a preselection with the available links (Fig. 22.2). GSOC then coordinates the final link selection and publishes the deconflicted links to all parties. The two control centers for TerraSAR and NFIRE then individually prepare their respective LCT operations with input from the instrument manufacturer TESAT (in the form of command input files) and their flight dynamics departments (Chebychev coefficients for LCT pointing). GSOC then produces the detailed sequence of events (SOE) and provides it to TESAT. After the links, GSOC and Orbital make a quick determination of the success of the operations and provide all the corresponding data to TESAT for evaluation. Results then flow into the input for the next links.

Occasionally, other partners (e.g., the DLR Institute of Communication and Navigation, IKN) request the opportunity to perform SGLs using their own optical ground stations. In this case, those partners present their objectives to

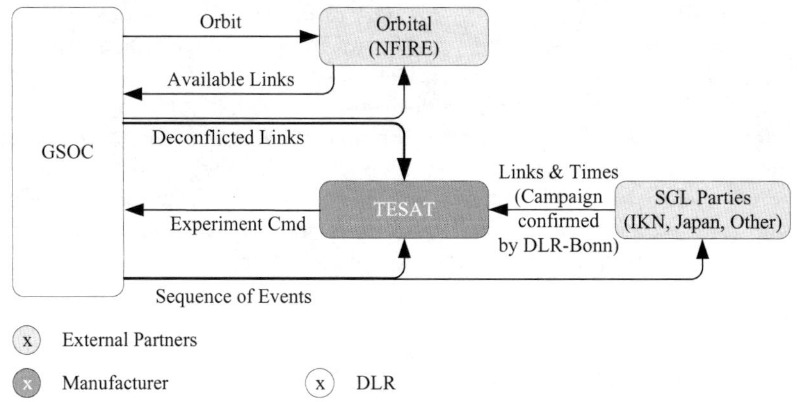

Fig. 22.2   TerraSAR LCT operations concept.

TESAT, which generates LCT configuration files, and GSOC again publishes the SOE and performs the operations.

B.   TECHNOLOGY DEMONSTRATION PROGRAM 1 (TDP-1) SYSTEM DESIGN

The next step is the project TDP-1 (Fig. 22.3). TDP-1 is an experimental mission giving the proof of concept of a relay for Earth observation data from LEO spacecraft via a GEO satellite using laser as a transfer medium. It is considered a precursor mission to the EDRS project.

GSOC's objective in the program is to establish a control center for LCT relay operations providing end-to-end service of data transfer from a LEO spacecraft via a relay satellite to the final data user. In theory, this means a handover of the data at the originating source (i.e., the LEO satellite) and delivery to a dedicated ground station or end-user. In practical terms GSOC will be able to execute all data transfer functions, including the operations of all participating laser terminals and RF equipment.

The TDP-1 payload hosted on a GEO satellite consists of an LCT, mainly for the ISL to a LEO spacecraft, and a Ka-band payload for the data downlink from the GEO satellite. GSOC began the preparatory design for the TDP-1 project in late 2011. The laser terminal can also be pointed to an optical ground station, a functionality that will be used during the commissioning phase for calibration purposes. The launch of the GEO relay satellite is currently scheduled for March 2013, with the LEO customers to follow several months later.

The participating agencies in TDP-1 are the German Aerospace Center (DLR) as the contracting entity or customer, with its institutes GSOC and DFD, INMARSAT hosting the GEO payload, TESAT as the LCT manufacturer, and the European Space Agency (ESA) as the first LEO customer.

The concept of operations for GSOC is that GSOC collects the orbital information of all participating spacecraft and possibly the ground contact information of the LEO satellites. It then performs visibility calculations and publishes visibility reports for periods of one week and collects link requests. On a specific day, GSOC forwards the final link selection to the TDP coordination office (TECO) at INMARSAT and receives feedback the next morning. GSOC then prepares the command information for all participating laser terminals. The downlink from the ALPHASAT is received via the Ka-band at DLR's DFD Institute and the data are distributed from there. GSOC receives a report about the success of the link and LCT diagnostic data for TESAT for evaluation of LCT performance.

The LCT terminal onboard ALPHASAT is also capable of contacting optical ground stations. This feature will be used for testing and calibration during the commissioning phase, when no LEO satellite is available. The first ISLs are planned for the first quarter of 2014.

The information being exchanged between the GSOC and the partners is as follows:

1. LEO SCCs to GSOC: orbit information, link requests, possible constraints, telemetry;

2. GSOC to LEO SCCs: link possibilities, SOEs, LCT command inputs;

3. INM/TECO to GSOC: orbit information, deconflicted links, telemetry, command logs;

4. GSOC to INM/TECO: link list, SOE, command inputs, telemetry requests;

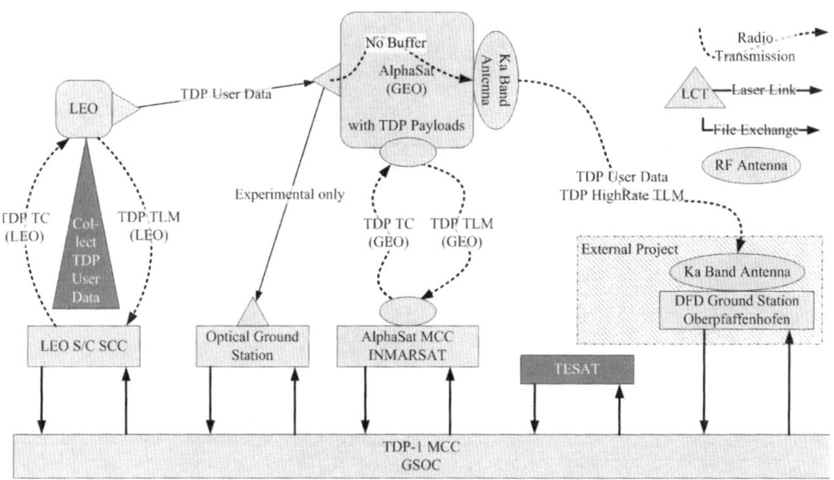

Fig. 22.3 TDP-1 project.

5. TESAT to GSOC: link requests, command inputs;

6. GSOC to TESAT: SOE, status reports, telemetry, command logs.

C. GSOC IN-HOUSE SYSTEM DESIGN [4]

The system designed will consist of several subsystems with specific tasks (Fig. 22.4):

1. *Exchange server.* Because of GSOC security policies the file exchange has to occur on an exchange server that is located outside of GSOC firewalls at the DMZ (Daten Management Zentrum). The exchange server is an FTP server that provides a data exchange point for the system. It is located outside the GSOC firewalls and the only bidirectional connection to the outside world is the SFTP service that enabled customers to send to and receive files from GSOC. It is also the only instance that is able to receive User Datagram Protocol (UDP) packet streams and so the Real Time (RT) check of this telemetry needs to be done there.

2. *Mission Planning System (MPS).* The MPS is responsible for optimizing the link planning based on input from the different customers and taking system-specific constraints into consideration.

3. *Flight Operations System (FOS).* The FOS manages and surveys the flow for the whole system. It is responsible for the routine and contingency operations for all LCTs supported by the system. In particular, it is responsible for

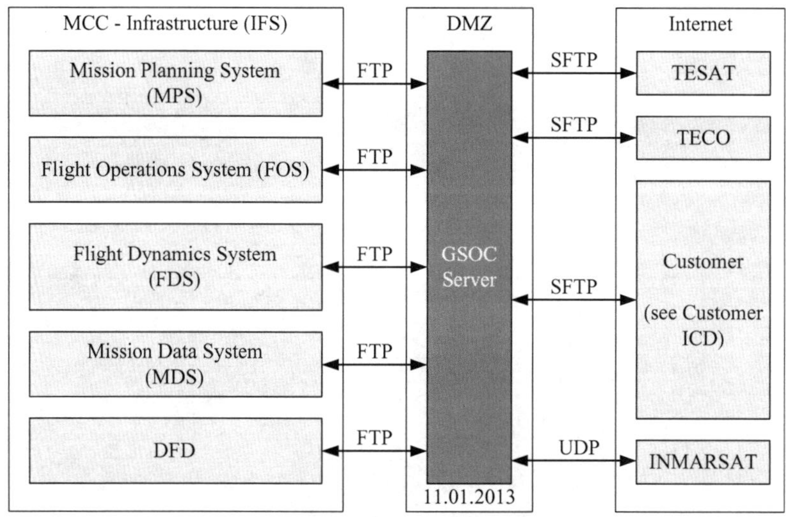

Fig. 22.4   TDP-1 functional system design.

maintaining and implementing all flight procedures in the system and handles almost all the outgoing traffic to the exchange server.

4. *Flight Dynamics System (FDS)*. The tasks of the FDS are to calculate the visibilities between the individual LCTs, generate the Chebyshev coefficients for laser acquisition and tracking, and perform alignment calculations on a periodic basis.

5. *Mission Data System (MDS)*. The MDS does the telemetry data processing for the system. It is also responsible for archiving and accessing data.

6. *Deutsches Fernerkundungs Datenzentrum, German Remote Sensing Data Center (DFD)*. The DFD is the receiving ground station for data from the relay satellite LCT (user as well as LCT diagnostic data). All data are then routed to the GSOC.

D. EDRS OPERATIONS

The final step in the development of EDRS is commercial operational transfer of data from LEO spacecraft via laser data links to a GEO relay satellite, and forwarding of the data to the end-user.

EDRS is an ESA program for data relay featuring one dedicated satellite (EDRS-C) and one piggyback payload (EDRS-A) (Fig. 22.5). Both will be positioned in GEO with visibility over central Europe. The dedicated satellite EDRS-C will be based on the new SmallGEO platform by Orbitale Hochtechnologie Bremen (OHB). The EDRS-A payload will be hosted on a EUTELSAT satellite. ESA will act as a major customer for Astrium, paying for relay services of data from SENTINEL satellites. However, the system is designed such that more customers can be integrated.

The link between the LEOs and GEOs will be established through a Ka-band high frequency (HF) transponder or an optical LCT. For these links, data rates of 300 Mbps (Ka-band) and 600–1800 Mbps (LCT) are targeted. The data transfer between the GEO and ground will be established through a high-bandwidth Ka-band link. Besides the main task, which is to relay Earth observation data to ground (return link), the system will also feature the possibility to relay data from ground to the LEO spacecrafts (forward link).

E. EDRS ROLES AND RESPONSIBILITIES

The MOC, operated by Astrium Services, is the core of the EDRS activities. Its responsibilities are the coordination of the overall system, planning and scheduling of the links, management of user data keys, providing a user help desk for customers and accounting and billing. Starting up to 75 days from a given T0, the MOC collects orbit predictions and operational constraints from all participating parties and publishes, based on the evaluations, possible link times. From T0–3

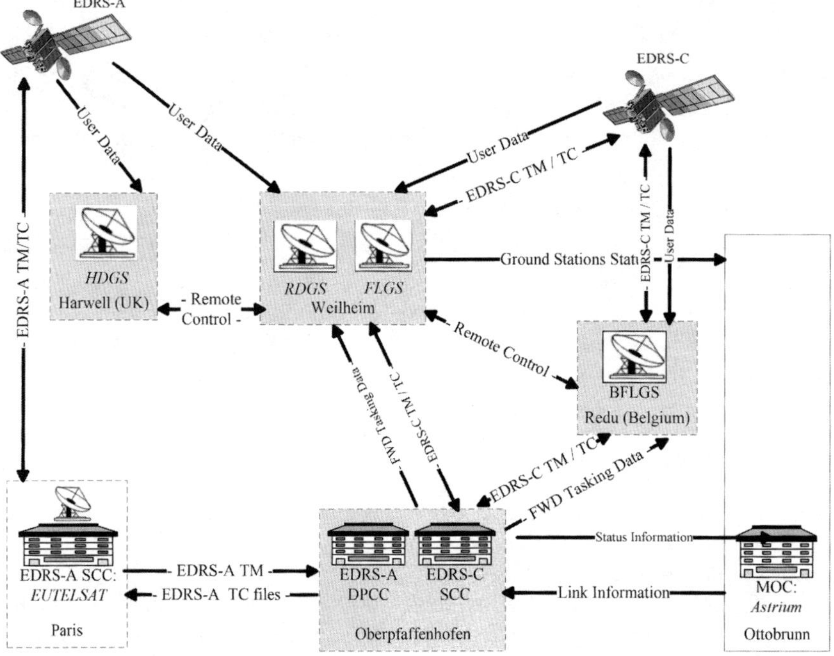

Fig. 22.5   EDRS system layout.

days to T0–8 hours, the actual link planning takes place and the relevant planning information is sent to the individual SCCs for command generation and uploads.

GSOC will be in charge of performing the following tasks:

1. Buildup and operations of the EDRS central ground stations in various places in Europe. All stations are operated from the DLR facilities in Weilheim. Stations located outside Weilheim are remotely operated using dedicated software developed at GSOC.

2. Buildup and operations of the control centers for the EDRS-A payload, called the Devolved Payload Control Center (DPCC) and the EDRS-C satellite (SCC). In the EDRS-A case, the TM/TC interface is connected to the Eutelsat SCC, from which the TM/TC data are transferred to the Eutelsat ground station controlling the EDRS-A host satellite. GSOC's role is to operate the EDRS payload onboard the host satellite. The EDRS-A GEO SCC is thus shared between Eutelsat and GSOC. In the case of the EDRS-C, the TM/TC interface is connected to the EDRS ground station in Weilheim with the Redu station as backup. In this respect, GSOC fulfils the role of the EDRS-C GEO SCC.

In the operational phase, GSOC will receive all link-related information from the MOC, with the exception of the pointing coefficients for the LCTs or the ILS Ka-band antenna on EDRS-A.

The SCC/DPCC at GSOC will be designed to process up to 120 link requests per day and satellite, calculate LCT pointing angles, generate commands, manage the spacecraft (S/C) time-tag buffers, enable correct user data-encryption keys, and log every activity for failure analysis.

For EDRS-A, specifically, the DPCC also has to calculate the Ka-band ISL antenna pointing angles, generate and send command information according to the interface control document (ICD) with Eutelsat, analyze the relevant host satellite telemetry, and coordinate all activities with Eutelsat.

V. CONCLUSION

At the present time, GSOC already has close to five years of experience operating LCTs for SGLs and ISLs. The TDP-1 project will incorporate the GSOC MPS into the operational concept and, with the implementation of the EDRS operations center, we will be the leading control center for the operation of satellite data relay systems based on laser data transfer.

ACRONYMS

ABLE	Atmospheric Boundary Layer Experiment
DLR	German Center for Aerospace
DPCC	Devolved Payload Control Center
DFD	German Remote Sensing Data Center (Deutsches Fernerkundungs Datenzentrum)
DMZ	Daten Management Zentrum (Data Mangement Center)
EDRS	European Data Relay Satellite
FDS	Flight Dydnamics System
(B)FLGS	(Backup) Feeder Link Ground Station
FOS	Flight Operations System
GEO	Geostationary (satellite)
GMES	Global monitoring for environment and security
GSOC	German Space Operations Center
HDGS	Harwell Data Ground Station
ICD	Interface control document
ISL	Intersatellite link
LCT	Laser communication terminal
LEO	Low Earth orbit (satellite)
MDS	Mission data system
MOC	Mission Operations Center
MPS	Mission Planning System

RDGS	Reference Data Ground Station
S/C	Spacecraft
SCC	Satellite Control Center
SGL	Space to ground link
TDP-1	Technology Demonstration Program 1
TECO	TDP ESA Coordination Office
TLM	Telemetry
UDP	User Datagram Protocol

REFERENCES

[1] ESOC GMES Team, "The Sentinels/EDRS Operations Constraints and Concept," ESTEC, 2010.

[2] Wallrapp, F., Ballweg, R., and Gataullin, Y., "The European Data Relay System (EDRS) Operational Challenges," IAC conference, IAC-11.B6.2.4, 2011.

[3] Gregory, M., "Tesat Laser Communication Terminal Performance Results on 5.6 GBit Coherent Intersatellite and Satellite to Ground Links," *International Conference on Space Optics*, 2010.

[4] Kuhlmann, S. "TDP-1 Planning Design Definition," GSOC, 2012.

Consideration of Space Debris Mitigation Requirements in the Operation of LEO Missions

Holger Krag,[*] Tim Flohrer[*] and Stijn Lemmens[†]
ESA/European Space Operations Centre, Darmstadt, Germany

I. INTRODUCTION

The low Earth orbit (LEO) altitude regime is the most frequently used region in space, and, today, it is the only region where there are manned spacecraft. As a consequence of this traffic, the global maximum in the spatial density of space objects is at an altitude of approximately 800 km, where the influence of the atmosphere on orbital lifetime is small. However, even in orbits used for human spaceflight, despite the denser atmosphere, the population of space objects is steadily increasing because of new fragmentations (approximately five per year) that are outpacing the decay of space objects.

A. STATUS OF THE ENVIRONMENT

Because most of these objects are not detectable by the available sensors, space debris models such as the MASTER (Meteoroid and Space Debris Terrestrial Environment) model of the European Space Station (ESA) are used to assess and describe the spatial distribution and physical properties of space objects in Earth orbits [1].

High impact velocities (which can reach 15 km/s for most missions in LEO) are the reason for the destructive energy of these objects, despite their small size. Most impacts will not cause any noticeable degradation of spacecraft function. However, micrometer-sized objects can generate pits on surfaces and optical sensors, degrading their performance. Such particles also generate charges during impact, which can influence electrical components. Particles larger than about 1 mm in diameter could, under special circumstances, terminate the function of payloads and thus endanger missions. The likelihood for this rises with increasing impactor diameter. A number of recorded incidents (NOAA 7 in 1997, Cosmos-539 in 2002, UARS in 2007, etc.) are believed to have been

---

*Space Debris Analyst, HSO-GR.
†Young Graduate Trainee, HSO-GR.

TABLE 23.1   MEAN TIME BETWEEN COLLISIONS OF OBJECTS WITH A SPHERICAL TARGET (OF 1 M^2 CROSS-SECTION) FOR SOME REPRESENTATIVE ORBITS AND SELECTED IMPACTOR DIAMETERS

Altitude	>0.1 mm	>1 mm	>1 cm	>10 cm
400 km	4.5 days	37 years	2,416 years	41,667 years
800 km	0.1 days	2.9 years	133 years	2,483 years
GEO	17.3 days	556 years	128,205 years	1,488,095 years

caused by physical damage from objects in space, which became obvious from a slight change in orbit. The events become more obvious and more critical to the environment when the kinetic energy reaches levels that shatter the satellite into several fragments. Such catastrophic collisions are believed to occur when an energy-to-target-mass ratio (EMR) of 40 J/g is exceeded. So far, there are four recorded examples in this domain (in 1991 between the inactive payload COSMOS-1034 and a fragment of the COSMOS-296 spacecraft; in 1996 between the active French CERISE microsatellite and a fragment of an Ariane-1 upper stage; in 2005 between a Thor Burner IIA upper stage and a fragment of a CZ-4B third stage; and in 2009 between the active Iridium-33 satellite and the decommissioned Cosmos-2251 satellite).

Table 23.1 compares the mean times between collisions of space objects with a spherical target with a cross-section of 1 m^2 for a few representative space missions. The numbers are a result of a collision flux analysis with MASTER-2009. We can see that Sun-synchronous orbits at an altitude of around 800 km suffer from higher exposures to debris flux.

With today's launch rates of 60 to 70 per year, the number of objects in space is steadily increasing. These rising object numbers also increase the probability of collisions in frequently used orbital regions. Today, it is a great concern that collisions could become the main future source of new debris, possibly causing a chain reaction in the space debris environment and rendering some orbital regions unacceptably risky for operations (an effect first postulated by NASA's Donald Kessler in 1978).

B.   COUNTERMEASURES

The problem was first noticed in the early 1960s, but the global dimension of this problem has only been understood recently. A first important step to an international application of debris mitigation measures was taken by the Inter-Agency Space Debris Coordination Committee (IADC), which was founded in 1993 as a forum for technical exchange and coordination on space debris matters. In 2002, the IADC published the "IADC Space Debris Mitigation Guidelines" and presented them to the United Nations Committee on the Peaceful Use of Outer Space (UNCOPUOUS) Scientific & Technical Subcommittee. In the meantime, space agencies in Europe have developed more technically specific guidelines, called the "European Code of Conduct," which was signed by ASI, BNSC (now UKSA), Centre National d'études Spatiales, Deutsches Zentrum für Luft- und Raumfahrt and European Space Agency in 2006, and which is building on the work of the IADC. At ESA, these guidelines have been translated into mandatory, technical requirements, the "Requirements on Space Debris Mitigation for Agency Projects" (ESA/ADMIN/IPOL (2008)2 Annex 1) [2], which apply to ESA missions procured after 1 April 2008. In parallel with these requirements, standardization of mitigation measures is important in achieving a common understanding of the required tasks leading to transparent and comparable processes. This is the task of normative international standardization bodies like the International Standards Organization (ISO) (Technical Committee 20 and Sub-Committee 14, e.g., ISO/WD 24113 Space Debris Mitigation) and the European Cooperation for Space Standardization (ECSS).

The major driver for future debris proliferation, besides the intentional and unintentional release of objects, is the continuance of objects with large masses and sizes in orbit that could be involved in catastrophic collisions. Mitigation measures thus concentrate on the prevention of object release [explosions, mission-related objects, solid rocket motor (SRM) exhaust products], the disposal of objects, and active collision avoidance. As ESA's simulations show, the most

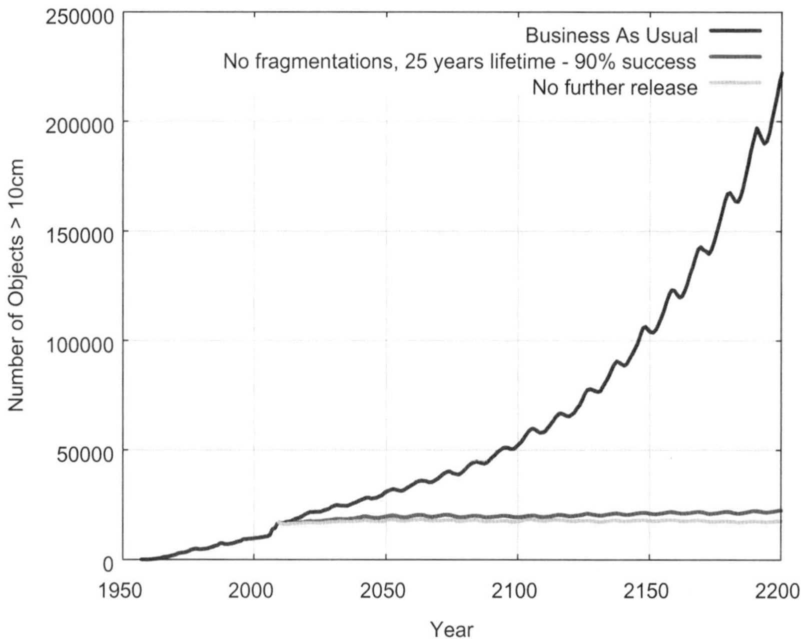

Fig. 23.1   ESA predictions for the evolution of the number of objects >10 cm in LEO.

effective means of stabilizing the space debris environment is the removal of mass from regions with high spatial densities. A limitation of the residence time of controlled objects in altitudes below 2000 km to 25 years followed by either atmospheric reentry or reboost to higher altitudes allows the growth of object number to be limited in the densely populated LEO environment. Figure 23.1 shows ESA predictions for the number of objects in LEO under assumptions of a business-as-usual scenario (no mitigation), 90% of success in implementing the 25-year lifetime rule, and no further release of objects. Active collision avoidance is another means to mitigate the generation of fragmentation debris, but less than 10% of the catalogued objects in LEO are under control. Accordingly, the efficiency of this measure is rather limited. However, collision avoidance is largely performed in any case to preserve the health of spacecraft.

II. COLLISION AVOIDANCE AT ESA

A. THE COLLISION AVOIDANCE PROCESS

The European Space Operations Centre (ESOC) has been providing collision avoidance services to ESA's major Earth observation mission for several years (ERS-1, ERS-2, Envisat, Cryosat-2, Cluster-II). The services and associated procedures in place are mission-independent and could be applied to all kind of missions in a similar manner. Different steps need to be performed to come to a decision for or against a collision avoidance maneuver. These steps, as explained in the following, are to a large degree implemented in the collision risk assessment software "CRASS" [3].

ESA's collision avoidance service was set up in 2004 and was based on the screening of publicly available Two Line Elements (TLE) sets provided by United States Strategic Command (USSTRATCOM), enriched by assessed covariance values [4]. Because of the relatively poor accuracy of TLEs, many alarms were raised. European independent tracking capabilites (such as the Fraunhofer-Institut für Hochfrequenzphysik und Radartechnik – Tracking and Imaging Radar) have been tasked to track the chaser object. With the subsequent orbit determination performed by ESA, the accuracy of orbit information could be improved by a factor of 10 or more. In most cases, events turned out to be false alerts.

Today, the collision avoidance process fully relies on the so-called conjunction summary messages (CSM) that have been published for operators by the Joint Space Operations Centre (JSpOC) since 2010. The CSMs report on conjunctions that have been defined by JSpOC based on their high-precision special perturbation (SP) data. These data include orbit information for all objects observed by the USSTRATCOM Space Surveillance Network and thus a much larger amount than is published in the form of TLEs.

Another advantage of the SP data with regard to the TLE sets is the accuracy of orbit information. Rough estimates suggest that the covariance associated with the SP data is on the order of up to 50 times smaller than that associated with TLEs for

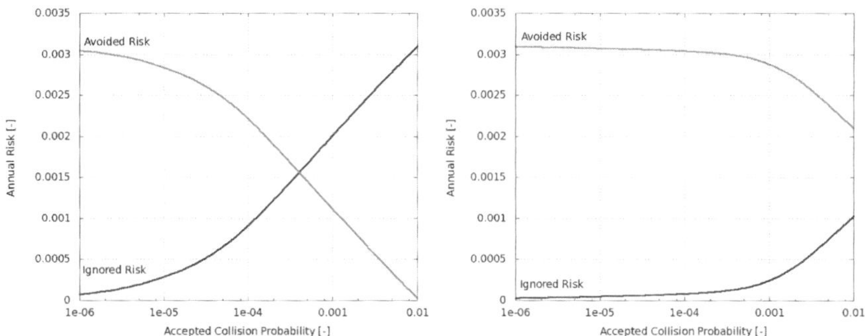

Fig. 23.2 Ignored and avoided risk for collision avoidance procedures based on collision probability levels, for estimated orbit covariances for a typical TLE and CSM.

orbits in 800 km altitude. High accuracy implies more reliable conjunction predictions, and high collision risks only occur for critical cases. This means that the number of false alarms decreases dramatically compared to conjunction assessments with TLEs. Figure 23.2 compares the avoided risk (i.e., risk accumulated from conjunctions that had risks above the reaction threshold) to the ignored risk (risk accumulated from conjunctions that had risks below the reaction theshold) as a function of collision risk threshold.This exercise is repeated for typical covariance levels of CSMs.

We can observe that, for the threshold levels for accepted collision probability, the use of CSMs allows for more efficient risk avoidance, which is a direct consequence of a more reliable identification of critical conjunctions through improved accuracy. As a side effect, the absolute number of alerts above the threshold is also significantly reduced through the CSMs.

CSMs also contain accompanying information on the orbit determination process and covariance matrices for both objects at the conjunction epoch. From this, the collision probability can be computed. The same software as for the previous TLE screening, CRASS, can be reused for this purpose. This software makes use of formulations provided by Alfriend and Akella [5]. It is assumed that the position uncertainties of both target and chaser can be described by separate and uncorrelated Gaussian distributions. Further, the problem is simplified to a linear one, that is, orbit curvatures and changes in the velocities are ignored.

The two error covariance matrices are combined into a common covariance matrix by simple addition of the two matrices. The 1σ position uncertainties expressed through the combined covariance span a three-dimensional error ellipsoid. In parallel, the combined target and chaser collision cross-section is determined. Both objects are assumed to have spherical cross-sections. A collision can only occur if these two spheres intersect [6].

To simplify the computation of the probability for this case (i.e., the collision probability), this three-dimensional problem is reduced to a two-dimensional one

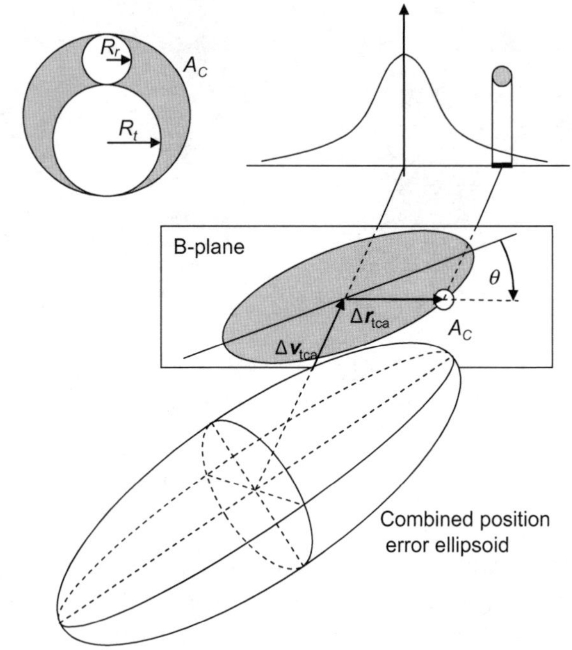

Fig. 23.3 Mapping of the combined position uncertainty onto the B-plane (θ) is the angle between the semi-major axis of the projected ellipse and the fly-by distance vector Δr_{tca}.

through projection of the conjunction geometry and uncertainties onto the B-plane (Fig. 23.3). The B-plane is perpendicular to the relative velocity Δv_{tca} at the time of conjunction and contains the conjunction range vector Δr_{tca}. The projection transforms the combined error ellipsoid into an ellipse and the spherical collision volume into a circular collision area of the same radius.

The two-dimensional Gaussian probability density for the combined position error in the B-plane is then integrated over the circular collision cross-section, which is centered in the predicted fly-by location, separated by the stand-off distance from the maximum of the Gaussian (representing the most probable position error), centered at the target location [6]. Information on the object radii and other object properties (cross-section, type, name, and identifiers) is taken from ESA's DISCOS database (Fig. 23.4) [7].

The collision risk assessment software automatically compiles a bulletin for the conjunction events and distributes it by e-mail. The bulletin contains all relevant information on the conjunction geometry and uncertainties in order to allow space debris analysts and mission operators to make a decision for or against a collision avoidance maneuver and to design the maneuver if necessary [8].

Before collision avoidance or orbit maintenance maneuvers, an ephemeris file that reflects the planned maneuver is supplied to JSpOC for a screening before maneuver execution. Only when sufficient clearance for the changed orbit is confirmed is the maneuver executed. The ability to ingest tracking data from European sensors is still maintained to improve chaser orbit information, but it is rarely used now because of the excellent quality of the data contained in CSMs.

B. STATISTICS

All data associated with the conjunction events have been archived for reporting reasons since initiation of the process in 2004. Statistical analysis of this long-term archive provides some interesting insights into the history of the environmental conditions to which ESA's satellites are exposed. For the first five years, CRASS analysis based on publicly available TLEs reported about 15 events per year where there were fly-by distances of less than 300 m; however, this changed dramatically in the following years.

An antisatellite test performed in January 2007 using the Chinese FengYun 1C satellite (860 km altitude) as a target, and the collision between the defunct Cosmos-2251 and the Iridium-33 satellite in February 2009, triggered two step increases in the object density in the vicinity of the operating altitude of ESA's satellites (near 800 km). With some delay, which is attributed to the time required by the US Space Surveillance Network to detect and correctly correlate the new fragment before publication, the number of conjunction events rose signifantly. With this increase in events, the number of radar tracking campaigns and avoidance maneuvers also increased.

The first CSMs, appearing in 2010, helped to reduce radar tasks, which then became superfluous in 2011. The number of collision avoidance maneuvers returned to a moderate level thanks to the better accuracy of the data contained in the CSMs.

Figure 23.5 shows the number of conjunction events in 2011 for the three satellites operating near an altitude of 800 km. The results labelled "CRASS" are for TLE screening. (Note that JSpOC bulletins have been available since 2010, and

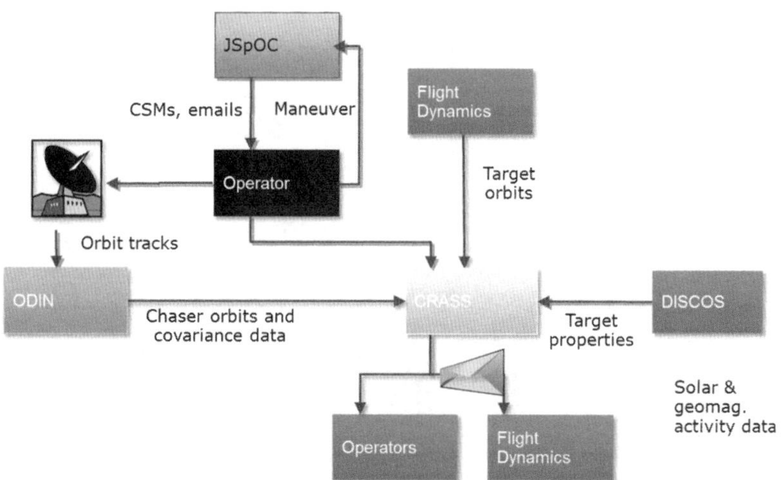

Fig. 23.4   ESA's collision avoidance process based on CSMs [8].

ERS-2 was deorbited in September 2011.) Because size is a major driver for collision probability, most avoidance maneuvers are attributed to Envisat (which has a span of 27 m, compared to the 5 m of Cryosat). To date, twelve maneuvers have had to be performed by Envisat to avoid critical conjunctions. One routine maneuver had to be rescheduled for the same reason. Four of these maneuvers were performed in 2011 alone. The following examples in 2010 underline the criticality of the current situation:

1.   On 21 January 2010, Envisat had a close conjunction with a CZ 2C rocket body (4 ton mass) that had been inserted just one month previously. Using refined orbit data (through radar tracking) the miss distance was 48 m and the collision probability 1/77.

2.   On 21 December 2010, Envisat underwent a series of three conjunction events with a fragment of the collided Iridium-33 spacecraft. The most critical of the three events had a total separation of only 47 m and a collision probability of 1/49 based on CSM information.

On the basis of TLE, the threshold of 1:10,000 for the collision probability is exceeded about 80 times per year for all three satellites. The conjunctions reported through CSMs and based on SP data seldomly reach such values. However, once the miss distance is small, very high values can be reached. The level of 1:100 has been exceeded twice for Envisat in the past two years.

Besides the quantity, the type of conjunction has also changed over time. In 2006, conjunction events were dominated by large, intact (i.e., physically intact, in one piece, but not necessarily operational) objects. After 2008, fragments began to dominate the conjunction events. One-quarter are attributed to fragments from the Chinese ASAT test alone. Because of the vicinity of the Iridium

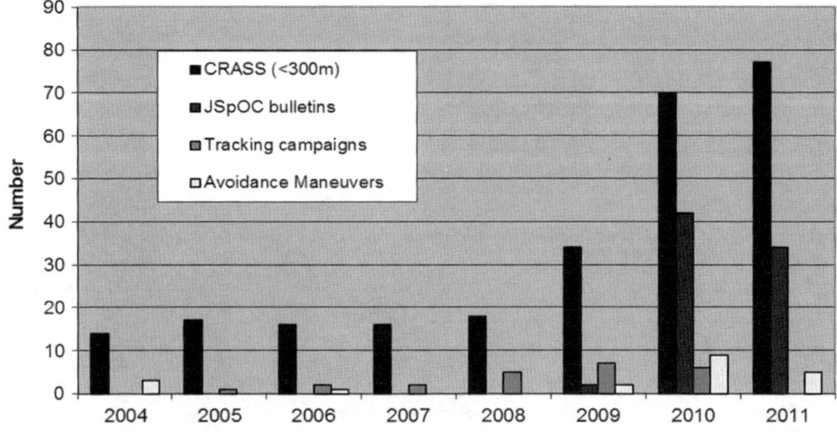

Fig. 23.5   Statistics on collision avoidance for ERS-2, Envisat, and Cryosat (since 2010).

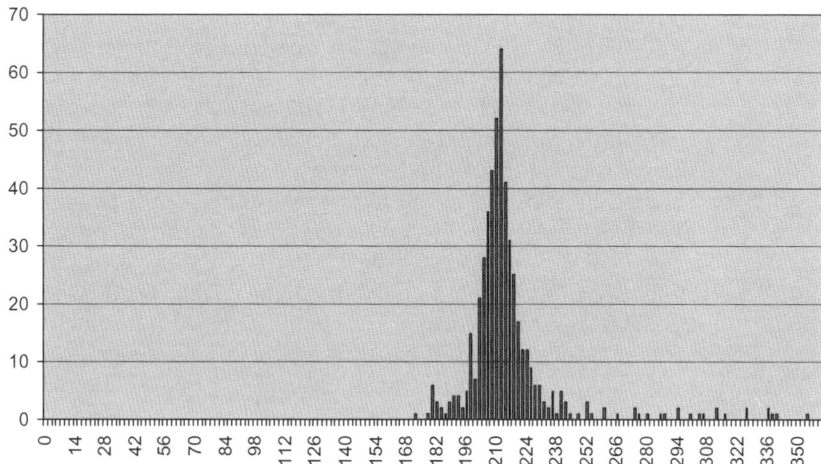

Fig. 23.6   Dispersion of the right ascensions of the ascending node (RAAN) of Iridium-33 fragments on 20 November 2010.

operational orbits, since 2010 about half of the conjunction events have been triggered by Cosmos-2251 and Iridium-33 fragments.

The distribution of the orbital nodes of the cloud of Iridium-33 fragments thereby generates a special problem to the conjunction assessment process. The distribution of the right ascensions of the ascending nodes is far from uniform due to the very low J2-induced perturbation (i.e. due to the flattening of the Earth) for orbits inclined at 86° (Fig. 23.6). As a consequence, the orbital planes are not yet far dispersed. Over periods of roughly nine months, the orbital planes of the ESA satellites operating in this vicinity drift into a co-planar but counter-rotating geometry with respect to these fragment orbits. This constellation offers the highest likelihood for the occurrence of conjunction events. This effect is clearly visible in the conjunction statistics and will persist over the next few years as a particular operational burden.

III.   ORBITAL LIFETIME REDUCTION

The active reduction of the orbital lifetime of LEO objects to at least 25 years is a firm requirement for all ESA missions procured after 1 April 2008 (the day when ESA's mitigation requirements entered into force). All missions procured before that date are encouraged to follow these requirements to the maximum possible degree. ERS-2 was launched in 1995, and after 16 years of successful operations was the first ESA object that lowered its orbit at the end of the mission in 2011 in order to comply with the rule.

ERS-2 had an on-orbit mass of 2080 kg and operated at an altitude of 790 km and inclination of 98.6°. The satellite concept was based on the reutilization of the Multimission Platform, developed within the French SPOT program. This platform provides the major services for satellite and payload operation, in particular attitude and orbit control, power supply, monitoring and control of payload status, and telecommunications with the ground segment. At the time of launch, ERS-2 and its sister spacecraft ERS-1 were the most sophisticated Earth observation spacecraft ever developed and launched by Europe. These highly successful ESA satellites collected a wealth of valuable data on Earth's land surfaces, oceans, and polar caps and were called upon to monitor natural disasters such as severe flooding or earthquakes in remote parts of the world. Both ERS satellites were built with a core payload of two specialized radars and an infrared imaging sensor. ERS-2 included an extra instrument to monitor ozone levels in the atmosphere.

In July 2011, ERS-2 was retired and the process of deorbiting began. In six weekly blocks of several maneuver pairs, a permanent decrease in altitude, while maintaining a certain circularity of the orbit, was achieved. A circular orbit was desired due to a combination of requirements related to clearing the operational orbit, ground-station coverage, and platform constraints. A delta-v per burn of up to approximately 2 m/s at the beginning of the deorbiting campaign led to a semi-major axis drop by about 40 km a week, based on a weekly pattern of five maneuvering days (Fig. 23.7).

The target altitude was then maintained while fuel depletion burns alternately increased and decreased eccentricity until the tank pressure dropped below operational levels. The decommissioning of the ERS-2 satellite was successfully

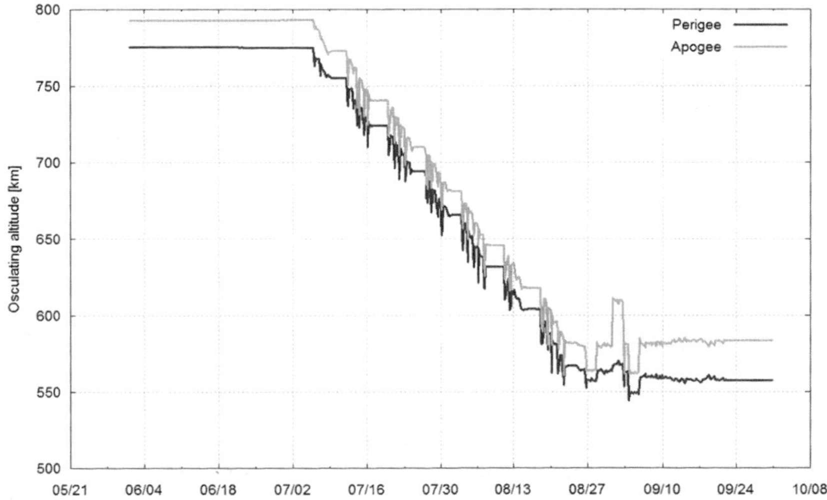

Fig. 23.7   Deorbiting of ERS-2 during July, August, and September 2011.

completed on 5 September 2011 after it reached the target circular orbit at an altitude of about 570 km. This ensured that reentry into the atmosphere would occur in less than 15 years. Although ESA's Space Debris Mitigation Requirements are only applicable for missions procured after 1 April 2008, ERS-2 fully complied with the associated rules. Once placed in its final orbit, ERS-2 was "passivated" (its batteries were disconnected and the communication system switched off once all the fuel was depleted). Since 5 September 2011, 1316 hrs UTC, no telemetry data have been acquired.

IV. GLOBAL COMPLIANCE WITH THE 25-YEAR RULE

Individual achievements are important as examples. However, a significant damping of the growth rate of the environment can only be achieved if effort is made globally. This section will therefore look into the degree of compliance achieved in the world with respect to the guideline of limiting orbital lifetimes in LEO to 25 years. Such an overview of the global level of implementing lifetime control is important so as to assess the acceptance of such rules in individual missions and to raise awareness by pointing out extreme cases of noncompliance. Overseeing global compliance is also important to predict the future evolution of the environment and, finally, to verify whether additional measures need to be applied.

A. APPROACH

This work plans to lay the grounds for a yearly report that concentrates on LEO missions that have reached end of life (EOL) in the past year. In this chapter, our analysis concentrates on missions decommissioned in 2010. It will be based on surveillance data, and lifetime estimations for the selected objects are performed independently. The major difficulty is to identify whether an object has reached its EOL and is therefore eligible for an analysis of its orbital lifetime compliance. A number of conventions and assumptions have to be established for this purpose:

1. Upper stages are analysed if
    (a) They were launched in 2010 (i.e., immediate EOL is assumed); and:
    (b) They have a perigee altitude of <6000 km at the time of reporting (to account for a possible lowering of the perigee of highly eccentric orbits due to orbital perturbation).

2. Payloads are analysed if they have a perigee altitude <2500 km at the time of reporting and were launched after 1990 (i.e., we do not expect any object launched before 1990 to be still operational at the time of the report) and are either
    (a) More massive than 50 kg (assumed to have orbit control capabilities) (according to DISCOS [7]);
    (b) Found to be maneuvring in 2010, but not in 2011;

(c) Not a reentry vehicle (STS, Dragon, Progress, Soyuz-TM, ATV, Shenzhou, etc.);

or

(d) Lighter than 50 kg (assumed to have no orbit control capability);
(e) Launched in 2000 (i.e., ten years of operational lifetime assumed);

or

(f) Heavier than 50 kg, launched in 2000 (i.e., ten years of operational lifetime assumed) and never maneuvered thereafter (assumed to have no orbit control capability);

or

(g) They were launched in 2010 and also decayed in that year.

For payloads, a maneuver detection mechanism will have to be implemented to determine the end of maneuverability (assumed to correspond to the EOL) and to detect deorbiting attempts.

The algorithm used to detect maneuvers in TLE-derived time series is essentially based on the moving window approach [9]. The time and orbital-parameter dimensions of the window are allowed to vary automatically while processing the time series, which makes this approach independent of the orbital parameters selected for the detection of maneuvers and reduces the fine-tuning effort required from an operator. The dimensions of the moving window are calculated directly from the time series by techniques from robust statistics and harmonic regression.

It has to be noted that the detection performance of the algorithm is a function of the altitude regime and the type of maneuver. Along-track maneuvers (i.e., typical orbit maintenance maneuvers) in high altitudes are the simplest case for detections, and maneuvers of a few mm/s can be identified. Figure 23.8 gives

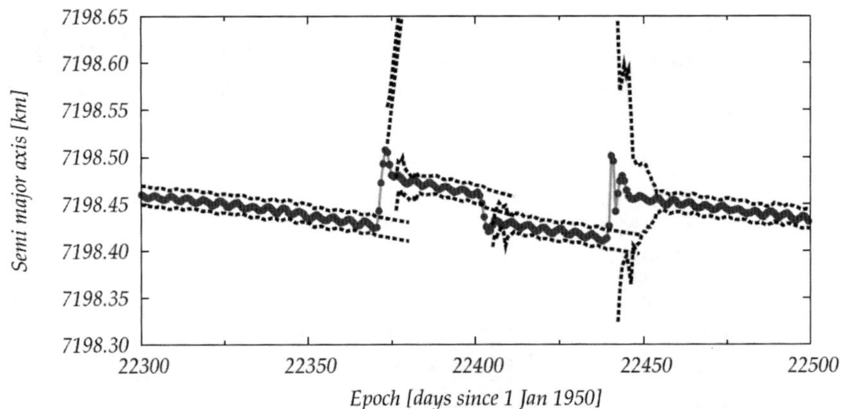

Fig. 23.8   History of the semi-major axis (km) of METOP-A between 21 January 2011 and 9 August 2011 (bold line) and the boundary of the moving window (small dots) [9].

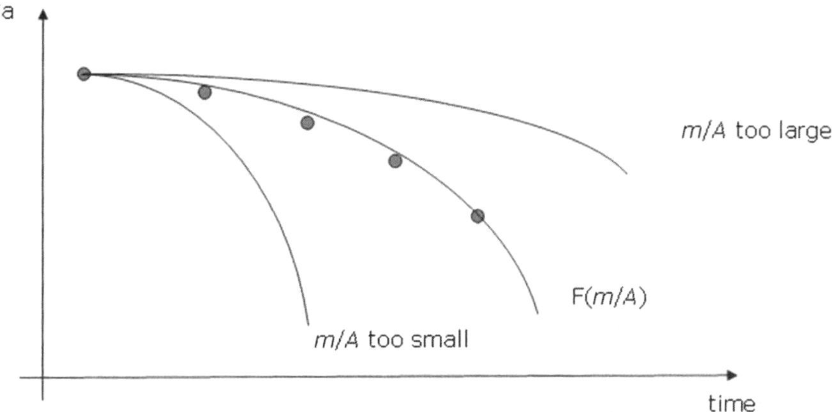

Fig. 23.9   Iteration to fit a mass/area (m/A) ratio into a history of orbital data.

an example for the semi-major axis of METOP-A between 21 January 2011 and 9 August 2011. The black lines mark the boundaries of the moving window. As some maneuvers are of the same order of magnitude as the expected noise level of the series, the algorithm will ignore them. Larger maneuvers, with a difference of a few meters in the semi-major axis, are correctly identified. Accordingly, manual checks are required [9].

The next step, after identifying the EOL, is to determine the remaining orbital lifetime. For this, the ballistic coefficient is fit to the orbit information using a propagator (Fig. 23.9). With the obtained ballistic coefficient the initial lifetime estimate is performed by straightforward application of King-Hele's formulations [10]. Based on this initial estimate, a refined analysis is performed with different fidelity (and computing power) according to the outcome. If the initial estimate leads to lifetimes of less than 100 years, King-Hele's formulations are applied iteratively. Semi-analytical propagation is applied if the initial estimate results in less than one year.

B.   RESULTS

The results will first be analyzed from the perspective of the upper stages, in a statistical way. A generous condition is applied to verify whether 25 years of orbital lifetime have been achieved. A total of 59 upper stages were injected into potentially LEO-crossing orbits in 2010. Of these, 29 will definitely decay (or have already decayed) within 25 years, another 8 are likely to decay in that time, and the remaining 22 will not. Figure 23.10 shows the accumulated object mass of the 59 upper stages for the different target injection orbits (LEO indicates medium Earth transfer orbit and GTO indicates geostationary transfer orbit). Obviously, most of the upper-stage mass is removed (directly or naturally) for

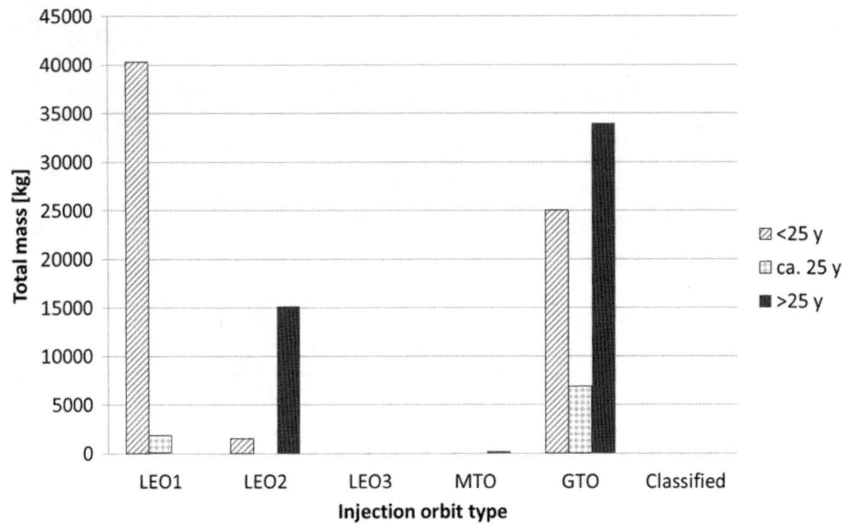

Fig. 23.10   Accumulated mass of upper stages injected in 2010.

most direct LEO injections. Stages injected into transfer orbits tend to accumulate more mass with longer orbital lifetimes. Transfer orbits mostly interfere with the LEO regions, but the interference is limited to a small fraction of the orbital period. The accumulated mass of the upper stages that will be removed from LEO interfering orbits within about 25 years or less is about 75.7 tons, while the mass of the upper stages with long-term LEO interference amounts to about 49 tons for 2010.

Regarding payloads, 67 LEO objects reached their EOL in 2010/2011. Of these, 14 seem to have lifetimes shorter than about 25 years. Five had already decayed. Four spacecraft in altitudes below 600 km stopped operating in 2010/ 2011. All will have shorter orbital lifetimes than 25 years. Seventeen spacecraft with orbit control capability have stopped operating in altitudes between 600 km and 1400 km. At these altitudes, an active reorbit maneuver (towards lower altitudes) is probably required to achieve a shortening of the remaining orbital lifetime to less than 25 years. Of these payloads, only one comes close to this remaining lifetime (although no particular lifetime shortening attempt seems to have been conducted). Two other satellites seem to have actively reduced their orbital altitude and have therefore shortened their orbital lifetime, although not enough to reach less than 25 years. Accordingly, performance in this critical altitude regime is poor. Figure 23.11 shows the apogee and perigee altitudes of the payloads ending operations in 2010. Payloads with orbit control capability at operational altitudes between 600 km and 1400 km are shown on the left, and payloads without orbit control capability at all altitudes on the right.

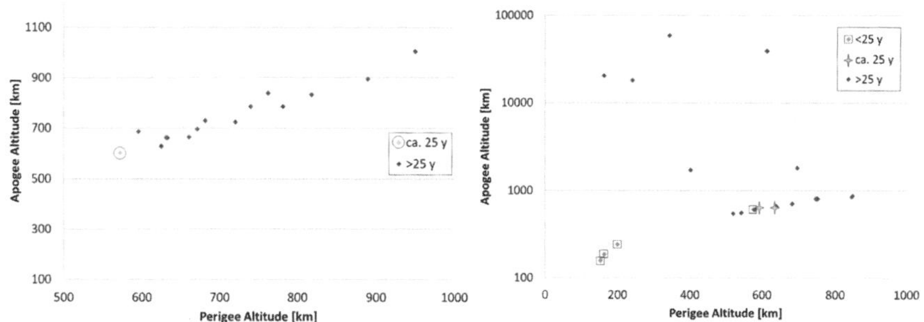

Fig. 23.11 Apogee and perigee altitudes of payloads ending operations in 2010.

Of the 67 LEO payloads, 31 do not have orbit control capabilities. A total of 28 are still in orbit. Of these, only five will decay within 25 years. As can be seen in Fig. 23.11, all these spacecraft, despite not having any maneuver capability, have been launched into orbit altitudes that do not allow for a natural reentry within 25 years. Also, 15 satellites have ended operations in LEO altitudes above 1400 km (where a reorbit to 2000 km is more fuel-efficient than a targeted natural decay within 25 years). All of these have moved away from their operational orbits towards higher altitudes. One has achieved an altitude slightly above 2000 km.

In summary, we can observe different performances in orbital lifetime shortening measures at different altitudes. In the most critical altitude band (between 600 km and 1400 km), the success rate in achieving limited orbital lifetimes is poorest. Furthermore, it should not be forgotten that it is not only the *number* of objects with long orbital lifetimes that has an environmental effect, but also their *mass* (and related to that, their size). Figure 23.12 underlines this by showing the remarkable amount of abandoned mass in the critical altitude band, which is only exceeded by the mass in the upper stages (where a

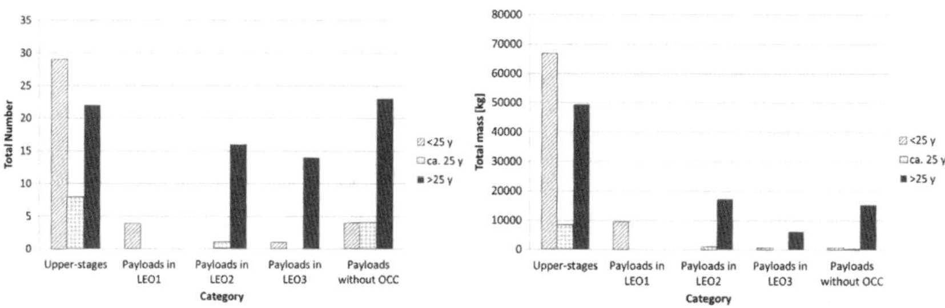

Fig. 23.12 Cumulated numbers (left) and mass (right) of objects that ceased operations in 2010, as a function of type and orbital regime and as a function of remaining orbital lifetime.

considerable amount has a short lifetime, and in most cases interferes with LEO only during fractions of an orbital period). (In Fig. 23.12, LEO1 indicates orbits <600 km, LEO2 orbits of 600–1400 km, and LEO3 orbits >1400 km.

V.   CONCLUSION

The current space debris environment has its highest density at altitudes between 800 and 1000 km. Collision avoidance is applied at ESA for all ESA-operated satellites in this regime. Recent major events have led to an increased risk and intensified action in the area of collision avoidance. Current estimates predict a considerable growth in the debris population if the countermeasures proposed by the IADC are not consistently applied. Of particular relevance is the rule for objects to clear the LEO region within 25 years. Following this rule, ESA successfully deorbited ERS-2 after 16 years of operational life. However, global application of this rule is required to preserve the environment. A method has been set up to analyze the degree of compliance achieved in this field per year. It has been found that upper stages tend to comply better than payloads. For payloads, the performance in critical regions (600–1400 km) is poor. About 88 tons of mass (payloads and upper stages) remained in orbits interfering with the LEO region for longer than 25 years. The degree of compliance has to improve considerably, or additional measures (e.g., active debris removal) will need to be taken into consideration.

REFERENCES

[1]   Lemmens, S., and Krag, H., "Maneuvre Detection Methods for Satellites in Low Earth Orbit Based on Two-Line Elements and Applications," *COSPAR Scientific Assembly*, Mysore, India, July 2012.

[2]   ESA, "Requirements on Space Debris Mitigation for ESA Projects," ESA/ADMIN/ IPOL(2008)2, Annexes 1, 2008.

[3]   Alarcon, J.-R., "Development of a Collision Risk Assessment Tool," Final Report, ESA contract 14801/00/D/HK, GMV, 2002.

[4]   Krag, H., Klinkrad, H., and Alarcón-Rodríguez, J. R., "Assessment of Orbit Uncertainties for Collision Risk Predictions at ESA," SP-645, *2nd IAASS Conference*, Chicago, Illinois, May 2007.

[5]   Alfriend, K., Akella, M., Lee, D., Frisbee, J., and Foster, J., "Probability of Collision Error Analysis," *Space Debris*, Vol. 1, 1999, pp. 21–35.

[6]   Klinkrad, H., "Space Debris – Models and Risk Analysis", Springer Praxis, Chichester, UK, 2006, ISBN 3-540-25448-X.

[7]   Hernandez de la Torre, C., Caballero, F. P., Sanchez Ortiz, N., Sdunnus, H., and Klinkrad, H., "DISCOS Database and Web Interface," SP-473, *3rd European Conference on Space Debris*, Darmstadt, Germany, March 2001.

[8]   Flohrer, T., Klinkrad, H., Krag, H., Bastida Virgili, B., and Merz, K., "Operational Collision Avoidance for LEO Satellites at ESA," *Proceedings of the 28th*

International Symposium on Space Technology and Science (ISTS), Okinawa, Japan, June 2011.

[9]   Flegel, S., "Maintenance of the MASTER Model," Final Rept., ESA contract 21705/ 08/D/HK, version 1.1, 2011.

[10] King-Hele, "Methods for Predicting Satellite Orbital Lifetime," Technical Report 87030, Royal Aircraft Establishment, Ministry of Defense, Farnborough, England, 1977.

CHAPTER 24

FAST: A New Mars Express Operations Concept, Quickly

Daniel T. Lakey,* **Jonathan Schulster**† **and Olivier Reboud**‡
SCISYS Deutschland, Darmstadt, Germany

Thomas Ormston§ **and Kees Van Der Pols**¶
VEGA Space, Darmstadt, Germany

Michel Denis** **and James Godfrey**††
European Space Agency, ESOC, Darmstadt, Germany

I. INTRODUCTION

Mars Express is Europe's first mission to Mars, and has been extremely successful. Since reaching Mars orbit in late 2003, the spacecraft has operated reliably and without significant anomalies. This consistency of operation allowed all its primary goals to be met (Mars surface, subsurface, and atmosphere), and more besides (Phobos observations, relay of NASA Lander data, to name but a few).

The eight-year mission, with sometimes serious anomalies, was halted in August 2011 when repeated errors in the solid-state mass memory (SSMM) resulted in the spacecraft transitioning to a "safe mode" and the suspension of nominal operations. The anomalies persisted and led to a switchover to the redundant mass memory controller unit. Science operations were restarted, but within a month the spacecraft was in safe mode again. A third safe mode a few weeks later forced the halting of science operations while the problem was diagnosed. Each safe mode consumed up to six months' worth of fuel and such frequent occurrences were unsustainable.

As a deep-space mission with long periods out of contact with Earth, Mars Express relies on an SSMM to store telemetry, science, and command data [1]. As these command data are critical to the continuation of operations, any interruption in the flow of commands from the SSMM would result in a safe mode, and a requirement to point the high-gain antenna at Earth and ensure a stable

---

*Spacecraft Operation Engineer (Onboard Data Handling), HSO-OPM, ESOC.
†Spacecraft Operation Engineer (Mission Planning, Payload & Thermal), HSO-OPM, ESOC.
‡Spacecraft Operation Engineer (AOCS and TT&C), HSO-OPM, ESOC.
§Spacecraft Operation Engineer (TT&C), HSO-OPM, ESOC.
¶Spacecraft Operation Engineer (Power & Thermal), HSO-OPM, ESOC.
**Deputy Spacecraft Operation Manager, HSO-OPM, ESOC.
††Deputy Spacecraft Operation Manager, HSO-OPM, ESOC.

attitude. All three safe modes experienced in 2011 were due to such a command interruption caused by SSMM anomalies. Specifically, the mission timeline command cache could not be filled with new commands from the SSMM, for various reasons, leading to an autonomous transition to safe mode.

II.   REVIEW OF PREVIOUS OPERATION CONCEPT

A.   OVERVIEW OF DRIVERS FOR THE MARS EXPRESS OPERATIONS CONCEPT

As a deep-space mission, Mars Express has a relatively low data rate and potentially long periods between ground-station passes. For telemetry and science data, this drives a look-and-store approach, recording to the SSMM for playback during the next contact with Earth. The same approach, in reverse, is applied for commands. An onboard mission timeline is populated with all the commands required for roughly three days' worth of commanding, which should provide adequate margin for ground-station problems and other factors that might inhibit daily command uplink.

The Mission Time Line (MTL) is a cyclic store in the SSMM of 3000 commands, up to 300 of which are copied to an in-memory cache by the data-handling processor for execution at their scheduled time. A daily command volume of roughly 1000 commands would normally be necessary for routine operations (Fig. 24.1). The Data Management System (DMS) maintains the cache of up to 300 commands, topped up by regular queries to the SSMM where the bulk of the MTL is stored. The ground keeps the MTL as full as possible via an Immediate-execution TeleCommand File (ITCF).

Another impact of being a deep-space mission is the significant two-way light time of between roughly 8 and 48 minutes. It is not feasible to wait for the acknowledgement of one command before sending the next. To address this, file-based operations based on the Telemetry and Telecommand Packet Utilization Standard PUS Service 13 [2] are used to send a "command file" containing fully encoded commands, which are executed one per second once all parts of the file have been received correctly. These command files can be either executed immediately upon completion (ITCF) or

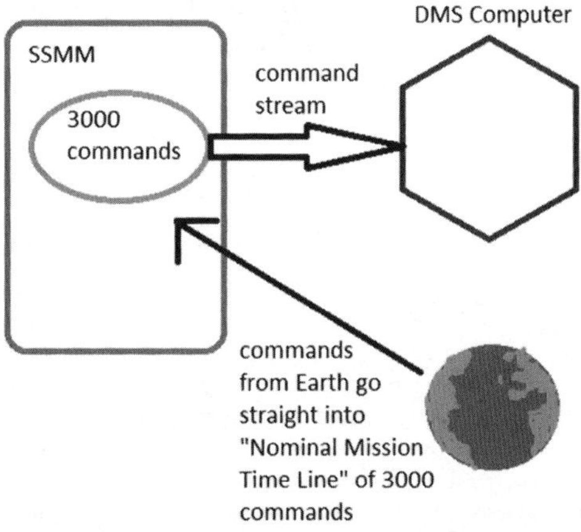

**Fig. 24.1   Nominal MTL SSMM/DMS
interaction.**

upon execution of a "trigger" command (Delayed-execution TeleCommand File, DTCF). To top up the MTL, command files containing "time tagged" commands are sent on an as-early-as-possible scheme. A three-day assumed margin gives ample time to resend any file that failed to be uplinked or executed properly.

One spacecraft constraint discovered post-launch was power on. Owing to an inconsistency between the power conditioning logic and the actual cabling, Mars Express cannot extract the maximum power from its solar arrays (it is limited to about 70%). The MTL therefore contains critical commands to the power system to optimize power usage and minimize battery wear.

Another constraint to the power system is spacecraft attitude. This is maintained by the Attitude and Orbit Control Management System (AOCMS), based upon commands issued to it via the MTL. The attitude is of vital importance to point the solar arrays correctly, but also to avoid thermal constraints from solar illumination on instruments (optics) and cooling radiators, and of course to correctly orient the spacecraft's high-gain antenna to Earth for communications. Getting "stuck" in a certain attitude could conceivably lead to drained batteries, violated thermal constraints, and loss of contact with Earth.

To manage the routine operations concept, a complex set of mission planning tools on the ground generated the commanding products one week at a time, split into files with the dual aim of keeping their number low and maintaining some independence between the various commanding streams for each subsystem and various instruments. Uplink planning tools then scheduled the uplink of these files as early as possible, to keep the MTL at the maximum possible fill level.

B.   REVIEW OF CONSTRAINTS

The constraints that drive the operations (ops) concept are significant two-way light time; limited commanding onboard; 1000 telecommands (TCs) per day commanding volume; irregular and time-limited ground-station pass schedule; and tight power, thermal, and pointing constraints, implying that safe attitude management is a priority. This ops concept relied entirely on a trustworthy MTL to store the commands and serve them without interruption, which implied a trustworthy SSMM. Following the string of SSMM anomalies, this could no longer be assumed. A new ops concept had to be found.

III.   FAST OPS CONCEPT

The SSMM anomalies caused safe modes because the DMS either failed to fill the MTL cache in a timely manner, or was unable to communicate with the SSMM entirely (Fig. 24.2). The autonomous recovery action to transition to safe mode could have been disabled, but that would not have solved the MTL cache-filling problem and would have potentially made critical the problem of a safe spacecraft

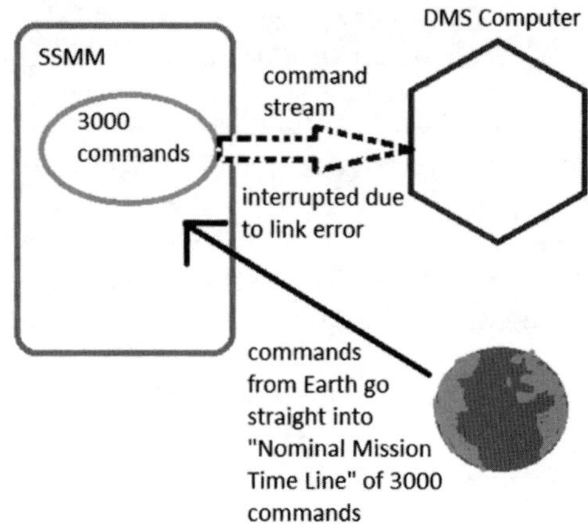

Fig. 24.2 SSMM link anomaly (interruptions to the refill of the MTL cache are detected by the DMS and result in safe mode).

attitude in the event of an interruption in AOCMS commanding. (Note that interruptions may be transient–only one or two packets affected–or may result in the SSMM being marked as "not-in-use" and the DMS severing the connection.)

Although the go-to-safe-mode action could be disabled, a loss of communication between the DMS and SSMM could lead to a partial file transfer execution. While Service 13 guarantees that file transfer from Earth to SSMM is complete (all or nothing), the once-per-second execution of the commands within a command file renders it vulnerable to an interruption.

An alternative, "short" in-memory MTL exists within the DMS RAM, for times when the SSMM in unavailable, such as after a safe mode. This is sized for 117 TCs, or about a tenth of a day's "routine" commanding volume. Between anomalies, the unit operated perfectly. This gave hope that a new ops concept could be found that was robust to intermittent SSMM failures.

A. ADDITIONAL CONSTRAINTS IMPOSED BY THE SSMM ANOMALIES

With a tiny MTL and an intermittent data connection, the new ops concept must therefore be:

1. *just-in-time*: there is no space for a long-term schedule;

2. *transactional*: leave no possibility for a partial operation;

3. *robust*: "expected" connection errors must not cause transition to safe mode;

4. *operator-friendly*: new concept must be as similarly practical in routine use as the previous concept;

5. *scientist-friendly*: avoid loss of mission data beyond the elementary operation ongoing at time of error;

6. *safe*: in no case should the new ops concept prevent a necessary safe mode nor result in a situation worse than with the previous concept.

B. OVERVIEW OF FAST OPS CONCEPT AND TRANSACTIONAL COMMANDING

The original ops concept interwove instrument and platform operations of different timescales. This ability needed to be retained, as short-term science operations and longer-term platform operations need to execute in parallel. For example, the transmitter switching ON and back to OFF may be spread over a 10 h period, but a pointing-plus-observation maneuver might last 1 h.

By paring down operations to the absolute minimum, it was conceivable to perform a science pointing and instrument operation in about 70 commands; indeed, this had already been considered as a mission optimization before the SSMM anomalies. Transmitter commands could be reduced to around 20 per pass, and other platform operations fitted into the remaining space. This was all very well, but would have reduced Mars Express operations to one per pass. However, it showed that something was possible.

For practical reasons, file transfer is the only reasonable way to uplink bulk commanding to the spacecraft. However, if each file could contain only the bare minimum of commands necessary to perform the next "activity," then a series of such activity command files could be scheduled to execute just-in-time, and keep the short MTL topped up directly (Fig. 24.3). This would emulate the SSMM-to-DMS-cache of the nominal MTL, on activity boundaries rather than on command number boundaries, and at the cost of MTL space for the "trigger" commands to execute the next file. In the flow chart shown in Fig. 24.3, a week's worth of time-tagged commands are stored in many DTCFs, each of which contains less than 100 commands. The command files are then executed just before the first command within them is scheduled, resulting in the short MTL being filled with the contents.

As the SSMM can no longer be relied upon for noninterrupted performance, the above could not form the entirety of the new solution. Furthermore, there are different timescales at which spacecraft activities run, meaning that different activities would need to run in parallel and independently. A failure to load the next activity of one type should not affect the loading of other activities, or interrupt already loaded ones.

Happily, the ECSS Telemetry and Telecommand PUS Service 11 "time tag" functionality provided the tools necessary to ensure both all-or-nothing

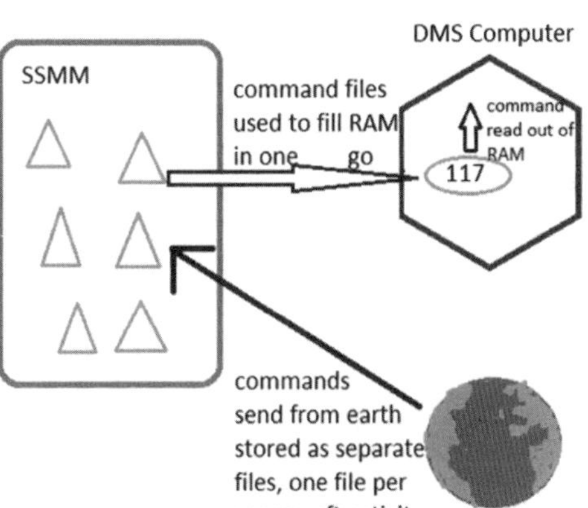

Fig. 24.3 File-based activities on short timeline (FAST) usage of SSMM and short MTL.

"transactional" execution of files and to manage different timelines within one MTL. The magic bullet was the "subschedule," up to 255 of which could be managed by the DMS. By assigning each activity type to a dedicated subschedule, they could be managed independently. The first command of an activity file would be an "execute immediate" command to disable its subschedule, and the last would be to enable it, ensuring that any partially loaded file would remain disabled and not affect any existing activity within the MTL.

This implied that only one activity of the same type Sub-Schedule Identifier (SSID) could be loaded at once, a core constraint for the new ops concept. Strictly speaking, any overlap of two activities of the same kind by even 1 s is now forbidden.

There have been many studies made of file-based operations (e.g., [3, 4]) concerning the transfer of data from ground to spacecraft as a discrete file, and ensuring its consistency during transfer. The problem we faced was one not dealt with by such schemes, as the data corruption happened *after* the file transfer had completed successfully. It was therefore necessary to devise a mechanism to ensure that a partial execution of the DTCF could remain "invisible," as if not started at all. Inspiration was drawn from the world of databases, whereby a series of critical and related operations are grouped into a "transaction," and only upon reaching a certain condition are the operations activated by "committing" the transaction.

As illustrated in Fig. 24.4, the existing SSID capability built into the MTL was used to approximate such a transactional capability on the spacecraft. By assigning each type of activity its own SSID, and using that ID as the transaction token, multiple activities could be "multiplexed" within the short MTL. The grouping of commands into discrete transactions also isolates any failures to only that transaction, ensuring that any already completed transaction is unaffected.

The utility of transactional commanding, and its application to file-based operations, is one of the key lessons learned from the recovery of Mars Express and the implementation of the FAST ops concept in general. We would recommend this be looked at for inclusion in a future standard as it is applicable to all critical operations spread over multiple commands.

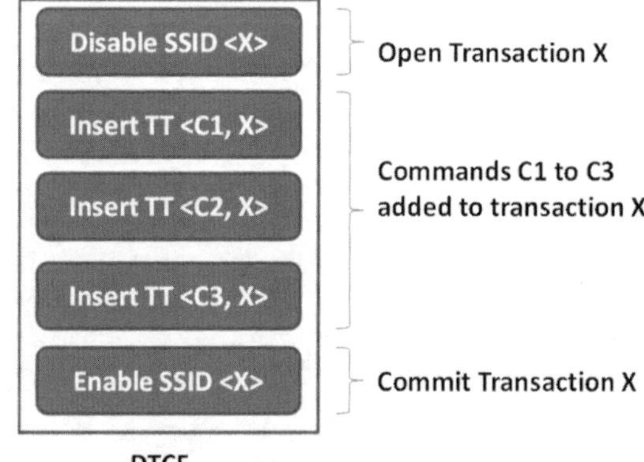

Fig. 24.4 Use of SSIDs to implement transactional commanding.

C. SAFE TO SAFE CONFIGURATION

Assuming each activity is constructed such that it will always leave the spacecraft in a safe and sustainable configuration (at least until the next intervention from ground), it is safe to have one or indeed all subsequent activities fail to load into the short MTL. The spacecraft would simply execute the remainder of the activity under execution in the short MTL, and then remain in its safe configuration until recovered by the ground. This is a more graceful safing strategy than the generic safe-mode, which covers practically all possible single failures but at the expense of fuel and complexity of recovery.

"Safe configuration" has a different meaning for each subsystem, but generally is a "leave it as you found it" principle and operates per activity. As the SSID disabling/enabling technique ensures an activity will either fully execute, or not at all, spacecraft safety is guaranteed by always ending any operation in the "safe" state:

1. Flight Dynamics (pointing & reaction wheel off-loading)–Earth pointed;

2. Transmitter (S- and X-band, independently)–transmitters off;

3. Thermal Control–"Instrument Non-ops." stable configuration;

4. Instruments–instruments off.

Generally, the "safe configuration" is that which would be set during a safe mode, so that should the SSMM become disconnected, the spacecraft will gradually converge to a safe-mode-style configuration as the already-loaded activities run to completion. The notable exception is the communications system, which remains configured for X-band and the high-data-rate capability it provides.

The concept thus assumes that all activity types are independent, and that each individual activity within a type is also independent. This is not always the case, but with smart nesting and cascading of the subschedule enable status, dependencies can be accommodated. Identifying and dealing with such inter- and intradependencies formed the bulk of the work in applying the new ops concept. Altogether, this gave rise to "File-based Activities on the Short Timeline Operations"–FAST Ops.

IV. DETAILED DESCRIPTION OF FAST AND ITS IMPACT ON SPACECRAFT AND GROUND SUBSYSTEMS

A. ONBOARD DATA HANDLING

Patches to the DMS's handling of the MTL and the DMS–SSMM interaction were considered, and have been investigated in some depth. These focus on implementing retry capability in the loading of the MTL cache, such than a transient interruption would not cause a safe mode. While promising, these would take time to implement, test, and apply onboard, with the risk attached to changing the original flight software in a few, very low-level, hardware-related critical routines. The

FAST ops concept offered a way to return to science within weeks rather than months, by combining advanced but already existing operational functionalities, and without touching heritage software onboard.

Because the SSMM would still occasionally issue error events, disabling the immediate safe mode path of the autonomous response to such errors was a necessity. By carefully considering the autonomous protection (Fault Detection, Isolation and Recovery (FDIR)) structure, it was concluded that such disabling was safe, as other protections would trigger if there was a further error. For example, an out-of-Earth-pointing timer exists, which will trigger a safe mode if the spacecraft does not return to Earth pointing within a certain period of time.

While a just-in-time strategy optimized the usage of the short MTL, 117 TCs is still a very hard target to hit. The use of Onboard Control Procedures (OBCPs) to reduce command volume was critical in allowing complex instrument operations to be fit into the available space. In the six months since resuming science operations, the number of OBCPs in use has grown by more than a factor of ten [5]. These have been developed in house by the Flight Control team, and are an ongoing focus of attention. Such a rapid growth in the number and complexity of OBCPs has required careful configuration management [6].

1. PRE-FDIR OBCP RECOVERY: PREFOU

The failure mode most frequently observed on SSMM-B is a transient communication error, which is detected by the DMS software and results in the SSMM being declared "not used" and thus halting all further DMS–SSMM interaction. Being transient events, the "SSMM In-Use" status can safely be immediately reset back to "used." In the event of a permanent failure, the FDIR would trigger again and again disable SSMM access. These events generally occur out of Earth contact, so an autonomous recovery technique was devised. Initially this consisted of simply scheduling a "Set SSMM Used" command following certain activities that were observed, or by analogy, suspected of being triggers for these transient errors.

A further refinement (following the Pre-FDIR concepts described in [7]) is now being implemented by the use of a combination of PUS Service 12b "Event Monitoring" and a recovery OBCP. This has several advantages:

1. Event Monitoring will react immediately upon the DMS FDIR triggering and raising the event–SSMM reset to "Used" within the same second as DMS FDIR triggering;

2. No MTL space is used for the "Set SSMM Used" safety command;

3. There is no longer the need to identify "at risk" activities in advance.

To avoid the case where a permanent failure causes an infinite loop of FDIR/ OBCP alternately switching the SSMM to "Used" and "Not Used," the OBCP saves the "last triggered" time in the DMS RAM. If the failure events are closer together than a configurable time delta (usually a few seconds), it disables the Event Monitoring and leave the SSMM in a not-used state.

2. MISSION SCHEDULER OBCP: MISO

The OBCP environment on Mars Express is sufficiently powerful to enable a "second MTL" to be run, via an OBCP, to schedule the execution of activity DTCFs. This removes the trigger commands from the short MTL, and frees up about 30 commands (25% of the short MTL) for more functional commands, for example, allowing more complex instrument operations. It also allows the uplink of all triggers at once, rather than spread throughout the week. This matches the once-a-week loading of the activity files and is a significant step towards control-room automation of all routine passes.

Owing to memory constraints, there is a strict limit on the number of OBCPs that can be executed in parallel, and MISO requires the use of one "execution slot" permanently. This is acceptable, considering the benefits in terms of capability and capacity that it offers (do more complicated things, and do more of them). As a fairly radical change to the way of commanding the spacecraft, this requires further changes to the ground software in different areas in order to support it:

1. Mission Planning: rather than creating the "trigger files" with commands to go into the short MTL directly, the data "when to call which file" now need to be communicated to the running OBCP by loading its internal schedule in advance, typically covering one week.

2. Mission Control System (MCS): a "second MTL" requires extensions to the ground modeling of command execution, so that the status of MISO can be modeled on the ground, and also so that the execution of DTCFs (and the commands therein) can be properly modeled.

3. Flight Control Team (FCT) Checks and Procedures: with the transfer of control of file execution away from the short MTL and to MISO, additional checks need to be defined and implemented, and associated procedures updated. With MISO, the short MTL has the remaining roles:
   (a) To host up to 117 commands loaded from activity files called by MISO;
   (b) To host manual commands (tests, special maintenance activities) not included in the weekly routine files.

B. ATTITUDE AND ORBIT CONTROL SYSTEM (AOCS)

The AOCS presented the biggest challenge in terms of partitioning operations into discrete, independent activities. This is because of the nature of attitude control: continuous regulation of physical analog parameters. Spacecraft attitude is controlled by varying the angular momentum stored in four reaction wheels. External torques on the spacecraft (solar radiation pressure, gravity gradient, etc.) cause an accumulation of momentum in these wheels that has to be dumped using the thrusters about every four orbits (each of 7 h duration). The buildup of angular momentum in the reaction wheels is directly affected by the number and type of spacecraft maneuvers.

The magnitude of wheel off-loadings (WOLs, momentum dumping using thrusters) is optimized by Flight Dynamics taking into account the momentum that has been accumulated since the last off-loading, and how much is expected to be accumulated before the next one. If a WOL were to be missed, subsequent science pointings could cause the stored momentum to exceed onboard mechanical limits, and the spacecraft would autonomously trigger a WOL. Should an autonomous WOL occur when the spacecraft was already slewing, a transition to safe mode would be triggered by the AOCS software. In Earth-pointed attitudes, autonomous WOLs do not risk triggering safe modes.

Leaving enough margin in the pointings/WOL calculations to account for a missed WOL activity was one approach, but it incurred too many constraints. Instead, a system of "chaining" was implemented, such that the trigger for pointing activities was dependent on the WOL SSID being enabled. A failure to fully execute the WOL activity file would lead to the SSID remaining disabled, and hence further pointings. Additional "Disable SSID" commands are inserted in the MTL just after the command triggering a WOL activity file, in case the file does not execute at all; in the nominal case this is then undone by the "Enable SSID" at the end of the activity file. See Fig. 24.5 for an illustration of this scheme. Here, the triggers for WOL (FW) and pointed activities (FD) use the same SSID as the FW activities, ensuring that an FW execution failure stops any more maneuvers of either kind. FD activities use a different SSID internally so that any failure there does not affect the subsequent FW/FD activity.

Accrued angular momentum due to the (non-WOL) "science pointing" activities is precalculated by the Flight Dynamics team to ensure that a missed one would not invalidate the following WOL. In this manner, the core FAST concept of "safe-to-safe" is maintained. WOLs are optimized so as to perform orbit maintenance in addition to angular momentum dumping. Should a missed WOL activity lead to an autonomous WOL when the spacecraft is close to pericenter, the orbit would be affected and require a later maneuver to correct it, costing extra fuel. To alleviate this, weekly deliveries of expected wheels levels during planned Earth pointings (computed by Flight Dynamics) have been set up. If a WOL is not performed due to a DTCF not executing correctly, a recovery WOL can be manually programmed by the Flight Control team at the next apocenter opportunity, ± 90 min. This recovery consists of restoring target wheel levels in Earth pointing and in reenabling the AOCS planned activities from that point onwards.

The possibility and latency to recover a WOL has to be considered at system level because it involves aspects such as coordination with the Flight Dynamics team, the available ground station coverage, their uplink allocation to Mars Express, and the room available in the short MTL. When a WOL is missed and

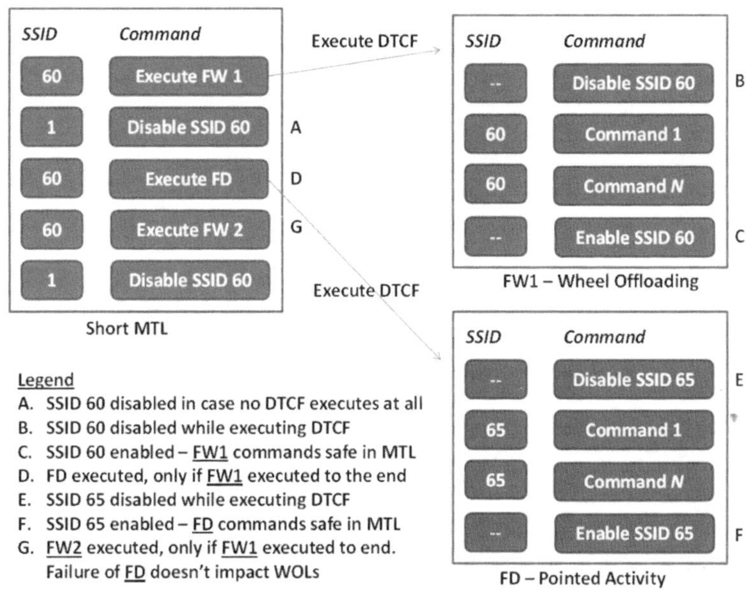

Fig. 24.5   Chaining of "Flight Dynamics" files.

the recovery cannot take place before the next planned WOL, then Flight Dynamics cannot guarantee that the slews leading to the pre-programmed WOL attitude are still safe (because of the momentum progressively built up by the external torques, like the gravity gradient at each apocenter with the spacecraft kept in Earth pointing). This necessitates the concept of disabling subsequent WOLs when a WOL is missed: a manual WOL designed with the expected target wheels levels needs to be performed in any case (preceding or replacing a planned one) once a WOL has been missed. Then, with safe levels restored in Earth pointing, the subsequent planned AOCS activities can be reenabled. Recovering as soon as possible from a missed WOL is important for both reference orbit maintenance and swift resumption of the routine science activities of Mars Express.

In April 2013 Mars will pass behind the Sun, as seen from Earth. During the month-long period of smallest Mars–Sun–Earth angle, commanding of Mars Express will be avoided because of radio interference from the solar corona. Furthermore, the solar conjunction necessarily means that the Mars–Earth distance is at its greatest, further hampering operations due to the long one-way light time of nearly 25 min. A safe mode during this period may have serious repercussions as spacecraft recovery may not be possible for a long period, as the S-band communications link used for safe mode recovery will be unusable for weeks around the solar conjunction. As a result, the usual FAST rule of "any activity can fail" is broken, with regards to the WOL.

A new scheme to maintain an Earth-pointed attitude throughout this phase and use a dedicated OBCP to allow one-command WOLs is being developed, so that the entire conjunction phase can be accommodated within the 117 commands of the short MTL.

C. COMMUNICATIONS: TRANSMITTER OPERATIONS

1. TRANSMITTER OPERATIONS

To minimize the commanding volume taken up by switching the X-band transmitter on and off during ground-station passes, the operations have been encoded into OBCPs. This results a 90% reduction of commands around passes, freeing up command slots in the short MTL for science dumps, time packet frequency increases for time correlation, and other in-pass activities. For spacecraft safety (especially due to power consumption of the transmitter), these OBCPs are loaded in the DMS RAM.

2. NO ACQUISITION-OF-SIGNAL PROCEDURE

Overall, minimal changes were made to the telemetry, tracking, and command procedures, which smoothed the transition to the new ops concept. The major shift in approach was that a new cause for no-acquisition of signal (AOS) was introduced: failure of a DTCF to execute, or execute incompletely, causing the X-band transmitter (X-Tx) to remain off during the pass. It was considered that such a failure would be the most likely cause of No-AOS (after ground issues such as planning or ground-station problems), ahead of safe mode or failure in the downlink chain. The response to No-AOS is therefore to send a "X-Tx ON" command in the blind. This is largely similar to the pre-FAST approach, except that it is sent immediately rather than waiting to see if a safe mode entry has occurred. This is safe, regardless of which of the following was the cause of the No-AOS:

1. *FAST DTCF execution failure*: X-Tx ON will reestablish the downlink and allow the issue to be confirmed and recovered.

2. *Safe mode*: X-Tx ON will not be received by the spacecraft because the uplink from ground station will be in the X-band and safe mode entry will have switched both receivers to the S-band.

3. *X-Band downlink failure*: X-Tx ON by OBCP will prompt a power cycle of the X-Tx, but leave the travelling wave tube amplifier ON. This follows the first stage of the expected recovery approach from ground to troubleshoot an X-band downlink chain failure.

In the case of FAST DTCF execution failure, the FAST concept "safe-to-safe" protection means that the entire X-Tx activity has failed to load, including the

X-Tx OFF command for the end of the pass. Therefore, a time-tagged X-Tx OFF needs to be loaded into the MTL to make sure that the transmitter is switched off at the end of the pass and thus prevent excess power consumption due to the transmitter remaining powered.

D. POWER AND THERMAL SYSTEMS

Constraints on the number of commands in the MTL have had several effects on the power subsystem. Routine periodic battery tests need to be spread out over several days as the commands cannot all be loaded into the short MTL in one go. Similarly, general power subsystem configuration settings are now done by immediate commanding rather than from the MTL. As with other areas of spacecraft operations, OBCPs are being developed to recover functionality that would otherwise be "too big" to fit. Owing to a wiring anomaly affecting power regulators, clamp voltage settings of the solar array power regulators (APR) need to be modified periodically for seasonal changes, and also updated back and forth every orbit to optimize power performance when the geometry of the planet fly-bys warms up the arrays too much.

Payload instrument thermal conditioning is provided via platform powered heaters for HRSC, the planetary Fourier spectrometer (PFS) and Mars Express Science Instrument Subsurface Sounding Radar Altimeter (MARSIS) instruments. These were previously operated on an instrument-specific basis, with two heaters used for HRSC, one for PFS and one for MARSIS. Each heater was reconfigured with the use of each instrument, from nonoperational control range or configuration to an operational range and back again after the observation was completed. This required eight commands per HRSC observation and four each for PFS or MARSIS. These were no longer affordable if we were to operate multiple instruments in a single pointing. To allow quick restart of science, all heaters were first turned ON permanently, but the additional power demand could not be maintained for the long term, for instance in eclipse seasons. A simple OBCP was written to configure all heaters to the operational range and another to set them back again, each running within 15 s.

Initially, the "set to op range" and "reset to non-op range" OBCPs were called each time the spacecraft was about to leave Earth pointing and shortly after the end of each science pointing, respectively. The total average heater demand is about 15–20 W higher with the payload interface conditioning heaters in the operational range (compared to the non-op range). Mitigating the power cost of this approach has been a driver for the recent eclipse season (May 2012). The new mission planning system (MPS 2012) allowed refinement of the approach such that the setting to operational range heater control is not made for science pointings that do not use HRSC, PFS, or MARSIS.

Further planned enhancements are to replace the existing, coarse "all or nothing" heater OBCPs with more configurable ones that allow the selection of a given heater configuration, such that heaters are only enabled as and when necessary, further saving power. Such a "set configuration" technique is also under consideration for application to the APR clamp settings, and possibly has other applications across all subsystems where configurations are currently applied by sending multiple low-level commands.

E. INSTRUMENT OPERATIONS

With the resumption of science operations it became clear that the number of TCs required to operate each instrument would be the driver for returning to a full science mission. In collaboration with the Science Ground Segment in the European Space Astronomy Centre (ESAC), it was determined which command sequences for payload operations could be combined or replaced one-for-one with OBCP equivalents. The first phase was to replace the "command-intensive" switch on and switch off command sequences with single-chain OBCPs. The initial development in the high-level procedure-like development environment led to inefficient coding of some commands. "Chains" of several OBCPs calling each other in sequence were therefore required to meet the 4 kb size limitation for a single OBCP. These are being replaced by recoding the OBCPs in the more efficient, low-level software-like environment.

With all the activation and deactivation OBCPs in place (January 2012), restoration of all science instrument operations was achieved. Phase two of this process has been to replace each instrument command sequence requiring more than four TCs (and no parsed parameters) with an equivalent 'macro' OBCP. With this in place, the total restriction on short-MTL size no longer has any significant impact on the number of instruments that can be operated in a single science pointing, and simultaneous operation of five instruments is now possible.

For occasions where the SSMM checksum error resulted in the SSMM being set to "not used," all OBCPs executed via load from SSMM would not execute until the unit was reconnected by the Flight Control team. This initially resulted in some instruments being left on after the failure event. The solution was to ensure that all OBCPs for instrument deactivation were directly stored into, and executed from, DMS processor RAM. Instrument switch off OBCPs have now been recoded using the more efficient low-level development environment, reducing their size by a factor of 5–10 and allowing all of them to reside directly in DMS RAM.

For ASPERA (Mars Express Science Instrument Energetic Neutral Atoms Analyzer), which is generally operated for a full day (between routine WOL "maintenance" slots), the process has been more comprehensive. The ops concept for the instrument has been redesigned from scratch and simplified into "switch on" high-voltage enabling, operation of the various subinstruments, disabling of the high voltages, and instrument switch-off. This has resulted in the development of eight completely new OBCPs based on the combination and reduction of several command sequences, with contents of up to 60 commands

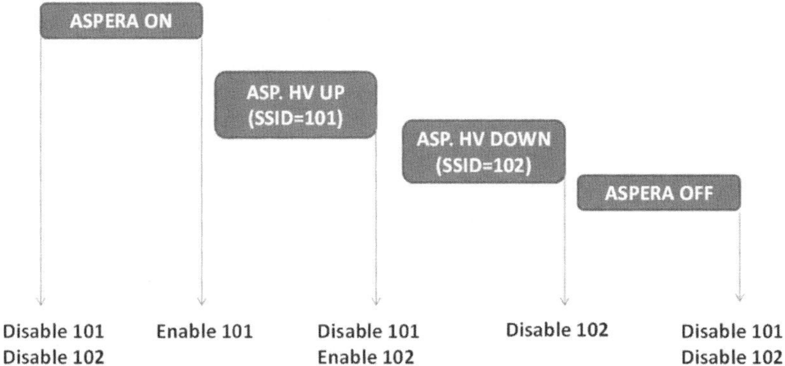

Fig. 24.6   ASPERA OBCP chain (the trigger commands for the next "Start OBCP" remain disabled unless enabled by successful execution of the previous step).

per OBCP. Since the high-voltage (HV) operations are safety-critical for the instrument, and shall only be used if ASPERA is completely switched on, the OBCPs for HV on and HV off have been assigned separate subschedules on the short MTL, which are enabled and disabled from within the ASPERA switch-on and -off OBCPs (Fig. 24.6). This protects the execution of the HV OBCPs such that switch on is only possible when ASPERA is on and HV off, and that HV switch-off OBCP only executes if HV is on.

Wherever possible, to reduce the total number of 'activities' that must be produced, checked, and uplinked each week, instrument operations have been included within the Flight Dynamics activity pointings. This is only feasible where the entire instrument operations (preheating, configuration, etc.) take place within the "pointing" and associated "slews." For some instruments such as the Mars Express Science Instrument Visible and Infrared Mineralogical Mapping Spectrometer (OMEGA), the duration of a full near-infrared observation (some 2.5 h of cryogenic cooling plus 3 h of scanner preheating) prevents this, but for all other "pointed" instruments (including OMEGA visible-range observations) this has or will soon be achieved. To make this possible, certain constraints on operations had to be extended, such as increasing the minimum slew duration to or from a science pointing from 10 to 20 minutes, to ensure all instrument conditioning heater operations were contained within the pointing activity. These constraints are purely technological and have no impact on science performance.

F.   GROUND/MISSION CONTROL SYSTEM

With one command file per activity, the number of command files to uplink to the spacecraft grew considerably, from about 25 to 125 for a week. Design limitations

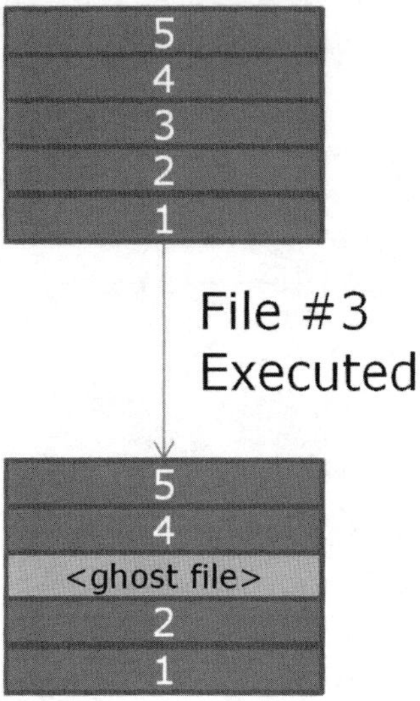

Fig. 24.7   SSMM ghost files.

on the SSMM also put constraints on the ordering of the files to avoid "ghost files" where space freed-up by executed files is not properly released (Fig. 24.7). In Fig. 24.7 it is shown how executing a DTCF causes its deletion, but the space cannot be recovered until the files "above" it (4 and 5) have also been deleted. This limitation means that files are to be uplinked in reverse chronological order, with the first to execute the last to uplink.

The manning overhead and potential for error incurred by this was great, and identified extremely quickly. An emergency patch to the ground control system was provided to automatically import a "manifest" of which DTCFs to load, in what order, and under what file name they should be stored into the SSMM, from a simple table generated by the planning tools together with the files themselves.

An unforeseen effect on the control system was incorrect command verification, due to incomplete modeling of DTCFs and their interaction with the MTL model. The MCS had difficulty in dealing with the embedded disable/enable SSID commands at the start and end of each activity file. This led to commands being incorrectly marked as deleted, disabled, or any number of other states other than correctly verified. Mars Express exclusively uses report-based acknowledgement for commanding, and failure reports are still flagged and raise alarms correctly, giving an alternative to the command verification model for failure identification. Moreover, slicing of activities in self-contained files shifts the focus from commands to files. A correction to the predictive and historical model of the onboard command status has been implemented.

G.   GROUND/MISSION PLANNING

As well as creating the activity files themselves, a tool was needed to schedule the execution of the activities and generate the trigger commands to execute the right DTCF at the right time. Associated to this is the creation of the Manifest file, whereby activity files can be loaded into the SSMM with a predefined file name, for the trigger command to then call.

The timing of the trigger commands was critical. Too early, and valuable MTL space is blocked for longer than necessary. Too late, and the commands would be rejected as past their execution time. Furthermore, an activity can only be loaded once the previous one of the same type has been fully executed from the MTL, to avoid interference from the Disable SSID command required for file-completeness protection.

An easy-to-check rule of 10 min before the first command of an activity was adopted for the scheduling of triggers. The DMS is only capable of executing one command file at a time, so the triggers also needed to be scheduled so as not to overlap each other.

An FCT-developed generic Finite State Machine already existed and formed the basis of a new trigger scheduling tool named DANIELS (for "Delayed Activity Numbering, Invoking, Evaluating and Listing Software"; Fig. 24.8), which ingested the command files generated by the MPS and calculated the optimal time for the trigger command to execute each one based on some simple rules, ensuring that each is scheduled about ten minutes before the first command of the activity. In Fig. 24.8, after the MPS has processed the requests from the Science Ground Segment, the command files are grouped by "activity type" and their loading scheduled on a just-in-time basis. These trigger commands are passed to the MCS for uplink, which is scheduled according to the "Manifest" file also generated by DANIELS.

The trigger generation capability has since been migrated to the central MPS, which uses a more refined algorithm to determine the time to execute the DTCFs.

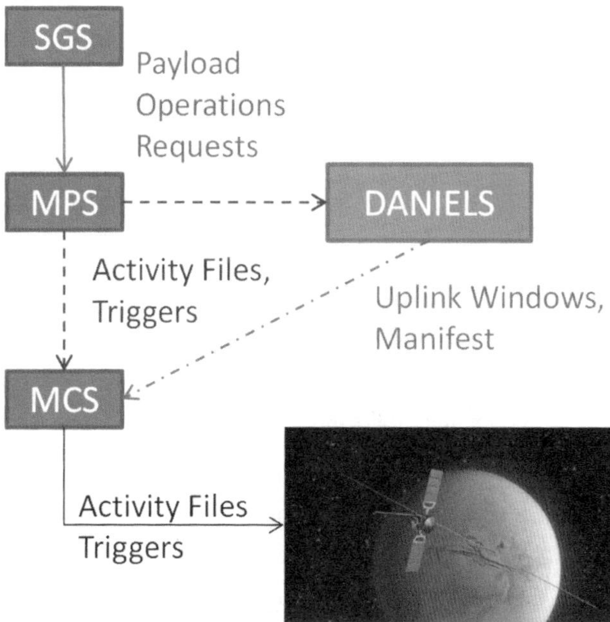

Fig. 24.8 FAST mission planning concept.

By taking into account the number of commands within a DTCF, a better approximation can be made for the execution duration, and precious space inside the short MTL is populated with the full activity only one minute before that activity starts.

As a validation tool, DANIELS ensures that no trigger, or the resultant execution, overlap. This functionality is still required now that the MPS generates the trigger files, such that the results can be cross-checked by an independent mechanism. The more complex MPS scheduling algorithm makes it more difficult for this previously manual check to be performed by hand.

In addition to the trigger schedule checking, a collection of checklists and reports are generated to help with product checking and uplink tracking. DANIELS is also responsible for creation of the Manifest file, which instructs the MCS to uplink the correct command files in the correct order with the correct file names. The reuse of existing tools and a rapid prototyping development approach meant the new tool could be adapted on a weekly (or more frequent!) basis as the ops concept evolved and expanded. The tools and concept mostly converged within two months, with minor tweaks when required thereafter.

Ultimately, the functionality of the ad hoc tool will be reduced to a secondary-level checking tool, as integration of additional functions is absorbed into the core MPS. The "MPS 2012" system has already taken over the process of adding the disable/enable SSID commands, and trigger command generation. The "upstream" mission planning tools are otherwise unaffected [8].

H.   GROUND/COMMAND PRODUCT CHECKING TOOLS

With roughly one hundred files to check per week and many constraints to respect, additional operations support tools were created. These have ranged from one-off shell scripts to modify generated command products until the formal MPS could be updated, to more complex modeling of the execution of activities and trigger commands. The generic Finite State Machine core has been reused for some of these tools, as most of the checks fall into the domain of state checking ("only allow transition X if model Y is in state Z . . .").

New tools have been written to automate the checking of commanding products as much as possible, following the general automation principles for Mars Express [9]. DANIELS performs checks on overlaps between triggers and activities, and also models the MTL fill level to ensure each DTCF can be fully contained within the short MTL (otherwise the remaining commands would be lost). On the principle that two independently written tools are unlikely to have the same bugs, these checks are performed by an additional tool run by the spacecraft operations engineer responsible for that week's command delivery and uplink (Fig. 24.9). Figure 24.9 shows that the timeline view, MTL fill level, and Product Checking & Uplink Confirmation tools

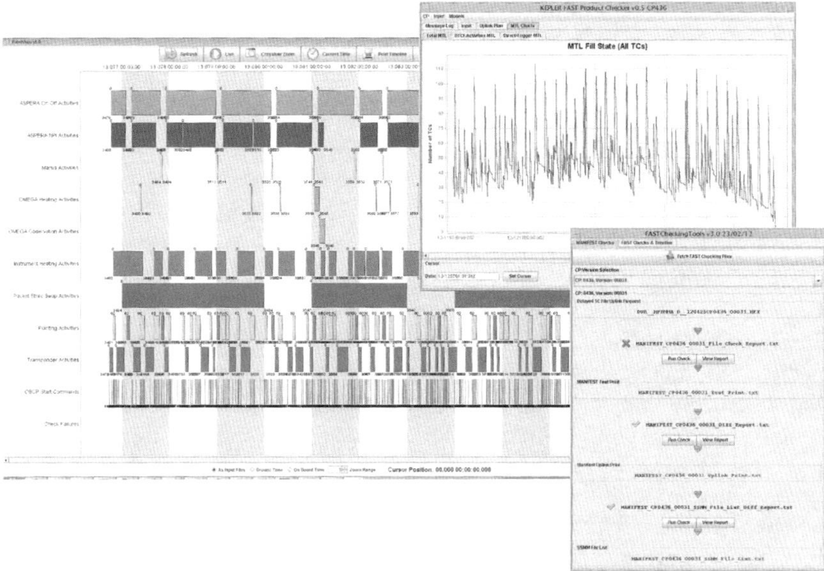

Fig. 24.9   FAST-related Checking & Uplink Verification tools.

give visual feedback to the operations engineers and automate much manual checking.

The necessity for new tools is driven by change in the ops concept that brought in new safety constraints that had to be adhered to, such as only one DTCF executing at a time. Mars Express's existing tool infrastructure [10] (generic timeline tool and generic Finite State Machine engine) meant that only minor modifications were necessary to integrate new rules, meaning that the main difficulty was in defining what needed to be checked to ensure safety of FAST ops. The key goal at all times was also to ensure this new checking did not lose quality and accuracy compared to the previous approach, and this although the ops concept and the checks themselves had evolved concomitantly.

I.   GROUND/SPACECRAFT MONITORING

New monitoring tools had to be created to monitor the progress of FAST operations (and any possible failures). These new tools complement the checking tools mentioned above, to ensure that what was planned was executed. The first step was to create definitions of what signs would indicate a problem with FAST operations: failed DTCF execution events, SSMM file count not as expected, SSID status not as expected, and so on. Because of the just-in-time nature of the new ops concept, these checks had to work on the earliest possible basis to be effective, rather than waiting for routine monitoring. Extra diagnostics are added onboard (such as various SSMM packets, diagnostic report packets) at the start of each pass to facilitate this, along with out-of-limit sets on the routine telemetry to monitor the process of the operations and alert the spacecraft controller to a problem as soon as possible.

V. CONCLUSION

The most significant conclusion is that it is entirely possible to run the full, rather complex, Mars Express science mission using only 117 TCs. Mars Express has already reached 100% of science operations, and future planning/OBCP developments hold much scope for increasing the science operations further still (Fig. 24.10). By implementing a just-in-time and transactional loading scheme, combined with careful planning on the ground, it was possible to work around the transient SSMM problems. Importantly, it is still necessary to rely on the correct functioning of the SSMM, but transient problems cannot spread beyond the initial impact.

It has been demonstrated that by judicious use of existing spacecraft capability and configurability of the existing ECSS-compliant onboard software, software patches at a lower, more critical level are not immediately required. Although the retry patch may allow resumption of the nominal MTL, we have reached a point where there is no value in doing so. Also, the new FAST approach brings additional benefits in terms of routine operability (more flexible via the activity granularity), potential for automation (more preprogrammable), and even safety (covers practically all possible failures that could come from the SSMM). That being said, the patch remains a backup solution, rather than a means to solve the already well-solved problems that occurred in 2011, and to provide more options in the future should further degradations or constraints arise on the aging Mars Express.

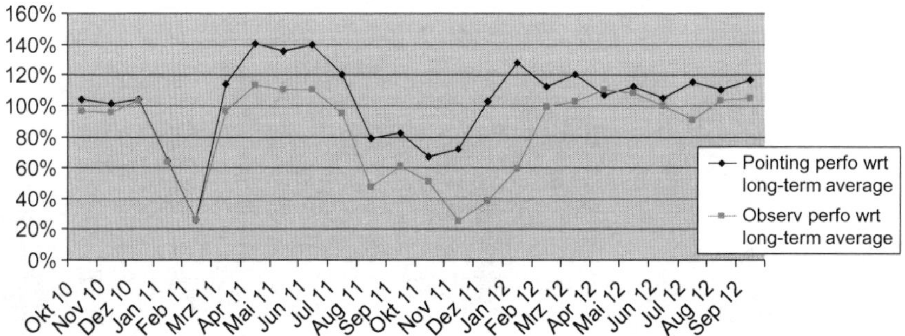

Fig. 24.10 Performance of Mars Express before and after SSMM anomaly.

A. MISSION REHABILITATION THANKS TO FAST: PERFORMANCE FIGURES

The Mars Express mission shows quite some dynamics in its performance figures along time, even in nominal phases, due to the variability of the conditions (Mars–Earth distance impacting data return, Mars–Sun distance impacting energy available, Mars surface illumination impacting the set of operated instruments, the presence of eclipse or radio-occultation, etc.). However the multiyear average performance can be summarized by two key figures that constitute a reference (100%) normally achieved or exceeded: 4.3 observations per 7 h orbit (measure of operations density) and 1.5 science pointings per orbit (measure of operations diversity).

Figure 24.10, normalized with respect to the multiyear average, illustrates the loss of performance following the SSMM problems in August–October 2011, the recovery by progressive introduction of the new FAST method from November 2011 (one instrument at a time, prioritizing the North Polar Cap observation campaign by the radar instrument MARSIS) until January 2012, and the stability thereafter.

The speed at which the new ops concept was conceived and implemented is remarkable. This involved developing, testing, and operationalizing the new ground software, onboard software, and commissioning units in Martian orbit (SSMM-B and HRSC-B) that had lain dormant since launch. To put it into perspective, a similar two-month dip in science operations occurs once every two years due to the solar conjunction severing communications with the spacecraft. An example of this is visible for the period January to February 2011 in Fig. 24.10. This shows the actual observation and pointing performance with respect to the long-term mission average, in terms of numbers of each successfully performed observation by the spacecraft. The big dip from September to November 2011 is due to the SSMM anomaly halting science operations. 100% represents the long-term mission average.

It should also be noted that by implementing the FAST approach, Mars Express has become significantly more robust to SSMM problems (the cause of many of Mars Express's 26 safe modes since launch). Such an anomaly no longer causes a safe mode and mission loss of three days, but instead causes a three-second data gap while the onboard protections reset and restore the link (using the event-driven recovery OBCP "PreFOU"). The FAST concept ensures that such a delayed SSMM reconnection is always safe, from a spacecraft commanding perspective. Three days to three seconds recovery time represents a 100,000-fold reduction in mission loss due to SSMM anomalies, resulting in increased science return and decreased stress on the Flight Control team.

B. LESSONS-LEARNED

Lessons to have come out of this experience affect all aspects of spacecraft operation. The importance of software flexibility in spacecraft operations for implementing workarounds and reconfigurations has been confirmed. This strongly advocates for a very careful assessment of the proposals for simplification of the onboard software during spacecraft development, as is often the case when schedule, complexity, or costs aspects become at stake.

It was surprising to find that the DMS software had apparently little robustness to transient communication errors between units. An explanation may be that the existing "too quick" retry mechanism relies on performance requirements as stated in the specification, but that the units involved may just marginally fulfill, leading to interlocks and failures on a few occasions (by definition, the ones that we see). It is certainly recommendable for future onboard software to provide a very solid functionality for "retry" if it relies on communication across a bus or network. A software patch will correct this for Mars Express in due course.

Development of ground tools, from the formal core Control and Planning Systems to informal checking tools and a somewhat grey area in between, has been key to the successful and speedy return to full science. The ability to quickly prototype new concepts and approaches in-house has certainly hastened the return to science, and the proven concepts have fed back into the formally supported software. Close support and flexibility from external software support teams also contributed greatly toward the success.

Transactional commanding, as a concept for critical groups of commands that should only be executed together as a logical entity, is a technique developed for FAST that has wider applicability to spacecraft operations. Future missions should consider whether they require a similar "all-or-nothing" technique to protect against incomplete file execution. This will become more critical as file-based operations gain popularity generally.

Finally, having an in-house development capability for OBCPs was vital. Mars Express no longer has industry support for onboard software developments, and having the development environment and the skills available to use it was a critical enabling factor. It is not anecdotal anymore that for project development reasons, Mars Express was originally conceived as a no-OBCP mission, although all involved actors have made efforts over time to keep or rehabilitate this functionality, in particular as it could be inherited from its precursor and sister spacecraft Rosetta. Without OBCPs, a FAST-based Mars Express mission today would still be possible but far less acceptable than has been achieved. The speed with which the science mission was recovered was greatly helped by the fact that preparation work on OBCPs to implement switch on and off of instruments had already been completed before the anomaly took place.

C. FUTURE WORK

Further development of OBCPs is the main focus for future work. It falls into two categories: command volume reduction and performance enhancement. The smaller the number of commands required to perform an activity the better, as it allows for more complex operations for the same number of TCs, and more flexibility in scheduling them, leading to better mission return for the same operational effort. Furthermore, the reduction in the number of TCs overall optimizes the space/ground command link by requiring shorter uplink time. The other side of the OBCP coin is to enhance the performance of the spacecraft. This can be in the frame of robustness/error handling, or utilizing onboard autonomy to permit more effective use of the spacecraft. These are not necessarily mutually exclusive goals. It can happen that having an OBCP to manage an operation might also do so more effectively, as well as in fewer TCs, as it can react much faster than the ground operators can.

Finally, we are considering the use of OBCPs to enhance the abilities of the onboard software specifically for FAST. An MISO, like a "second MTL," is being implemented by using an OBCP to schedule the execution of trigger files and release that space in the short MTL. It has been proposed to extend the scheduler to be able to process other commands than "Execute TC File," such as "Start OBCP" or "Set Thermal Configuration." This would remove further commands from the short MTL, and reduce the size (and possibly number) of certain types of activity. However the trade-off between OBCP complexity, execution time, and schedule capacity must be examined closely to ensure that no "over-optimization" takes place that actually reduces the mission flexibility by introducing more strict operational constraints.

ACKNOWLEDGMENTS

We are grateful for the effort, cooperation, and flexibility of the entire Science Ground Segment team at ESAC, Spain, the various instrument teams, and also the software support teams at ESOC.

REFERENCES

[1]   Shaw, N., Moorhouse, A., Denis, M., Porta, R., and Mounzer, Z., "File Transfer, Mass Memory and Mission Time Line–Providing Spacecraft Remote Commanding at Mars," *AIAA-2006-5843, SpaceOps 2006 Conference*, Rome Italy, June 2006.

[2]   "Telemetry and telecommand packet utilization," ECSS-E-70-41A, http://www.ecss.nl/ [retrieved 24 April 2012].

[3]   Haddow, C. R., Pecchioli, M., Montagnon, E., and Flentge, F., "File Based Operations–The Way Ahead?," *AIAA-2012-1294297, SpaceOps 2012 Conference Proceedings*, Stockholm, Sweden, June 2012.

[4]   Blake, R., "Deployment of File Based Spacecraft Communication Protocols," *AIAA-2012-1291500, SpaceOps 2012 Conference Proceedings*, Stockholm, Sweden, June 2012.

[5]   Choukroun, P., Denis, M., Schmitz, P., and Shaw, M., "Evolving ESA Mars Express Mission Capability with On-Board Control Procedures," *AIAA-2010-2186, SpaceOps 2010 Conference Proceedings*, Huntsville, Alabama, April 2010.

[6]   Lakey, D., Eiblmaier, M., Denis, M., de Sousa, B., Porta, R., Shaw, M., and Francisco, T., "Multi-Mission End-to-End OBCP Configuration Control," *AIAA-2012-1291059, SpaceOps 2012 Conference Proceedings*, Stockholm, Sweden, June 2012.

[7]   Ormston, T., Denis, M., and Peschke, S., "Redundancy or Retry? Expanding FDIR in Flight," *AIAA-2010-1910, SpaceOps 2010 Conference Proceedings*, Huntsville, Alabama, April, 2010.

[8]   Rabenau, E., Denis, M., and Peschke, S., "Mars Express Mission Planning–Expanding the Flight Box in Flight," *AIAA-2010-1967, SpaceOps 2010 Conference Proceedings*, Huntsville, Alabama, April, 2010.

[9]   Shaw, M., Denis, M., Rabenau, E., Lakey, D.T., and Schulster, J., "Mission Automation and Autonomy: In-flight Experience Derived from More than 8 years of Science Operations in Orbit about Mars," *AIAA-2012-1284058, SpaceOps 2012 Conference*, Stockholm, Sweden, June 2012.

[10]   Ormston, T., Lakey, D., and Denis, M. "Product Verification on Mars Express–Routine Validation to Ensure Routine Success," *AIAA-2012-1293945, SpaceOps 2012 Conference Proceedings*, Stockholm, Sweden, June 2012.

CHAPTER 25

Emergency Scheduling of Multiple Imaging Satellites with Dynamic Merging

Jianjiang Wang,[*] Xiaomin Zhu[†] and Dishan Qiu[‡]
National University of Defense Technology, Changsha, China

I. INTRODUCTION

Imaging satellites are the platforms equipped with optical sensors that orbit the Earth to take photographs of special areas at the request of users [1]. Recently, using such imaging satellites to take photographs in emergencies has become a critical means for obtaining first-hand information [2]. For example, when an earthquake occurs, images of stricken areas can be obtained. Importantly, the images are expected to be acquired within a few hours or even within dozens of minutes so that damage assessment and planning of rescue policies can be carried out promptly. From this scenario, it is easy to conclude that emergency tasks have short user-expected finish times but not strict restrictions, rather than the deadlines of real-time tasks; that is, although the finish time of a task is a bit later than its user's expectation, the task execution is still valuable. As the imaging of stricken areas, missing users' expected finish time may result in inferior rescue efficiency, but it is still worth getting the satellite images.

There are some uncertainties when imaging satellites take photographs, including users' requirements, weather conditions, satellite states, and so on [3]. For example, the arrival times and task numbers submitted by users are uncertain, which intensifies the problems for imaging satellite resource management during emergencies. Thereby, providing a novel planning and scheduling algorithm for emergency tasks is mandatory due to the critical natures of the tasks and dynamic environment.

To date, many scheduling algorithms have been developed for imaging satellites (see details in Sec. II). Unfortunately, to the best of our knowledge, less work

---

*PhD, candidate, Science and Technology of Information Systems Engineering Laboratory.
†PhD, Science and Technology of Information Systems Engineering Laboratory.
‡Professor, Science and Technology of Information Systems Engineering Laboratory.

has been done regarding dynamic emergency scheduling of multiple imaging satellites. The difficulties are threefold:

1. The features of time windows, users' expected finish times, and the dynamic environment have to be considered, which makes task scheduling on multiple satellites more complicated than traditional schemes.

2. Dynamic emergency scheduling has no fixed horizon (e.g., one day for daily planning, as shown in [4]), making the scheduling of imaging satellites difficult in modeling and solving.

3. Multiple conflicting objectives need to be considered while scheduling, such as schedulability, user satisfaction ratio, stability, and so on.

The aforementioned arguments present a big challenge to the design and implementation of novel and fast dynamic scheduling algorithms for emergency tasks submitted to imaging satellites, motivating us to develop an efficient dynamic emergency scheduling strategy to solve the issue:

1. We established a multi-objective mathematic programming model to formulate the dynamic emergency scheduling problem of multiple imaging satellites.

2. We proposed a task dynamic merging (DM) policy to enhance satellite resource utilization.

3. With the DM in place, we designed a novel dynamic emergency scheduling (DM-DES) algorithm for multiple imaging satellites.

4. Through extensive simulation experiments, we discovered that the DM-DES algorithm could efficiently improve the scheduling quality of conventional scheduling algorithms.

The reminder of this chapter is organized as follows. The next section reviews previous work in the literature. Section III formally models the DES problem. The following two sections describe the task DM policy, as well as DES algorithm (DM-DES). This is followed by extensive simulation experiments and performance analyses in Sec. VI. Finally, Sec. VII concludes the chapter with a summary.

II. RELATED WORK

Over the past decades, many studies have worked on the scheduling of imaging satellites, most focused on a static scheme (i.e., making scheduling decisions in an offline planning phase). Bensana et al. [4] investigated the daily management problem for imaging satellite SPOT-5, and formulated the problem as a Variable Valued Constraint Satisfaction Problem or as an Integer Linear Programming Problem. Several exact methods such as Depth First Branch and Bound or a Russian Dolls search were proposed to find the optimal solution, as well as approximate methods such as Greedy or Tabu searches to find a good solution [4].

Cordeau and Laporte described the problems in selecting a subset of requests for each orbit to yield a maximum profit within constraints, that is, the satellite orbit problem (SOP), and presented a Tabu search heuristic [5]. Again to solve the scheduling problem of imaging satellite, Lin et al. used the Lagrange Relaxation technique to integrate with other approaches, like Tabu or Liner searches [6, 7]. In addition, Bianchessi et al. described an improved Tabu search algorithm to solve the multisatellite, multi-orbit and multi-user scheduling [1]. Globus et al. analyzed and discussed the multisatellite scheduling issue and developed an evolutionary algorithm to solve the problem. Moreover, they compared the evolutionary algorithm with other existing algorithms including hill climbing (HC), simulated annealing (SA), two variants of genetic algorithm, and so on [8, 9]. It should be noted that the aforementioned static scheduling methods have definite horizons. Once a scheduling decision is made, it cannot be changed, which is obviously not feasible in a dynamic environment.

There have also been a few studies directed towards the dynamic scheduling of imaging satellites. Pemberton and Greenwald discussed the dynamic scheduling problem and analyzed contingency conditions [3]. Also, central to the max-flexibility retraction heuristic, Kramer and Smith suggested a repair-based search method for an oversubscribed scheduling problem [10]. With a similar idea, Wang et al. proposed a heuristic to solve a dynamic scheduling problem focusing on multiple imaging satellites [11]. Unfortunately, all the above dynamic scheduling schemes only focus on tasks with no timing requirements, and lack a guarantee to complete emergency tasks within their expected finish times.

For task merging, only Cohen et al. considered the context of multiple targets in a single scene, in what was viewed as a preliminary investigation of task merging [12]. In this chapter, we concentrate on designing a novel task DM policy, and apply it in our proposed DES algorithm of multiple imaging satellites.

III. PROBLEM FORMULATION

In contrast with static scheduling, dynamic scheduling mainly handles aperiodic tasks for which arrival times are not known a priori; this is the most significant feature of emergency tasks besides users' expected finish times. Without loss of generality, we formulate the dynamic emergency scheduling of multiple imaging satellites with a multi-objective mathematic programming model that contains five basic objects: tasks, resources, available opportunities, operational constraints, and objectives.

A. TASKS

In this section we focus on dealing with targets, which can be photographed in the scene of a sensor. A set $T = \{t_1, t_2, \ldots, t_n\}$ of emergency tasks are dynamically submitted for execution. Each task $t_i \in T$ has a priority p_i, an arrival time a_i, an expected finish time e_i, and a resolution requirement rs_i. Considering the practical condition that users often submit numerous tasks at once in emergency, we therefore assume the new tasks arrive in batch style.

B. RESOURCES

A set $R = \{r_1, r_2, \ldots, r_m\}$ of resources (or sensors) are available for assignment to tasks. Each resource $r_j \in R$ is denoted by $r_j = (d_j, \sigma_j, s_j, b_j, o_j, as_j, m_j, \rho_j, \pi_j, M_j, P_j, M_j', P_j', \varepsilon_j, msg_j, bs_j)$, where $d_j, \sigma_j, s_j, b_j, o_j, as_j, m_j, \rho_j, \pi_j, M_j, P_j, M_j', P_j', \varepsilon_j, msg_j,$ and bs_j are the duration of task execution, field of view, slewing pace, start-up time, retention time of shutdown, attitude stability time, memory consumption per unit observation time, energy consumption per unit observation time, energy consumption per unit slewing angle, maximum memory, maximum energy, remaining memory, remaining energy, supplementary energy every time, maximum slewing angle, and best ground resolution, respectively. Because each target is imaged in a scene and the size can be neglected, all tasks on each resource $r_j \in R$ have the same duration denoted by a small constant d_j. In general, the best ground resolution will be obtained when the satellite takes a photograph of the target on the ground track without slewing. It is worth noting that slewing will degrade the image resolution when the resource deviates from the ground track (see Fig. 25.1).

The target resolution in Fig. 25.1a is bs_j, and the target resolution in Fig. 25.1b is calculated by the following equation:

$$observation\ resolution = bs_j \cdot L/H_j, \tag{25.1}$$

where H_j denotes the orbit height of r_j, L is the distance between the resource and ground along the direction of sensor slewing, $L = (H_j + R) \cos \theta - \sqrt{R^2 - (H_j + R)^2 \sin^2 \theta}$, in which R is Earth's radius, and θ is the observation angle.

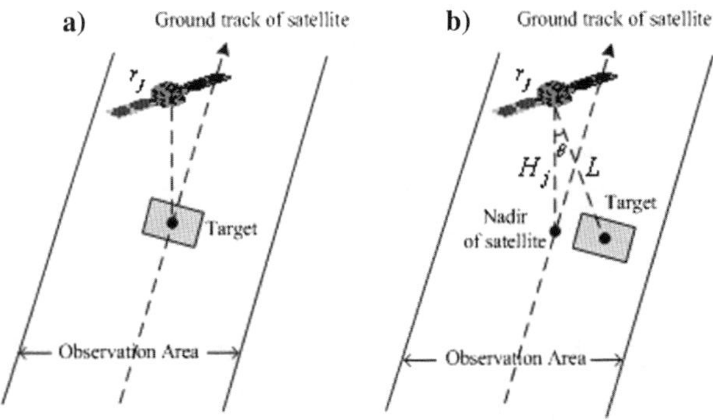

Fig. 25.1   Resolution and slewing angle.

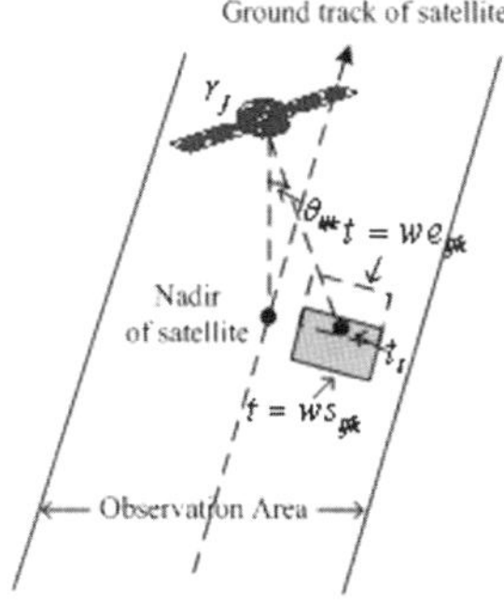

Fig. 25.2 Available opportunity ao_{ijk}.

C. AVAILABLE OPPORTUNITIES

We denote $AO_{ij} = \{ao_{ij1}, ao_{ij2}, \dots, ao_{ijK_{ij}}\}$ as a set of available opportunities of task t_i on resource r_j. For a given available opportunity $ao_{ijk} \in AO_{ij}$, it is represented by $ao_{ijk} = \{[ws_{ijk}, we_{ijk}], \theta_{ijk}\}$, where $[ws_{ijk}, we_{ijk}]$ is the time window and θ_{ijk} is the ideal slewing angle (depicted in Fig. 25.2).

Based on the expected finish time of t_i, we divided AO_{ij} into three subsets: valid available opportunity set $v\text{-}AO_{ij}$, critical available opportunity set $c\text{-}AO_{ij}$, and delay available opportunity set $d\text{-}AO_{ij}$. For any available opportunity ao_{ijk}: 1) if $we_{ijk} \le e_i$, $ao_{ijk} \in v\text{-}AO_{ij}$; 2) if $ws_{ijk} \le e_i < we_{ijk}$, $ao_{ijk} \in c\text{-}AO_{ij}$; 3) if $ws_{ijk} > e_i$, $ao_{ijk} \in d\text{-}AO_{ij}$. It is defined that the amounts of elements in AO_{ij}, $v\text{-}AO_{ij}$, $c\text{-}AO_{ij}$, and $d\text{-}AO_{ij}$ are K_{ij}, K_{ij}^v, K_{ij}^c, and K_{ij}^d, respectively. Apparently, K_{ij}^v is a finite integer, K_{ij}^c is equal to "1" or "0," and K_{ij}^d is an infinite integer. In practice, tasks cannot wait to be executed for a long time (e.g., several days), so we assume there exists a due date dd_i for each task t_i in T, and $ws_{ijk} < dd_i$. As a result, K_{ij}^d is also a finite integer, as shown in Fig. 25.3 ($K_{ij} = K_{ij}^v + K_{ij}^c + K_{ij}^d$).

An allocation matrix $X = (x_{ijk})_{n \times m \times K_{ij}}$ is used to reflect a mapping of tasks, where element x_{ijk} equals "1" if task t_i is allocated to the kth available opportunity on resource r_j; otherwise, x_{ijk} is "0." Additionally, bt_{ij}, ft_{ij}, ϕ_{ij}, and or_{ij} represent the beginning time, finish time, observation angle, and observation resolution of task t_i on resource r_j, respectively, where $ft_{ij} = bt_{ij} + d_j$ and $or_{ij} = bs_j \cdot L/H_j$ [see Eq. (1)].

To facilitate understanding the DES of imaging satellites, we define in our study four sorts of tasks: finished tasks (FT), executing tasks (ET), waiting tasks (WT) and new tasks (NT), which are illustrated in Fig. 25.4. a_1 and e_1 represent the arrival time and expected finish time of task t_1, respectively; $bt_{1,2}$ and $ft_{1,2}$ denote the beginning time and finish time of t_1 on resource r_2; as well as $ws_{1,2,1}$ and $we_{1,2,1}$ are the start time and end time of the time window in the 1st available opportunity $ao_{1,2,1}$ of t_1 on r_2, respectively. Task t_1 is allocated to the 1st available

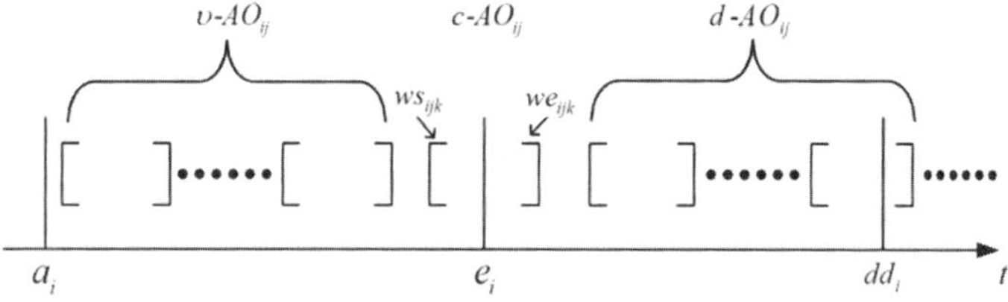

Fig. 25.3 Available opportunity sets.

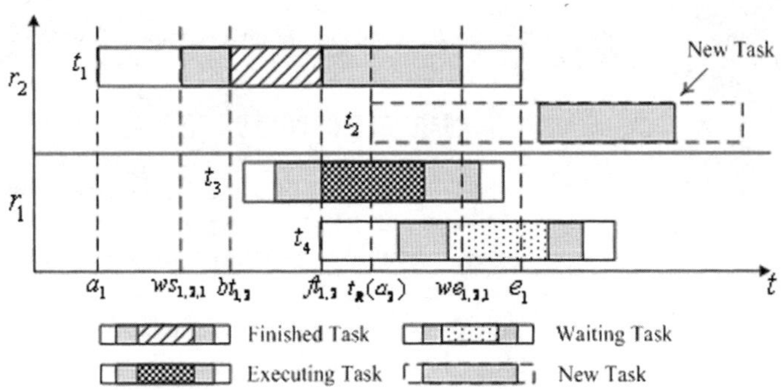

Fig. 25.4 Four sorts of tasks while dynamic scheduling.

opportunity on resource r_2, $x_{1,2,1} = 1$. Labels for other tasks are omitted for simplicity.

The type of a task is determined by the timing instant t_R when a scheduling event is triggered. For any task t_i the following apply: 1) if $ft_{ix} < t_R$, $t_i \in FT$; 2) if $bt_{ix} < t_R < ft_{ix}$, $t_i \in ET$; 3) if $bt_{ix} > t_R$, $t_i \in WT$; 4) if $a_i = t_R$, $t_i \in NT$. Because the nonpreemptive scheme is used in our study, the scheduling decisions for finished tasks and executing tasks cannot be changed. However, the waiting tasks and newly arrived tasks can be taken into account together for scheduling optimization. Therefore, we consider dealing with tasks in WT and NT.

D. OPERATIONAL CONSTRAINTS

Because each task is neither disjunctive nor preemptive, a task can only be allocated to one resource, and executed once. We therefore have the following constraint C_1:

$$C_1: \quad \sum_{j=1}^{m} \sum_{k=1}^{K_{ij}} x_{ijk} \leq 1 \quad \forall\, t_i \in T \tag{25.2}$$

Each task t_i must be executed in an available opportunity ao_{ijk}, $ao_{ijk} \in \bigcup_{r_j \in R} AO_{ij}$.

Hence, we have the available opportunity constraint as follows:

$$\forall t_i \in T, \quad r_j \in R, ao_{ijk} \in AO_{ij}, \quad \text{if } x_{ijk} = 1$$

$$C_2: \quad \begin{cases} bt_{ij} \in [ws_{ijk}, we_{ijk} - d_j] \\ \varphi_{ij} \in [\max\{\theta_{ijk} - \sigma/2, -msg_j\}, \min\{\theta_{ijk} + \sigma/2, msg_j\}] \end{cases} \tag{25.3}$$

To satisfy the requirements of users, the observation resolution of each task must be better than or equal to the resolution requirement. So we have the following resolution constraint:

$$C_3: \quad or_{ij} \leq rs_i, \quad \forall t_i \in T, \quad r_j \in R, \quad \text{if} \sum_{k=1}^{K_{ij}} x_{ijk} = 1 \tag{25.4}$$

We consider a set $T_j = \{t_{1j}, t_{2j}, \ldots, t_{Hj}\}$ $(T_j \in T \backslash FT)$ of tasks waiting for execution with order on resource r_j; that is, task t_{1j} will be executed first and t_{Hj} last. If task t_{ij} is finished successfully, resource r_j requires sufficient setup time to prepare for executing the next task, $t_{i+1,j}$. Thus, the ready time constraint must be considered.

We first introduce some relevant definitions.

Definition 1: Setup Time. Setup time, say $c_{i,\,i+1,j}$, is the minimum needed duration from the finish time of t_{ij} to the beginning time of the next task $t_{i+1,j}$ in T_j, which is denoted by

$$c_{i,i+1,j} = b_j + o_j + as_j + |\phi_{i+1,j} - \phi_{ij}|/s_j \tag{25.5}$$

where ϕ_{ij} and $\phi_{i+1,j}$ are the observation angles of tasks t_{ij} and $t_{i+1,\,j}$, respectively.

Definition 2: Ready Time. Ready time, say rt_{ij}, represents the time from which t_{ij} can be executed on r_j, which is defined as

$$rt_{ij} = ft_{i-1,j} + c_{i-1,i,j} \tag{25.6}$$

Therefore, we can obtain the ready time constraint formulation as

$$C_4: \quad \forall t_{ij} \in T_j, \quad r_j \in R, rt_{ij} \leq bt_{ij} \tag{25.7}$$

The following memory constraint of resource must also be satisfied:

$$C_5: \quad \forall r_j \in R, \sum_{i=1}^{H} (ft_{ij} - bt_{ij}) \cdot m_j \leq M_j' \tag{25.8}$$

Finally, the energy constraint is presented as

$$C_6: \quad \forall r_j \in R, \sum_{i=1}^{H} (ft_{ij} - bt_{ij}) \cdot \rho_j + \sum_{i=1}^{H-1} |\phi_{i+1,j} - \phi_{ij}| \cdot \pi_j \leq P_j' \tag{25.9}$$

E. OBJECTIVES

In this section, we give first priority to scheduling benefit; thus, the primary objective is to maximize the sum of priorities of accepted tasks under the constraints

$$\max_{t_i \in T, r_j \in R} \left\{ \sum_{j=1}^{m} \sum_{i=1}^{n} \sum_{k=1}^{K_{ij}} p_i x_{ijk} \right\} \qquad (25.10)$$

Moreover, to make the scheduling stable, the perturbation of the whole tasks should be minimized. Before introducing this objective, we first define the perturbation in our study.

Definition 3: Perturbation. Perturbation δ_u is the measurement of the distance between a new schedule and the initial one in the uth scheduling.

Generally, the distance results from the following three types of variances of tasks: 1) variance of finish time within the expected finish time; 2) variance of finish time resulting in delay, that is, dissatisfaction with expected finish time but not rejected; and 3) rejection.

Assume there are s batches of tasks in total. Because the scheduling event is triggered by the arrival of a batch of tasks, the scheduling times is equal to s. Thus,

$$\delta = \sum_{u=1}^{s} \delta_u = \sum_{u=1}^{s} \sum_{i=1}^{n} \sum_{v=1}^{3} \omega_v disturb_v(i, u) \qquad (25.11)$$

where δ is the total perturbation in the scheduling of the whole tasks, ω_v, $v = 1, 2, 3$ represent the influence measurement for users of type v variance, and generally $\omega_1 < \omega_2 < \omega_3$:

$$disturb_v(i, u) = \begin{cases} 1, & \text{if type } v \text{ variance happens on task } t_i \text{ at the } u\text{th scheduling} \\ 0, & \text{otherwise} \end{cases} \qquad (25.12)$$

Consequently, the minimum perturbation objective is depicted as

$$\min \sum_{u=1}^{s} \sum_{i=1}^{n} \sum_{v=1}^{3} \omega_v \, disturb_v(i, u) \qquad (25.13)$$

Finally, we prefer that more tasks be finished within their expected finish times, so we have the following objective:

$$\max_{t_i \in T, r_j \in R} \left\{ \sum_{j=1}^{m} \sum_{i=1}^{n} \sum_{k=1}^{K_{ij}^v} x_{ijk} \bigg/ \sum_{j=1}^{m} \sum_{i=1}^{n} \sum_{k=1}^{K_{ij}} x_{ijk} \right\} \qquad (25.14)$$

Accordingly, the scheduling problem in our study can be formulated as a multi-objective mathematic programming problem with the aforementioned constraints.

IV. DYNAMIC MERGING

Because a scene imaged by a sensor is of a certain size, some adjacent targets in the same scene can be observed simultaneously (simultaneous observation), as shown in Fig. 25.5a. Our DM refers to selecting a waiting task $t_{i'j}$ in T_j to merge with a new task t_i for simultaneous observation, generating a new waiting task t_{i*j}. Assume $t_{i'j}$ is allocated to its lth available opportunity on r_j, then the DM must satisfy the following constraints:

1. Available resource constraint:

$$K_{ij} > 0 \qquad\qquad (25.15)$$

2. Visibility constraints:

$$\exists k \in [1, \dots, K_{ij}]$$

$$\text{s.t}\begin{cases} (a): & \theta_{ijk} \in [\phi_{i'_j} - \sigma_j/2,\ \phi_{i'_j} + \sigma_j/2] \\[4pt] (b): & \min\{we_{ijk}, we_{i'l}\} - \max\{bt_{i'_j}, ws_{ijk}\} \ge d_j \\[4pt] (c): & \max\{bt_{i'_j}, ws_{ijk}\} + d_j + c_{i',i'+1,j} \le bt_{i'+1,j} \\[4pt] (d): & or_{i^*,j} \le rs_{i^*,j} \end{cases} \qquad (25.16)$$

The available resource constraint means that the resource r_j must be available for the new task t_i. With respect to visibility constraints, the following apply. Constraint (a) is the slewing angle constraint. As illustrated in Fig. 25.5b, the target of task t_i must be in the observation scene of resource r_j when executing task $t_{i'j}$. The observation angle of $t_{i'j}$ is immutable, due to the constraint of not resulting in conflicts with other tasks in T_j. Constraint (b) is the time window constraint. It is shown in Fig. 25.5c that tasks t_i and $t_{i'j}$ must be executed to satisfy their own available opportunity constraints, respectively, and the beginning time cannot be brought forward so as not to conflict with earlier tasks. Constraint (c) is the setup time constraint. It is implied that the setup time should be enough to execute the next task after merging. Constraint (d) is the resolution constraint of composite task

$$t_{i*j}.\ or_{i*j} = bs_j \cdot L/H_j,\ L = (H_j + R)\cos\phi_{i*j} - \sqrt{R^2 - (H_j + R)^2 \sin^2\phi_{i*j}}.$$

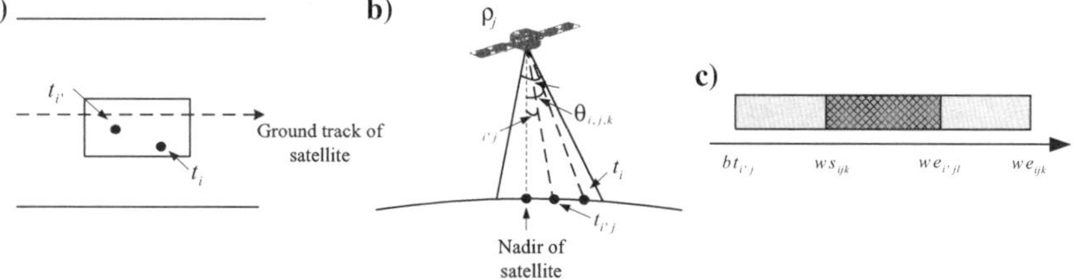

Fig. 25.5 Dynamic merging constraints.

The memory constraint [see Eq. (8)] and energy constraint [see Eq. (9)] must be satisfied for task DM. The beginning time of the new waiting task t_{i*j} is max $\{bt_{i'j}, ws_{ijk}\}$, the finish time $ft_{i*j} = bt_{i*j} + d_j$, the observation angle $\phi_{i*j} = \phi_{i'j}$, and the observation resolution $or_{i*j} = or_{i'j}$. Moreover, the priority of task t_{i*j} is $p_{i*} = p_i + p_{i'}$, the expected finish time $e_{i*} = \min\{e_i, e_{i'}\}$, and the resolution requirement $rs_{i*j} = \min\{rs_{ij}, rs_{i'j}\}$.

All the tasks in WT satisfying the merging constraints with t_i comprise an alternative task set ATS_i for DM. The role of the following alternate task set establishment (ATSE) algorithm is to establish the set ATS_i.

Algorithm 1. ATSE Algorithm

1 Initialize the set $ATS_i = \phi$, available opportunity set $AO_i = \bigcup\limits_{r_j \in R} AO_{ij}$;

2 **for** each available opportunity ao_{ijk} in AO_i, **do**

3 **for** each task t_{ij} in T_j, **do**

4 **if** t'_{ij} and ao_{ijk} satisfy the constraints (15) $\sim$ (16) and mermory, energy constraints **then**

5 $ATS_i = ATS_i \cup t'_{ij}$;

6 **end if**

7 **end for**

8 **end for**

It is denoted in Algorithm 1 that ATSE first explores all the available opportunities of t_i on the entire resources in R. For each available opportunity ao_{ijk}, ATSE seeks the entire tasks in T_j that can be merged with t_i from ao_{ijk}, and then puts them into the alternate task set ATS_i. Finally, ATS_i contains all tasks that can be merged with t_i.

If the maximum number of available opportunities of a task is m, and the total number of tasks for scheduling is n, the worst-case time complexity of ATSE is $O(nm)$.

V. EMERGENCY SCHEDULING ALGORITHM

The DES of multiple imaging satellites is an Non-Polynomial (NP)-complete problem [3], which motivated us to use heuristic approaches to allocate emergency tasks to achieve a close-to-optimal solution. In our study, the dynamic scheduling can be viewed as task insertions from NT to the current schedule.

A. TASK INSERTION CONDITIONS

Task insertion is the insertion of a task from NT into the waiting task set WT within operational constraints. To avoid complicated constraint checking, in this section we give the straightforward task insertion conditions.

A new task is usually inserted into the idle time slot between two tasks in T_j, $j \in [1, \ldots, m]$. A special case is that the new task is inserted before the first task or after the last task in T_j. Without loss of generality, we add two virtual tasks (i.e., a primal task t_{sj} and a terminative task t_{ej}) in each T_j that have the following properties. 1) The beginning time of t_{sj} is t_R, which is the timing instant of scheduling, and ft_{sj} is the finish time of the task that is being executed at t_R. If there is no task being executed, $ft_{sj} = t_R$. 2) Both the beginning time and finish time of task t_{ej} are ∞. 3) The setup times of t_{sj} with other tasks are 0, like t_{ej}.

Based on the above construction, the task insertion can be described as inserting a task into an idle time slot of r_j between task $t'_{i'j}$ and $t_{i'+1,j}$, denoted by $<r_j, t'_{i'j}, t_{i'+1,j}>$. The insertion conditions are depicted as follows:

1. Available resource condition:

$$K_{ij} > 0 \tag{25.17}$$

2. Time conditions:

$$\exists k \in [1, \ldots, K_{ij}]$$

$$\text{s.t.} \begin{cases} (a): & rt_{ij} \leq we_{ijk} - d_j \\ (b): & rt_{i'+1,j} \leq bt_{i'+1,j} \end{cases} \tag{25.18}$$

The available resource condition indicates that the resource r_j must be available for t_i. As for the time conditions, condition (a) denotes that the beginning time of task t_i must satisfy the available opportunity constraint C_2 in Sec. III, and condition (b) represents that the insertion of the new task t_i cannot affect execution of the subsequent tasks.

The memory constraint [see Eq. (8)] and energy constraint [see Eq. (9)] must be satisfied for task insertion. If task t_i satisfies the conditions above, it can be inserted into $<r_j, t'_{i'j}, t_{i'+1,j}>$ to yield a new waiting task t_{ij} with beginning time $bt_{ij} = \max \{rt_{ij}, ws_{ijk}\}$, finish time $ft_{ij} = bt_{ij} + d_j$, and observation angle $\phi_{ij} = \theta_{ijk}$.

B. SOME RULES

There are two steps that are critical in the DES of multiple imaging satellites: selecting new tasks and selecting tasks for DM. To address this issue, we develop two corresponding rules: task requirement degree and optimal task merging.

1. *Task Requirement Degree*
    Task requirement degree TRD_i represents the urgency to be allocated to t_i:

$$TRD_i = p_i \Big/ \sum_{r_j \in R} K_{ij} \tag{25.19}$$

Thus, a task with higher priority and less available opportunities should be allocated preferentially. This rule is for selecting new tasks in *NT*.

2. *Optimal Task Merging*

Definition 4: Best Merging Task, Nonperturbation Merging Task, and Non-delay Merging Task. For a given task t_i, $t_i \in NT$, if $\exists t_{i'_j}$, $t_{i''_j} \in ATS_i$, that is able to be merged with t_i to generate a new waiting task $t_{i'_j}$:

If $ft_{i*j} = ft_{i'j}$, and $ft_{i*j} \leq e_{i*}$, the task $t_{i'j}$ is called a best merging task of t_i.

If $ft_{i*j} = ft_{i'j}$, and $ft_{i*j} > e_{i*}$, the task $t_{i'j}$ is called a nonperturbation merging task of t_i.

If $ft_{i*j} \geq ft_{i'j}$, and $ft_{i*j} \leq e_{i*}$, the task $t_{i'j}$ is called a nondelay merging task of t_i.

From Definition 4, it is easy to find that if task t_i can be merged with a best merging task, there will be neither perturbation nor delay. Thus, we should first select a best merging task for DM if it exists. If there are multiple best merging tasks in ATS_i, we select one randomly. If there is no best merging task in ATS_i, we arbitrarily choose a nonperturbation merging task. If no nonperturbation merging task exists either, a nondelay merging task with minimal $ft_{i*j} - ft_{i'j}$ will be selected. Finally, if all the above tasks do not exist in ATS_i and $ATS_i \neq \phi$, we select one task with minimal $ft_{i*j} - e_{i'}$.

C. ALGORITHM DESCRIPTION

In this section we present the novel DES algorithm with DM (i.e., DM-DES) of multiple imaging satellites. As described before, our scheduling algorithm can be triggered by the following events:

1. *Data transmission.* At the time of data transmission, we assume that all data in memory is transferred, so we should update the remaining memory to be the maximum memory.

2. *Energy supplement.* We need to update the remaining energy according to the supply of energy.

3. *Task execution.* If a task, say t_i, is being executed on r_j, the remaining memory and the remaining energy will be updated by the following procedure.

4.

Algorithm 2 Update (t_i, r_j)

$$1 \; M'_j = M'_j - (ft_{ij} - bt_{ij}) \cdot m_j;$$
$$2 \; P'_j = P'_j - (ft_{ij} - bt_{ij}) \cdot \rho_j - |\phi_{ij} - \phi_{i-1,j}| \cdot \pi_j$$

5. With the arrival of a batch of new tasks, our scheduling algorithm will schedule the new tasks.

The pseudocode of DM-DES is shown in Algorithm 2, in which *Flag* represents the trigger of our algorithm.

Algorithm 2 DM-DES Algorithm

1   **if** $Flag = 1$ **then**

2    $M'_j = M_j$;   //Update the remaining memory at the time of data transmission of r_j

3   **end if**

4   **if** $Flag = 2$ **then**

5    $P'_j = \min\{P_j, P'_j + \varepsilon_j\}$;   //Update the remaining energy when r_j supplies the energy.

6   **end if**

7   **if** $Flag = 3$ **then**

8   Update (t_i, r_j);   //Update the remaining memory and the remaining energy of r_j when task t_i is being executed on r_j.

9   **end if**

10   **if** $Flag = 4$ **then**

11   **while** $NT \neq \phi$ **do**

12   Select task t_i in NT with maximum TRD_i, $NT \leftarrow NT \setminus t_i$;

13   Establish ATS_i for t_i with **ATSE Algorithm**

14   **if** $ATS_i \neq \phi$ **then**

15   Select one task in ATS_i, say t'_{ij}, according to optimal task merging rules;

16   Merge task t_i and t'_{ij} to generate a new waiting task t^*_{ij};

17   break;

18   **end if**

19    $AO_i \leftarrow \bigcup_{r_j \in R} AO_{ij}$;

20   **while** $AO_i \neq \phi$ **do**

21   Select an available opportunity ao_{ijk} with minimum we_{ijk}, $AO_i \leftarrow AO_i \setminus ao_{ijk}$;

22   **for** each task t'_{ij} in T_j **do**

23   **if** ao_{ijk} satisfies the conditions (17) $\sim$ (18) with $<; r_j, t'_{ij}, t'_{i+1, j} >$ **then**

24   Insert the task t_i into $<r_j, t'_{ij}, t'_{i+1, j}>$;

25   Go to line 2;

26   **end if**

27   **end for**

28   **end while**

29   **end while**

30   **end if**

When a batch of new tasks arrives, the DM-DES algorithm first picks out a new task t_i from NT according to task requirement degree (line 12). The DM-DES then calls the ATSE algorithm to obtain all the tasks in WT that can be merged with t_i (line 13).

If the alternate task set ATS_i is not empty (i.e., some tasks in WT can be merged with t_i), DM-DES chooses one task from ATS_i in terms of the optimal task merging rule to merge with t_i, which implies that t_i has been allocated successfully, and then DM-DES goes back to line 12 to select another task until no tasks are in NT (lines 14–18).

If no tasks are capable of being merged with t_i, DM-DES explores the available opportunities of t_i on all the resources in time order. For an available opportunity ao_{ijk}, DM-DES seeks after the available time slots. If there is an available time slot $<r_j, t'_{ij}, t'_{i+1,j}>$, which represents that t_i can be scheduled successfully, DM-DES inserts t_i into $<r_j, t'_{ij}, t'_{i+1,j}>$ and goes back to line 12 to select another task. Otherwise, if there are no available time slots, DM-DES explores the next available opportunity. After the while loop, if task t_i has not been inserted into a time slot, it is noted that t_i fails to be scheduled. DM-DES then goes back to line 12 to allocate another task until no tasks are in NT (lines 20–28).

We now evaluate the time complexity of the DM-DES algorithm. The time complexity of updating the remaining memory and energy is much less than scheduling numerous newly arrived tasks, which can be ignored in our study. Assume there are s batches of tasks for scheduling, the maximum number of tasks in a batch is n, and the maximum number of available opportunities of a task is m. When a batch of new tasks arrives, DM-DES first sorts the tasks according to task requirement degree and selects the task with maximum task requirement degree (line 12), for which the time complexity is $O(n \log n)$. According to the above analysis, the time complexity of the ATSE algorithm is $O(nsm)$. In addition, the computational complexity of selecting DM tasks (lines 14–18) is $O(m)$. DM-DES will consume $O(nm)$ time to insert a new task into the current schedule. Hence, the time complexity of scheduling a batch of new tasks is $O(n \log n) + O(n)[O(nsm) + O(m) + O(nm)] = O(n^2 sm)$. Finally, the time complexity of scheduling s batches of tasks is $O(n^2 s^2 m)$.

VI. PERFORMANCE EVALUATION

We evaluate in this section the performance of the proposed DM-DES algorithm. To demonstrate the performance improvements of DM-DES, we quantitatively compare it with Retraction-Based Heuristic Algorithm (RBHA), which was proposed in [11], and a baseline algorithm, a dynamic emergency scheduling algorithm without dynamic merging (DES).

1. *RBHA*. The basic idea of RBHA is to carry out some amount of iterative repair search inside each new task's available opportunities. Within the repair search

for a given new task $t_i \in NT$, some rules are adopted to decide which tasks to "temporarily" retract, making room for incorporating t_i.

2. *DES.* DES is a variant of the DM-DES algorithm. The difference between DES and DM-DES is that DES does not consider DM. The goal of introducing DES is to evaluate the effectiveness of task DM.

To make the comparison fair, we slightly modify the RBHA in such a way that it chooses the available opportunities in $\cup_{r_j \in R} v\text{-}AO_{ij}$ first. If no available opportunity is obtained, we then select the other available opportunities.

The performance metrics by which we evaluate the system performance include the following:

1. *Total task priorities (TTP).* This is defined as $TTP = \sum_{j=1}^{m} \sum_{i=1}^{n} \sum_{k=1}^{K_{ij}} p_i x_{ijk}$.

2. *Satisfaction ratio (SR).* This is defined as $SR =$ total amount of tasks within their expected finish times/total amount of tasks being executed.

3. *Perturbation measurement δ.* This is used to measure the scheduling stability.

A. SIMULATION METHOD AND PARAMETERS

In the simulation experiments, targets were randomly generated (latitude $-30°$ to $60°$ and longitude $0°-150°$). The numbers of targets were 200, 400, 600, 800, 1000, and 1200, respectively. Without loss of generality, the priorities of tasks were uniformly distributed in [1, 10]. According to extensive literature [3–5, 12], three sensors on different satellites are considered in this chapter. The sensor parameters are presented in Table 25.1; orbit models of the satellites were obtained from STK (where the parameters with * denote designed values based on the literature). In our simulation experiments, we also assumed that the remaining memory and energy were updated once every day, at 12:00:00 hrs.

The arrival rate of a batch is denoted as $a_i = a_{i-1} + intervalTime$, where *intervalTime* is a random positive real number, uniformly distributed in [a, b].

The expected finish time e_i of task t_i is chosen as follows: $e_i = a_i + Time$, where *Time* is subject to normal distribution, $Time \sim N(baseTime, baseTime/10)$.

The due date dd_i of task t_i is described as $dd_i = a_i + DueDate$, where *DueDate* $\sim N(baseDueDate, baseDueDate/10)$.

TABLE 25.1   PARAMETERS OF SENSORS

Satellite	*bs*	*msg*	σ	d^*	s^*	b^*	o^*	as^*	*m*	ρ	π	*M*	*P*	ε
IKONOS-2	4	45	0.931	2	1	3	3	5	5	3	0.5	5000	3000	1000
QuickBird-2	2.44	25	2.1	2	1	3	3	5	5	3	0.5	3000	2500	1000
SPOT-5	10	27	2.09	2	1	3	3	5	5	3	0.5	3000	3000	1000

Resolution unit, m; angle unit, °; time unit, s; memory unit, MB; energy unit, W.

TABLE 25.2   PARAMETERS FOR SIMULATION STUDIES

Parameter	Value(Fixed) – (Varied)
Task Count	(800) – (200,400,600,800,1000,1200)
Priorities	([1,10])
Resource Count	(3)
IntervalTime (h)	([0,4]) – ([0,4],[4, 8],[8,12])
baseTime (h)	(6) – (3,6,9,12)
baseDueDate (h)	(24) – (12,24,36)

Table 25.2 gives the simulation parameters and their values. In addition, regarding minimal perturbation objective, the parameters ω_1, ω_2, ω_3 are defined as $\omega_1 = 0.5$, $\omega_2 = 1$, $\omega_3 = 2$, respectively.

B.   PERFORMANCE IMPACT OF TASK COUNT

In this experiment, we investigated the performance impact of task count. Figure 25.6a–c shows comparisons of DM-DES, RBHA, and DES in terms of total task priorities, satisfaction ratio, and perturbation measurement.

Figure 25.6a shows that with the increase in task count, the total task priorities of all algorithms increases. This is because in the condition of sufficient resource capacity, the amount of accepted tasks will increase with the increment of coming task count. In addition, DM-DES has a better scheduling quality than RBHA and DES. This confirms that DM-DES benefits from task DM, which reduces conflicts, so scheduling quality is improved. RBHA slightly outperforms DES, because RBHA considers "temporarily" retracting some allocated tasks to accept some higher-priority tasks.

From Fig. 25.6b it is found that the satisfaction ratios of all algorithms decrease with increasing task counts. This can be attributed to the fact that the intervals between arrival times and expected finish times are finite. When the task count increases, more tasks cannot be accepted within their expected finish times, making the satisfaction ratios decrease. In addition, the satisfaction ratios of both

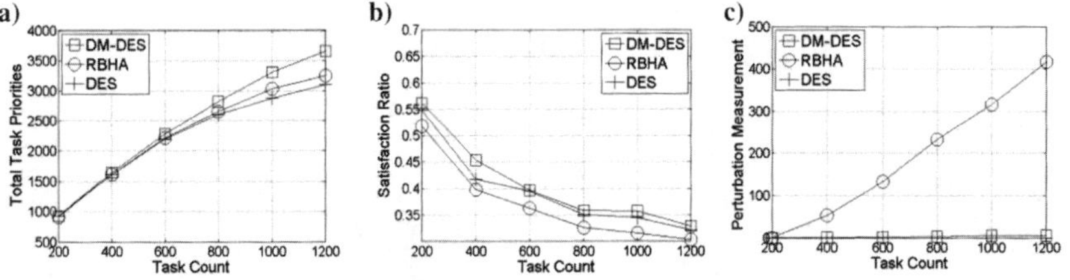

Fig. 25.6   Performance impact of task count.

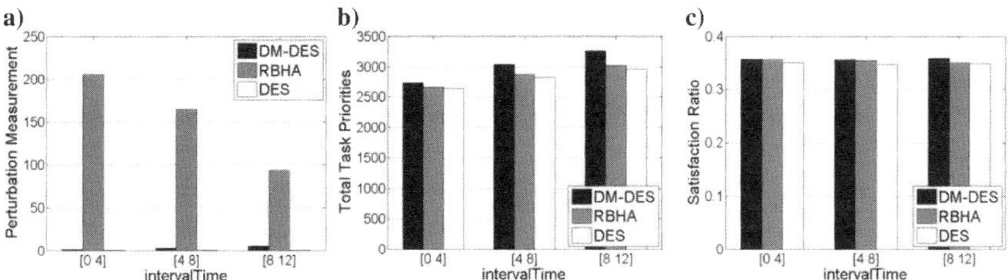

Fig. 25.7   Performance impact of arrival rate.

DM-DES and DES are higher than that of RBHA, which proves that task retraction may result in more tasks not being accepted within their expected finish times.

Based on Fig. 25.6c, it can be seen that, as a result of the task retraction strategy, RBHA may greatly change the schedule, and even some accepted tasks may be rejected in rescheduling; thus, RBHA produces the highest perturbation measurement. With DM-DES, perturbation can only be induced by DM; hence, the perturbation measurement for DM-DES is much less than that of RBHA. The perturbation measurement of DES always remains "0," because it considers neither DM nor retraction; that is, the allocated tasks cannot be adjusted in rescheduling. In addition, the perturbation measurements of both DM-DES and RBHA become larger with an increase in task count.

C.   PERFORMANCE IMPACT OF ARRIVAL RATE

We carried out a group of experiments to observe the impact of arrival rate on all algorithms. The experimental results are presented in Fig. 25.7a–c.

Figure 25.7a shows that, with an increase in *intervalTime*, the total task priorities of all algorithms increase. This is because the arrival rate of emergency tasks reduces with increasing *intervalTime*, so fewer tasks are waiting on resources, leading to tasks having fewer conflicts. Hence, more tasks will be accommodated. DM-DES is superior to the other algorithms in terms of schedulability for the reasons described previously.

From Fig. 25.7b it is found that the satisfaction ratios have no obvious variance. This confirms that the counts of tasks were accommodated within expected finish times and did not increase with increasing *intervalTime*, so the satisfaction ratios remain essentially invariable.

Figure 25.7c shows that RBHA always has the highest perturbation measurement, and DES always remains "0" (as explained for Fig. 25.6c). Moreover, the perturbation of RBHA decreases with a reduction in the arrival rate, which can be explained by the fact that, with a lower arrival rate, the system load becomes lighter, and fewer tasks will be adjusted in rescheduling. In contrast, the perturbation measurement of DM-DES increases, proving that a lower arrival rate may induce more opportunities for DM.

D. PERFORMANCE IMPACT OF EXPECTED FINISH TIME

The objective of this experiment was to investigate the impact of task expected finish time on the performance of the three algorithms. Figure 25.8a – c illustrates the experimental results.

From Fig. 25.8a it is clear that the total task priorities of the algorithms show no obvious change in whether the expected finish times become longer. This can be attributed to the fact that the total number of accepted tasks will not change because the due dates of tasks are invariable. The increase in expected finish times enhances the number of tasks accommodated within their expected finish times, but reduces those outside their expected finish times.

Figure 25.8b shows that the satisfaction ratios of the algorithms have ascending trends with increase in expected finish times. This is because the increase in expected finish times causes more tasks to be accepted within their expected finish times, but the total amount of accepted tasks is invariable. Hence, the satisfaction ratios increase as the expected finish times go up.

Figure 25.8c shows that the perturbation measurements of the algorithms have no obvious variance. This is because the variances of expected finish times only affect the count of accepted tasks within their expected finish times, and have no obvious influence on the other metrics.

E. PERFORMANCE IMPACT OF TASK DUE DATE

In this experiment we evaluate the performance impact of task due date. The parameter *DueDate* is a normal distribution, where the expectation *baseDueDate* varies from 12 to 36 with an increment of 12. Figure 25.9a – c plots the performances of the algorithms.

Figure 25.9a shows that all the algorithms produce larger task priorities as the due dates of tasks are prolonged. This is because, with due dates being prolonged, tasks will have more available opportunities in scheduling, so more tasks can be accommodated. The total task priorities of DM-DES are larger than those of RBHA and DES, which can be explained as for Fig. 25.6a.

Figure 25.9b demonstrably shows that the satisfaction ratios of all algorithms decrease when *baseDueDate* varies from 12 to 36. This is because the total count of

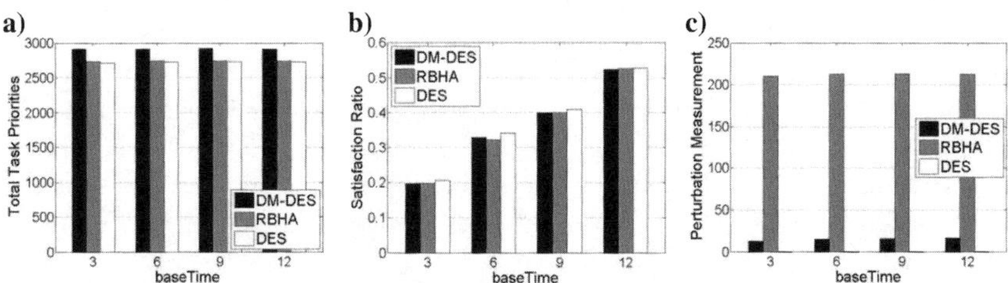

Fig. 25.8   Performance impact of expected finish time.

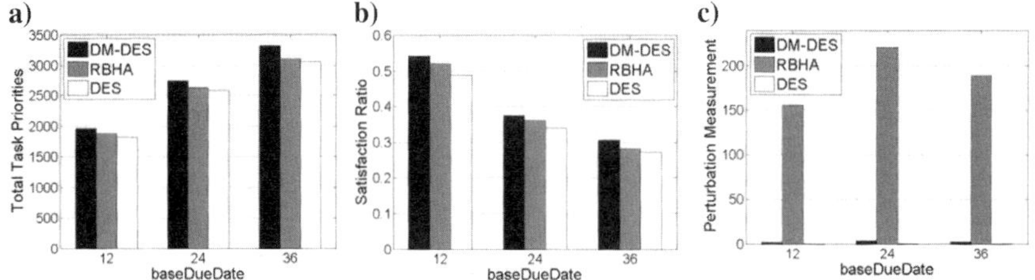

Fig. 25.9 Performance impact of task due date.

accepted tasks increases with the task due dates being prolonged. However, the number of tasks within their expected finish times is invariable due to the invariance of expected finish times. So, the satisfaction ratios decrease. In addition, the descending rate becomes smaller as the value of *baseDueDate* becomes larger. This is because the number of accepted tasks tends to saturation gradually, and the total number of tasks ascends more slowly.

Figure 25.9c indicates that RBHA induces many more perturbations than the other algorithms, for the same reasons stated for the former experiments (see Figs. 25.6c and 25.7c). This can be attributed to the fact that when the task due dates are prolonged, the increase in available opportunities will cause more tasks to be accommodated, which may result in more task DM in DM-DES and task retraction in RBHA. As a result, the perturbation measurements will increase first. However, with a further increase in task due dates, time conflicts will be reduced, leading to more tasks being accepted without retraction. Thus, perturbation measurements will decrease.

VII. CONCLUSION

This chapter establishes a multi-objective mathematic model to formulate the DES problem of multiple imaging satellites. We have proposed a novel task DM strategy, and developed an ATSE algorithm to be used for DM. Moreover, we present a novel DES algorithm DM-DES, which adopts DM to improve the scheduling quality in terms of schedulability, satisfaction ratio, and stability. We have conducted extensive experiments to compare this with its variant DES and also RBHA [11]. The experimental results prove that the DM-DES is an excellent scheduling algorithm and is suitable for dynamic emergency scheduling of imaging satellites.

ACKNOWLEDGMENTS

This research was supported by the National Natural Science Foundation of China (grant nos. 61104180 and 71271213) and the National High-Tech Research and Development Plan of China (grant no. 2008AA7070412). The authors also thank the editors for their constructive comments.

REFERENCES

[1] Bianchessi, N., Cordeau, J., Desrosiers, J., Laporte, G., and Raymond, V., "A Heuristic for the Multi-satellite, Multi-orbit and Multi-user Management of Earth Observation Satellites," *European Journal of Operational Research*, Vol. 177, 2007, pp. 750–762.

[2] Zhu, K. J., Li, J. F., and Baoyin, H. X., "Satellite Scheduling Considering Maximum Observation Coverage Time and Minimum Orbital Transfer Fuel Cost," *Acta Astronautica*, Vol. 66, 2010, pp. 220–229.

[3] Pemberton, J., and Greenwald, L., "On the Need for Dynamic Scheduling of Imaging Satellites," *ASPRS-2002-0068, American Society for Photogrammetric and Remote Sensing Conference*, Denver, Colorado, 2002.

[4] Bensana, E., Verfaillie, G., Agnese, J., Bataille, N., and Blumestein, D., "Exact and Inexact Methods for the Daily Management of an Earth Observation Satellite," *AIAA-1996-23454 SpaceOps 1996 Conference*, Munich, Germany, September 1996, pp. 507–514.

[5] Cordeau, J., and Laporte, G., "Maximizing the Value of an Earth Observation Satellite Orbit," *Journal of the Operational Research Society*, Vol. 56, Issue 8 August 2005, pp. 962–968.

[6] Lin, W. C., Liao, D., Liu, C., and Lee, Y., "Daily Imaging Scheduling of an Earth Observation Satellite," *IEEE Transactions on Systems, Man, Cybernetics–Part A: Systems And Humans*, Vol. 35, Issue 2, March 2005, pp. 213–223.

[7] Lin, W. C., and Liao, D. Y., "A Tabu Search Algorithm for Satellite Imaging Scheduling," Proceedings of the IEEE International Conference on Systems, Man and Cybernetics, IEEE, The Hague, The Netherlands, 2004, pp. 1601–1606.

[8] Globus, A., Crawford, J., Lohn, J., and Pryor, A., "A Comparison of Techniques for Scheduling Fleets of Earth-Observing Satellites," *Journal of the Operational Research Society*, Vol. 54, Issue 9, September 2003, pp. 962–968.

[9] Globus, A., Crawford, J., Lohn, J., and Pryor, A., "Scheduling Earth Observing Satellites with Evolutionary Algorithms," *IEEE-2003-14532, Space Mission Challenges for Information Technology Conference*, IEEE, Pasadena, California, July 2003.

[10] Kramer, L. A., and Smith, S. F., "Maximizing Flexibility: A Retraction Heuristic for Oversubscribed Scheduling Problems," *Proceedings of the Eighteenth International Joint Conference on Artificial Intelligence*, Acapulco, Mexico, August 2003, pp. 1218–1223. Morgan Kaufmann Publishers, San Francisco, CA.

[11] Wang, J. M., Li, J. F., and Tan, Y. J., "Study on Heuristic Algorithm for Dynamic Scheduling Problem of Earth Observation Satellites," Proceedings of the 8th International Conference on Software Engineering, Artificial Intelligence, Networking and Parallel/Distributed Computing, , IEEE, Qingdao, China, 2007, pp. 9–14.

[12] Cohen, R., "Automated Spacecraft Scheduling the ASTER Example," *GSAW-2002-1423, Ground System Architecture Workshop*, El Segundo, California, March 2002.

CHAPTER 26

Planning and Execution of Tele-Robotic Maintenance Operations on the ISS

M. Caron[*]
Canadian Space Agency, St-Hubert, Quebec, Canada

I. Mills[†]
NASA, Johnson Space Center, Houston, Texas

I. INTRODUCTION

Canada's primary contribution to the International Space Station (ISS) consists of the Mobile Servicing System (MSS). The MSS includes the Space Station Remote Manipulator System (SSRMS, or Canadarm2), the Mobile Base System (MBS), and the special purpose dexterous manipulator (SPDM, or Dextre), shown in Fig. 26.1. The Canadarm2 is a 17-m-long, seven-degree-of-freedom (DOF) robotic manipulator that has been used extensively in the assembly of the ISS. The MBS is equipped with four power data grapple fixtures (PDGF), which can each be used as a base by the Canadarm2. The Canadarm2 can "walkoff" from one MBS PDGF to another, in addition to two other PDGFs located on the ISS Destiny (Lab) and Harmony (Node 2) modules. Attached to the U.S.-built mobile transporter (MT), the MBS can be relocated to any one of eight predefined locations along the ISS truss (worksites 1 to 8), thereby extending the effective MSS reach. Dextre is designed to be operated as a dexterous extension of the Canadarm2 for ISS maintenance. Astronauts can control the MSS from two redundant robotics workstations (RWS) on the ISS. The MSS has also been increasingly commanded directly from ground workstations located at the Johnson Space Center's (JSC) Mission Control Center Houston (MCC-H), and from the Canadian Space Agency's (CSA) mission control facility in St-Hubert, Canada.

With ISS construction now complete, the focus of MSS operations has shifted from assembly support to resupply and maintenance activities. In September 2009, ISS astronauts used the Canadarm2 to catch the first ISS free-flyer [1]. It is anticipated that the MSS will be required to capture, offload, and release up to six unmanned resupply vehicles per year. The continued need to use the MSS for ISS maintenance was also vividly illustrated in August 2010 with the

---

[*]Supervisor, Mission Control.
[†]Lead Robotics Operations Officer, Johnson Space Center – DX26.

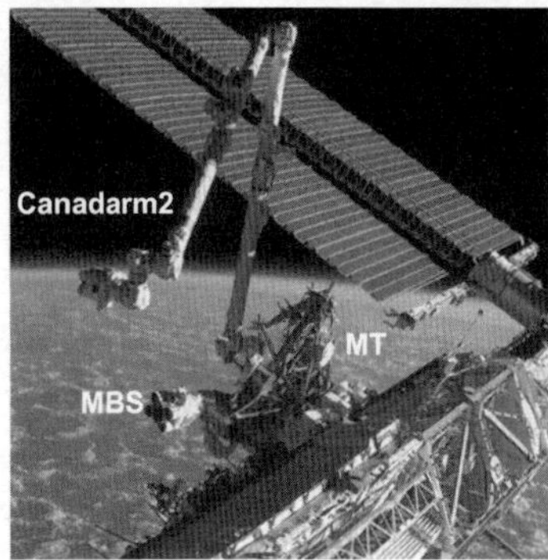

Fig. 26.1   The mobile servicing system.

failure of a critical cooling pump located outside the ISS pressurized volume. The Canadarm2 was used extensively over the course of three extravehicular activities (EVA) to assist the astronauts in the replacement of the failed component [2]. It should be noted that ISS systems include hundreds of such external electronics components, intended to be replaceable on-orbit. Approximately 250 of these on-orbit replaceable units (ORU) were designed to be replaced by Dextre in an effort to reduce the projected EVA workload.

Dextre began supporting logistics and maintenance operations in late 2010. First, it demonstrated its capability to transfer a spare ORU from one ISS location to another. It was then tasked with the stowage of new ISS spare parts and science payloads, delivered by the HTV-2 (2011) and HTV-3 (2012) resupply capsules, on ISS external stowage platforms. In 2011, Dextre also replaced a faulty remote power controller module (RPCM) with a spare unit. In 2012, Dextre was involved in a series of on-orbit tests aimed at demonstrating that robots such as Dextre could be used to repair and refuel existing satellites. This chapter summarizes these first on-orbit telerobotic operations and shares the technical challenges encountered during the planning and execution of these tasks. It begins with a brief description of Dextre's architecture, before presenting the current operational philosophy. The chapter then describes the planning and execution of typical Dextre on-orbit activities, drawing examples from the HTV transfer and RPCM replacement operations. It concludes with a discussion on how the lessons learned over the past year of Dextre operations can be applied to future ISS maintenance tasks, but also to more general robotic exploration and satellite-servicing missions.

II. DEXTRE OVERVIEW

Dextre, shown in Fig. 26.2, comprises upper and lower body structures. The upper body includes a PDGF, avionics platforms, as well as two identical seven-DOF arms. The PDGF allows Dextre to be operated at the end of the Canadarm2, as shown in Fig. 26.1. Dextre's lower body consists of a latching end effector (LEE), a tool holder assembly (THA), two cameras, and a three-sided enhanced ORU temporary platform (EOTP). Multiple ORUs can be temporarily stowed on the EOTP, which can be rotated by Dextre's arms to facilitate ORU access. Dextre's LEE is a grappling effector, similar to those on the Canadarm2, which allows Dextre to be based on an ISS or MBS PDGF when not operated at the end of the Canadarm2. The upper and lower body structures are connected via a body roll joint (BRJ). The BRJ allows Dextre's arms to reach the THA and EOTP by rotating the upper body structure relative to the lower one.

Each of the two 3.5-m-long arms comprises seven revolute joints and an ORU/tool changeout mechanism (OTCM). The OTCM (Fig. 26.3) uses grippers to mechanically grasp specially designed fixtures on the ISS and robotically compatible ORUs (Fig. 26.4). There are six different types of grasp fixtures and six different types of alignment targets in use on the ISS. Once grasped to a fixture, the OTCM's 7/16 in. drive socket, located between its grippers, can be extended to engage bolts located in the center of the fixtures. These bolts actuate mechanisms that fasten (or unfasten) an ORU to its receptacle, or that mate (or demate) its electrical connectors. In addition to its gripping and bolt-driving capabilities, each OTCM is equipped with power/data/video umbilicals for communicating with and/or powering attached payloads. Finally, OTCMs include a bore-sight camera designed to align the OTCM with a grasp fixture using its associated target. Dextre is also equipped with three different types of tools, which can each be grasped and actuated by an OTCM. These tools allow Dextre to interface

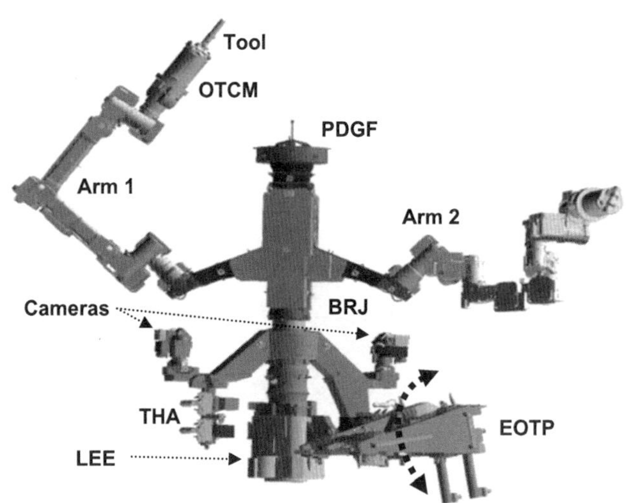

Fig. 26.2   Dextre components.

with fixtures and bolts not designed to be handled directly by an OTCM. When not in use, the tools are stowed in the THA.

Because of the delicate nature of its insertion tasks, Dextre relies on a force–moment sensor (FMS), located at the back of each OTCM, for force control. FMS measurements are used by a force–moment accommodation (FMA) control algorithm, which simplifies the alignment process and limits the forces and moments exerted on ORUs (and their surroundings) during insertion.

III.　DEXTRE OPERATIONAL PHILOSOPHY

A.　EVOLUTION OF DEXTRE'S OPERATIONS CONCEPT

Early Canadarm2 operations relied extensively on human-in-the-loop (HIL) modes of operation, in which operators on the ISS use hand-controllers to maneuver the manipulator. However, it was quickly revealed that the use of operator-commanded autosequences, where an operator specifies the desired final manipulator configuration, either as a set of target joint angles or as the desired position/orientation of the payload, could considerably streamline robotic operations when compared to HIL modes.

As further experience was gained with the Canadarm2, concerns were raised that the duration of complex Dextre maintenance operations would not meet ISS astronaut workday constraints, which is normally limited to 6.5 h, or would exceed maximum unpowered periods for ORUs. A 2002 CSA–NASA technical interchange meeting (TIM; Special Purpose Dexterous Manipulator Task

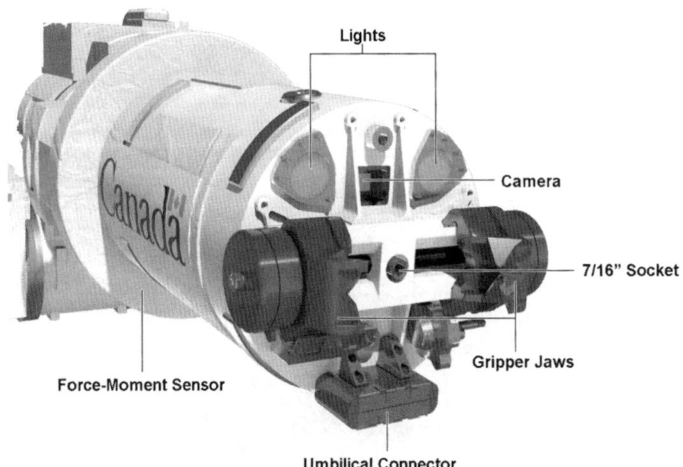

Fig. 26.3　ORU/tool changeout mechanism.

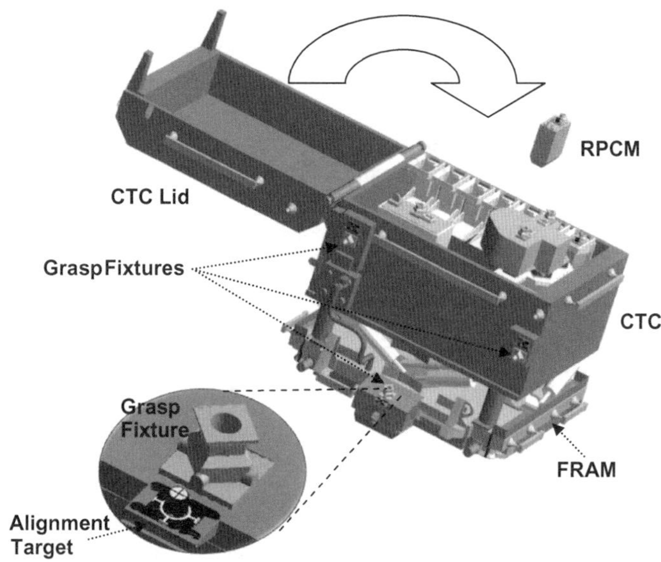

Fig. 26.4   Dextre-compatible cargo transport container and flight releasable attach mechanism (FRAM).

Timeline TIM, Johnson Space Center, Houston, 6–8 August 2002) proposed three key solutions to mitigate these concerns:

1. Development of a ground control capability that would allow for the more mundane portions of robotic maintenance to be performed and monitored from the ground;

2. Replacement of the original ORU temporary platform (OTP) with an EOTP, providing multiple storage areas for ORUs as well as the ability to provide power to their heater circuits. The EOTP would not only simplify the operational sequences, but also allow them to be broken up into smaller segments that could be scheduled over longer periods. As a result, Dextre's OTP was replaced with the EOTP in 2010.

3. Development of a cargo transport container (CTC), a box containing smaller ORUs and compatible with Dextre's EOTP and OTCM, which could provide thermal conditioning to the replacement units during maintenance operations (Fig. 26.4).

Following this TIM, a limited ground control capability was first introduced for the Canadarm2 in 2005. Previously, ground commanding had been limited to MSS activation, diagnostics, and camera surveys. The ability to safely perform small maneuvers in free space was demonstrated first, and this capability progressively grew to include coordinated joint motion without any restriction on maneuver size. When Dextre arrived on orbit in 2008, a large proportion of Canadarm2 maneuvers were operator-commanded autosequences initiated from the ground. Consequently, it was proposed that Dextre's on-orbit checkout and commissioning activities be designed as a demonstration and validation of its ground-based telerobotics concept of operation. The commissioning process spanned nearly two years, and involved progressively more complex demonstrations of free-space maneuvering, OTCM functionality, fine-alignment tasks, and contact operations involving Dextre's tools [3]. Hence, all Dextre operations discussed in this chapter were controlled entirely from the ground.

B. MISSION PLANNING

The replacement of an ISS component by Dextre involves a complex, meticulously choreographed series of robotics operations. Typical removal and replacement operations begin with the retrieval of Dextre from its stowage location, which can be one of several MBS or ISS PDGFs, by the Canadarm2. The desired spare is then retrieved from one of the Station's external stowage platforms (ESPs) or express logistics carriers (ELCs) (Fig. 26.5). Spare units covered in this chapter were either contained in a CTC or integrated directly onto a flight releasable attach mechanism (FRAM). A FRAM, shown on Fig. 26.4, is a type of ORU carrier compatible with Dextre's OTCM and EOTP. The failed component is then replaced in situ with the spare unit. The operation concludes with the return of the failed component to a suitable storage area, which is typically the initial location of the spare unit. Dextre must then be stowed on the ISS to allow the Canadarm2 to support other robotics objectives. For all operations, the

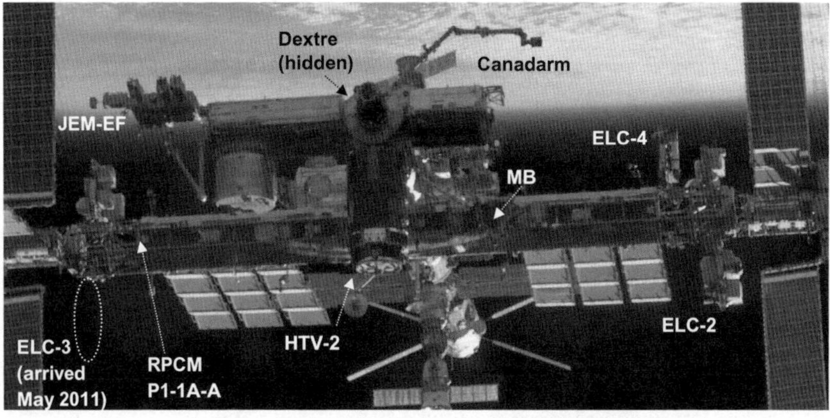

Fig. 26.5   Dextre work areas on ISS for 2011–2012.

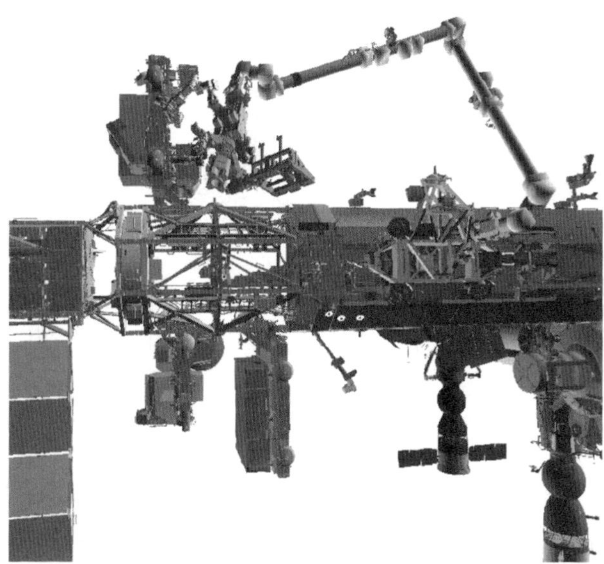

Fig. 26.6   RPS graphics simulation of FRAM relocation demonstration.

MT/MBS and Canadarm2 must bring Dextre into the vicinity of the desired worksites (i.e., the stowage area for the spare unit and location of the failed component).

MSS mission planning can be summarized into the following six major aspects: 1) trajectory design; 2) procedure development and validation; 3) command script development and certification; 4) generation of task-specific software configuration files; 5) dynamic and thermal analyses; and 6) integration into ISS activities.

Once an MSS logistics or maintenance operation has been identified, Canadarm2 and Dextre trajectories are designed using the Robotics Planning System (RPS). The RPS is a kinematic graphics simulator that replicates the external ISS environment, as well as the motion of the Canadarm2 and Dextre, with a high degree of fidelity (Fig. 26.6). Mission designers use this tool to divide the trajectory into a series of autosequences and verify that the resulting operational sequence meets its intended objective, while remaining clear of singularities, self-collisions, joint limits, and structural clearance concerns.

Mission designers then specify the desired autosequences and MSS commands into a specialized syntax made of keywords, each specifying a particular MSS function or control feature. This customized language is used by an automated procedure generation tool to produce the required operational procedures and command scripts [3]. The procedures, developed in MS Word format, are used as the primary document for the review, approval, and execution of the operation by the Flight Control team. The command scripts generated are text files used by MCC command applications to implement the directives listed in the procedures.

Once generated, procedures are subject to a careful independent validation process involving multiple flight controllers using both simulator validation and desktop reviews. This process ensures that the procedures meet all documented constraints, standards, and flight rules. This includes ensuring continuity from one procedure to the next, that no maneuver exceeds maximum travel constraints, that clearance concerns are properly identified and mitigated, that the MSS will not interfere with the operation of external ISS systems or payloads, and that documented system behavior, previous flight history, and anomalies have been suitably addressed. In parallel, flight controllers also verify the predicted maneuver durations and resulting communication link requirements. They also perform command script certification to ensure that the script accurately reflects the procedure directives.

Dextre's performance must also be optimized for each specific task by using customized thresholds, gains, and limits contained in software parameter files. Because FMA plays a critical role throughout Dextre's contact operations, extensive dynamic analyses are carried out before each operation. Thermal analysis is also needed to establish how much time an ORU can remain unpowered during transfer operations before reaching its lower temperature limit.

Procedures must be completed and validated one week before the scheduled start of operations to allow their integration into the overall ISS plan. MSS activities are highly interdependent with several other ISS systems. For instance, a failed component must be electrically isolated from its own system before Dextre can remove it. In addition, several robotically serviceable ORUs are located in the vicinity of articulated structures such as solar arrays and radiators. These must locked in position before the MSS enters the area of concern to ensure that safe clearances are maintained. The extensive use of FMA, in addition to the desired positional accuracy (of the order of a few millimeters), requires ISS thrusters to be inhibited during contact operations. Even the small accelerations generated by some astronaut exercises may affect MSS operations and must be constrained. Use of video downlink channels, the selection of suitable telemetry formats, and the availability of relay satellite coverage must also be negotiated with other flight control disciplines.

Finally, contingency procedures and analyses must be developed to return the ISS to a safe configuration and recover all required functionality in the presence of an ISS or MSS anomaly. These include Canadarm2 procedures to extract Dextre from its worksite following a failure, use of Dextre's backup drive unit to complete an ORU insertion or extraction, and crew procedures to recover from an ISS loss of attitude control.

C. MISSION EXECUTION

Every on-orbit MSS operation is supported by a team of three robotics flight controllers (the "ROBO" team). The lead flight controller sits in the primary ISS flight control room in MCC-H, and is assisted by two other flight controllers who are located in separate rooms ("backrooms"), either in MCC-H or at CSA. All MT relocations, Canadarm2, and Dextre operations described in this chapter were controlled entirely by this team from the ground.

Ground-controlled MSS operations typically begin with MSS activation, which can take up to 90 minutes. A video survey of the ISS exterior follows, using MSS and ISS cameras, which is downlinked to the ground. The survey covers the entire space the Canadarm2 and Dextre are expected to traverse and verifies there are no clearance concerns. The primary objective is to confirm that the ISS configuration along the planned trajectory matches the RPS models used to develop and validate the procedures. The survey is also required to protect against the possibility of loss of video downlink during ground-controlled MSS motion, thereby affecting the ability of ground operators to monitor clearances.

The MSS ground control protocol is a methodical, highly structured, and scripted commanding sequence that requires two flight controllers to execute any command. During execution, the procedures are followed by all team members to ensure that the correct step of the correct procedure is being executed. For each step in the procedure, the team agrees on the step to be executed, confirms that the system is in the expected configuration, and verifies there is adequate communication coverage for the command execution. Before each command uplink, the lead flight controller and one other agree on the command to be uplinked. The command is then uplinked to the ISS and the team monitors its execution. Initiating Canadarm2 or Dextre joint motion requires the uplink of three successive commands (using the above process) in a "Ready–Arm–Fire" protocol. It should be noted that the MSS software only allows one of its manipulators to be in motion at any given time. Here, the "Ready" command corresponds to loading the desired kinematic configuration, while the "Arm" command confirms the desired configuration, and the "Fire" command initiates motion. Before uplinking the "Load" command, both the commander and verifier are required to verify that the contents of the command (destinations) match the procedure being executed. Once motion is initiated, the ROBO team monitors the Canadarm2/Dextre trajectory and clearances using downlinked views (when available), telemetry displays, and telemetry-driven RPS models. If unexpected motion is observed, the lead flight controller issues a single "Safing" command to arrest motion.

IV. FRAM RELOCATION DEMONSTRATION (DECEMBER 2010)

Dextre's commissioning was scheduled to be concluded in July 2010, when it would carry out the complete replacement a partially failed RPCM. Aside from commissioning considerations, the primary objective of this operation was to swap two operational RPCMs. The "RPCM swap" would place the inoperative output channel of the failed RPCM in an unused location at the other RPCM's location, thereby regaining full functionality on both RPCMs. Dextre performance was nominal, but it was not able to extract the RPCM from its worksite, despite pulling with the maximum force rated for this ORU (100 N). It was noted that each unsuccessful attempt to remove the RPCM resulted in significant Canadarm2 deflections. Although Dextre could have applied a higher pull force (up to 200 N), concerns that the resulting Canadarm2 oscillations could cause the RPCM to recontact the ISS once it broke free prevented further attempts. Further ground analysis isolated the issue to an unexpected mechanical interference from the RPCM's electromagnetic shielding. The RPCM swap was deferred to allow the determination of a valid higher pull force, and the generation of procedures to safely extract the RPCM.

The additional analysis and potential operational concept modification required to complete the RPCM replacement drove the decision to shift the focus to the preparation of the upcoming HTV-2 ORU transfer, which was deemed a higher priority. In its unpressurized section, the HTV-2 was bringing a new flex hose rotary coupler (FHRC) to be used as a spare for the Station's external thermal control system, and CTC-2, which carried spare RPCMs and an mss video distribution unit (VDU). Because both the CTC-2 and FHRC were installed onto a FRAM, Dextre was assigned the task to transfer these ORUs to ELC-4 (Fig. 26.5). To ensure readiness for the HTV-2 mission and verify Dextre's ability to overcome the FRAM's magnetic soft-dock force (estimated to be 80 N), it was decided to perform a FRAM extraction and insertion before the mission. This "FRAM relocation" would use Dextre to relocate CTC-3 from Site 1 to Site 2 on ELC-2 (Fig. 26.7). As a secondary objective, the EOTP's FRAM mechanical and electrical interfaces would also be verified by temporarily stowing CTC-3 on the EOTP before its final installation to Site 2. Although the CTC-3 relocation would improve EVA access to the CTC's contents, its success was not critical to ISS operations, making this particular task well-suited for such a demonstration.

The FRAM relocation took place over 6 days, from 20 December 2010 to 4 January 2011. The Canadarm2 was first walked from Node 2 over to the MBS, before retrieving Dextre, which had been stowed on the Lab PDGF. The MT then relocated the entire MSS to the vicinity of ELC-2. Once maneuvered into position by the Canadarm2, Dextre removed CTC-3 from ELC-2 Site 1, before placing it on the EOTP. The next day, CTC-3 was removed from the EOTP and installed on ELC-2 Site 2. The operation concluded when Dextre was put down on MBS PDGF2 and the MSS was translated back towards the center of the Station's main truss. Table 26.1 shows the detailed sequence of events associated with this operation. The times for each subtask are also shown in hours and minutes to illustrate typical activity durations.

The successful relocation of CTC-3 from one storage site to another was representative of many of Dextre's future tasks. Although the Canadarm2 had been previously used to maneuver Dextre from one area of the ISS to another, this was the first end-to-end Dextre task involving complex contact operations and the

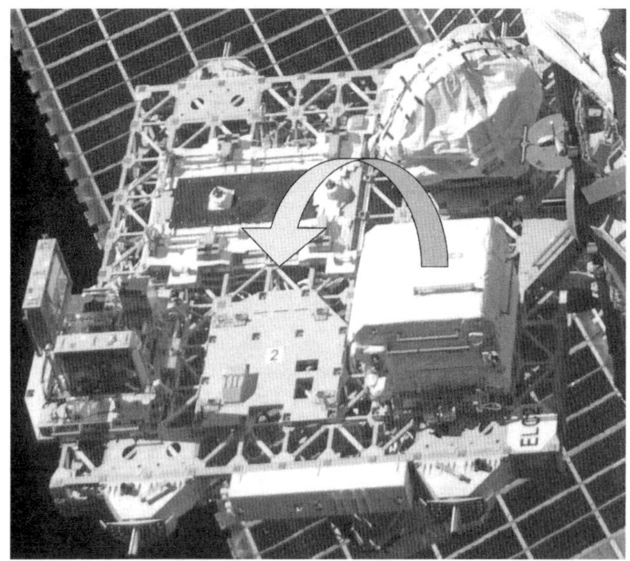

Fig. 26.7   CTC-3 relocation on ELC-2.

extensive use of FMA. The FRAM relocation not only validated Dextre's concept of operation and completed its on-orbit commissioning, but also led to the following observations.

Despite concerns that the required extraction force would exceed robotics requirements (and potentially cause the Canadarm2/Dextre to bounce back towards structure), no such issues were associated with the extraction from either ELC-2 or the EOTP. In both cases, the OTCM demated successfully the FRAM connector with the nominal torque, and the Canadarm2 pulled Dextre holding CTC-3 smoothly. The FRAM's magnetic soft dock forces remained below 27 N as the Canadarm2 began motion approximately 3 s after the command was sent.

The insertion maneuver on the EOTP was performed using FMA. During the final insertion maneuver, ground operators concluded that the maneuver had completed and issued a command to terminate the motion. Although real-time FMS readings and the absence of motion on the video downlink seemed to confirm the completion of the maneuver, post-operation analysis revealed that FMA was still working at reducing the moment in the yaw axis at that time, which was still relatively high. The premature interruption prevented the FMA from fully seating the interface and resulted in the need to perform several additional alignment corrections. The insertion challenges were compounded by a temporary loss of FMS data, which caused the insertion to be completed without FMA. The FMS issue was induced when power was deliberately

TABLE 26.1 FRAM RELOCATION OPERATIONAL SEQUENCE

Day	Major activity	Detailed activities	Duration (h:min)
20 Dec.	MT relocation	MT is translated from Worksite 3 to 5.	3:35
	Canadarm2 "walkoff" from Node2 PDGF to MBS PDGF3	MSS activation and pre-motion survey.	1:00
		Canadarm2 maneuvers to grapple MBS PDGF3.	1:45
		Canadarm2 grapples MBS PDGF3, changes its base, before releasing Node 2 PDGF.	0:55
		Canadarm2 reconfiguration.	0:50
21 Dec.	Dextre relocation from Lab PDGF to Canadarm2 and reconfiguration for translation	MSS activation and pre-motion survey.	1:00
		Canadarm2 maneuvers to grapple Dextre.	0:50
		Canadarm2 grapples Dextre, Dextre changes its base to the Canadarm2, before releasing the Lab PDGF.	2:05
		Canadarm2 and Dextre reconfiguration.	2:30
	MT relocation	MT is translated from Worksites 5 to 2.	3:15
22 Dec.	CTC-3 extraction from ELC-2 (Site 1) and installation on EOTP	MSS activation and pre-motion survey.	2:00
		Dextre reconfiguration (BRJ and Arm1).	0:20
		Canadarm2 moves Dextre into position near ELC-2 Site 1.	0:25
		Dextre Arm 1 moves in to grasp CTC-3 FRAM fixture.	0:25
		Arm 1 alignment and final maneuver for grasping.	0:50

		Arm 1 grasps FRAM fixture and unfastens FRAM from ELC-2.	0:40
		Canadarm2 maneuvers Dextre and CTC-3 FRAM away from ELC-2.	1:00
		Dextre BRJ rotation to permit EOTP access.	0:10
		Dextre Arm1 maneuvers CTC-3 FRAM over EOTP Side 2.	1:05
		CTC-3 FRAM alignment, insertion, and fastening onto EOTP using Arm 1 (four attempts were required).	4:40
		Canadarm2 maneuvers to overnight configuration.	0:30
23 Dec.	CTC-3 removal from EOTP, installation on ELC-2 (Site 2), and Canadarm2/ Dextre post-activity reconfiguration	MSS activation and pre-motion survey.	1:00
		Dextre Arm 1 releases CTC-3 FRAM from EOTP.	0:25
		Dextre Arm 1 extracts CTC-3 FRAM from EOTP.	0:45
		Dextre BRJ rotation to move EOTP clear of the worksite.	0:10
		Canadarm2 maneuvers Dextre into position near ELC-2 Site 2.	0:30
		Dextre Arm 1 positions FRAM for alignment maneuvers.	0:15
		CTC-3 FRAM alignment/insertion attempts without FMA.	2:30
		CTC-3 FRAM alignment/insertion with FMA, and fastening onto ELC-2 (Site 2).	0:50
		Dextre Arm 1 releases the FRAM microfixture.	0:02
		Dextre Arm 1 maneuvers away from CTC-3.	0:10
		Canadarm2 reconfiguration.	1:00
		Dextre arm reconfiguration for stow.	0:30

(Continued)

TABLE 26.1 FRAM RELOCATION OPERATIONAL SEQUENCE *(CONTINUED)*

Day	Major activity	Detailed activities	Duration (h:min)
3 Jan.	MT relocation	MT translation from Worksites 2 to 5.	4:00
4 Jan.	Dextre stowage on MBS PDGF2	MSS activation and pre-motion survey.	1:30
		Canadarm2 maneuvers Dextre to grapple MBS PDGF2.	0:55
		Dextre grapples MBS PDGF2, Dextre base change, Canadarm2 releases Dextre.	0:40
		Canadarm2 reconfiguration.	0:30

removed from Dextre's lower body in order to safely mate the FRAM electrical connectors to the EOTP power distribution system. Overall, it was concluded that future final insertion maneuvers must be allowed to run for a longer duration to give FMA the opportunity to completely correct misalignments. In addition, future FRAM attachments to the EOTP should be performed as a two-stage operation, during which power to the SPDM lower body would only be removed once the FRAM is structurally attached to the EOTP, but before electric connectors are mated. This would prevent any loss of FMS data.

The insertion maneuver on ELC-2 was first attempted without using FMA to demonstrate the feasibility of a non-FMA installation in the event of a failure of the FMS. Despite several attempts, it was not possible to complete the installation without FMA. Timeline constraints prompted the operators to revert to an FMA approach. Capitalizing on the previous day's experience, FMA was given a longer time to relieve loads at the interface and successfully complete the installation. It was concluded that installation without FMA is a "best effort" approach and can require considerable additional time to accomplish.

Finally, scheduling this operation proved to be surprisingly difficult because of the frequently changing launch date of the STS-133 mission, resulting in conflicting configuration requirements for MT and Canadarm2 support. In addition, the timing of MT translations was made difficult by the requirement to have an EVA capability (which is not available throughout the year due to ISS crew rotations) to protect against stranded MT contingencies requiring EVA support.

V.   HTV-2 ORU TRANSFER (FEBRUARY 2011)

Dextre's first official mission consisted of unloading two newly delivered ORUs, a spare FHRC and CTC-2, from the HTV-2's exposed pallet (EP). The HTV-2 was launched from Japan on 22 January 2011 and rendezvoused with the ISS five days later. ISS astronauts used the Canadarm2 to capture the vehicle and berth it to the Station. The following day, the astronauts commanded the Canadarm2 to remove the EP from the HTV-2's external cargo trunk and hand it over to the Japanese Experiment Module's (JEM) robotic arm, which then attached it onto the JEM Exposed Facility (JEM-EF), seen in Fig. 26.5. Both the FHRC and CTC-2 were integrated onto FRAMs, themselves securely fastened to the EP. The extraction of both ORUs followed a sequence similar to the one presented in Table 26.1. Dextre Arm 1 was used to transfer the FHRC from the EP to its EOTP, where power could be provided to the FHRC heating circuits. Dextre Arm2 then extracted CTC-2, handling it by the FRAM fixture, while Arm1 grasped another fixture directly on the CTC lid, through which power could be provided for thermal regulation. The FHRC would remain on EOTP Side 2, while the CTC-2 would be held at the end of SPDM Arm1 until the arrival of ELC-4, scheduled for the following month.

The approach previously developed for the FRAM relocation was used: Canadarm2 maneuvers were used to remove both the FHRC and CTC from the EP once they had been grasped and unfastened by Dextre's arms (Fig. 26.8). Again, the extraction was executed smoothly for both ORUs. Applying lessons learned from the December FRAM relocation, the FHRC installation on the EOTP was completed more efficiently by letting FMA operate longer until all loads had been relieved, and by completing the FRAM's mechanical engagement before powering down Dextre's lower body.

Nevertheless, unanticipated difficulties arose when grasping the FRAMs for both the FHRC and CTC. The ROBO team was surprised to discover that the OTCM had drifted by approximately 7 cm relative to the microfixture after standing down for a planned 30 min communication outage (Fig. 26.9). This drift was later attributed to the length of the kinematic chain between Dextre's OTCM and payloads attached at the end of the JEM–EF. The chain was highly susceptible to the relative motion of the Station's modules and structures induced by the orbital environment. However, experience from the FRAM demonstration had led the ROBO team to schedule additional shifts of longer durations to account for such unexpected issues.

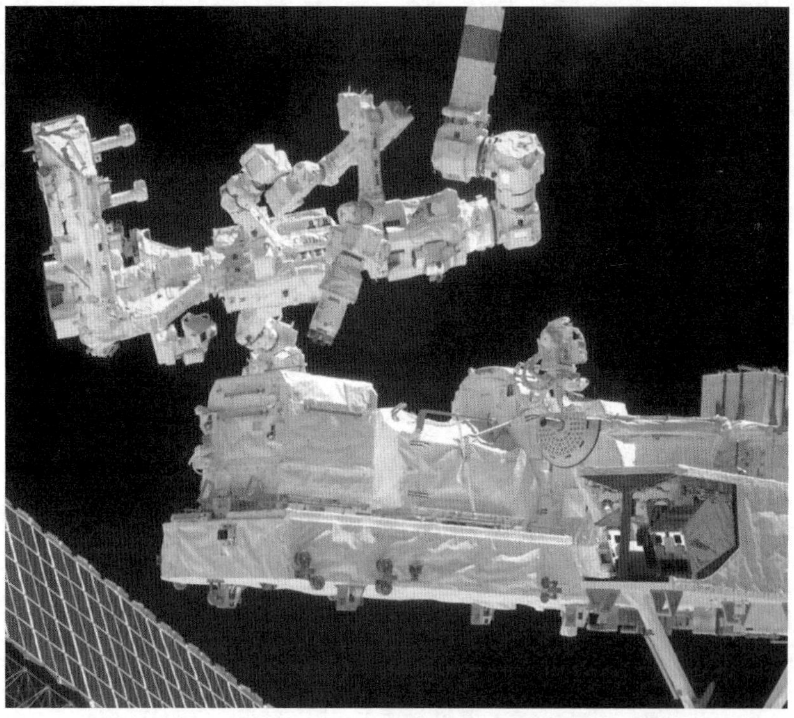

Fig. 26.8   FHRC extraction from EP.

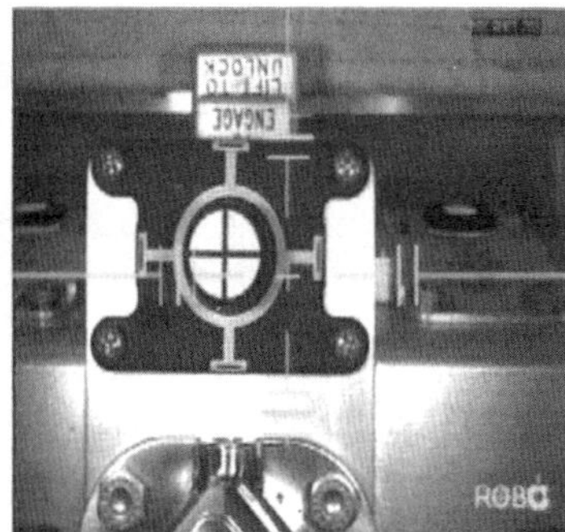

Fig. 26.9   OTCM drift during communication outage.

VI.   RPCM P1-1A-A REMOVAL AND REPLACEMENT (AUGUST 2011)

In August 2011, the focus for Dextre operations shifted back towards the removal and replacement (R&R) of RPCM P1-1A-A, which had been unsuccessfully attempted the previous year. The plan was revised to replace the failed RPCM with a new spare unit contained in CTC-2, instead of swapping it with another RPCM currently in operation. Additional analysis had been performed over the previous year to ascertain the highest force Dextre could apply to safely extract the RPCM from its enclosure. Several options were considered to overcome the high extraction force, but the final decision was to have Dextre's arm pull with a force of 125 N for several minutes, relying on FMA to control the magnitude of the applied force and minimize any misalignment. Ground analysis and hardware tests had indicated that this force was likely to free the RPCM, without completely extracting it from its alignment guides, thus ensuring that the Canadarm2 deflection would not cause the RPCM to escape the worksite and hit the ISS. An alternate technique, combining straight pulls with small positive and negative pitch adjustments, was also developed as a backup method for extraction.

Many aspects from the RPCM R&R differed from previous FRAM transfers. First, RPCM's are equipped with micro-conical fixtures, which are not designed to be grasped directly by an OTCM. This meant that both of Dextre's robotics micro-conical tools (RMCT) would be used to grasp and actuate the RPCMs fixtures and co-located bolts. Second, this maintenance operation would mark the first time that Dextre would actuate a CTC mechanism to unlatch, open, and close its lid (Fig. 26.10). Third, this would be the first time that two payloads would be placed simultaneously on the EOTP. Astronauts had installed the Robotic Refueling Mission (RRM) demonstration module onto Dextre's EOTP (on the now vacant Side 2, because the FHRC had previously been installed on ELC-4 in April 2011) during an EVA on the final Space Shuttle mission in July 2011. The RPCM R&R also called for CTC-2 to be installed on the EOTP. This would require an EOTP rotation to install CTC-2 onto EOTP Side 1. Table 26.2 summarizes the main execution steps associated with the RPCM R&R.

The RPCM extraction out of the P1-1A-A from its site was successful on the first pull, with a maximum force of 128 N (Fig. 26.11). The dynamic response of the on-orbit system matched well that predicted by ground analyses; after 60 s at maximum force, the RPCM broke free, but did not escape its receptacle on the ISS. However, the elastic energy stored in the Canadarm2 deflection did cause the RPCM to bounce back within its alignment guide in an over-damped fashion. The insertion of the new RPCM into the P1-1A-A site was also successful on the first attempt. The RPCM insertion proceeded smoothly, requiring a 160 N push force to overcome initial worksite resistance. However, maintaining the required preload during the RPCM fastening proved difficult. Four unsuccessful attempts to bolt the new RPCM were required before Dextre's brakes were applied to maintain an adequate preload on the bolt during its initial thread engagement. Bolting was then completed nominally. The insertion of the failed RPCM into the CTC was completed without any issues. Actuation of the CTC lid was also performed without any problems. In the days following the RPCM R&R, the CTC-2 and RRM module were both removed from Dextre's EOTP and installed on ELC-4 Sites 1 and 3, respectively.

Fig. 26.10   Dextre opening CTC-2 lid.

TABLE 26.2 RPCM R&R OPERATIONAL SEQUENCE

Day	Major activity	Detailed activities
	Initial configuration	MT/MBS located at Worksite 6, near location of failed RPCM P1-1A-A Canadarm2 based on MBS PDGF3 Dextre located at the tip of the Canadarm2, holding CTC-2 at the end of Arm1, and supporting the RRM module on EOTP Side 2.
1–2 Aug.	CTC-2 stowage on EOTP	Dextre Arm 2 rotates the EOTP 120° clockwise to facilitate the installation of CTC-2 on EOTP Side 1.
		Dextre Arm2 grasps the CTC-2 FRAM fixture.
		Arm 1 releases the CTC-2 lid fixture.
		Arm2 installs CTC-2 on EOTP Side1 (multiple attempts over two days).
10 Aug.	Tool unstow and CTC lid checkout	Arm2 grasps and unstows RMCT2 from the THA.
		Arm1 grasps the CTC-2 lid latch fixture, actuates the mechanism to unlatch the CTC lid, and releases the latch fixture.
		Arm1 grasps the CTC-2 lid hinge fixture, and actuates the mechanism to fully open and close the CTC lid.
28 Aug.	RPCM R&R day 1	Arm1 actuates the mechanism to fully open the CTC-2 lid and releases its fixture.
		Arm1 grasps and unstows RMCT1 from the THA.
		Arm2 grasps the spare RPCM's fixture in CTC-2 and engages its collocated bolt.
29 Aug.	RPCM R&R Day 2	Arm2 unfastens the spare RPCM and extracts it from the CTC.
		Arm1 captures the RPCM P1-1A-A fixture.

(Continued)

TABLE 26.2   RPCM R&R OPERATIONAL SEQUENCE *(CONTINUED)*

Day	Major activity	Detailed activities
		Failed RPCM P1-1A-A is first powered down, before Arm1 unfastens and extracts the RPCM from its site.
		Arm2 inserts and fastens the spare RPCM in the P1-1A-A site, before the new RPCM P1-1A-A is powered up. Arm2 then releases the RPCM.
		Arm1 inserts and fastens the failed RPCM into the CTC.
30 Aug.	RPCM R&R Day 3	Arm1 releases the failed RPCM and stows RMCT1 back in the THA.
		Arm1 grasps, and closes the CTC-2 lid, before releasing it and maneuvering clear.
		Canadarm2 and Dextre are reconfigured for translation.

Fig. 26.11   Dextre extracting failed RPCM from P1-1A-A site.

VII.   ROBOTIC REFUELING MISSION (SEPTEMBER 2011 AND ONGOING)

Since September 2011, several Dextre's activities have been directed towards supporting the RRM. The RRM is a joint NASA–CSA collaboration aimed at demonstrating that robots such as Dextre can successfully repair and refuel existing satellites utilizing specialized tools. The NASA-built RRM module provides Dextre with four new tools and multiple test beds, allowing the robot to cut safety tethers and thermal blankets, remove caps, and install a refueling nozzle (Fig. 26.12).

Shortly after installing the RRM on ELC-4, Dextre released the RRM's tool stowage bay and hose management launch locks. A vision test was then performed using the OTCM camera to gather imagery of various RRM features at different distances and rates for ground algorithm post-processing. The next session of RRM-related operations took place in March 2012. Over a period of three days, Dextre retrieved and checked out three RRM tools to ensure they had remained functional after the launch: the safety cap tool (SCT), the wire cutting tool (WCT), and multi-function tool (MFT). Dextre then released seven additional launch locks that secured four small tool adapters using the MFT, before maneuvering the WCT to cut two thin wires fastening valve caps to the module. One of

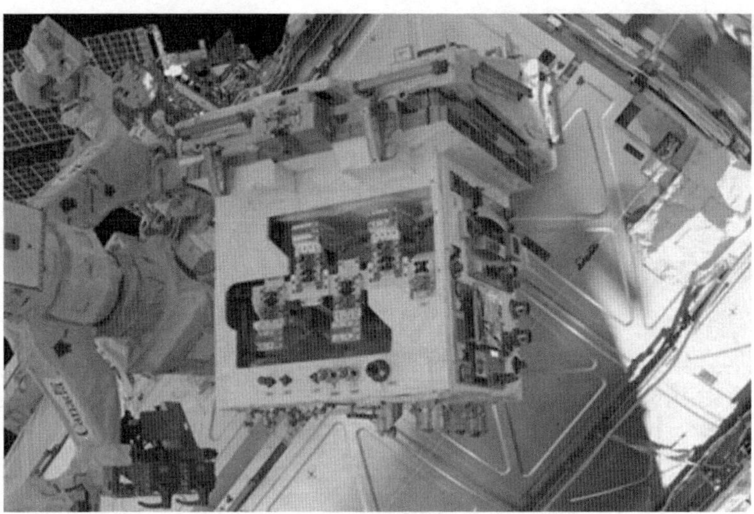

Fig. 26.12   RRM module on Dextre EOTP.

these cuts required the robot to slide a small hook under a wire with a precision of approximately 1 mm. Later, in June 2012, the MFT was used to successfully retrieve and manipulate various RRM adapters, remove caps and valves, and stow away all components once the operations were complete. The next series of RRM operations, completed in January 2013, effectively integrated all previous RRM tasks to simulate the entire refueling process for a satellite. This involved the use of RRM tools by Dextre to cut three wires, remove and stow several caps, connect a robotically compatible nozzle tool to a fill/drain valve, before flowing a simulated fuel through the nozzle.

VIII.   HTV-3 PAYLOAD TRANSFER (AUGUST 2012)

RRM operations had to be scheduled around competing requirements for the MSS to support visiting vehicle operations and ISS maintenance activities. One such activity was the HTV-3 mission, which began with the arrival of the resupply capsule on the ISS on 27 July 2012. As was done for HTV-2, ISS astronauts used the Canadarm2 to capture the vehicle before installing it on the Station. The EP was then extracted from the HTV's external cargo trunk and attached to the JEM–EF. Two new payloads were delivered on the EP, one of which was the Space Communications and Networks (SCAN) testbed. The SCAN payload was integrated onto a FRAM, itself securely fastened to the EP, and was to be installed on ELC3 by Dextre. Because of the relative proximity between the JEM-EF and ELC-3 (see Fig. 26.5), it was possible to transfer the SCAN payload directly from the EP to its final location on ELC-3 without having to relocate the MT or use the EOTP as a temporary stowage area. The same extraction and insertion approaches used for the HTV-2 ORU transfers were applied again for the HTV-3 payload transfer. The experience gathered from HTV-2 led to an efficient grasp of the FRAM's microfixture and smooth extraction from the EP. Nevertheless, achieving proper alignment of the FRAM with the ELC-3 site proved difficult, and more than 3 h was required to complete the insertion.

IX. KEY LESSONS-LEARNED

The transfer of several FRAM-based ORUs and the RPCM replacement clearly demonstrated that complex on-orbit robotic operations could be controlled safely and successfully from the ground. Ground operators were able to perform the required fine-alignment maneuvers and corrections using the available system telemetry and downlinked video views; to recover from various system anomalies, unmodeled dynamics and thermal effects; and to adapt to changing plans, conflicting requirements, and unforeseen circumstances. On-orbit RRM tests have also shown that Dextre's operational concept could be readily applied to tasks involving payloads not designed to be handled robotically.

Table 26.3 gives the time required to secure each FRAM to its mating interface, from an initial hover position, for all FRAM insertions taking place from December 2010 to August 2012. Although improving tools and operational techniques have reduced insertion times to approximately 1 h, these times were found to vary significantly for different mating interfaces. For instance, the August 2011

TABLE 26.3   FRAM INTERFACE ALIGNMENT EFFORT

FRAM insertions	Duration (h:min)
CTC-3 on EOTP side 2 (Dec. 2010)	4:40
CTC-3 on ELC-2 (Dec. 2010) • Without FMA (with FMA)	2:30 (0:50)
FHRC on EOTP side 2 (Feb. 2011)	1:04
FHRC on ELC-4 (April 2011)	2:05
CTC-2 on EOTP side 1 (Aug. 2011) • 1st day (2nd day)	5:05 (2:30)
CTC-2 on ELC-4 (Sept. 2011)	0:58
RRM on ELC-4 (Sept. 2011)	1:05
SCAN on ELC-3 (Aug. 2012)	3:12

FRAM insertion onto EOTP side 1 was surprisingly difficult, requiring a total of 7 h and 35 min over two days of operations. Pre-operation analyses of this case had not identified any issues and Dextre's performance was nominal. Although a root cause was not determined, it was suggested that manufacturing variations of the tight clearance between the fine alignment features might have been a contributing factor. Both of the following FRAM insertions in September 2011 were completed without any issues using the same approach, suggesting that the issue may have been interface-specific. Nevertheless, use of FMA was found to considerably facilitate insertion tasks in all cases.

The ability to continuously improve the on-orbit software played an important role in the success of ground control. Key features required for ground-controlled contact operations were designed and uplinked after Dextre had been launched. In early 2012, Dextre's performance was further enhanced with the ability to automatically transition from a positioning control mode to a force control mode upon detection of applied forces or moments. This removed the need for operators to stop and manually transition from one control mode to the other while approaching the worksite for future operations.

Ground control of the MSS has resulted in considerable savings in ISS crew time. For instance, the RRM module transfer from the Space Shuttle's payload bay to Dextre's EOTP by EVA monopolized four astronauts (two Canadarm2 operators and two spacewalkers) for 1 h and 10 min each. Although ground-controlled MSS operations for similar tasks can take significantly longer, they offer the net advantage of not requiring any astronaut time. They also optimize the use of ISS consumables by reducing the number of required EVAs. This explains why the reliance on ground control has increased tremendously since this capability was first commissioned for the Canadarm2 in 2005. In 2011, the proportion of MSS Stage operations (operations taking place between Space Shuttle visits) controlled from the ground reached 80%.

Use of the EOTP allowed large-scale logistics and maintenance operations to be broken up into smaller phases that could be executed over a longer period of time. For instance, CTC-2 was temporarily stowed on Dextre for more than six months, whereas the FHRC and RRM both remained on the EOTP for approximately one month. This would not have been possible with Dextre's original OTP, because all three payloads required thermal regulation. The segmenting of MSS tasks increased operational flexibility by allowing Dextre to support multiple concurrent objectives and permitted operations to fit within reasonable flight control shifts. The EOTP's capability to support multiple payloads simultaneously was also found to simplify operational sequences significantly, as was the case for the RPCM R&R.

The processes and tools used to plan telerobotic operations must be flexible to account for replanning challenges. The delay of Space Shuttle mission STS-133 prevented the complete transfer of the HTV-2 ORUs to their final destination on ELC-4 and resulted in significant work to redesign Dextre's operations. A single day of ORU transfer necessitated the generation and validation of hundreds of procedure pages and the uplink of a correspondingly large number of commands. This prompted the ROBO team to automate the initial development of flight procedures and command scripts. Such an automated product development tool, using consistent and validated procedure templates, not only increases the efficiency of the development of an operation, but also reduces the likelihood of an error during the generation of long procedures. Efforts are currently under way to further automate both the generation and execution of MSS procedures, with the ultimate goal of having the ground operator simply monitor the execution, ready to take action if required. In addition, the experience gathered from the on-orbit activities discussed here have led to the development of new ground commanding tools, such as a "software hand controller" application, designed to facilitate alignment corrections, and tools aimed at streamlining OTCM and RMCT operations.

Finally, the generation, certification, and delivery of Dextre software parameter updates in the presence of quickly evolving requirements, together with the challenges of attempting to simulate ill-defined dynamic conditions, were also found to be one of the key areas affecting the planning and scheduling of Dextre operations.

X. CONCLUSION

MSS ground-controlled operations were used to successfully complete four logistics and maintenance robotics tasks on the ISS: 1) relocating a CTC from one ISS location to another; 2) offloading spare parts delivered by the HTV-2; 3) replacing a failed RPCM; and 4) installing an external science payload delivered by HTV-3. Through these examples, this chapter demonstrated that ground-controlled robotic systems could successfully perform complex tasks on-orbit, which involved the extensive use of robotics tools, the actuation of several mechanisms, as well as extraction from and insertion into intricate worksites. These representative tasks have also demonstrated the benefits of using ground-controlled telerobotic operations to optimize the use of ISS resources. These benefits will become even more pronounced with the decision to extend ISS operations until 2020 and the corresponding increase in maintenance requirements. Efforts are currently under way to ensure that all required mission planning products will be in place to face this unprecedented challenge.

Approximately 250 ISS ORUs were originally deemed to be robotically compatible. However, several of these do not completely meet all requirements for robotic serviceability. Nevertheless, Dextre will frequently be solicited to attempt their replacement should they fail, regardless of their robotics limitations. For instance, the ISS Program enquired in November 2011 whether Dextre could be used to replace a suspect main bus switching unit (MBSU), despite the presence of thermal blankets obstructing the spare's fixtures. Dextre's ability to interact with components which are themselves not designed to be handled robotically will be an accurate demonstration of telerobotic capabilities for satellite servicing. Because most existing satellites were launched without any consideration for on-orbit servicing, it will be important to demonstrate the robotics capability to capture, maintain, refuel, and release these satellites to extend their useful life. Dextre's early successes show that its technology and operational concept are well suited to satellite servicing and exploration applications beyond low Earth orbit.

REFERENCES

[1]   Lucier, L., and Smith, C., "Cosmic Catch: Canadarm2's First Capture of a Free-Flying Vehicle – Operational Risks, Considerations, and Results," *IAC-10.B6.1.3, 61st International Astronautical Congress*, Prague, Czech Republic, October 2010.

[2]   Dyer, J., and Lucier, L., "Lessons Learned in Robotic Support of the Maintenance and Repair of the International Space Station," IAC-2010-2345, International Space University, *ISS Symposium*, Strasbourg, France, September 2010.

[3]   Aziz, S., "Development and Verification of Ground-Based Tele-Robotics Operations Concept for Dextre," IAC-10.B6.1.1, *61st International Astronautical Congress*, Prague, Czech Republic, October 2010.

CHAPTER 27

Efficacy of the Dawn Vesta Science Plan

Carol A. Polanskey[*]
Jet Propulsion Laboratory, California Institute of Technology, Pasadena, California

Steven P. Joy[†]
University of California, Los Angeles, California

Carol A. Raymond[‡]
Jet Propulsion Laboratory, California Institute of Technology, Pasadena, California

I. INTRODUCTION

The Dawn spacecraft began its Vesta science phase on 3 May 2011 when it acquired its first optical navigation image of asteroid (4) Vesta. The spacecraft was captured into Vesta orbit on 17 July 2011, where it remained in orbit until 5 September 2012 until escaping from Vesta on its way to asteroid (1) Ceres. Dawn is a NASA Discovery Mission managed by the Jet Propulsion Laboratory with the Principal Investigator and Science Operations Center located at the University of California, Los Angeles. The Dawn spacecraft payload consists of two redundant framing cameras (FC1 and FC2) developed and built under the leadership of the Max Planck Institute for Solar System Research, Katlenburg–Lindau, Germany, with significant contributions from the DLR (German Aerospace Center), the Institute of Planetary Research, Berlin, and in coordination with the Institute of Computer and Communication Network Engineering, Braunschweig, a visible and infrared mapping spectrometer (VIR) contributed by the Agenzia Spaziale Italiana in Italy, and a gamma ray and neutron detector (GRaND) developed by the Los Alamos National Laboratory and currently operated by the Planetary Science Institute. Details of these instruments and their operations have been published previously [1–3].

Before the arrival of the spacecraft at Vesta, the Dawn Science Plan was developed to establish an architecture for acquiring the necessary datasets to satisfy the science objectives of the mission. The science observations in each of the three different science orbit altitudes were designed to take advantage of the unique

[*]Dawn Science Operations Manager, Planning and Execution Systems.
[†]Dawn Science Center Manager, Department of Earth and Space Sciences.
[‡]Dawn Deputy Principal Investigator, Solar System Exploration Directorate.

characteristics of that orbit (Fig. 27.1). Additional observations were designed for the approach to and departure from Vesta. Table 27.1 lists the dates and durations of the science phases of the Vesta mission as executed on the flight system. Although all three instruments were operated during each science orbit phase, one instrument was designated as the primary observer, with the spacecraft attitude designed specifically to meet the objectives of that instrument. The highest science orbit, Survey, provided the opportunity to observe multiple rotations of Vesta in a single orbit because the orbit period of 69 h was much greater than the 5.3 h rotation period of Vesta. The VIR instrument was the primary observer during this science phase and was able to take advantage of the orbit characteristics to obtain near-global coverage with a combination of push-broom mapping and use of the scan mirror. After Survey, the spacecraft transferred to the high-altitude mapping orbit (HAMO), which was the optimum science phase for acquiring global topography and color datasets with FC2. The spacecraft ground track of this approximately 12 h orbit provided complete coverage of Vesta in 10 orbits, allowing a self-contained mapping cycle in 5.5 days. The spacecraft passed through the HAMO altitude twice, taking similar datasets at two different Vesta seasons. Six mapping cycles were included in the first HAMO imaging phase (HAMO-1). Two cycles were at a nadir attitude and color filter images were also acquired to clear filter data for the topography analysis. The remaining four cycles were imaged at different off-nadir attitudes to obtain clear imaging

to support both stereo analysis as well as stereophotoclinometry [5]. Following another transfer period, the spacecraft arrived at the low-altitude mapping orbit (LAMO). LAMO was dedicated to acquiring data to determine elemental composition with GRaND and obtain gravity science measurements using coherent Doppler tracking with both the high-gain and low-gain antennas. The spacecraft operated in LAMO for over 4.5 months, orbiting Vesta nearly six times a day with communications to Earth only three times per week. After completing the

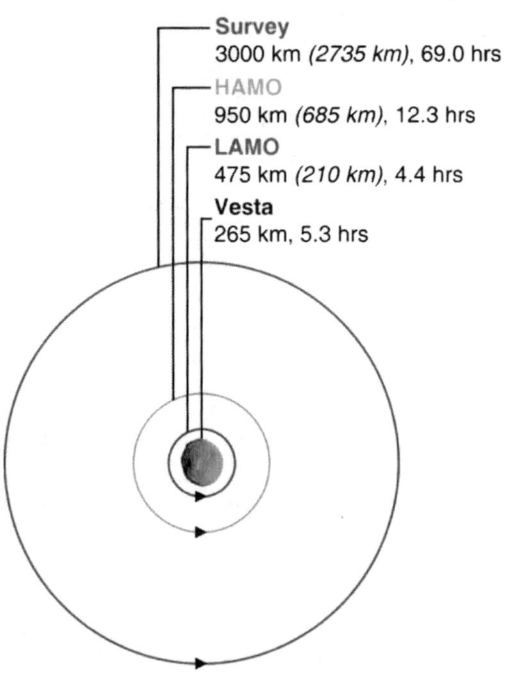

Fig. 27.1   Vesta science orbits [4].

TABLE 27.1 VESTA SCIENCE PHASE DATES

Science phase	Start date	End date	Duration (days)
Approach	3 May 2011	10 Aug. 2011	100
Survey	11 Aug. 2011	31 Aug. 2011	21
HAMO-1	29 Sept. 2011	2 Nov. 2011	34
LAMO	12 Dec. 2011	1 May 2012	141
HAMO-2	15 June 2012	25 July 2012	40
Departure	26 July 2012	5 Sept. 2012	42

LAMO observations, Dawn was transferred back to the HAMO altitude (HAMO-2) to obtain a complementary set of six topography cycles with better illumination in the northern latitudes, including an additional color imaging cycle. Following HAMO-2, the remote sensing instruments acquired a final set of observations of Vesta during departure before starting on a cruise to Ceres.

II. DAWN SCIENCE PLAN CONCEPT

It is the job of the science operations team to provide a high-quality dataset to the science team for their analysis. The observations needed to be planned, and the spacecraft and instrument sequences built, many months in advance of arrival at Vesta. As such, the strategy had to be flexible enough to adjust to the final details of the orbits once they were defined, and robust against the inevitable losses of science data either during acquisition or transmission. Very little was known about the surface of Vesta before arrival, so the observation strategy had to capture the interesting features regardless of what or where they were on the target body. The approach taken was to systematically map the surface of Vesta as the resolution increased with each lower science orbit, relying on global coverage to capture all interesting aspects of the body and minimizing reliance on observations targeted to specific features. The technique that evolved during the development of the Dawn Science Plan is referred to as functional redundancy. Functionally redundant datasets are not truly identical in nature but are designed to provide essentially the same data where either dataset would satisfy a given requirement in a unique way [6]. The specific science planning constraints and uncertainties that led to the adoption of this approach are fully discussed in [6] and included knowledge of when the spacecraft would arrive at Vesta, the fixed length of time that the spacecraft could remain in orbit, Vesta season, Vesta pole orientation and spin rate, and details of the science orbits such as period, inclination, and phasing. The science operations team developed a set of orbit requirements that would enable the observations, but those requirements had relatively large tolerances for error compared to the exact timing required for building spacecraft and instrument sequence products. The methods that were developed to obtain functional redundancy included repetition of particular observations over different longitude ranges, different off-nadir attitude, changing seasons, all with judicious deployment of margin. In general, a complete set of observations was packaged into interval named cycles, and cycles were repeated under varying conditions to acquire the functionally redundant datasets.

To illustrate some aspects of the science planning, examples of FC2 footprints on the surface of Vesta are shown in Fig. 27.2 for each science orbit. The views were generated using the science opportunity analyzer (SOA) planning tool [7], which was used by the science operations team to develop the Dawn Science Plan. Vesta is displayed using the best shape model available at the time that the observations were sequenced. Note that during the Survey orbit planning, no data were available to influence observation design. Instead eleven 1×3 mosaics were taken over a full Vesta rotation to provide global coverage. The mosaics were repeated four times (two from a southern viewing position and two from an equatorial position) for redundancy. In the HAMO example, several orbits of footprints are displayed, showing how the ten-orbit mapping cycle would provide global coverage. The footprints are displayed on the Vesta shape model derived from Survey and Approach data. (Vesta shape models were provided for planning by Robert Gaskell of the Planetary Science Institute.) While features are now visible in the shape model (note the prominent set of three craters informally referred to as "the Snowman"), it is unnecessary to alter the previously determined mapping plan because all surfaces can be imaged. In LAMO there were many more nadir orbits than data volume available to return images, so the camera could only acquire data from about one-third of the orbits. The ten orbits shown comprise most of what could be accomplished by the camera in one week of LAMO. Because there were no requirements to obtain any imaging during LAMO, the plan to collect camera data using a fixed

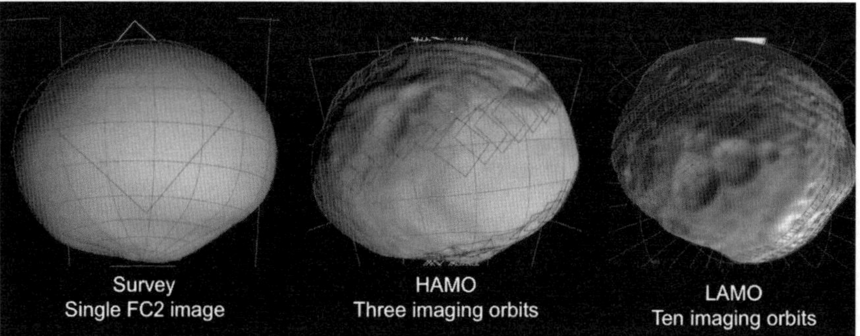

Fig. 27.2   FC2 footprints on Vesta are shown for each science orbit (Vesta is represented by more mature versions of the shape model at each science orbit).

template of orbits between data playback passes was sufficient. The remainder of this chapter will describe some of the as-flown results and explore the effectiveness of the Dawn Science Plan strategy.

III. PERFORMANCE OF THE SCIENCE PLAN

There are multiple ways to assess the efficacy of the Dawn Science Plan. The performance metrics used are specific to the individual datasets acquired. In this section we will report on a subset of the science data that Dawn returned from Vesta, including surface coverage for color imaging and topography, imaging opportunity in LAMO, and the deployment of margin in LAMO to achieve the GRaND objectives. One of the most straightforward measurements to quantify involves a comparison of the planned surface coverage in a given imaging campaign against the achieved coverage. The tool used by the Dawn science operations team to compute surface coverage is CKVIEW. CKVIEW is a tool that was developed at the Institute of Planetary Research at the DLR. Development started in 2000 when planning started for the Cassini Jupiter fly-by. CKVIEW was later adopted for planning and image processing purposes for the Venus Express, Dawn, Rosetta, and MESSENGER missions. SOA was developed for Cassini after CKVIEW and incorporated the ability to output the SOA imaging observation plan in a format that could be directly ingested into CKVIEW for evaluation. CKVIEW was used in conjunction with SOA in the development of the Dawn Science Plan and it was also used during flight operations to monitor data gaps and gores. While CKVIEW can now ingest the Vesta shape model, this model was not available during the planning stages so the results will be presented using the triaxial-ellipsoid model of Vesta. Two different example FC2 datasets from HAMO-1 are described.

The first analysis is the performance of the FC2 color filter image acquisition. In Cycles 1 and 6 of HAMO-1, FC2 acquired global coverage of the illuminated surface in the clear filter plus the seven color filters. Two cycles were planned because it was known that there would be small gores between orbits at the HAMO-1 altitude. The expectation was that the images from Cycle 6 would fill the gores in Cycle 1 coverage. The planned coverage for the FC2 clear filter (F1) for Cycle 1 is shown in Fig. 27.3. Vesta is displayed with a sinusoidal projection, with the grey area representing surface that was covered by one or more images with incidence angle less than 90°. The northern hemisphere was in shadow above a latitude of 50° north and could not be observed. The gores between ground tracks are black. During planning it was estimated that Cycle 1 and Cycle 6 would each provide coverage of 85% of the surface of Vesta for each filter.

Figure 27.4 shows a sample of the FC2 color coverage as-flown results for filters 2, 3, and 4. The data losses display a similar pattern for all filters in a given cycle. This was not necessarily an expected result, but in the case of this dataset, the data losses were related to anomalies that impacted more than one image at a time. The gores between orbit ground tracks are clearly visible in all

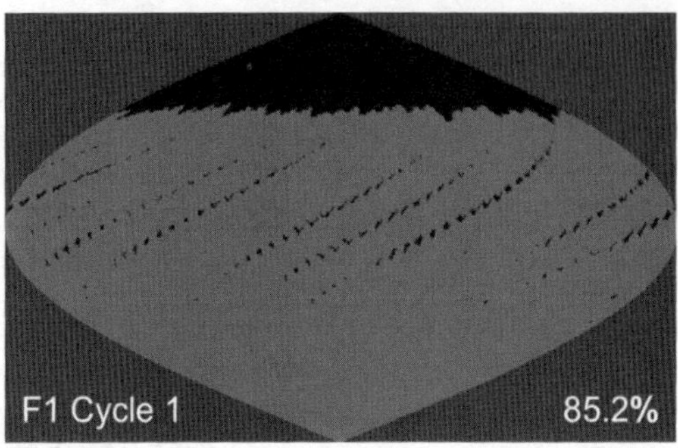

Fig. 27.3 FC2 planned imaging coverage for HAMO-1, Cycle 1, clear filter.

filters for the individual Cycle 1 and 6 coverage maps, but are significantly reduced in the combined coverage map. It was anticipated that the gores between orbits would disappear entirely, but deviations in the actual trajectory from what was planned resulted in minor residual gores. However, the prominent gores stemming from various unexpected data losses were almost completely eliminated in the combined dataset. In general, the FC2 color coverage imaging campaign in HAMO-1 was very successful, covering 87% of the total surface of Vesta in at least three filters and exceeding the goals for this dataset. An additional cycle of color imaging was acquired in HAMO-2. Although the primary goal of HAMO-2 was to obtain coverage in the newly illuminated northern latitudes, the planned color coverage shown in Fig. 27.5 also extended to southern latitudes and provided yet another opportunity to fill gores from HAMO-1. In this view, the

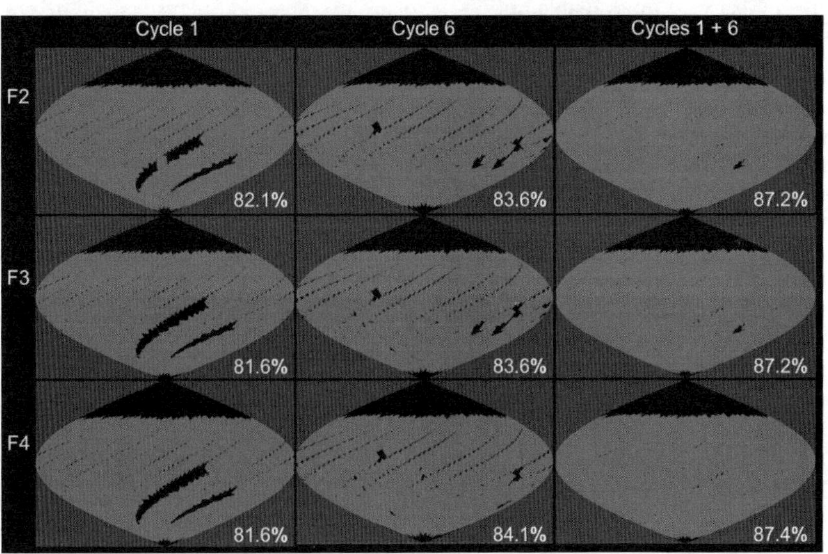

Fig. 27.4 FC2 actual coverage for HAMO-1 three color filters in Cycles 1 and 6.

footprints for the planned color imaging are white and the clear filter imaging is grey.

Although a preliminary shape model had already been derived from Survey data [8], a key objective of HAMO-1 was acquiring the optical topography dataset. This dataset consisted of the FC2 clear nadir images in Cycles 1 and 6, as well as clear off-nadir images in Cycles 2 through 5. Each topography cycle placed the spacecraft at a unique off-nadir attitude. This allowed the commanding for each orbit in the cycle to follow a similar template. Two off-nadir attitudes combined with one nadir view were required for stereo processing, so the additional two off-nadir cycles provided a functionally redundant dataset for margin. Images from any two of the four off-nadir attitudes could be used at each point on the surface to derive the topography, which allowed flexibility to accommodate minor data losses. Figure 27.6 shows the results of the clear imaging campaign for HAMO-1. The clear filter coverage in Cycles 1 and 6 display similar results to those described previously for the color filters. Aside from Cycle 2, which lost one full orbit of imaging, the off-nadir imaging cycles were essentially complete. The width of the gores between individual orbits varied with the off-nadir angle selected, but were generally reduced compared to the nadir cycles. No attempt was made to fill these gores because it was antici-pated that different off-nadir images could be used in those areas.

The topography campaign was repeated again in HAMO-2 with slight vari-ations in the off-nadir angle selections. The primary improvement in the

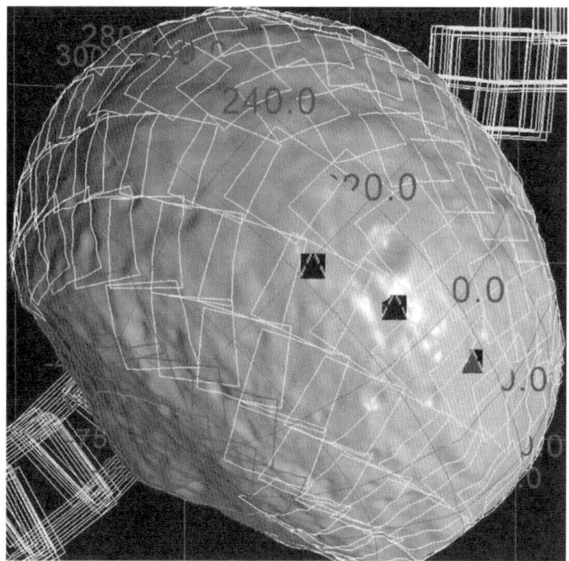

Fig. 27.5 FC2 planned color imaging for HAMO-2.

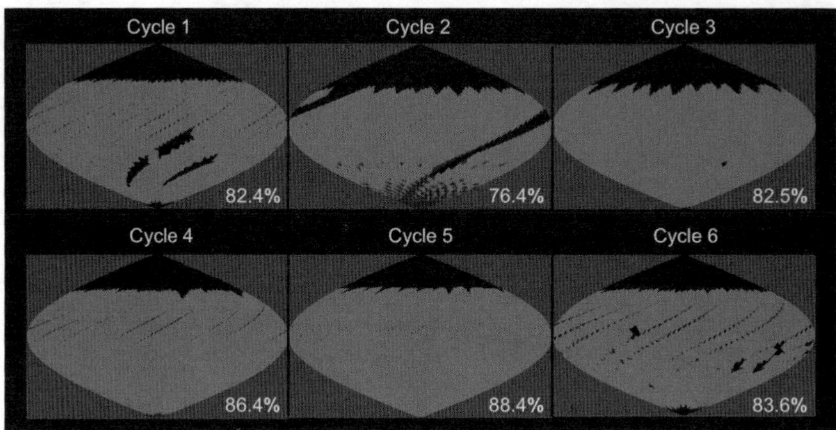

Fig. 27.6 Actual FC2 coverage for HAMO-1 clear imaging in all six cycles.

topography model obtained from HAMO-2 was increased visibility into the northern hemisphere because the subsolar latitude had moved from approximately 18° south of Vesta's equator at the end of HAMO-1 to only 3° south at the end of HAMO-2. Therefore, not only did HAMO-2 provide additional nadir and off-nadir opportunities to fill in data gaps from HAMO-1, it was also conducted with significantly better overall illumination conditions for Vesta. The improved illumination resulted in much better surface coverage for roughly the same number of images per orbit. Figure 27.7 shows the FC2 clear images acquired in Cycles 1 through 6 of HAMO-2. For these plots the incidence angle criteria were relaxed to 100° to show coverage that was intentionally acquired past the predicted dark-to-lit terminator crossing in the northern hemisphere. Early data return in HAMO-2 had indicated that there were some areas of higher elevation beyond the terminator that were illuminated beyond what had been predicted from a smooth triaxial ellipsoid model, so images were acquired earlier than the dark-to-lit terminator crossings in Cycles 5 and 6.

In addition to better illumination, HAMO-2 benefited from the lessons learned during the first HAMO orbit. During nadir cycles in HAMO-1, the spacecraft axis along which the camera is mounted was pointed towards the center of Vesta; however, this vector was sufficiently different from the surface normal vector that the images were not truly nadir. To mitigate this effect in HAMO-2, the spacecraft was pointed slightly off-nadir in the northern and southern regions in Cycle 1 to better align the viewing angle with the surface normal. This technique also reduced the gores between orbit swaths in this cycle, as is evident in Fig. 27.7. Additional improvements in the management of the images within the camera and the spacecraft data storage subsystem improved the ease of operations in HAMO-2. Fewer images were lost during HAMO-2 execution than for HAMO-1, with short data interruptions at the deep space network being the only source of image loss.

Figure 27.8 shows the six HAMO-1 cycles plotted in CKVIEW, where the colors indicate how many views are available for each part of the surface that meet the criteria specified for stereo imaging of usable quality. In particular, the incidence angle for each image must fall between 5° and 75°, and the stereo angle between images must be between 15° and 65°. A minimum of three views (one nadir and two off-nadir) is required. Any area that is green, blue, or purple meets the stereo criteria. The only portions of the surface that do not meet the criteria are in the far northern latitudes where illumination is poor, or at the south pole where the incidence angle criteria cannot be met even though those regions were imaged repeatedly in each cycle. For HAMO-1, the usable stereo coverage is 76% of the body, with resolution between 60 and 70 m/pixel. The topography observation strategy proved robust to the incidental data losses. Even the worst case of the missing orbit in Cycle 2 did not result in any areas with less than three qualifying images.

The results for HAMO-2 were even more successful. Figure 27.9 shows the usable stereo coverage for HAMO-2, with 92% of Vesta covered with at least three qualifying views. Note that these projections use a triaxial ellipsoid model rather than the actual Vesta shape model and give more optimistic results because this model does not account for regions where the topography creates shadows that do not meet the incidence angle criteria. The stereo coverage extends farther to the north than in HAMO-1; however, there was still a small area of Vesta left in shadow at the end of HAMO-2. A final set of observations was taken near the end of August 2012 during departure, at lower resolution.

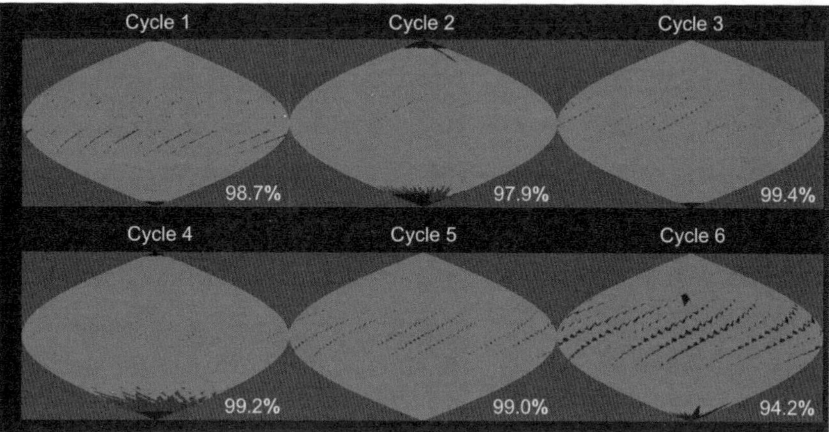

Fig. 27.7 Actual FC2 coverage for HAMO-2 clear imaging in all six cycles.

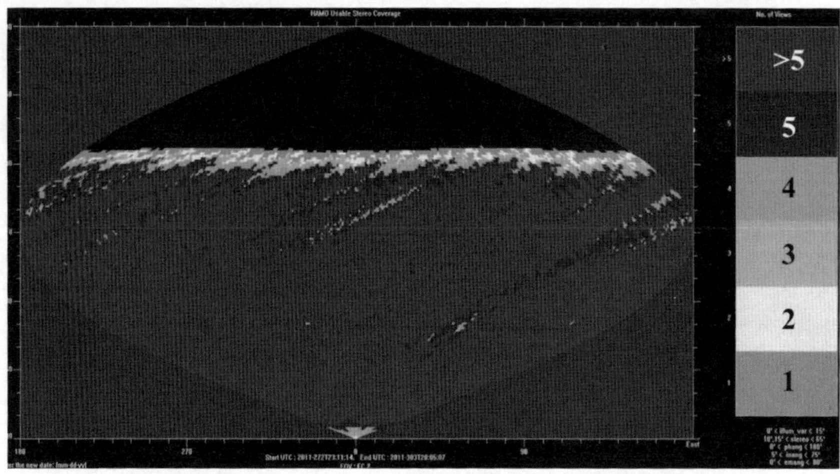

Fig. 27.8   Actual FC2 usable stereo coverage for topography analysis during HAMO-1 (incidence angle cutoff is <75°; colors display the number of available stereo views).

These images filled in most of the remaining northern terrain, although there were still some small areas at the north pole that have remained unobserved.

The stereophotoclinometry technique can take advantage of the combined HAMO-1 and HAMO-2 datasets, as shown in Fig. 27.10. The combined coverage of 94% for three views is not a significant improvement over HAMO-2 coverage

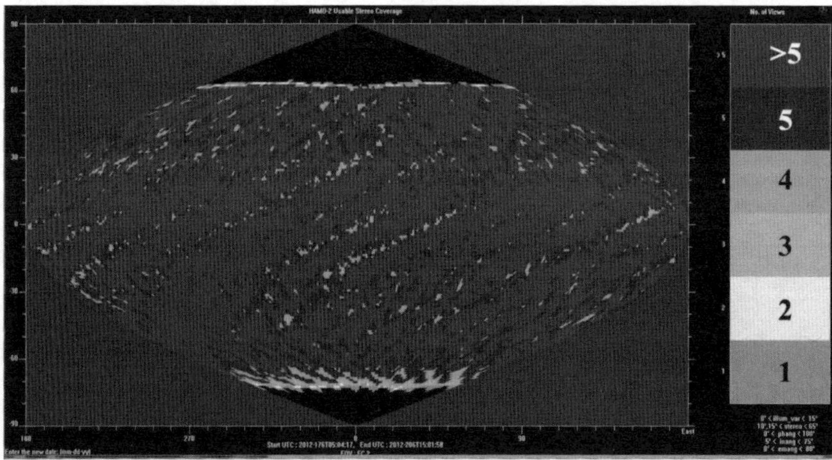

Fig. 27.9   Actual FC2 usable stereo coverage for topography analysis during HAMO-2 (as in HAMO-1, the incidence angle cutoff is <75°).

alone, but the region covered by more than five views increased from 67% for HAMO-2 to 85% for the combined dataset.

Accumulation of gamma ray and neutron data in the LAMO science orbit required a different strategy than used in the imaging campaigns. The GRaND instrument required a nadir attitude over a long period of time to collect enough data to detect the major rock-forming elements. The baseline operational requirement for the GRaND dataset was 70 days of observations at the LAMO altitude with the spacecraft within $5°$ of nadir for 70% of the time. The remaining 30% of the time was spent at an attitude where the high-gain antenna could communicate with the Earth so that gravity science could be obtained. This period was also used to return the GRaND science data accumulated while pointing nadir. The other two science orbits preferred a roughly 50/50 split between time at nadir and time returning data to Earth. The goal of the science plan was to put as much available time into the LAMO science orbit phase as possible to improve the GRaND results; however, the project also required that the science plan carry margin against unexpected anomalies that could impact any science orbit phase. The compromise was to carry 40 days of operations margin that could be used at any time to lengthen a science orbit phase to recover from a severe anomaly that was not already covered by functional redundancy. This time would not be sequenced into the plan unless needed. Whatever margin was remaining at the start of LAMO would be used to delay the end of LAMO. With a nominal duration of 70 days there was sufficient planning time available to add the extra time to the end without a negative impact to the flight team schedules. Although there were three safe-mode entries between the start of Approach and the start of LAMO, they all occurred during periods when the

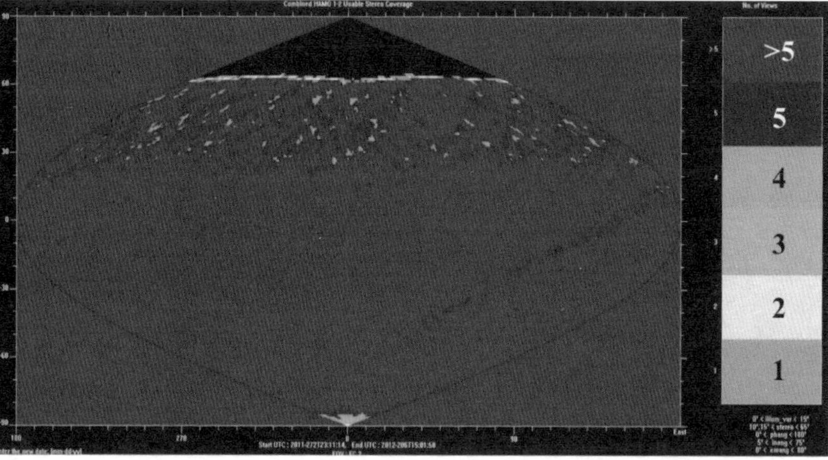

Fig. 27.10   Combined topography imaging coverage using data from HAMO-1 and HAMO-2.

flight team was able to recover nominal operations without any loss of time to a science orbit phase. The full 40 days of margin were therefore added to the end of LAMO. Subsequently, the navigation team revised the time necessary to arrive at Ceres from Vesta departure, allowing an additional 30 days to be added to LAMO. During LAMO there were two spacecraft safe-mode entries at a cost of 15 days of GRaND data acquisition. In the end, the LAMO duration was 80% longer than the baseline plan, resulting in a significant improvement to the GRaND dataset.

The extra time spent in LAMO orbit provided a unique opportunity to deviate from the template-driven observation strategy to plan some rare FC2 targeted observations. However, the targets were image gaps, not Vesta features. This additional flexibility had been anticipated for periods near the end of the Vesta mission when the orbit behavior had been fully characterized. For 16 weeks, FC2 had acquired data on orbits dictated by the standard template. However, when the final four-week mission extension was added, the available orbits for acquiring data were evaluated against the remaining gaps in coverage. While the trajectory would not provide opportunities to fill all remaining gores, careful selection of the remaining imaging orbits significantly improved overall coverage. Using a conservative selection criteria of images with less than 708 incidence, the CKVIEW plot in Fig. 27.11 shows 66% coverage of the surface at a LAMO resolution of less than 30 m/pixel (plot and shape model coverage analysis provided by Thomas Roatsch, DLR, personal communication, 28 April 2012). The actual image gores are visible as thin stripes. Note that, in Fig. 27.11, the most recent Vesta shape model is used to display the coverage. Resolution is indicated with color for the two ranges of 10–20 m/pixel and 20–30 m/pixel. The other black areas in the illuminated region have been imaged, but are on slopes that do not meet the incidence angle criteria. This nearly global coverage is a

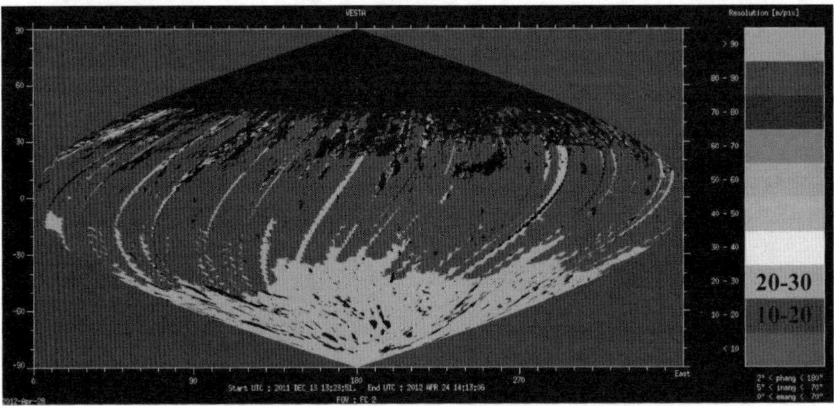

Fig. 27.11   Actual FC2 coverage obtained during the LAMO phase.

spectacular result for "bonus" imaging that was not required for this science phase, and will be used to improve Vesta age determination and mapping fidelity.

There was some flexibility built into the science plan architecture that was not exercised to the extent expected. Starting a year before arrival, the Vesta command sequences were all generated relative to geometric epochs related to Vesta. The most commonly used epoch was the time of the dark to lit terminator crossing for each orbit, because the remote sensing instruments would typically observe only during the dayside portion of the orbit. All instrument activities were commanded relative to that single epoch. The Vesta arrival date had been uncertain by several weeks, so command sequences built a year earlier were shifted to align with the final reference trajectory. Transfer times between science orbits were also unknown until shortly before each transfer started, so the sequence epochs were updated once at the start of each science phase. A mechanical process would update the epochs without requiring changes to the delivered instrument sequences. This process was executed without incident, and all observations retained their required attributes as expected. However, there was also an expectation that the orbits would drift within a science phase or that the orbit periods would be substantially different from the original specifications, requiring timing updates between observation cycles. For Survey and HAMO-1, the orbit delivery was extremely precise. The Survey orbit requirement was a 3000 km radius with a range between 2800 km and 3200 km, and the orbit-phasing requirement was ± 45 min change in the dark-to-lit terminator crossing time. The as-flown orbit results are reported in [9]. The Survey orbit radius was 2997 km and generally varied by less than 4 km, with the phasing error in the dark to lit terminator crossing never exceeding 9 min. The HAMO-1 experience was similar. Following the epoch update to the initial sequence delivery, the deviation of the dark to lit terminator crossing was found to be significantly less than the 10 min tolerance, and the orbit radius variation between 934 km and 960 km was well within the allowed range of 925–975 km. There was a change to the LAMO planned orbit radius, but it was identified well in advance. Before the delivery of the first set of LAMO sequences, the mean orbit radius was shifted higher by 10 km, from 465 km to 475 km, to improve orbit stability. This change was acceptable to the GRaND instrument team, so no further sequence adjustments were required. Although a sequence epoch update process was put in place to accommodate the expectation of mid-phase adjustments, the navigation performance exceeded expectations and the sequences were acceptable without additional intervention. The flight team was relieved of the work involved in updating the sequences, which simplified the overall uplink process.

IV. CONCLUSION

The Dawn mission has revealed Vesta and all her mysteries to the world. Despite the challenges that are inevitable during the height of scientific discovery with an interplanetary spacecraft at an unknown body, a wealth of data has been returned to Earth, achieving the goals that were set many years before launch. The principal of functional redundancy in science planning has proven to be effective and efficient for developing an observation strategy that is robust to the many varied sources of data losses experienced during a year of flight operations. Although real-time commanding was often required to respond to instrument and spacecraft anomalies, real-time intervention was not needed to change the observation plans when data were lost, because no given image or spectra was uniquely critical data. In the cases where the data return was complete, as expected, the variations in acquisition conditions allowed all data to contribute to the scientific analysis. At completion of the Dawn mission at Vesta, all the datasets designed to meet the scientific objectives and requirements of the mission were acquired at a level of quality and quantity that exceeded the planning expectations. Despite the many successes, there are still challenges ahead for the Dawn mission. After departure from Vesta, planning begins for the Ceres mission, where the same principals will be exercised in developing the Dawn Ceres Science Plan. Many of the details of the Ceres mission will differ from Vesta, but the goals of redundancy and adaptability will be applied to the global mapping campaigns of the three Dawn instruments at their next asteroid rendezvous.

ACKNOWLEDGMENTS

The success of this endeavor depended greatly on the support of the Dawn Flight Operations team and their ability to respond to the many surprises and challenges during Vesta operations. Specific recognition is given to Thomas Roatsch, Klaus-Dieter Matz, and Ralf Jaumann at DLR for development of the CKVIEW tool, which was critical to evaluation of the observation plans as well as the as-flown performance evaluation. The authors also thank the instrument operations leads—Pablo Gutierrez-Marques (FC, MPS), Tom Prettyman (GRaND, PSI), and Sergio Fonte (VIR, IFSI)—and their teams for their support of operations and flexibility in responding to the challenges of the science phase of the mission. A portion of this work was performed at the Jet Propulsion Laboratory, California Institute of Technology, under a contract with NASA.

REFERENCES

[1] DeSanctis, M. C., Coradini, A., Ammannito, E., Filacchione, G., Capria, M. T., Fonte, S., Magni, G., Barbis, A., Bini, A., Dami, M., Ficai-Velotroni, I., and Preti, G., and the VIR Team, "The VIR Spectrometer," *Space Science Reviews*, Vol. 163, 2011, pp. 329–369.

[2] Prettyman, T. H., Feldman, W. C., McSween, H. Y., Jr., Dingler, R. D., Enemark, D. C., Patrick, D. E., Storms, S. A., Hendricks, J. S., Morgenthaler, J. P., Pitman, K. M., and Reedy, R. C., "Dawn's Gamma Ray and Neutron Detector," *Space Science Reviews*, Vol. 163, 2011, pp. 371–459.

[3]   Sierks, H., Keller, H. U., Jaumann, R., Michalik, H., Bubenhagen, F., Büttner, I., Carsenty, U., Christensen, U., Enge, R., Fiethe, F., Gutiérrez Marqués, P., Hartwig, H., Krüger, H., Kühne, W., Maue, T., Mottola, S., Nathues, A., Reiche, K.-U., Richards, M. L., Roatsch, T., Schröder, S. E., Szemerey, I., and Tschentscher, M., "The Dawn Framing Camera," *Space Science Reviews*, Vol. 163, 2010, pp. 263–327.

[4]   Russell, C. T., and Raymond, C. A., "The Dawn Mission to Vesta and Ceres," *Space Science Reviews*, Vol. 163, 2011, pp. 3–23.

[5]   Raymond, C. A., Jaumann, R., Nathues, A., Sierks, H., Roatsch, T., Preusker, F., Scholten, F., Gaskell, R. W., Jorda, L., Keller, H.-U., Zuber, M. T., Smith, D. E., Mastrodemos, N., and Mottola, S., "The Dawn Topography Investigation," *Space Science Reviews*, Vol. 163, 2011, pp. 487–510.

[6]   Polanskey, C. A., Joy, S. P., and Raymond, C. A., "Dawn Science Planning, Operations and Archiving," *Space Science Reviews*, Vol. 163, 2011, pp. 511–543.

[7]   Streiffert, B. A., and Polanskey, C. A., "Science Opportunity Analyzer: Not Just Another Pretty Face," *AIAA-2002-12342, SpaceOps 2004*, Houston, Texas, October 2002.

[8]   Jaumann, R., Williams, D., Buczkowski, D., Yingst, R., Preusker, F., Hiesinger, H., Schmedemann, N., Kneissl, T., Vincent, J., Blewett, B., Buratti, B., and Carsenty, U., "Vesta's Shape and Morphology," *Science*, Vol. 336, 2012, pp. 687–690.

[9]   Rayman, M., and Mase, R. A., "Dawn's Exploration of Vesta," *IAC-12-A3.4.3, 63rd International Astronautical Congress*, Naples, Italy, October 2012.

Simplify ISS Flight Control Communications and Log Keeping via Social Tools and Techniques

Hugh S. Cowart[*] and David W. Scott[†]
NASA—Marshall Space Flight Center, Huntsville, Alabama

Daniel J. Stevens[‡]
Barrios Technology, NASA—Johnson Space Center, Houston, Texas

I. INTRODUCTION

Space flight control requires a river of observations, decisions, and actions involving multiple information streams, which may be parallel, divergent, or redundant. The heart of the control operation requires a) communicating effectively in real time with other controllers in the room and/or in remote locations and b) tracking significant events, decisions, and rationales to support the next set of decisions, provide a thorough shift handover, and troubleshoot/improve operations. For real-time communications, International Space Station (ISS) flight controllers speak with each other via multiple voice circuits or "loops," each with a particular purpose and constituency. Controllers monitor and/or respond to several loops concurrently. The primary tracking tools are console logs, traditionally kept by a single operator and not visible to others in real time. Information from telemetry, commanding, and planning systems also plays into decision making. E-mail is very secondary/tertiary because of timing and archival considerations.

As noted above, voice circuits are a primary tool for real-time communication. A flight controller's "juggling capacity" can be saturated by excessive voice traffic and/or demands from other systems, especially during high-tempo operations when saturation is least affordable.

Console logs have been used since the beginning of the Space Program to document significant observations, conversations, and actions of flight controller console positions. From the 1960s through the 1990s, pen and paper were used. Voluminous physical storage was required and document search was mentally exhausting. During that era, terminals for mainframe/microcomputer-based

---

[*]Data Management Coordinator, Mission Operations Laboratory, MSFC/EO30; hugh.s.cowart@nasa.gov.
[†]Computer Engineer, Ground Systems Development & Integration, MSFC/EO50; scotty@nasa.gov.
[‡]ISS Flight Controller, Mission Operations Directorate, JSC/DI2

flight control and planning systems took up the real estate that was not required for communications gear. Application switching on the terminals was awkward and slow. Owing to reliability requirements for the control and planning systems, the evolution of workstation and PC servers/clients took quite a while. Figure 28.1 shows a typical Spacelab-era console log.

When desktop computers with graphical user interfaces (GUIs) became commonplace in control rooms at the beginning of ISS operations, log-keeping based on office automation applications evolved rather quickly. Word processor or spreadsheet programs were typically used and had several advantages over handwritten logs, in that they were more legible, easier to search, and cut-and-paste from other sources could be applied to save typing time. However, automation also brought the new challenges:

1. Search was serial, and all entries related to a particular topic could not be gathered easily.

2. It was hard to search across files from different console positions or time intervals.

3. Cut-and-paste could produce a glut of information.

Whereas Shuttle-based missions could last for a few weeks, the ISS is permanently manned. From 2000 to the present, log-keeping approaches evolved to handle years of data. Mission Control Center-Houston (MCC-H) uses a Microsoft Word template and macros so that log entries can be harvested into a centralized database archive that can then be searched. Payload Operations and Integration Center (POIC) has been using Filemaker Pro to maintain position-specific logs with flexible capabilities for search, selection, one-click setting of status flags, and so on. Figure 28.2 shows a recent excerpt from a Filemaker Pro-based log at POIC. Examples from MCC-H's current approach appear in Sec. IV.

A time-honored principle in operations is "better and best are the enemy of good enough." While this has merit, the last few years have seen massive growth in communications and planning activity due to the increase in crew size and the transition to post-assembly utilization. Yesterday's "good enough" is today's "hanging in there" and could well be tomorrow's "coming up short." Because a gram of prevention is worth a kilogram of cure, the authors

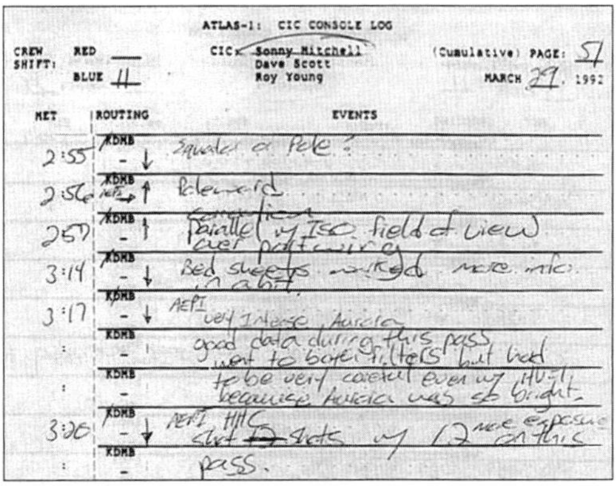

Fig. 28.1 Spacelab console log, 1992.

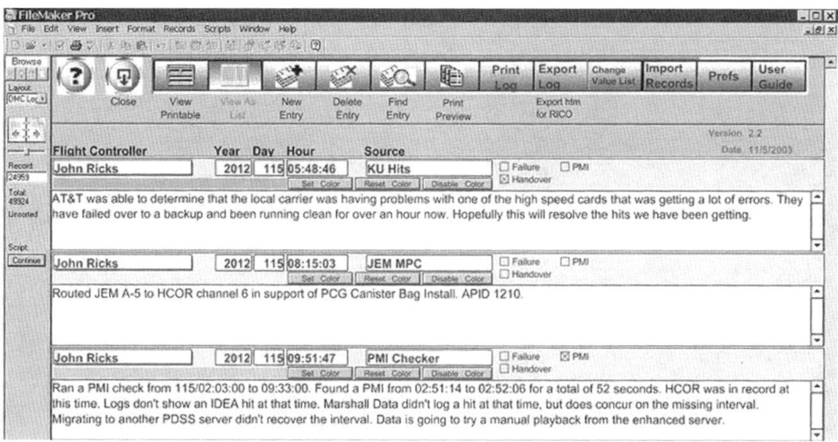

Fig. 28.2 POIC Filemaker Pro-based log, 2012.

offer three log-keeping systems (one recently deployed and two seed concepts) that should be good enough to grow with operations for quite some time:

1. A Web-based Console Log Tool (CoLT) that a) with a log owner's permission supports real-time access and editing across operator/discipline logs by on-duty and off-duty personnel whether on-site or off-site, and b) could links to other processes.

2. A scheme that uses commonly accepted microblog syntax (e.g., @targetid, #hashtag) to enable text conversation among the logs themselves (only portions of the log that the originating console operator wishes to share would be viewable).

3. A communications dashboard that integrates a console log, focused text chat channels for certain types of conversations, status updates of general and discipline-specific interest, and some aids for handling short discussions and vectoring longer ones.

The following sections assume that the reader is familiar with ISS Operations Protocol, that is, standard phrasing conventions for verbal and written communications in the real-time flight control environment. Translations of acronyms referring to flight controller positions, systems, payloads, and statuses are generally not given, as they are not central to describing how the concepts work.

II. HOW SOCIAL MEDIA METHODS COULD EMPOWER LOG-KEEPING SYSTEMS

When people think of social media, recreational friendship typically comes to mind, but the term "social" refers just as readily to "interactions among organisms and their systems." In that context, social interactions merit just as much attention as we pay to interactions among our electronic systems. Earlier papers have explored the functional characteristics of social media that can benefit operations [1–3]. Figure 28.3, which was adapted from presentation charts, reviews some principles and behaviors that relate particularly well to this chapter.

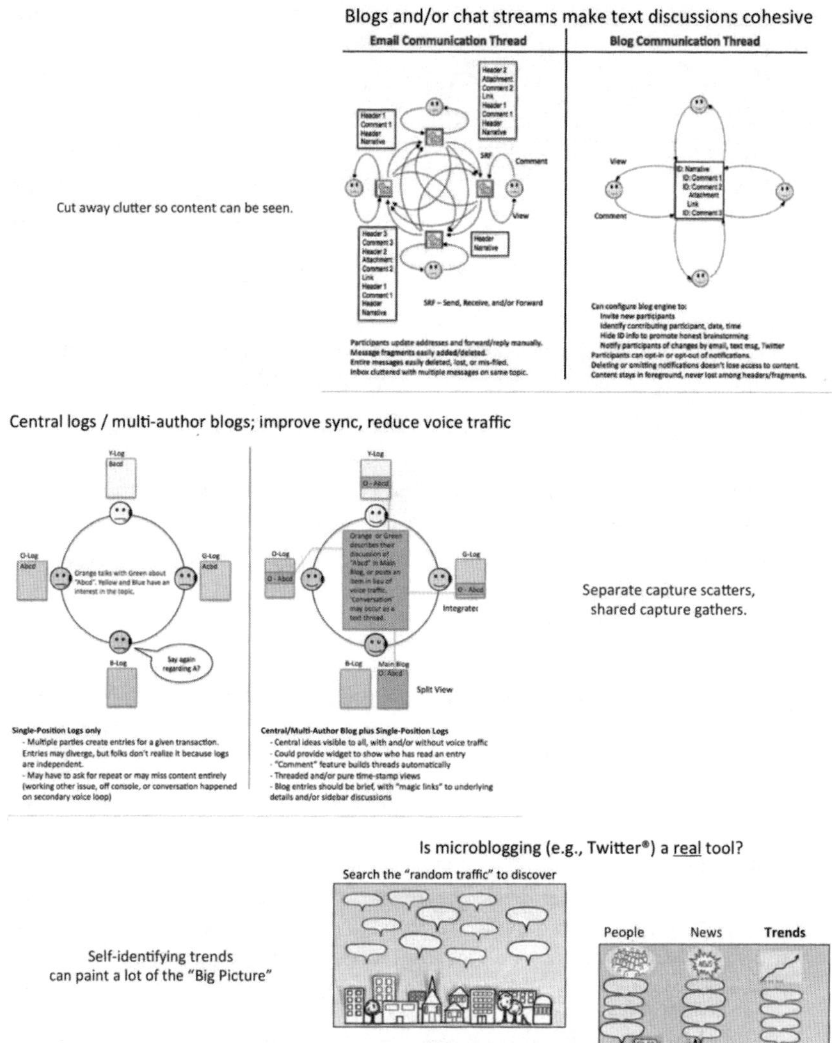

Fig. 28.3   **Some social media functional characteristics and implications for log keeping [3].**

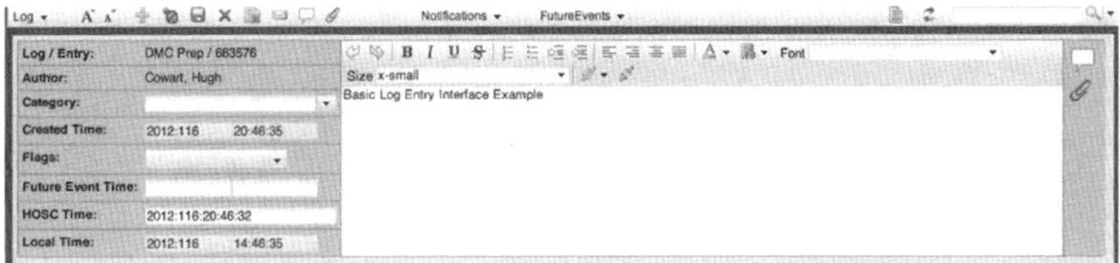

Fig. 28.4   CoLT log entry in "Edit" mode.

III.   CONSOLE LOG TOOL (CoLT)

A.   CoLT INTRODUCTION

The POIC software developers and operations cadre recently deployed a new Web-based console log utility to replace the legacy Filemaker Pro-based system. First and foremost, CoLT provides a streamlined interface for producing flight operations log entries. In addition to this fundamental capability, it provides functions that directly support bidirectional/multipath flow of operational information between a real-time operator and "back room" or other support. As a result, CoLT has the potential to morph "The Log" into a central communications hub that produces higher-fidelity logs of real-time support activity, yet requires less log-keeping effort by the real-time operator. The following sections describe major CoLT features and how they enhance communications.

B.   ACCESSIBILITY

A primary issue with legacy POIC logs was limited access. Because of firewall issues and the use of local clients, direct viewing of logs was restricted to locations inside the POIC. Information logged during the day often has a very short shelf-life. With daily export going out after normal business hours, back office support could experience up to 24 h delays in information receipt unless the console operator makes additional/duplicate efforts to send out information via e-mail or phone. CoLT is far more accessible. In basic terms, anyone with the proper credentials and a virtual private network connection can view console logs in real time. In addition to providing real-time access to logs on a team-by-team basis, CoLT permits viewing multiple logs concurrently in separate windows/tabs or in an interleaved presentation. This is very advantageous for grasping a "big picture" view of operational progress or post-ops analysis.

What about remote editing of the log? Although the idea is somewhat controversial in terms of traditional operations concepts, it could, if well managed, significantly streamline front-room communications processes (Fig. 28.4). The tool has the following capabilities:

1.   Remote and local creation, editing, and/or commenting to logs;

2.   Control on a per log basis, administered by assigned log owner(s).

In the future, console applications could be enabled to autogenerate log entries based on telemetry or user-defined triggers, for example, commanding, with or without a confirmation dialog. Autogenerated entries could be embedded in the operator's log so that "it's all one story" or placed in a separate log for interleaving or hiding with respect to the operator's log as needed.

C. FLAGS

In CoLT, each log has a "Flags" field with a customizable pull-down value list (Fig. 28.5). Typical flags are Failure, Anomaly, Handover Item, and so on (Fig. 28.6). When editing a log entry, the user can set multiple flags as appropriate. A user with Group administrative privilege can, in addition to modifying flag names, associate automatic actions with a given flag's status, such as sending a notification e-mail of the entry to a flag-specific distribution list when the flag is set. Flag status also serves as a filtering criterion for CoLT's search engine.

Future versions of CoLT might allow other actions, for example generating a linked action item to be dispositioned by back-office processes. Consider how this might simplify anomaly processing. In the legacy log-keeping system, an operator determines an event is an anomaly, notes it as such in the log, and hopes the back room sees it when reviewing the daily log export. If back-room processing needs to begin before the daily log export, the operator has to open an e-mail, copy-and-paste the log entry, determine who it goes to (which may require several phone calls), then address and send it. In CoLT, when the Anomaly flag is checked, the operator would only have to review the autogenerated e-mail/text message that will notify the back room and press OK. The idea is to reduce the amount of coordination and actions needed to pass the ball to someone off-console who can run with it.

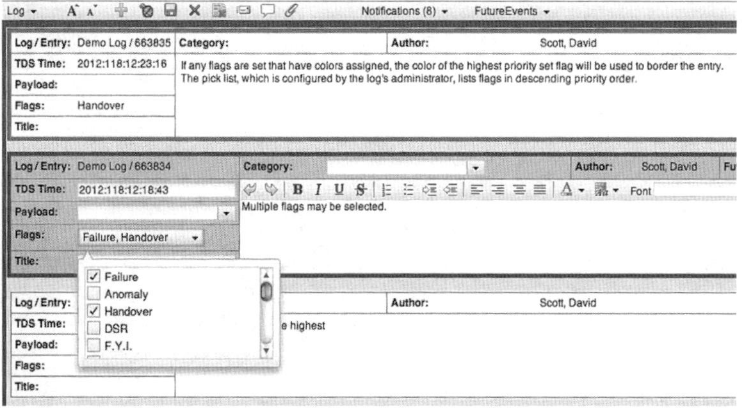

Fig. 28.5   Flags: behavior during log keeping.

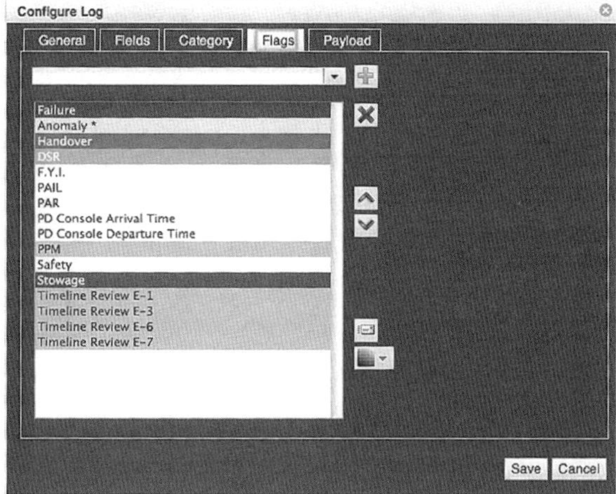

Fig. 28.6   Defining flags.

D.   FUTURE EVENTS

Many log entries pertain to a future event. This information would be useful to whoever is on duty when the event actually occurs. In traditional logs, this information often becomes buried as time passes and new log entries are created. One could carry the information in a shift handover log, but this is unwieldy if the future event is a long way out. CoLT allows an author to include a future event date/time in a given entry. When that date/time arrives, CoLT notifies anyone viewing that log (which, at a minimum, consists of the on-console operator) and provides a link to retrieve the given entry for reference.

Of course, discussions about a significant upcoming real-time activity also happen in back rooms (and halls). Useful information from these discussions can now be recorded as a log entry, authored by the back room, with an assigned future event date/time. Additional conversations related to the same activity will logically lead to new entries and/or updates to existing ones, so that the log now carries the activity as a "pre-loaded" event. Although entries pertaining to the activity may originate at different times, they can be tied to a common future event time that is early enough for the pre-loaded information to be reviewed and acted on before the actual activity.

This scheme is a little bit like having an alarm clock that wakes you up *and* gives you a copy of written "to do" lists. This improves communications between real-time operations and back-room support in both directions. The log becomes more than a serial set of text entries regarding operations events and evolves towards an integration tool. After all, if we are not integrating, we are probably disintegrating!

E.   QUALITY RECORDS

POIC is required to maintain an uneditable archive of console logs for record and/or investigation purposes. These fulfill the role of "Quality Records" (QR) in the International Organization for Standardization's (ISO 9000) quality management system. Traditional and current practice is for each console position to e-mail a Hypertext Markup Language (HTML) file of each day's console log entries to an archivist. CoLT supports manual and automatic QR generation, with options to minimize the human effort needed.

F.   RICH TEXT FORMATTING (RTF)

A common challenge in the legacy tool was unfriendly formatting. The most popular layouts only showed five lines of text, and using scroll bars became old after a while. Alternative layouts could show the entire content of large entries, but then displayed vast amounts of blank space for small entries. Text formatting options were limited and did not include common features such as bullet lists, numbered lists, and hyperlinks. CoLT provides a robust Rich Text Formatting (RTF) suite and auto-adjusts the sizes of individual entries as needed to display all the content. This makes it much easier to view and edit detailed entries.

G.   ATTACHMENTS

Log entries are often associated with documents or other forms of communication. CoLT allows information to be attached to an entry just as attachments are added to an e-mail. The tool should evolve to include seamless integration with document repositories and/or other mission-related systems.

H.   COMMENTS

During CoLT requirements development, the concept of external participation in the log spawned a requirement for a non-intrusive means of interacting with an existing entry. The result was an integrated commenting system that works very similarly to blog sites that let users comment to an article (Fig. 28.7).

Commenting allows someone to provide information or request clarification regarding a specific log entry without changing its main content. During pre-deployment CoLT testing, comment discussions often brought out

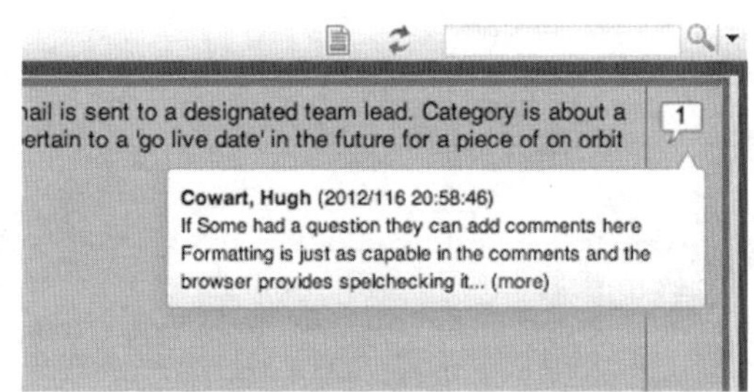

Fig. 28.7   Viewing a comment.

more useful information than the main entry. One team plans to update main entries occasionally based on the essence of the comments, retaining or deleting comments as appropriate. This could be very beneficial when a back-office process stores actionable information in a log entry with a future event time, then updates that information as needed until the future event time is reached. Examples include the following:

1. A back-office team discussion results in a log entry containing actionable information and a future event time.

2. Comments to an entry and/or changes in the plan result in updates to the entry.

3. At the future event time, CoLT notifies the operator that this entry pertains to the related activity.

4. An operator edits the entry and/or adds/edits comments to best capture the performance of the event.

5. This "cradle to grave" log entry could be "reincarnated" by duplicating it into a new log entry and editing for the next instance of the same event.

I. NOTIFICATIONS

In a traditional log, information is provided by one operator at a time. Any new information, such as new entries or updates to existing entries, is introduced by their own action. In CoLT, multiple sources can edit entries and/or add comments (depending on assigned privileges). It is important to advise a given user (particularly the operator on console) of changes made by other parties. CoLT's Notifications feature functions similarly to that seen in Facebook:

1. A user may subscribe to an entry of interest and is automatically subscribed to entries they create, edit, or comment on.

2. A user may unsubscribe to any entry.

3. Change notifications appear in a pull-down menu and include a link to the entry or the ability to pull up a filtered search list returning all entries with related notifications.

4. A user may opt in to receive e-mails for each notification. (A future build may offer daily/weekly digests.)

J. E-MAILING OF INDIVIDUAL ENTRIES, STANDARD REPORTS, AND SEARCH RESULTS

If a user selects an entry and wants to send it to someone, a single click on an envelope icon converts that entry, with attachments and comments, into an e-mail and presents it for addressing and review. A compact dialog makes it very easy to generate and e-mail standard reports (per POIC operational requirements) and customized reports (via CoLT's search engine). Access to global address lists and predefined mailing lists is being improved. Future development will include other ways of streamlining communications processes and eliminating the tedium of manual cut-and-paste.

K. HOW IT FITS TOGETHER

While CoLT supports traditional narrative log keeping as practiced for decades of space operations support, its true strength lies in power-boosting the flight team's efforts:

1. "Anywhere" access allows back-room support, off-duty personnel, and management to maintain situational awareness without waiting for daily log export. Search capability across the entire log is no longer limited to on-console personnel.

2. Console operators can push information, as needed, far more easily (e-mail options, report generation) and sometimes CoLT takes care of it automatically (e-mail alerts based on Flag status).

3. Comments capability gathers additional knowledge about a log entry at the same "location" without interfering with the entry.

4. Notifications feature keeps users abreast of changes in items of interest.

5. Future Events scheme puts accumulated support information for certain activities where it is needed (right in front of the operator), when it is needed (shortly before the activity), and without multiple transfers.

6. Future builds could include automatic linkages between CoLT and other communication, coordination, and/or planning systems, reducing the manual keying load on console operators.

L. EARLY OPERATIONAL EXPERIENCE

Most of the discussion above is based on design, testing, and limited operations simulation experience. POIC began using CoLT for actual ISS flight operations in June 2012, three months after the *Space Ops 2012* conference. Overall, the usage experience has been positive, with a relatively smooth transition into operational use. Of course, any new system has its challenges:

1. Owing to facilities schedules, servers were not yet in a "highly available" configuration, so transition to an alternate instance of the tool required use of a different URL. It has been easy to inform on-console personnel of transitions as they occur. Off-console users, especially those working outside an operations context, may not always get the word in a timely manner.

2. The future events capability described above was incomplete at the transition time so we have not yet been able to test its efficacy for flight support.

3. When an off-console participant received an e-mail advisory concerning an activated flag, they would have to a) manually enter CoLT to add a comment to the log entry or b) generate an e-mail thread that would have to be manually re-incorporated into the log comments, perhaps by a different operator. This can splinter the conversation somewhat and we are looking at ways to make it much easier to respond directly into CoLT.

We have seen indicators of significantly improved communications with CoLT:

1. The "Flags" feature has vastly improved the flow of information from real-time operations to back-room support. This flow will improve even more when discussion via e-mail threads (as noted above) is replaced with discussion via Log comments.

2. Automated daily reports are far more consistent than manually mailed log exports.

3. External access to the Log and ability to create specific use logs is being explored by several teams. This allows running logs of operations to be used in more contexts.

4. Within the first month of operations, a console team that did not participate in the legacy Filemaker Pro logs scheme asked to be included in CoLT. Some entities within the European Space Agency (ESA) have inquired about availability.

Work is continuing to evolve CoLT into a flexible, multimission tool. We hope that we can open up an environment for other centers and agencies to test this capability, especially for any related ISS use.

IV. COMMUNICATING AMONG LOGS VIA SOCIAL TECHNIQUES

With the advent of Facebook, Twitter, and other social media networks, the typical knowledge worker is increasingly comfortable with data sharing and expects such capabilities. Data storage is shifting from local storage to distributed servers and/or clouds. Applications are moving from local client-based software to server-side applications with Web-based or mobile client interfaces. NASA could benefit greatly by adopting and adapting these techniques throughout the Agency.

For the first 35 years of the Space Program, console logs were kept on paper, which required large amounts of physical storage and could be searched only by the human eye. The rise of the PC spawned a gradual evolution to electronic logging. A major challenge with early efforts was that logs might be spread over hundreds of individual Word files, so searches were quite cumbersome. Thanks

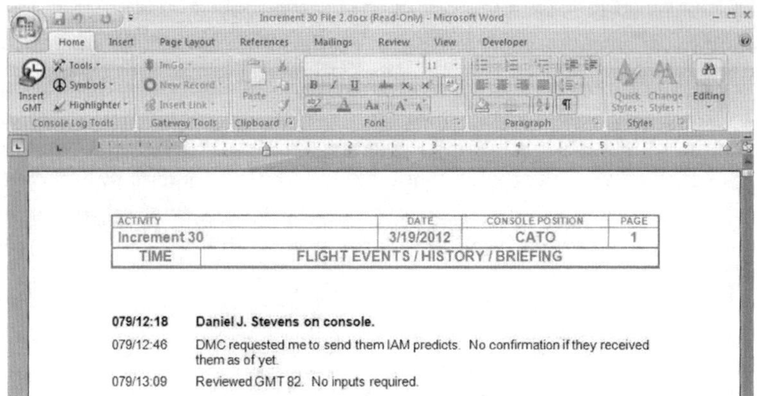

Fig. 28.8   MCC-H console log template.

to advancing technologies (and some enterprising flight controllers!), most current Johnson Space Center (JSC) ISS Flight Controllers use a Microsoft Word-based template with specialized macros, as shown in Fig. 28.8.

A major function of the template-and-macro scheme exports log entries to an SQL Database, which allows for a quick and thorough search across all of a specific discipline's logs via a convenient Web-based client, as depicted in Fig. 28.9. Although this log-keeping system has been stable for around eight years and works well, traffic is increasing and it is time to think and act forward.

A.   SOCIAL MEDIA MEETS CONSOLE LOGGING

The techniques and mindsets made popular by social media could be harnessed to dramatically improve Log accessibility, cross-discipline communication, and the dissemination of information. For example, microblog sites such as Twitter use the following "tag" protocols:

1.   To refer to a person, place the "at" symbol (@) in front of their username to reference that person. The post "MyUserID – @JohnDoe see you tonight." will appear in @JohnDoe's stream of posts and will be visible to others who are following @JohnDoe.

2.   To refer to an object or topic, place a "hash" (#) in front of the item name. The post "MyUserID – is attending the #AIAA #SpaceOps2012 conference this week" will appear in MyUserID's stream. Anyone searching for #Space-Ops2012 will see a stream containing MyUserID's post and anyone else's post referencing #SpaceOps2012. These streams are sometimes called trends when they are receiving a lot of traffic. There are several ways to generate and display lists of top trends.

How can these microblogging techniques be translated for use in the control center? The @ tag would be used to refer to other disciplines' call signs, entire control centers, other organizational units, or even a particular person; for examples, @FlightDirector would reference the Flight Director on console at the time; @MCC-H would refer to all disciplines within the MCC-H control center; @Org would be used to reference all users who are members of a particular organizational unit; @myAUID would be used to reference a particular person by their Agency User IDentifier (AUID). The #tag would be used to reference objects, such as a piece of hardware or software, procedures, a timelined activity, Flight Notes, Anomaly Reports, and many other frequently used console products.

Using @tags and #tags within a log would cause entries from one log to appear in other logs. For example, Flight Controller X has created a new Flight Note to document XYZ.

1. Currently this exchange of information would require:
    (a) Flight Controller X logs "Created new Flight Note that is ready for Flight Director approval."
    (b) Flight Controller X calls Flight Director – "FLIGHT, CONTROLLER X, new Flight Note created and ready for approval." "CONTROLLER X, FLIGHT, copy."
    (c) Flight Director has to search in Flight Note application, then opens Flight Note, reviews, and approves.
    (d) Flight Director calls Flight Controller X – "CONTROLLER X, FLIGHT, Flight Note approved." "FLIGHT, CONTROLLER X, Roger, thanks."
    (e) Both positions create log entries discussing Flight Note's approval.

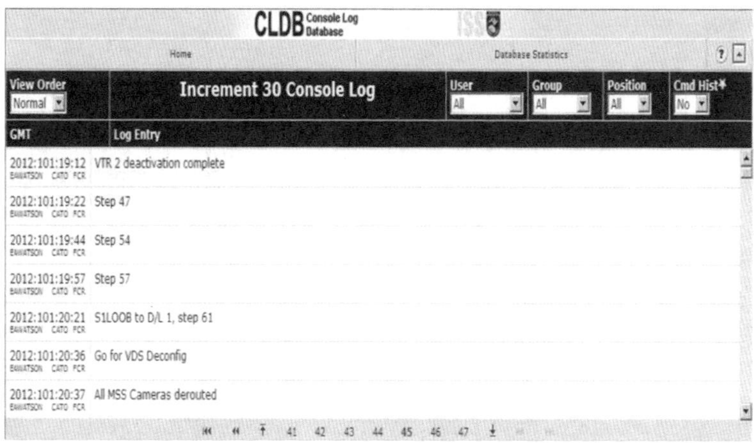

Fig. 28.9   MCC-H console log database.

2. Using microblogging techniques, voice loop traffic could be significantly reduced:

   (a) In Flight Controller X's log–"GMT 300/10:30 @FlightControllerX has created #FNxxxx to document XYZ and it is ready for @FlightDirector approval." Entry would automatically appear in Flight Director's log and would include a direct link to the Flight Note.

   (b) Flight Director takes direct link to Flight Note, reviews, and approves.

   (c) In a manner similar to a Facebook comment, Flight Director responds in his log–"GMT 300/10:35 @FlightDirector has approved #FNxxxx." Entry would appear in both logs as a child to Flight Controller X's original entry.

   (d) Taking this concept a step further, the Flight Note system might be enhanced to generate appropriate @tag and #tag traffic, sometimes automatically, sometimes with dialogs for Proceed/Cancel decisions. This would free up both Flight Director and Flight Controller from logging the event by hand.

This communication would not have to be restricted to just the Flight Control team. A Flight Controller could easily make a similar log entry to contact a Subject Matter Expert (SME) or other personnel in the office, for example, "@jdoe equipment #XYZ is showing an error X. How should console proceed?" The SME would then receive an e-mail or a client Web application pop-up with the question. Then @jdoe could respond to that question. Likewise, if a SME or other person from the office would like to inform the console of some update or new product, they could then create an entry "@FlightControllerX the following Anomaly Report is ready for closure: #ARxxxx (Title of the AR)."

With the stream and trends concept, each user would have access to a list of "popular" references. For example, when several controllers make multiple log entries about an observed failure, the associated #tag would quickly become visible to all users as a top trend (Fig. 28.10). This would allow office members to keep up with important issues as they evolve, without requiring console operators to send out additional e-mails and with less third-party reporting.

Streams could dramatically simplify shift handovers. Currently, the Flight Director calls each control center and each discipline individually for reports on significant events of the previous shift and upcoming activities on the timeline. If @tags are adopted by all team members, the Flight Director could initiate a log entry to the entire team and each position would respond with their status:

1. "@ISS ready for #Handover items";

2. "For #Handover, @FlightControllerY reports all systems functioning nominal and nothing on the timeline";

3. "For #Handover @FlightControllerY needs @ISS to review #SomeProduct".

All logs would show the original post from the Flight Director and all replies as children of the original post. In all likelihood, there would still be a need and desire for some handover-related voice traffic, but the existence of the stream would

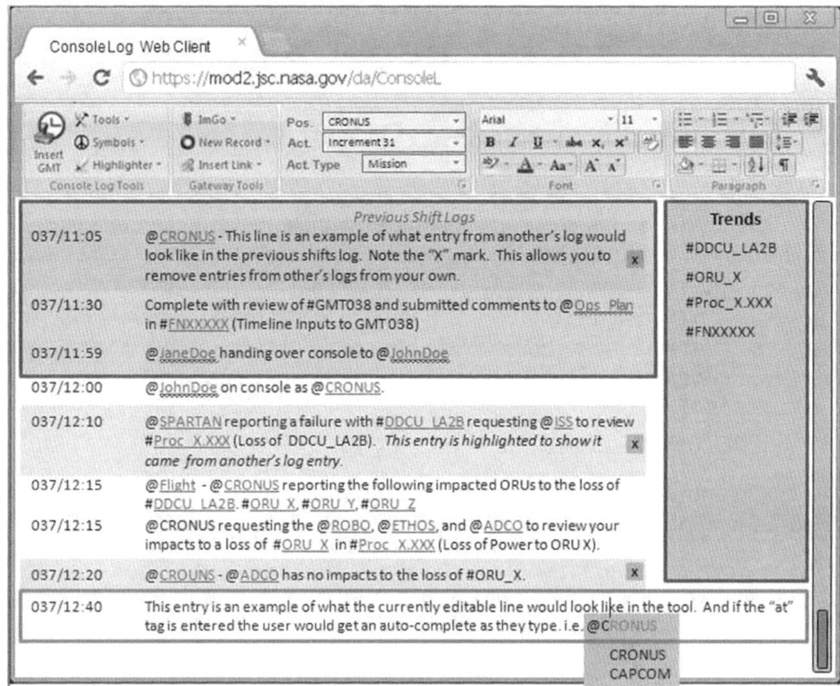

Fig. 28.10 **Console log with @tag and #tag features.**

mean 1) less additional log keeping to document the content of the handover, 2) shorter voice handover briefing, and 3) talking points are visible throughout and after the briefing (very handy if someone is working an urgent issue and has to miss the briefing).

B. WHAT WOULD THE TOOL LOOK LIKE?

The first step in this shift would be to move away from file-based Microsoft Word Templates to a Web client that accesses a more robust database. This move alone would reduce the support burdens and redundant storage associated with use of local clients. The end product would resemble a chat window with all the added functionalities of the current Visual Basic Word Macros (e.g., Timestamps, Hyperlinks to Flight Products). The user would define the number of old entries to view and the client Web interface would only maintain those entries in the viewable screen. Figure 28.10 is a conceptual image of this type of Console Log.

The first 3 entries have a gray background to show that they are from a previous shift. Colored shading of the 1st, 5th, and 8th entries shows that they came from other positions' logs. The "X" boxes hide/reveal their content. A bright "active box" around the 9th entry means it is selected for editing, commenting, or other actions. The "Trends" frame shows which topics are receiving the most attention.

C. AUTOCOMPLETE

Because this system would be working with an established list of Discipline names and userIDs, it is possible to have autocomplete functionality on @ and # tags. As shown at the bottom of Fig. 28.10, if a user began typing "@C", a window of auto-complete options would appear with the closest completion selected. The user could press Enter to autocomplete @CRONUS or press the down arrow followed by Enter to autocomplete @CAPCOM. This could also be extended to AUIDs, Procedure numbers, or various other Flight Product references.

D. UNIQUE CONSIDERATIONS

The flight control environment has some more stringent performance needs than better known forms of social media:

1. The client-side interface must be able to work offline in the case of network outages and then synchronize with the Database when connectivity is restored.

2. Access privileges must be managed with extreme care. Sensitive material may exist within a log entry that must not be shared outside a specific group. For example, the Biomedical Engineer (BME) reports to the Flight Director via an @tag post in the Biomed log that a crew member has fallen ill. Because medical information is sensitive, the BME would need to be able to turn pro-tected posting on so that his post would only be available to his group and to anyone he references in that post. Protected posts would not be visible to anyone else, for example via #tag or @tag searches or following. For posts that are not sensitive, the BME could turn protected posting off. Adminis-tration of protected posting merits much additional discussion.

E. WHERE THIS MIGHT LEAD

This concept could port very easily to mobile devices and even other control centers around the world. The flexibility of server-side processing would allow for thin-client applications and ease of distribution. "Anywhere" access to detailed log information would make it easier for a manager or on-call flight controller to maintain situation awareness or answer calls for assistance. (ISS core system flight control operations have been streamlined with multiple console positions often consolidated into a single Operator who is trained to safe malfunctioning systems until a certified Specialist can resolve the issue. With log details literally "in their hand," a remotely located Specialist would be in a much better position to talk the Operator through a quick resolution or to assess the need to go on-site.)

By harnessing microblogging's simple yet novel tools to log keeping, NASA would "grease the machines" for the next generation of flight controllers who have grown up Tweeting and Facebooking. Given that our forebears in space operations did amazing things wielding slide rules, just imagine what tomorrow's flight controllers could accomplish with tools built around a technique that is as natural to them as breathing. Those of us currently in the business might also find some fresh air!

V. COMMUNICATIONS DASHBOARD

Over the years, electronic messaging and documentation have lightened the paper burden and made searching easier, but have also spawned the unintended consequence of a text explosion! Just because we *can* generate more words more quickly does not mean we should; nonetheless, we do. Much effort goes into transferring text from one communications or planning system to another. With the increased pace of operations and addition of more console operators, all kinds of voice discussions wind up being repeated to bring folks "up to speed" and/or get transcribed or summarized into text and injected into the text explosion, making the blast even bigger. At times, the barrage of background can cause the foreground to go underground.

If we could integrate presentation and exchange of flight control communications in intracenter and/or intercenter contexts, we might drastically reduce the repetition of both text and voice information, thus empowering flight controllers to manage more operations with fewer words and with greatly reduced stress. These techniques could apply to a wide variety of real-time control environments, such as aircraft flight control, factory operations, military operations, 911 call centers, and inter-agency responses to large-scale disasters.

A. DASHBOARD COMPONENTS

A communications dashboard could 1) use social networking service (SNS), micro-blog, and/or fortified text chat methods to supplement voice loop communications and 2) display various information streams in an arrangement that requires less human processing to grasp the "big picture" of operational events and context. Typical SNS applications, such as friends, games, and free-form chatter, would naturally give way to more operations-oriented services:

1. Console log contents.

2. Text chat channels analogous to traditional voice loops and/or text chat streams created "on the fly" and dedicated to work a particular issue. Text chat *is* a bit slower than voice, but when something is "said" it remains visible. New or existing participants can catch up or keep up with the conversation just by reading, and the conversation is self-documenting.

3. Mapping capability with a quick, intuitive interface so that console log entries can be kept brief by providing links to access underlying details stored in text channels or streams. This also reduces duplication of cut-and-pasted text.

4. Consolidated indexing of mapped text conversations and streams for easy browsing or searching.

5. Individual/group "hailing" capability that would advise operators of someone else's need to communicate. Unlike a voice loop call, which may not be heard or may have to be repeated, a dashboard hail would persist visually until acknowledged.

6. Announcement capability. As for hailing, the advantage over voice loop transmission is visual persistence.

7. High-level status indicators, such as Comm-link status with ISS, time to major events, and other items of general and/or position-specific interest.

8. Scrollable status updates from manual (console operator) and/or automated (telemetry monitoring) sources.

These services could be displayed integrally to slash redundancy among multiple threads. Also, if people look at the same instance of a conversation instead of creating separate accounts of it, there will be fewer disconnects. Those that do appear will be more obvious and can be quickly resolved.

B.   HOW SOCIAL MEDIA TECHNIQUES COULD IMPROVE INFORMATION DISPLAY

Social media methods present intriguing potential for powering a communications dashboard. Well-known schemes with a personal, professional, or institutional focus (e.g., Facebook, LinkedIn, and Yammer) are clearly not suited for direct application to real-time mission operations. If we examine *how* these services engage users, we find some visually simple yet extremely powerful system behavior:

1. A few very small icons, representing events of keen interest, change to a brilliant color when those events happen, indicating how many instances have happened and easily attracting the operator's attention. A simple click on the icon summarizes the new events and a simple highlight-and-mouse-release opens a specific instance's details.

2. Multiple streams of dynamic information appear in frames proportionate to their "importance" (based on assumptions about a typical user and some user-configurable preferences). When moving about the workspace, some frames change while others remain the same, just as scenery outside our car changes as we drive, yet certain items in the car stay in place, although their values may fluctuate.

C. CONCEPTUAL MOCKUP

Figure 28.11 illustrates the concept of integrating several text streams, a console log, hailing, and status indicators to present a "big picture" to the console operator. A description of related voice loop traffic appears above the mocked-up display window.

An added benefit of building a communications dashboard around a SNS model is that, because SNSs are common in our personal and office cultures, training to use one in an operations setting would be relatively easy.

D. WHAT GAIN WOULD A COMMUNICATIONS DASHBOARD PROVIDE FOR VOICE COMMUNICATIONS?

With more robust text communications and display of appropriate status items, some voice traffic from announcements and detailed discussions could be moved to dashboard media. Voice is vital but is serial in nature. Text is

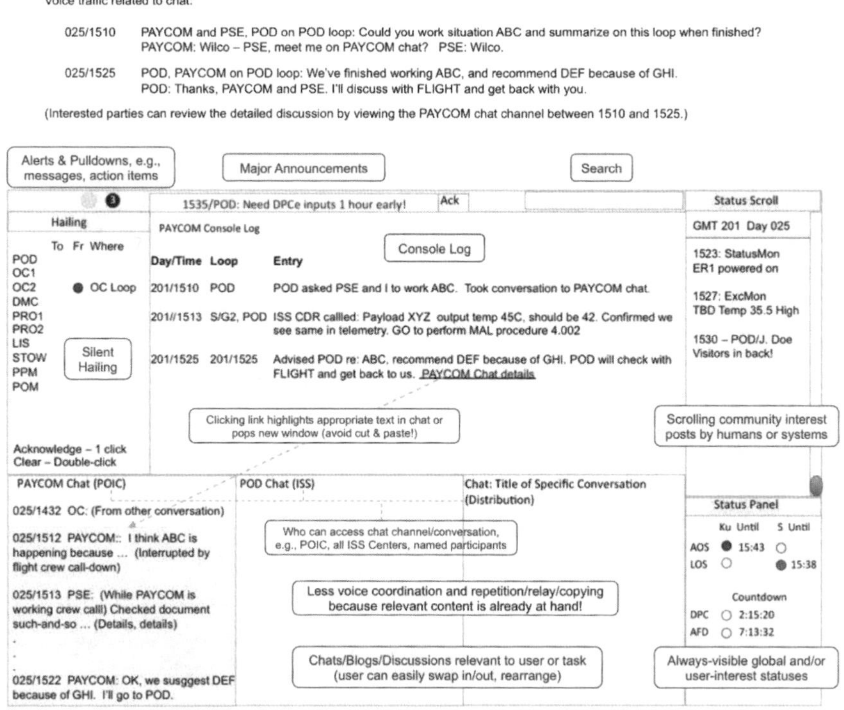

Fig. 28.11   Communications dashboard concept.

semiparallel and graphics is parallel, and both have the advantage of visual persistence. If the most essential voice traffic is retained due to time and/or criticality and underlying details are worked elsewhere with less repetition, aural saturation and competition between audio and visual stimuli are reduced. This *will* lower stress on the console operator. A communications dashboard lab could

1. Develop and demonstrate dashboard capabilities and their effects.

2. Allow flight controllers to a) experiment with new representations and communication methods to see how they might like to evolve their operations concepts and b) generate ideas for additional capabilities and features.

3. Develop ways to access appropriate existing mission support software, e.g., CoLT, via either their traditional interface or a dashboard. Existing SNS schemes are essentially entities unto themselves. Applications running inside the SNS are not available outside it. The author knows of no SNSs that integrate with real-time operations systems.

4. Explore the effects of various layouts in a mission operations context and draft guidelines for console operators to optimize their setups.

5. Qualify and quantify what voice traffic might be moved to dashboard media.

E. DEVELOPMENT EFFORTS

As of September 2012, the Principal Investigator (PI) is developing a Communications Dashboard via the "IT Labs" (Information Technology) program sponsored by NASA's Chief Technology Officer. This approach consists of a series of 90- to 120-day phases joined by update–evaluate–approve–perform–evaluate–decide processes, as shown in Fig. 28.12.

"CommDash" is currently in the Idea/Issue phase. A one-minute "elevator pitch" video summarizing the CommDash concept and objectives is available online at http://tinyurl.com/CommDashElevatorPitch.

The PI has identified collaborators at multiple NASA centers and in some commercial firms, such as social network/enterprise providers and software innovators, and welcomes inputs from others. The current effort includes looking for mashup potential and may expand into visual interface prototyping. Hopefully, a working prototype will emerge sometime in 2013.

Fig. 28.12   NASA IT labs phases.

VI. CONCLUSION

Innovation is the art of the "adjacent possible" [4, 5]. This chapter offers three innovative, Web-based technology approaches to console log keeping (one in initial deployment, two in concept stage) that could reduce stress on space flight controllers while increasing their productivity. The three approaches are

1. *Console Log Tool (CoLT)*: a recently deployed log-keeping application at MSFC's POIC that provides "anywhere" access, comment, and notifications features similar to those found in SNSs, report generation, and e-mail transmission at several levels of automation, and a future events tracking capability that could foster appropriate bidirectional communication via the Log. Future versions might include connections to other ISS communications and/or planning systems. Initial use in support of real-time operations has been successful. Ongoing development may lead to offering CoLT as a service to multiple missions and/or agencies, or as licensed or open-source software.

2. *Cross-Log Communication via Social Techniques*: a concept being explored at JSC's MCC-H that would use microblogging's @tag and #tag protocols to make information/requests visible and/or discoverable in logs owned by @Destination addressees. Some #Object designations would also serve as links to products/locations in other systems, for example, Flight Notes. As the PI has made a career change, this concept eagerly awaits its next champion.

3. *Communications Dashboard*: an MSFC concept that proposes an SNS approach to integrate console logs, text chat streams analogous to voice loops, text chat streams dedicated to particular conversations, generic and position-specific status displays/streams, and a graphic-based hailing capability. Further development is in work via the NASA IT Labs program, with hopes of a working prototype or system in 2013.

Such approaches can reduce flight controller stress because of the following goals and principles:

1. Information is represented in fewer places, so there is less opportunity for disconnects.

2. Multiple parties look at the same instance of information, reducing disconnect opportunities even further.

3. Manual cut/copy-and-paste operations are minimized.

4. Information is discoverable when needed and disappears when it is not.

5. Less voice traffic is needed because the log-keeping system can now carry many conversations with excellent retention, organization, and visual persistence.

There is a wiki at http://simplifyopscomm.wikispaces.com/home for continuing the discussion and starting new ones.

REFERENCES

[1] Scott, D. W., "Using Social Media in Engineering Support and Space Flight Operations Control," *IEEE Aerospace Conference*, 2011, http://ieeexplore.ieee.org/xpl/articleDetails.jsp?arnumber=5747662 [last accessed 23 May 2012].

[2] Scott, D. W., "Using Web 2.0 (and Beyond?) in Space Flight Operations Control Centers," *SpaceOps*, 2010, 210896, http://ntrs.nasa.gov/archive/nasa/casi.ntrs.nasa.gov/20100021135_2010020435.pdf [last accessed 29 October 2011].

[3] Scott, D. W., "How to Boost Engineering Support Via Web 2.0 (Seeds for the Ares Project ... and/or Yours?)," Unpublished (available on request from scotty@nasa.gov).

[4] Johnson, S., "The Genius of the Tinkerer," *The Wall Street Journal*, 25 September 2010, http://tinyurl.com/76aeqqd [last accessed 23 May 2012].

[5] Johnson, S., "Where Good Ideas Come From," http://www.youtube.com/watch?v=Mb0ssmoXG1I [last accessed 23 March 2011].

CHAPTER 29

TDX-TSX: Onboard Autonomy and FDIR of Whispering Brothers

A. Schwab,[*] C. Giese[†] and D. Ulrich[‡]
EADS Astrium GmbH, Friedrichshafen, Germany

I. INTRODUCTION

The TerraSAR-X and TanDEM-X satellites (TSX and TDX, Fig. 29.1) were launched in June 2007 and June 2010, respectively. They are key elements of the German national radar Earth observation mission and were developed in the frame of a public–private partnership between the German Aerospace Centre (DLR) and Astrium Satellites. TSX and TDX are both versatile synthetic aperture radar (SAR) satellites with active phased-array antenna technology operating in the X-band. The two satellites orbit in a closely controlled formation and form a single-pass SAR interferometer with adjustable baselines (typically below 250 m) across- and along-track. The satellites operate individually for TerraSAR-X mission product generation as well as jointly for Digital Elevation Model product generation, thus forming an integrated system of systems. They enable Earth observation independent of the time of day or cloud cover. Operating together in close formation, TSX and TDX allow 3-D observation, that is, derivation of a Digital Elevation Model (DEM). Combined operation for three years allowed data collection for a complete and consistent DEM of the Earth surface with a height resolution of about 2 m [1]. The TDX mission was envisioned during the development of TSX, so only absolutely necessary design extensions were implemented on TSX to support future synchronized operation of both radars. Individual satellite onboard failure detection, isolation and recovery (FDIR) is based on a distributed, hierarchical, multiple-level concept, in which each level performs its dedicated failure detection, (if possible) failure isolation, and (if possible and desired) failure recovery. If a failure cannot be recovered on a particular level, a corresponding event report will be performed and the next higher level will take care of the predefined actions. On application software (SW) level, the concept is mainly realized by standardized configurable services supported by FDIR preprocessing within onboard algorithms. A special characteristic of each individual satellite is the extremely high power consumption and

*MPC Manager Operation & FDIR, AET2.
†TDX Operation & FDIR Architect, ACE72.
‡TSX-2 Project Manager, AED7.

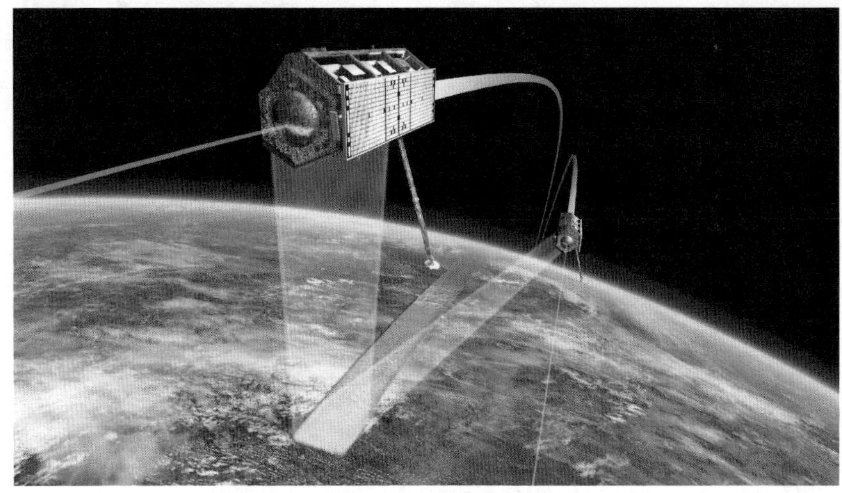

Fig. 29.1   TSX and TDX satellites.

dissipation of the SAR instrument during operation (i.e., Data Take). For economic reasons a design to cost and budget approach had to be chosen, with the consequence that the energy resources and thermal control system of the satellite were not sized to support a continuous operation of the SAR instrument. Consequently, improper operation of the instrument, violating given resource limits, can endanger the satellite. So, onboard FDIR must not only cover hardware failures as identified in the failure mode, effect and criticality analysis, but also functional operational errors and transient failures, such as those caused by single event upset-induced bit flips in memories or electromagnetic interference-induced corruption of data during transmission, because all these cases could lead to resource limits being exceeded. The special characteristic of the TDX mission is the close-formation-flying constellation of TDX and TSX satellites. The bus design of TDX has been extended with respect to TSX to allow formation flight of both satellites. The software changes relevant for the TDX mission were verified during the TDX on-ground tests and then also uplinked to TSX in preparation for the formation flight.

II.   KEY SATELLITE DESIGN CHARACTERISTICS

The 5-m-long and 2.4-m-wide satellite structure has a hexagonal cross-section that fits the rocket fairing efficiently. One side carries the SAR antenna, which is 80 cm in elevation and points to the Earth on the right side of the subsatellite ground track. The SAR instrument, operating in the X-band (9.65 GHz), allows for weather-independent Earth imaging, day or night. The design is based on an active antenna using an array of transmit–receive modules that provides rapid electronic beam steering and programmable antenna patterns. Each module feeds a subarray comprising two slotted waveguides, one each for vertical and horizontal polarizations. Rapid switching of polarizations and other radar parameters, even pulse-to-pulse, results in a very flexible SAR design that provides a large variety of experimental modes in addition to the modes required for standard product generation. StripMap is the standard mode of bistatic operation of the TDX mission. It provides a resolution on ground of 3 m and a measurement swath width of 30 km. Although, originally, an average SAR operation time of 170 s per orbit was specified, which corresponds to a measurement swath width of about 1200 km, the system in orbit has demonstrated that an extended operation profile can nominally be operated for an average SAR transmit time of 210 s per orbit and additional 130 s operation time in receive-only mode. Tests have proven that an operation time of up to 600 s is possible under favorable conditions. In Spotlight mode, azimuth (along track) scanning of the radar beam allows the resolution to be improved to 1 m for measurement scene "patches" of about 10 km × 10 km. The active antenna design ensures small long-track separation of the patches. In a special instrument configuration called dual receive antenna (DRA), the nominal and redundant instruments of one satellite can be operated in parallel, essentially as two individual instruments, including data recording on the SSMM, with each controlling one half of the antenna.

The Sun-synchronous dusk–dawn orbit allows the use of a fixed solar array, mounted on the left side of the satellite. The energy is stored in a lithium-ion battery system that provides a capacity of more than 100 Ah. The battery design is based on the use of the lowest capacity cells integrated in a high number of parallel strings. An open failure in one cell, that is, in one string, will only lead to the loss of a small amount of capacity, without degradation of the overall battery voltage, so there is no need for a bypass. The battery is designed to provide the peak load during operation of the radar instrument in eclipse, while keeping the minimum voltage at the battery terminals higher than the cutoff voltage of the input converter of the instrument power distribution unit (IPDU). The worst-case case of voltage drops due to radar operation, from begin of life (BOL) to end of life (EOL) conditions, as expected during the design phase, is shown in Fig. 29.2. Evidently it has been a key challenge of the power system design to trim the FDIR logic and particularly the hardware (HW) protection elements of the power system to fit any of these conditions. Although it is difficult to determine the state of charge (SOC) of a lithium-ion battery system, analysis has shown that the voltage drop at peak load currents is much more dependent on the actual battery impedance than it is on the SOC of the battery. As neither the SOC nor the battery impedance can be known accurately enough onboard the spacecraft, an alternate ground-based method for power system surveillance was introduced. As a direct monitor of battery health during ground contacts, the onboard system reports the power characteristic during peak load conditions for a selectable worst case, for a consistent set of

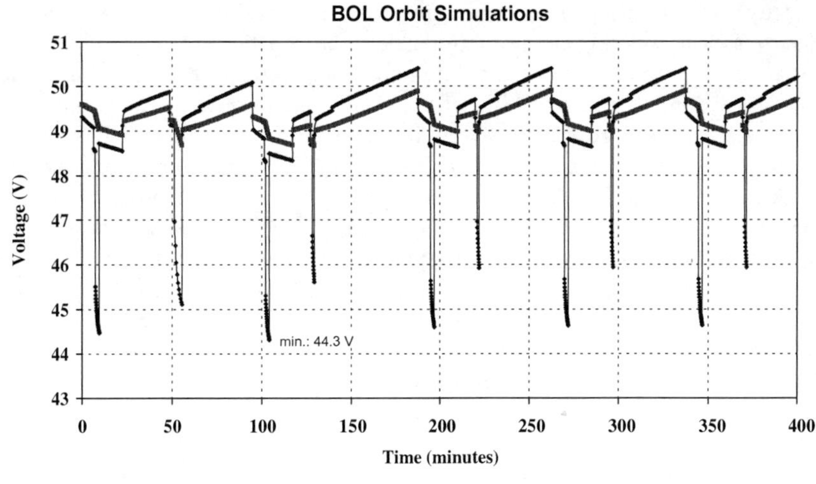

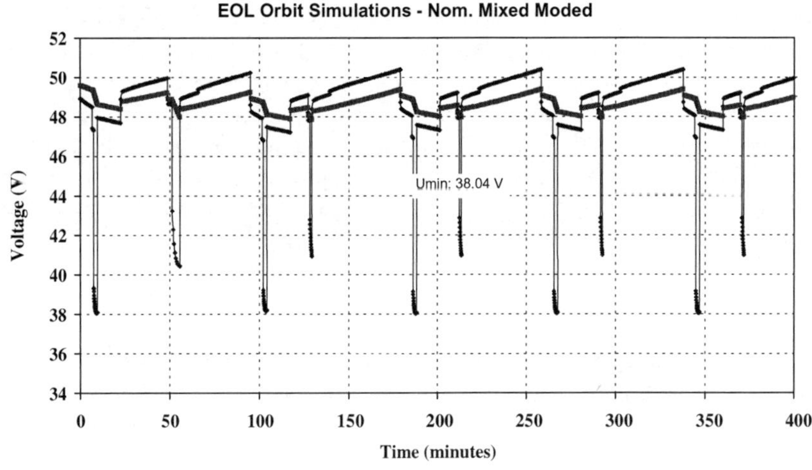

Fig. 29.2   Worst-case primary voltage behavior at BOL and EOL for a nominal operation scenario (design phase).

power system data (several temperatures, voltages, and currents) since the last reset of the reference conditions. This minimum battery condition monitor supports simple ground monitoring of power system health.

Both TSX and TDX carry an IGOR dual-frequency GPS receiver connected for zenith precision orbit determination (POD), as well as aft- and forward-looking occultation antennas, as well as a laser retro reflector to support the tracking, occultation and ranging (TOR) mission of the Geosciences Research Centre (GFZ) in Potsdam. Ground processing of the IGOR POD data verified by high-precision laser ranging measurements from both satellites allows accurate and reliable determination of the interferometry baseline (i.e., the effective SAR antenna separation) required to ensure DEM height accuracy. Besides this synergetic contribution to the SAR mission, the TOR mission allows determination of atmospheric pressure, temperature, water vapor content, and electron density, which can be used for weather prediction or research on climate change from radio occultation measurements of signals through the atmosphere from settling and rising GPS satellites at Earth's horizon [2].

SAR measurement data are transmitted to ground stations in the same frequency band as used by the radar. To avoid mutual disturbance, the X-band downlink horn antenna is mounted at the tip of a 3.3-m-long boom. This boom is deployed towards nadir by a reliable passive spring mechanism soon after launch.

For TDX DEM data collection, formation flight is necessary, with separations between the two satellites down to 150 m and synchronized operation of the SAR instruments on TSX and TDX. For TanDEM-X the SAR design was extended to allow exchange of so-called Sync pulses to support coherent operation of both SARs during bistatic operation. Six Sync horns on each satellite provide quasi-omnidirectional coverage (Fig. 29.3). These are fed by transmit–receive modules that were already present as intermediate amplifiers with spare input/output ports. To support the close-formation-flying constellation, TDX is equipped with an additional propulsion system based on high-pressure nitrogen gas. This cold gas system (CGS) provides smaller impulse bits than the primary hydrazine system that is used for major orbit maintenance on both satellites. Last, but not least, an intersatellite link receiver and decoder is accommodated as additional hardware on TDX, which allows TDX to "listen" to the TSX low-rate S-band telemetry. The S-band telemetry contains all housekeeping data from the satellite. This one-way intersatellite link operates the nominal formation-flying separation range. Because the S-band antennas are mounted in nadir and zenith directions and the ground-station contacts are nominally performed at high rate, there are contact gaps depending on the relative formation geometry [3].

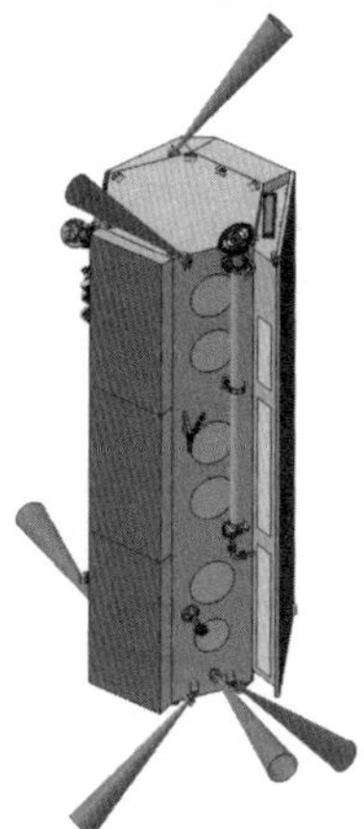

III. INDIVIDUAL SATELLITE FDIR

The functional operational architecture of the TSX was therefore defined to provide an optimum modular and

Fig. 29.3 Sync horn accommodation.

flexible design to support easy integration and update of key functions as well as adaptation and subsequent increase of onboard autonomy without endangering spacecraft health. Although the basic system configuration focused on the preservation of spacecraft health, several autonomous onboard recovery procedures were implemented, which have subsequently been enabled, thus ensuring a higher operational state. In addition, the spacecraft resources provide the potential for a further increase in onboard autonomy and support a relatively easy upgrade from TSX to TDX with substantial additional functionality. System modularity is achieved by a functional breakdown of the overall system into application processes, each individually providing its own dedicated functions, which are embedded into a common set of operational services for standard functions. This basic set of operational services provides full commandability and observability for individual applications as well as for the application-specific FDIR. Communication between applications, including the flight operation system on the ground, is based on an application and service concept following the ECSS E-70-11 standard. The hierarchical relations of the applications (Fig. 29.4) ensure an efficient and transparent routing of command and telemetry packets and efficient handling of failures at the lowest possible level and earliest time. Where a failure cannot be resolved at lower levels, a more stringent reaction at a higher level is defined.

As the top-level onboard application, system control is responsible for command distribution from and telemetry routing to the mission operation segment (MOS) on ground as well as for the overall system FDIR. The Bus

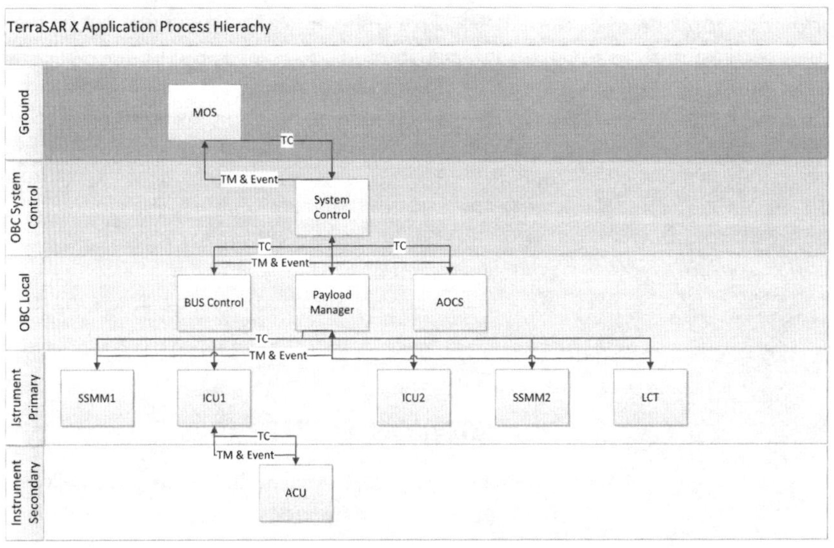

Fig. 29.4 Application process hierarchy.

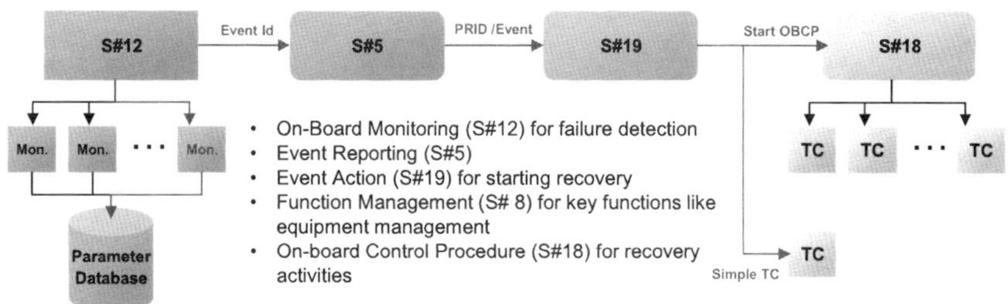

Fig. 29.5   PUS service application process hierarchy.

Control and Attitude and Orbit Control System (AOCS) application provides all services and functions required for operating the platform and supporting the overall satellite system. The Payload Manager application provides spacecraft support functions including FDIR for the payload compartment of the satellite, which is composed of further applications distributed on several different HW units that are responsible for operation of the SAR instrument (standard cold redundant SAR as well as hot parallel dual receive antenna configuration), Instrument Control Unit 1 (ICU1) and ICU2, payload data recording on a solid-state mass memory (SSMM), as well as data transmission via laser communication terminal (LCT). All applications support a common service set ensuring command acknowledgment, observability, function management, as well as local FDIR. Higher-level applications are only responsible for resolving problems, which are identified in failure signatures visible in directly acquired telemetry information or are reported via event reports, which cannot be resolved on an individual application level. The FDIR within each application comprises a configurable set of services based on the ECSS-E-70-41A Packet Utilization Standard (PUS), as depicted in Fig. 29.5. This distributed hierarchical onboard FDIR system on the SW level is accompanied by HW-embedded FDIR functions.

As an example, the power system FDIR comprises several primary bus voltage monitors on the SW level, so-called software disconnect non-essential load (SW DNEL) monitors on the platform and satellite levels. In addition to these SW based surveillances of the primary power bus, several levels of power shedding functions for non-essential equipment, essential equipment and ultimate battery health protections are implemented. The SW also provides for complementary consistent recovery startup from depleted energy resources as soon as minimum solar input is available. This FDIR is extensively verified during system-level tests, as shown in Fig. 29.6, which includes the famous Astrium "Munchhausen" low power contingency recovery test. The name of this test is drawn from "The Adventures of Baron Munchausen," and signifies being able to recover with minimum external support, fully on one's own resources. At (1), the Sun simulator is switched off. The fully charged battery is discharged, if necessary by an external load. Subsequently, the installed SW and HW primary power bus

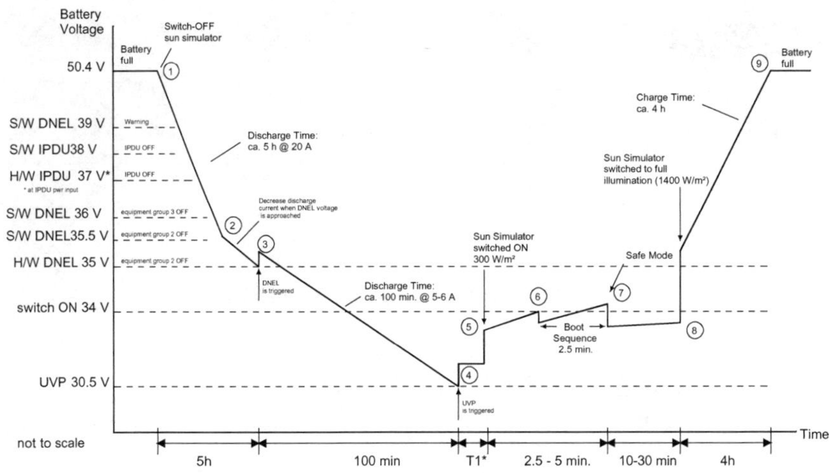

Fig. 29.6   Voltage profile.

monitors trip. At each level, correct occurrence of the related FDIR is checked. At (3), all non-essential loads are disconnected, which results in a small increase in the primary bus voltage. Because of the power consumption of the remaining essential units and heaters, which are still ON at some time, the latching current limiter under-voltage protection is reached (4), causing the power control and distribution unit (PCDU) converters and all remaining loads, including heaters, to switch off. Without external load the battery voltage increases slightly. After some time, the Sun simulator is switched ON with a faint stimulus $(300 \ W/m^2)$ (5). Because of the charge current, the battery voltage rises immediately; as soon as the switch-on levels for essential loads are reached, the PCDU and essential equipment such as onboard computers 1 & 2 (OBC1 & 2), RX1, RX2, and TX1 are switched on, the system reconfiguration sequence in the OBC is triggered, and the system initialization sequence of the central software starts, recovering the system into its satellite safe mode (7). Finally, the Sun simulator stimulus is switched to simulating full illumination. In this condition the satellite is at its nominal position with respect to the Sun. The battery is charged to its full charge state. The satellite is fully recovered. As a last resort, the mission operations center performs regular monitoring of the long-term system behavior. This FDIR design ensures the most robust system, even for multiple-failure cases.

IV.   TaNDEM-X MISSION FDIR

For DEM data collection formation flight is necessary, with separations between the satellites as small as about 150 m. The close-formation-flying constellation is primarily safeguarded by introducing a special orbit control approach, ensuring a helix formation under all circumstances (Fig. 29.7). By combining an out-of-plane (horizontal) orbital displacement via different ascending nodes with a radial (vertical) separation via different eccentricities and arguments of perigee, an elliptic relative motion is achieved between the satellites. Because the orbits never cross, autonomous control is not necessary for safe satellite operation. In other words, the helix orbit provides passive safety. Nevertheless, the very close formation suggests a risk of the satellites colliding and the risk of mutual illumination by the main beams of the radar antennas.

A number of new functions and extensions of the original TSX FDIR were introduced into the satellite design to safeguard against these risks. The major changes with respect to the original TSX implementation are as follows:

1. An additional Acquisition and Safe Mode magnetic torquer safe mode (ASM-MTQ);

2. A Reaction Control System (RCS) thruster safe mode (ASM–RCS) as fallback safe mode in the case of dynamic or power/thermal problems during ASM–MTQ;

3. Surveillance of close formation flight by exclusion zones, intersatellite link (ISL), and Sync warning data takes;

4. System-level reactions, comprising safety measures in several subsystems, in parallel;

5. FDIR for automated onboard relative navigation via the TDX Autonomous Formation Flight (TAFF) algorithm.

The mechanisms were realized by a combination of new hard-coded SW functions and additional monitoring, event action, and onboard command procedure (OBCP) definitions, which were installed via configuration data update of the individual services in the affected applications.

Fig. 29.7   Helix formation.

A.   SAFE MODE ASPECTS

In a robust, simple, system approach, the original Safe Mode of TSX and TDX was based on the propulsion system, as a prime actuation system supported by magnet torques to minimize fuel consumption. Analyses have shown that the use of thrusters in Safe Mode leads to an unacceptable risk of collision for certain satellite formation geometries. During close formation flight, that is, vertical separations of <300 m, the collision risk was evaluated to be 1:500 when one spacecraft experiences a Safe Mode drop. Use of magnetic torque actuation only avoids a change in altitude in the case of a drop to the AOCS ASM. Thus, on both satellites, in the AOCS application an AOCS ASM sub-mode was introduced that is solely based on use of the magnet torquers. These actuators are powerful enough to regain attitude and pointing stability for moderate rotation rates of the satellite, but under the constraint of a significantly longer convergence time. Owing to this constraint, the original more powerful Safe Mode based on the use of hydrazine thrusters is maintained as a safety net to handle residual risks of significant low-power conditions, as well as unfavorable thermal conditions of particularly exposed critical spacecraft elements. To maintain simple orbit injection in launch and early orbit phase for the TDX satellite, RCS-based ASM has also been used for fast initial acquisition. Because of the longer convergence times to a power and thermal safe attitude during the MTQ-controlled safe mode and the transitions between MTQ and RCS control, ASM impacts on the power and thermal system required an extension of the related FDIR settings at the platform and system level.

The logic for handling safe mode recoveries focused on the AOCS mode and submode impacts is shown in Fig. 29.8. The entry point in the case of problems in one of the operational modes is always the rate damping in the magnet torquer safe mode (ASM-MTQ-RD). An onboard autonomous sequence brings the spacecraft into Earth- and Sun-oriented acquisition mode (ASM-MTQ-AQ) once the rotation rates are below a certain threshold. If the rates are not below the specified limit within a few hours, an onboard computer (OBC) warm boot is triggered to recover potential sensor, actuator, or software problems. Thereafter, the spacecraft starts again in ASM-MTQ-RD. The exception to this scenario occurs in the case of severe thermal or power problems, in which case the transition into ASM (RCS) occurs immediately. The total amount of thruster activities upon first entry of ASM (RCS) is limited, so that the maximum orbit change does not lead to a substantially increased collision risk. The actual limits depend on intersatellite distance and current tank pressure.

After transition to ASM (RCS), thruster on-time is monitored to realize limited thruster operation. Depending on the formation and tank pressure, this limit is adapted from the ground. If the maximum thruster limits are reached, an OBC reboot will be performed and the ASM-MTQ is reactivated. If the limited thruster activities cannot recover the previous problem, the ASM (RCS) is triggered a second time with unlimited thruster activities.

B. EXCLUSION ZONE FDIR

To handle the risk of mutual illumination by the radar main beams when in the close-formation flying constellation, "exclusion zone" logic is introduced to both satellites (Fig. 29.9). Bistatic operation of the radars is used for DEM data collection, where only one SAR instrument transmits but both radars receive. Radar

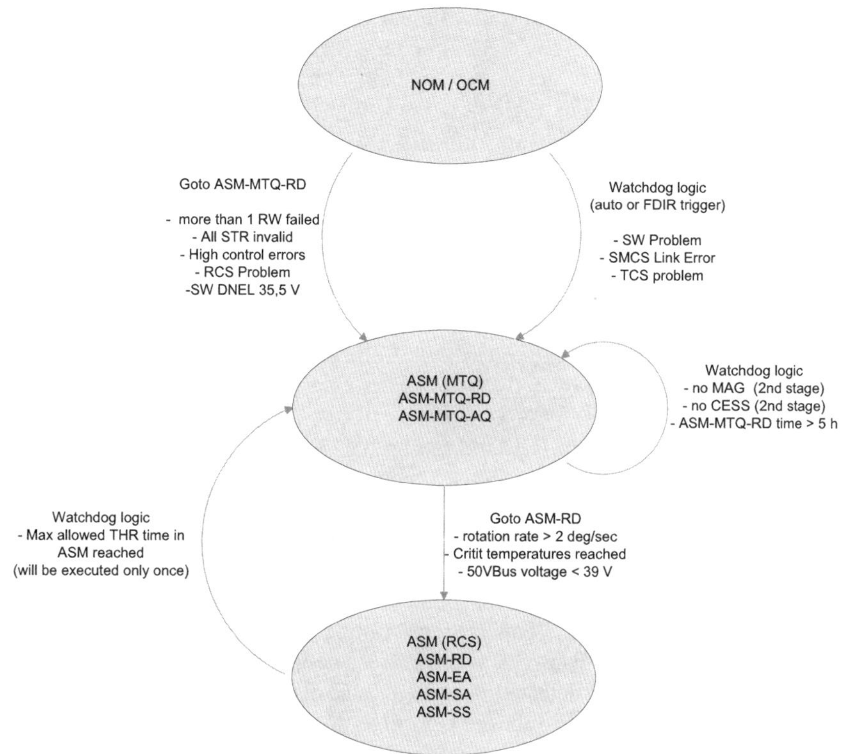

Fig. 29.8   Safe mode transition logic.

transmission is allocated to the satellite for which there is no risk of illuminate of its partner satellite with the main beam of its radar antenna for part of the orbit. Radar transmission by the other satellite is prevented within this part of the orbit by onboard logic – its "exclusion zone." The onboard exclusion zone logic is implemented by a new Payload Manager (PLM) application software function and updates of onboard FDIR settings (monitoring, event/action, OBCP execution) in the OBC Central SW and ICU application SW, which interact according to the following scheme:

1. The Payload Manager application generates nominal events that indicate the start and end of the exclusion zone and triggers a PLM OBCP.

2. Within the instrument, SAR transmission is prevented for data takes that begin inside the exclusion zone. For ongoing SAR data takes, active transmission is aborted at the beginning of the exclusion zone.

On TSX, the PLM application can already request, via onboard data access functions, various onboard data to establish a satellite ancillary data service to support individual payloads and instrument operation, as well as embedding of related information in the TSX mission data stream. For SAR transmission permitting operational AOCS modes, that is, stable right-looking or Sun-side-looking attitude, the exclusion zone logic uses the actual orbit position, which is determined by the AOCS application onboard from the navigation solution of the Astrium Mosaic GPS receiver, to release individual normal progress severity event reports (EZ_TX_OFF and EZ_TX_ON) for indication and triggering of the start and end of the exclusion zones. For each satellite and operating mode (i.e., right- and left-looking), individual lower and upper limits for the exclusion zone are defined. All exclusion zone limits are configurable and given as an argument of latitude values. When the helix orbit is changed, the exclusion zone limits have to be updated via flight procedures from ground. Figure 29.10 shows the exclusion zone logic and related FDIR implementation.

The detection of an EZ_TX_OFF event triggers a PLM OBCP, which enables an event/action entry in the ICU and sets an ICU onboard parameter, controlling the exclusion zone behavior of the instrument. Thanks to the efficient operational progress reporting of the instrument data take preparation and execution functions, as well as their comfortable configuration capabilities, the enabled event action triggers an appropriate preventive reprogramming step in the antenna control electronic setup upon occurrence of an ICU data take processing start event (DT_PROC_START). This reprogramming step disables the transmission of radar signals for the front-end transmit receive modules (TRMs). As soon as an EZ_TX_ON event is detected, a PLM OBCP disables the related ICU FDIR reaction and resets the ICU onboard parameter to indicate unconstrained instrument operation. For data takes that start outside the exclusion zone and are continuing into the exclusion zone, reprogramming of the antenna control electronics

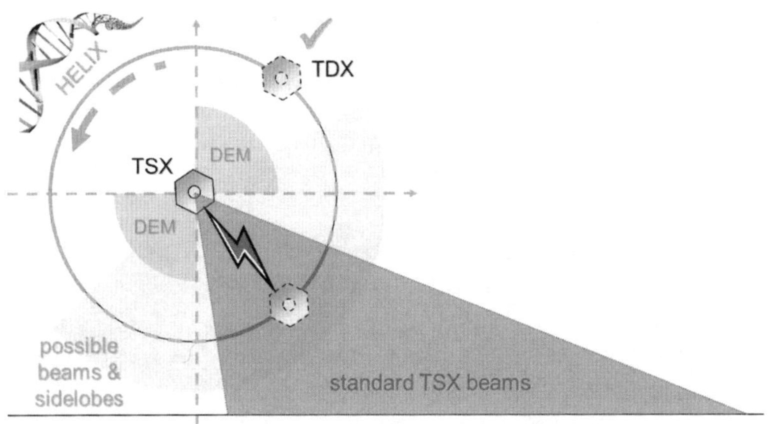

Fig. 29.9 Relative orbit geometry and exclusion zone.

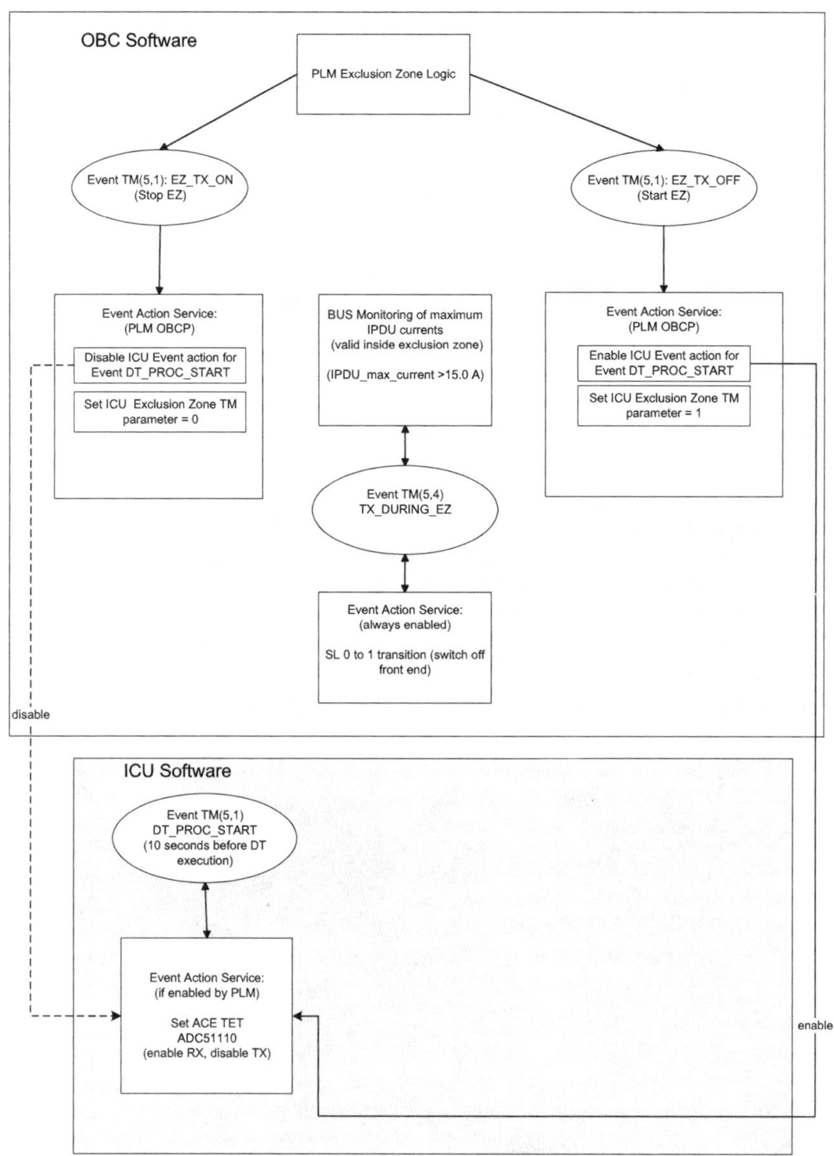

Fig. 29.10   Exclusion zone FDIR.

is not possible. Therefore, an exclusion zone-dependent monitoring of the high-power instrument supply current is implemented, which shuts down the SAR front end in the case of active RF transmission in forbidden zones.

Passive bistatic, as well as sync warning, data takes are not influenced by exclusion zone reaction. The onboard logic provides online protection for the individual satellites. Primary protection is provided by a similar logic, implemented in the ground segment to check and prohibit dangerous operation using the planned orbit geometry during mission plan preparation and generation.

C.   SYNC WARNING MECHANISM

While the above mentioned exclusion zone FDIR addresses the top-level interactivity coordination between the two brothers, we are now looking into the whispering communication exchange between them. The synchronization antenna system comprises six circularly polarized X-band horn antennas arranged in such a way that spherical coverage is obtained (Fig. 29.11). Bistatic data takes for DEM generation make use of the exchange of RF pulses between the radar instruments of the two satellites via selectable horn antennas as part of a synchronization scheme. This allows monitoring of the phase difference and minimization of the interferometric phase error. Besides these nominal exchanges of calibration sequences between the two brother satellites during synchronous TDX data take operations on the two satellites, this exclusive bidirectional communication channel between the two satellites has been identified as an excellent means for a minimum information exchange between the two, whenever no operational data take is executed. Like two brothers prowling through a dangerous area

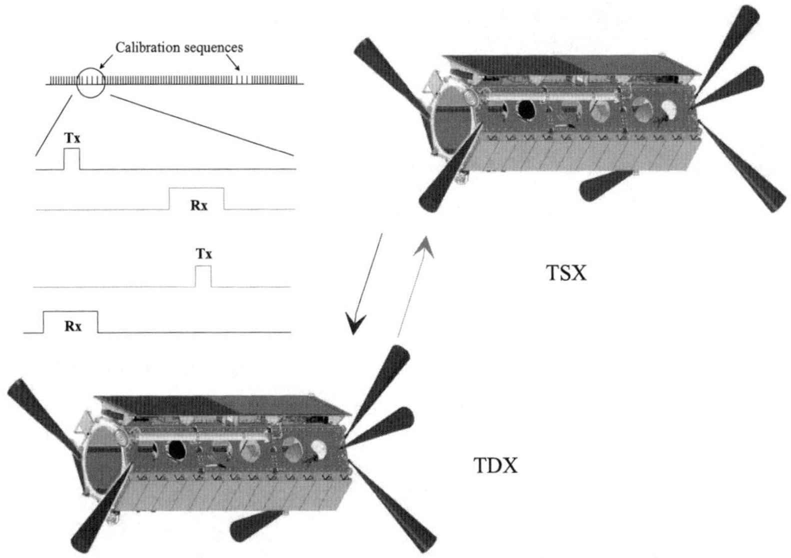

Fig. 29.11   Sync warning exchange.

without visible connection, at predetermined times each is whispering "I'm okay," while waiting for the same message from the other.

Technically this is achieved by dedicated Sync warning data takes, which are performed synchronously by TSX and TDX using deterministic Sync horn pairs. Typically two Sync warning data take pairs per orbit are planned from the ground based on a fully predictable satellite orbit position and attitude. An ICU onboard logic has been added to compare the received signal-to-noise ratio (SNR) with an adjustable threshold value. A value below the threshold is interpreted as a potential discrepancy in the planned relative formation geometry and a high severity event is generated by the ICU. The reaction mechanism is realized by the exclusion zone FDIR as described. After a failed Sync warning data take event, SAR transmission is suppressed to avoid the risk of mutual illumination by the call of exclusion zone OBCP and the exclusion zone function is disabled thereafter. Recovery from a Sync warning failure can be performed either by ground operation or automatically after the next successful Sync warning data take.

D. CONSTELLATION MAINTENANCE AND AUTONOMOUS ORBIT CONTROL FDIR

As a second, but only unidirectional, means in the whispering communication exchange of the two satellites, an additional S-band receiver and decoder have been installed on TDX, allowing it to "listen" to TSX S-band telemetry data. Thanks to this information link, TDX gains enhanced and real-time knowledge about TSX. Utilizing the configuration capability of the housekeeping and diagnostic report service capability, a special diagnostic telemetry packet is defined on TSX, which contains GPS navigation and TSX overall satellite status data. This one-way intersatellite link works for low data rates within the nominal separation distance range of the constellation, but leaves some contact gaps depending on communication system settings as well as on the relative formation geometry. The S-band antennas are mounted in the nadir and zenith directions; gaps therefore occur at equator crossing when the two spacecraft are next to one another. These gaps depend on the helix orbit and will not exceed 15 min per orbit.

Although the nominal formation flying is under ground control, TDX embarks on an autonomous formation flying algorithm (TAFF) able to conduct autonomous onboard constellation maintenance via the additional cold gas system on TDX. Data received from TSX telemetry via the ISL as well as internal TDX AOCS parameters are pre-processed within a specific module of the AOCS software (constellation safety preprocessing) to derive onboard parameters that can be monitored by the related onboard services. Furthermore, the received TSX GPS navigation solution data is used as input for the TDX TAFF algorithm running on the onboard computer. The TAFF algorithm was tested in navigation and open-loop mode during the commissioning phase, and has also been successfully operated in closed-loop mode.

Obviously, this novel approach of autonomous onboard constellation maintenance deserved an extension of the TDX FDIR. Based on the above mentioned preprocessed data, the configuration tables of the FDIR service suite in the AOCS application have been amended to cope with feared events for constellation and orbit maintenance operations. As an example, the constellation maintenance FDIR terminates and disables all autonomous constellation activities in the case where the TAFF algorithm indicates a collision risk but also if the time since the last valid data update expires beyond a given threshold. To ensure coordinated and synchronous orbit maintenance maneuvers on TSX and TDX, the execution of orbit control maneuvers on TDX is prohibited if TSX is not ready to transition to the orbit control maneuver. Last, but not least, all orbit control or constellation maneuvers are disabled and the SAR transmission function is prevented if the received TSX status via ISL indicates an OBC reboot or a switch to AOCS ASM.

V. CONCLUSION

The ambitious demands and risks of the close formation flight of TSX and TDX, forming the first configurable SAR interferometer in space, are handled by a variety of failure management mechanisms implemented on-ground, intrinsic to the constellation management, as well as onboard. The latter, being the last resort of satellite health preservation as well as key for high satellite availability, have been outlined within this chapter. The considerable challenges of the system upgrade from TSX to TDX could be efficiently implemented thanks to the modular functional architecture and consequent software design. The very few cases where experience could be gained during nominal operation are

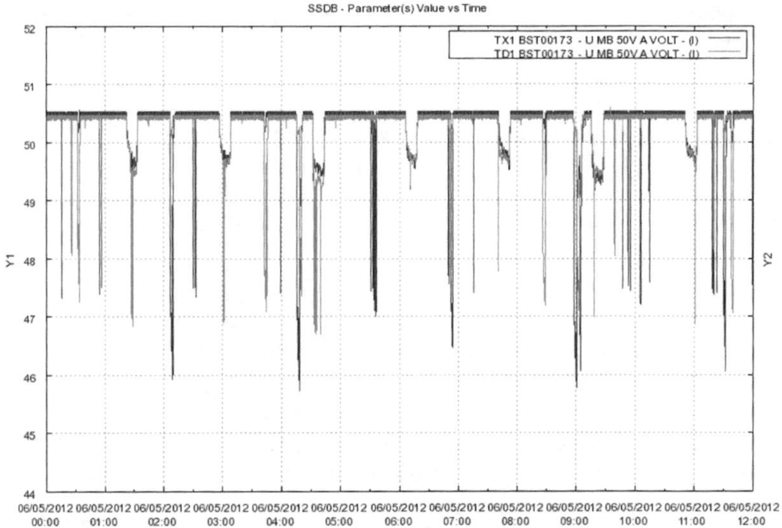

Fig. 29.12 Battery main bus voltage during SAR operation.

already showing the operational benefit of the implemented failure management mechanisms. Fortunately, no safe mode has occurred since the satellites have been in close formation. All data takes have been reliably planned with respect to exclusion zone limits, and have been uploaded to the satellites by the mission operations center, so the onboard exclusion zone mechanism has never needed to prevent SAR transmission or abort an active data take. The same applies for energy resource management, which is an integrated part of the mission planning system as well as for orbit control performed by the flight dynamics system of the DLR ground segment, supporting an efficient and safe operation of this ambitious mission [4–7].

More than 16,000 Sync warning data takes have been successfully exchanged between the satellites during the last 18 months of mission. Once, due to a small difference in Sync warning data take start times, a failed Sync warning was detected during a short period without full onboard time synchronization to the GPS time reference. During the next ground contact immediately after detection of the event, the related recovery action was performed by ground operations. If the installed event action entry, which is installed to further increase the availability of the SAR instrument and reduce ground operation interaction, had been enabled, after the next successful Sync warning a fully autonomous recovery back to full operational state would have been performed without ground interaction.

The long-term monitoring of satellite main bus voltages at the specified lifetime of TSX shows excellent margins according to the defined SW and HW DNEL levels. Figure 29.12 shows the TSX and TDX main bus voltages during a typical SAR operation scenario, comprising individual single-satellite TSX mission data takes as well as combined TDX mission data takes on both satellites. The chosen period includes a periodic discharge phase of the batteries during summer solstice eclipses. The slightly smaller minimum battery voltage of TSX compared to TDX is consistent with the longer in-orbit life of TSX. As reported for other spacecraft resources, the TSX power system characteristic, being far better than the original predictions, will allow a mission extension far beyond the specified as well as original planned mission lifetime.

ACKNOWLEDGMENTS

The TDX project is financed and implemented as a public–private partnership between DLR and Astrium, with partial funding by the German Federal Ministry for Economics and Technology (Förderkennzeichen 50 EE 0601).

REFERENCES

[1]   Zimmermann, R., Fischer, D., Giese, C., and Miller, D., "Safety Mechanisms for Operational Formation Flight of TerraSAR-X and TanDEM-X", *4th Spacecraft*

Formation Flying Missions & Technologies Conference, St. Hubert, Canada, May 2011, pp. 351–367.

[2] Giese, C., and Ulrich, D., "The TANDEM-X Space Segment,", *IEEE-2011-3895, International Geoscience and Remote Sensing Symposium*, Vancouver, Canada, July 2011.

[3] "TanDEM-X: Design & Interface Document," Document # TDX-AED-DD-0001, Astrium, Suresnes Cedex, France, 2008, pp. 408–410.

[4] Ardaens, J. S., D'Amico, S., and Fischer, D., "Early Flight Results from the TanDEM-X Autonomous Formation Flying System," *4th Spacecraft Formation Flying Missions & Technologies Conference*, May 2011, pp. 201–221.

[5] Herman, J., Fischer, D., Schulze, D., Loew, S., and Licht, M., "AOCS for TanDEM-X – Formation Flight at 200 m Separation in Low-Earth Orbit," *AIAA-2010-2375, SpaceOps*, Huntsville, Alabama, April 2010.

[6] Herman, J., Schulze, D., and Loew, S., "TanDEM-X Close Formation Flight – AOCS Safety Measures and Operations," *AIAA-2010-2195, SpaceOps*, Huntsville, Alabama, April 2010.

[7] Schättler, B., Kahle, R., Metzig, R., Steinbrecher, U., and Zink, M., "The Joint TERRASAR-X/Tandem-X Ground Segment," *IEEE-2011-2001, International Geoscience and Remote Sensing Symposium*, Vancouver, Canada, July 2011.

CHAPTER 30

Innovative Rover Operations Concepts–Autonomous Planner (IRONCAP): Supporting Rover Operations Planning on Ground

R. Steel,[*] **A. Hoffman**[†] **and M. Niézette**[‡]
Telespazio VEGA Deutschland, Darmstadt, Germany

A. Cimatti[§] **and M. Roveri**[¶]
Fondazione, Povo, Italy

K. Kapellos[**]
TRASYS, Hoeilaart, Belgium

A. Donati[††] **and N. Policella**[‡‡]
European Space Operations Centre, Darmstadt, Germany

I. INTRODUCTION

This chapter outlines the aims of the Innovative Rover Operations Concepts–Autonomous Planner (IRONCAP) study, gives a brief background as to why this study is needed, and presents the architecture for the prototype detailing the relation and positioning within a ground segment. Included are the expected interactions with the various components of a ground segment; contrasts between rover science operations planning and engineering operations planning illustrating the common or conflicting requirements; planning and scheduling techniques that are used within the prototype; details of the two demonstration test cases;

---

[*]Project Manager, Technology Division; robin.steel@vega.de.
[†]Software Engineer, Technology Division; alexander.hoffmann@vega.de.
[‡]Practice Leader, Technology Division; marc.niezette@vega.de.
[§]Project Manager, Embedded Systems Unit; cimatti@fbk.eu.
[¶]Software Engineer, Embedded Systems Unit; roveri@fkb.eu.
[**]Project Manager, Robotics; Konstantinos.Kapellos@trasys.be.
[††]Technical Officer, HSO-OS; alessandro.donati@esa.int.
[‡‡]Deputy Technical Officer, Future Studies Section; nicola.policella@esa.int.

and finally the synergies with other ESA projects currently under development. We conclude with a look at the results of the project to date and present the final steps needed to bring the project to completion. Through this study, new techniques, concepts, and technologies will be explored that will potentially benefit current and future rover missions planned by the European Space Agency (ESA).

IRONCAP is an ESA-run study project started in January 2011 and is being performed by a consortium of three members. The prime contractor is Telespazio-VEGA Deutschland-GmbH, with its two partners FBK and TRASYS. Each provides its own specific expertise to the study. VEGA brings its expertise in operational ground segments and flying space missions. TRASYS has extensive knowledge of rover operations and simulation. FBK provides the planning and scheduling experience of model synchronization and planning with uncertainty. This constellation provides for an effective and diverse knowledge base on which the study will develop, providing a fruitful result.

The key objectives of the study are as follows:

1. Assess and summarize the state-of-the-art concepts and technologies for operations of both orbiting spacecraft and rovers.

2. Define advanced concepts for controlling and monitoring rover operations, considering the presence of autonomous planning and execution capabilities in the rover segment. Enabling cutting-edge technology shall be considered during the course of the study because the focus is on future rover missions.

3. Identify possible engines and languages to handle the different types of planning data, such as occurrences, event, activities, and resources.

4. Identify optimum ways to synchronize onboard and ground planning processes.

The ultimate result of the study will be the development of a general-purpose proof-of-concept prototype providing a coherent and complete working implementation of an Automated Ground Activity Planning/Scheduling and Validation System for rover operations for ESA.

A. BACKGROUND

Before diving into the core of the work it is necessary to first understand some of the drivers for this. Planning of operations within ESA is usually distributed between two entities, one providing the science planning inputs and the other supplying the engineering planning inputs to the overall planning activities of any given spacecraft. Nominally, the interactions are at a planning level between these two entities, usually with each entity having their own set of tools to perform analysis of previous activities and to facilitate further planning tasks. These tools commonly have different knowledge bases built into them, with no real synchronization. One of the foremost rationales for this study is to investigate and define ways to harmonize these interactions, developing concepts and techniques that are applicable to both teams.

Another important driver is the prescribed levels of autonomy used to describe the capability of a spacecraft. There are four different levels of autonomy defined in the ECSS-E-70-11 standard for space applications:

1. E1: Execution under ground control;

2. E2: Execution of pre-planned mission operations onboard;

3. E3: Execution of adaptive mission operations onboard;

4. E4: Execution of goal-oriented mission operations onboard.

As can be expected, for most space rover operations only the E2 to E4 levels are really applicable due to the time delays that are nominally involved and synchronization of the daylight hours at the rover's location with those of the human controllers on Earth. Moon rovers may well be able to cope with the small latency to allow for E1 control of a rover, although simulated motion techniques would need to be used to guide the operator in control.

Currently, for most missions, autonomy level E2 is used. These are plans that consist of time-tagged schedules of activities. The onboard controller will execute these schedules and monitor the status of the execution. If some activity fails, the execution will be aborted immediately. This usually results in the controller ensuring that the payload and platform are in a safe state and then waiting for further instructions from the ground.

The next level of autonomy that is increasingly used in new and already operational missions is level E3. This level prescribes that the execution of activities onboard can be triggered and driven by events. The controller monitors the status of the system and environment, starting or selecting activities to be performed on the basis of conditions of the onboard telemetry. In this case IRONCAP is required to generate conditional plans that would integrate uncertainty regarding time, resource consumption, and environment/system state. The controller monitors the execution of the plan together with the environment and system state, triggering the execution of the plan branches based on their enabling conditions. If the plan execution fails the controller will react in the same way as at the lower level of autonomy and wait for further instruction from the ground (although non-critical operations can continue to run in the background). This could happen, for example, in the case that none of the precompiled options in the conditional plan are valid.

At the highest level of autonomy, the E4 level, the onboard controller expects a high-level plan of goals with constraints, which is expanded and mapped down to lower-level activities by the onboard planning system. The goals and constraints that are passed to the onboard system will generally be expressed at a lower level of abstraction than the goals that are defined as input to

IRONCAP. The on-ground planning process is therefore not limited to a process of merging and checking the consistency of possible conflicting goals, but also compiles the on-ground timeline to the level of abstraction required by the onboard controller. A mandatory requirement of the project is to be able to demonstrate the IRONCAP concepts using the goal-oriented autonomous controller (GOAC).

For most ESA missions the E2 level of autonomy is currently the preferred choice, even though this is typically applied to spacecraft and not rover missions. The E3 autonomy level is foreseen for future ESA missions but has already been used in a limited context within some missions in the form of onboard control procedures. For the time being, the E4 autonomy level has been addressed only at a prototyping level within ESA, although plans are already being discussed for operational missions.

B. CHALLENGES OF ROVER OPERATIONS

Rover operations in a foreign environment have many uncertainties that must be taken into account when planning tasks for the rover to achieve. When drilling into an object to take a sample it may not be known exactly how hard the object is, which has an impact on the length of drilling time needed to achieve the correct sampling depth. This then has an impact on the amount of power required to perform the drilling operation, which may then have further impacts on future operations. Another example is the simple task of driving the rover to a new location. This task may at first seem pretty simple, but there is an element of uncertainty that must be considered. The rover, as it traverses a terrain, may slip and slide as it is travelling. This slippage can slow down or, in some cases, even speed up the travel to the intended location. It could also require a non-straight line route to get to its target location due to this slippage or due to objects the rover encounters that must be avoided. All these situations require some type of uncertainty to be taken into account during the planning process. This is one of the challenges faced by the IRONCAP project.

Together with the inherent uncertainty involved in any rover mission there is the need to trade off where the actual planning of the rover operations is performed due to the differences in processing capabilities onboard and on-ground. Usually, as can be expected, rovers' capabilities in this area are weaker than those possible on-ground, and a trade-off is made as to which planning activities are to be done where. Processor- or memory-intensive tasks are usually performed on-ground to alleviate the load on the rover and save time during the planning process. This is not to say that the rover itself cannot perform such tasks onboard, but it may well be that in performing such a task the rover would need more computational time than that required on-ground and use necessary resources to achieve the task, possibly requiring additional recovery time on top of the task itself before continuing operations. The IRONCAP study aims to support this by providing on-ground planning tasks to support rover operation at the different levels of autonomy.

C. WHERE DOES IRONCAP FIT IN?

IRONCAP studied the concepts needed to define the operations of autonomous rovers and the systems required on-ground to support these operations. In relation to ground aspects it complements two parallel studies, the Autonomous Controller study completed by ESA in October 2011 and the Man–Machine Interface (MMI) for Exploration Missions (MMI4EXPL) study completed by ESA in September 2012. The Autonomous Controller study aimed to define and prototype an onboard autonomous controller capable of supporting the levels of autonomy up to E4. The end result provided an integrated onboard goal-oriented replanning functionality (GOAC) [1] that could be used on future ESA rover missions. Meanwhile, the MMI4EXPL study aimed to define and prototype the MMI concepts needed to support the different aspects of future and present space exploration missions [2].

This study project enables ESA to bridge the gap between the science planning and engineering planning for rover operation, combining both concepts into a single prototype tool that can be used in both environments. We provide concepts and techniques that can be used for current and future rover missions, as well as providing the necessary integration with GOAC by supporting the E2 to E4 levels of autonomy prescribed by the ECSS standard.

In the context of the ground segment, IRONCAP is positioned primarily in the Rovers Operations and Control Center of a given mission. Its aim is to enable the planning and scheduling of rover operations by both the scientific and engineering teams involved in the mission by providing the level of support needed for the autonomy level on the rover itself. To support this IRONCAP interfaces with the Mission Control System, obtaining telemetry data in the form of science/housekeeping data, and provides the schedules to be uploaded to the spacecraft in the form of time-tagged commanding and/or goal-oriented plans. For the planning of communication windows it also interfaces with the ESTACK Management System, receiving plan view files that contain station allocations to the rover.

II. ARCHITECTURE

The architecture of the system, shown in Fig. 30.1, takes into account the situational assessment required by the science and engineering planners and foresees an integrated 3-D visualization component. The study has already highlighted the reuse of 3DROV for this purpose, which is a comprehensive system developed to visualize and simulate rover activities [3].

IRONCAP allows the planner to generate plans of different kinds to fulfill the requirements imposed by the three different autonomy control levels, E2 to E4. During the course of the project we have identified the following four classes of

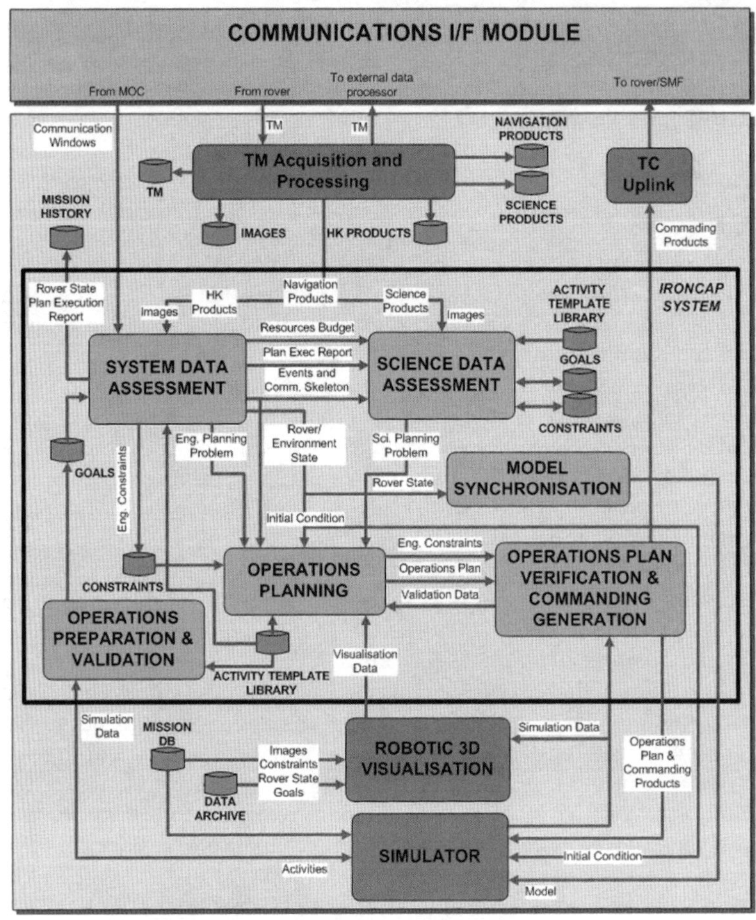

Fig. 30.1 IRONCAP architectural overview.

plans derived from the three levels of autonomy that are considered applicable to the project:

1. *Class A1.* This kind of plan is a simple sequence of activities with no conditional branches and no flexibility regarding the duration of activities, but allows for parallel activities. Typically, this is a "time-tagged schedule" within the ground segment. This kind of plan will be suitable for representing the E2 level of autonomy.

2. *Class A2.* This is an extension of the class A1 plan but with two aspects:
   (a) Conditions are introduced with only one branch (e.g., "if (battery_level_good) then goto(x,y); experiment(1)"). When the condition does not hold

during the execution of this type of plan, then the execution is halted, requiring ground-segment intervention/guidance.

(b) We also introduce flexibility on the start time and duration of activities, thus allowing for dependency on activities encoded in relationships between activities.

3. *Class A3.* This plan class further extends A2 by allowing for conditions with multiple branches, providing event-based autonomous operations to be executed onboard. For this class the execution of the plan is not necessarily halted if a condition is not met, as there may be a recoverable path still available in the execution.

4. *Class A4.* At the highest level of autonomy we derive the class A4 plan. This consists of a set of goals to be uploaded onboard to achieve goal-oriented mission replanning. The execution of this type of plan is decided onboard rather than on-ground, in contrast to the other classes of plan. These kinds of plans are compliant with those goal plans that the GOAC system is able to support.

From these definitions we can see that plans belonging to classes A1 to A3 allow for the execution of more than one activity in parallel. Plans in all four classes are also associated with the set of assumptions used for plan generation, which can be monitored during execution of the plan. These plans are not only tagged with timing information, they are also annotated with other situational checkers such as checking the battery level.

III. PLANNING AND SCHEDULING

An Operations Planner is responsible for the generation of possible solution plans of a given planning problem if such a solution exists. There may be times when no solution can be found and this is reported back to the user, who can then adjust the planning problem or initial constraints based on the reported debugging information. During refinement of the planning problem, several actions can be taken, such as the relaxation/strengthening of planning goals or assumptions, modification of constraints, or the introduction of additional goal and constraints. The appropriate action to take is governed by the debugging information provided to the operator for analysis.

Realization of this processing is provided within three elements: the Planner and Scheduler, the Planning Problem Editor, and the Operations Plan Editor. The Planner and Scheduler is used to determine and generate the actual planning and scheduling activities. In combination with this, the Problem Editor facilitates the refinement process, taking into account the debugging details produced by the Planner and Scheduler and/or validation data from a simulation facility when a solution to the planning problem cannot be found. Finally, the Operations Plan Editor makes last-minute adjustments or refinements that can be revalidated using the Planner and Scheduler.

To aid in the assessment of a valid solution, should one be found, additional components are used, such as 3-D visualization and simulation tools. These provide graphical feedback to the user, allowing a more thorough assessment of the solution. The level of autonomy has an influence on the procedure used to produce and analyze the solution. In the case of autonomy levels E1 to E3, the plan is returned to the operator for further refinement/analysis if a solution is found or processed further to obtain debugging information in the case a solution cannot be found, and this is then provided to the operator. For the highest level of autonomy, E4, a slightly different procedure is used whereby a selected subset of goals can be used to find a specific specialized solution. To do this, the feasibility of the goals is checked along with their controllability with respect to how strongly or weakly controllable they are. If the selected goals all provide a solution, then the operations plan is produced and returned to the operator; otherwise, debugging information is obtained, similar to that obtained from the other levels, which the operator can use to descope goals or modify the subset selection of goals to be achieved. These different procedures are necessary to cater for the different levels of autonomous behavior a rover supports and to successfully, efficiently, and effectively perform rover operations.

A.   SCIENCE AND ENGINEERING ASSESSMENT AND PLANNING

One of the most challenging aspects of the project was the understanding, definition, and analysis of the various angles that make up the science assessment/planning and the engineering assessment/planning. As with any rover mission, a situational assessment of the location of the rover has to be performed to establish the context in which the planning of operations can be performed. This situational analysis is performed on an engineering level and on a science level, both with their own goals and objectives. From an experience point of view, there are few rover missions around from which to really study the interactions and demands of the science and engineering teams. Accordingly, some assumptions have had to be made during the study based on experience from non-rover missions when devising possible operational scenarios.

With this in mind, the science assessment is mainly concerned with the evaluation and assessment of what science has been achieved since the last assessment, what exciting new science could be done from what we see now, the science observations already planned to be performed, and how to maximize the scientific return to the benefit of the community.

In stark contrast to this, the engineering assessment is aimed at the state of the space vehicle, constantly checking its health with respect to the last assessment, and tweaking parameters to better utilize the platform but in the safest way possible. This would usually involve an evaluation of any energy sources on the space vehicle (i.e., batteries, solar panels, etc.) and their performances, power consumers from the payload/platform, evaluation of all moving/mechanical parts on the vehicle (such as wheel motors, camera arms, internal relays, etc.), noting and reacting to any degradation in performances that could indicate a potential future failure on the spacecraft. This is similar to the assessments made for orbiting satellites.

This means that both assessments provide goals and objectives for the next planning stage, and these may or may not conflict with one another. It is therefore important to cater for both assessments when performing planning operations and provide a mechanism whereby conflicts of interest between the teams can be resolved and harmonized. The study investigated the collaboration and combination of these two situational assessment analyses with the ultimate aim of producing a prototype that will support both necessary approaches and facilitate interaction between them at the planning domain level.

B. MODEL REPRESENTATION AND SYNCHRONIZATION CONCERNS

Within the IRONCAP model, synchronization is responsible for the update and synchronization of the models used for planning and for plan validation, and for the synchronization of the model used by an external simulator with the model used for reasoning. Formal method techniques are used for the first kind of synchronization between planning and plan validation, and can be further broken down into the following:

1. Update of the initial state used for successive plan generation and validation;

2. Update of the model used for all formal reasoning;

3. Update of the assumptions used for plan generation.

For the synchronization of the initial state we will simulate the plan previously executed and downloaded from the rover. The simulation starts from the previously known state derived from telemetry and used for the previous plan generation, driven not only from the new plan but also from the information coming from this initial telemetry. Within this phase it is also possible that we will discover problems with the model used for simulation, so synchronization of the initial state is tightly integrated with the update of the reasoning model and of the assumptions. For the update of the model and of the assumptions under which planning takes place, we use techniques developed within the Onboard Model Checking Autonomous Reasoning Engine (OMC-ARE) [16] project for fault detection and identification. We identify possible faults and/or wrong assumptions from the telemetry information received, using this information to revise the model (e.g., introduce new fault behavior, strengthen the assumptions). Within IRONCAP we therefore exploit and extend techniques defined and used within the OMC-ARE project (OMCARE), and those discussed in [4, 5].

IV. VALIDATION AND VERIFICATION

Model and plan validation and verification are based on symbolic model checking techniques exploiting Satisfiability Modulo Theory (SMT) and abstraction refinement [6–12]. As described previously, we have formulated four types of plan definitions (A1–A4) that cover the three autonomy levels (E2–E4) covered within this study project. These four plan types cannot be treated in the same way because of their structure/content, so two algorithms were devised to cater for them.

Model verification and validation can be seen as a two-sided process, one validating the domain model against a set of expected behaviors, and the other validating the domain model against a set of unexpected behaviors. When validating using the expected behavior approach the operator is verifying that the model is sufficiently defined to allow safe operation of the spacecraft. When validating using the unexpected behavior approach the operator verifies that the domain model will not allow dangerous situations to occur due to faults in the model. The combination of these approaches allows the operator to perform what-if analysis by modifying one or more of the initial conditions, the expected or unexpected behaviors, the planning domain, the planning problem, or the operations plan. Observing the effect of the changes allows the operator to determine the validity of the models used and determine if the safety of the spacecraft would be jeopardized.

During plan verification and validation a two-algorithm approach is used. When dealing with plans of type A1 to A3 we use specialized verification algorithms that are based on model checking methods (Fig. 30.2). These algorithms

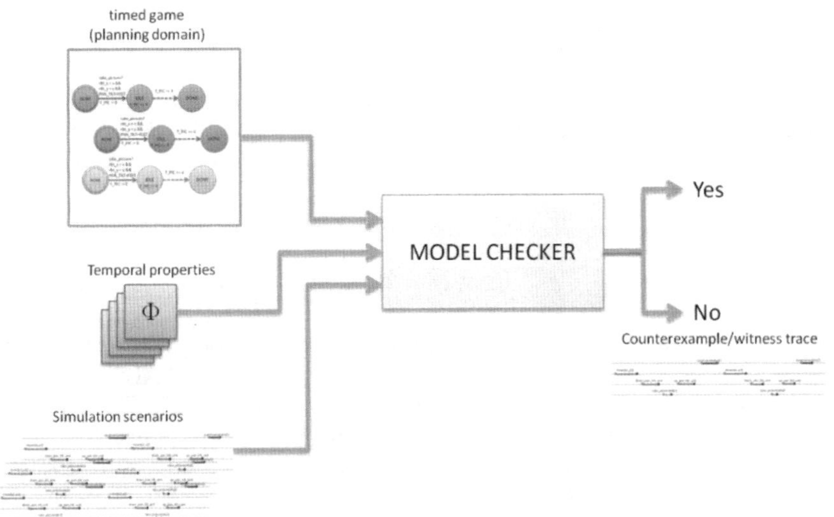

Fig. 30.2 Basic model checking approach.

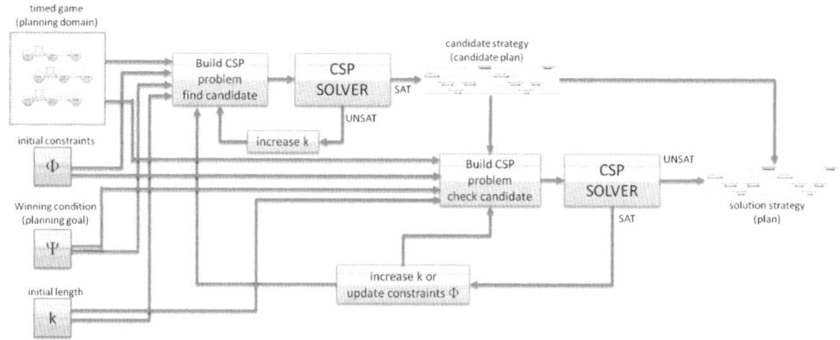

Fig. 30.3   Goal-based approach using the time-game principle.

aim to find paths of the operations plan that violate the goals set by the planning problem. If such a path exists then the operation plan is not a solution to the defined planning problem. Counter to this, if no such path exists then the operations plan is validated and is a solution to the planning problem.

The situation is different when dealing with plans of type A4, because these plans can contain a mixture of sequences, similar to A3 plan types, as well as goals. An initial processing of the operations plan needs to be performed first to elaborate the goals into sequences on the plan using knowledge of the goal planner mechanism. Once this has been achieved the resulting plan can be viewed and treated as a type A3 plan, meaning that the same verification algorithms as used for A1 to A3 plans can now be invoked to validate the operations plan. No further processing is performed for the parts of the original A4 operations plan that are already at the granularity of an A3 class. Instead, these are integrated with the results of the goal expansion, where the important feature of this expansion is that it should match the expansion that would occur onboard the actual spacecraft, meaning that the on-ground and onboard models have to be synchronized and the same processing mechanisms have to be available to the goal expansion routines.

For planning and scheduling we exploit hybrid-game approaches based on a mechanism of generate and test, but extended to the hybrid domain case, where we use SMT Model checking techniques to generate a candidate solution [13, 14]. We check, also with model checking techniques, that the candidate solution is a real solution for the considered planning and scheduling problem. Figure 30.3 illustrates the overall reasoning strategy that is used within the system.

V.   REUSE OF EXISTING SOFTWARE/CONCEPTS

As part of the exploration of software that could be reused for this development, the initial Advanced Planning and Scheduling Initiative (APSI) framework was considered in the context of this study, as well as the more recent version of this framework (APSI v2) [15]. APSI offers a structured and flexible library for effective modeling and solving of the planning and scheduling problems in different domains. The framework is based on the concept of timelines representing the temporal evolution of the system and of the environment. An initial analysis shows that the framework should be extended to allow the expression of nondeterministic effects at the discrete level, specifying uncertainty on the duration of activities, and uncertainty on resource consumption/production within an activity. Moreover, we see an extension to the framework to explicitly model the sensors (i.e., the observations). As far as a goal language is concerned, the possibility to associate goals with some sort of preference scheme (e.g., mandatory, or nice to achieve if resources allow) will also be assessed during the course of the study for its applicability.

The main reasoning capabilities of the prototype are provided by the New Symbolic Model Verifier (NuSMV) framework, which provides symbolic model checking in the form of pure binary reasoning. This is complemented by the MathSAT tool, which is an efficient SMT solver capable of handling conjunction constraints along with arbitrary Boolean combinations of theory atoms supporting a wide range of theories. To complete the process the algorithms of the NuSMV Extentions (NuSMT) tool will be exploited, providing user-friendly mechanisms and dealing with hybrid systems.

For the graphical MMI side of the prototype, the use of the Eclipse Rich Client Platform has been used to provide a common look and feel to that already being used within the ESA ground segment and other projects of the Agency. This is augmented by the Eclipse Modeling Framework (EMF), the Extended Editing Framework (EEF), Log4J, and components of the Eclipse Nebula Project.

Graphical representation of the simulated plan and rover activities is provided by the 3DRov environment and is the basis of one of the two demonstration cases. The tool provides end-to-end simulation for planetary rovers via 3-D visualizations. It provides models for the mechanical, electrical, and thermal subsystems of the rover, as well as the planetary environment. Scientific instrument models are also included to allow simulation of science-based scenarios. ESA's SIMSAT 4.1 simulator is the basis of the tool, with its models designed to comply with the Simulation Model Portability (SMP) 2.0 standard.

VI.   DEMONSTRATION VIA CASE STUDIES

A key aspect of the study is to demonstrate the capabilities and concepts defined by the study within two demonstrative test cases, both of which make use of different onboard controllers, the generic controller contained within 3DROV and the more specific one of the GOAC.

The first test case performed is based on the use of and interaction with the generic control provided by 3DROV, focusing on the planning of rover activities at the E2 level of automation. It is based on the international "Mars Sample Return" mission, which features at least two surface elements: a lander hosting a Mars ascent vehicle (MAV) and a sample-fetching rover (SFR). The SFR will collect soil samples from the Martian surface or subsurface while travelling from its landing place, and will return those back to the MAV. A robotic arm attached on the lander, the so-called Mars surface sample transfer and manipulation system, transfers the samples from the rover drill to the MAV for transfer back to Earth. The recent Phoenix mission has demonstrated that this type of system needs to be flexible and the operations have to be carefully planned to cope with unforeseen outcomes of the activities. This handling of uncertainty is a key feature of the IRONCAP study.

For the second case we used the controller developed by the GOAC study to operate the rover, focusing on the E4 level of automation. The scenario used is one that was also used to demonstrate the functionally of the GOAC. It consists of the Sample Acquisition part of the ExoMars experiment cycle, which is composed of two subtasks, Site Panoramic Investigation and Drill Region Characterization. Initially, the rover acquires a panoramic view of the site using the PanCam instruments to provide the scientists on-ground with the necessary information that allows them to select an outcrop of rocks to be further investigated by the rover module instruments. The region selected for drilling is selected ether with ground intervention or autonomously onboard. This is typically performed after a long journey when a new site is reached to evaluate the local characteristics. The rover travels closer to the selected area, performs the Drill Region Characterization, and drills at the selected area. The Drill Region Characterization (also called the WISDOM pattern) is an experiment where the rover covers a 5 m × 5 m area, taking a WISDOM measurement at appropriate points. This provides the scientists on-ground with the information necessary to identify the most interesting point for drilling within that area. Usually in the next sol, this target point is communicated to the rover to allow it to approach the point and accurately position the drill for drilling.

VII. CONCLUSION

IRONCAP is setting the grounds for the planning and scheduling of operations and activities of future ESA interplanetary rover missions. It is developing and evolving the concepts required to efficiently and successfully carry out rover operations at the three main levels of autonomy. It provides a prototype that intends to bring the science and engineering situational assessments together into a common tool.

Future extensions/usage of IRONCAP could include the assistance of human space flight and surface operations on foreign worlds through autonomous robotic control via goal-oriented commanding. Further functionality could include the addition of plug-ins to enhance the communications between the teams and team members by means of twitter or RSS-like interfaces reporting to mobile devices, or even being used directly on mobile devices to issue additional goals to be achieved by a remote agent already executing the current schedule.

These exciting aspects will need to be studied further to establish their feasibility and real-world applicability outside the scope of this work. The project was concluded in the first half of 2013.

REFERENCES

[1] Fratini, S., Cesta, A., De Benedictis, R., Orlandini, A., and Rasconi, R., "APSI-based Deliberation in Goal Oriented Autonomous Controllers," *Proceedings of 11th ESA Symposium on Advanced Space Technologies in Robotics and Automation, ESA/ESTEC*, Noordwijk, The Netherlands, April 2011, pp. 345–360.

[2] Man–Machine Interface (MMI) for Exploration Missions (MMI4EXPL) Final Report, ESA Report 2012-3421, Paris, France, June 2012.

[3] Engelmore, R., and Morgan, A. (eds.), *Blackboard Systems*, Addison-Wesley, Reading, MA, 1986, pp. 134–136.

[4] Bozzano, M., Cimatti, A., Guiotto, A., Martelli, A., Roveri, M., Tchaltsev, A., and Yushtein, Y., "On-Board Autonomy via Symbolic Model Based Reasoning," *10th ESA Workshop on Advanced Space Technologies for Robotics and Automation*, ESA/ESTEC, Noordwijk, The Netherlands, 2008.

[5] Cimatti, A., Guiotto, A., and Roveri, M., "On Board Model Checking for Space Applications," *Proceedings of ESA Workshop on Avionics Data, Control and Software Systems (ADCSS)*, ESA/ESTEC, Noordwijk, The Netherlands, 2008, pp. 654–670.

[6] Audemard, G., Cimatti, A., Kornilowicz, A., and Sebastiani, R., "Bounded Model Checking for Timed Systems," *Proceedings of FORTE*, Vol. 2529, 2002, pp. 243–259.

[7] Bozzano, M., Bruttomesso, R., Cimatti, A., Junttila, T., van Rossum, P., Schulz, S., and Sebastiani, R., "An Incremental and Layered Procedure for the Satisfiability of Linear Arithmetic Logic," *Proceedings of TACAS*, Vol. 3440, 2005, pp. 317–333.

[8] Bozzano, M., Bruttomesso, R., Cimatti, A., Junttila, T., van Rossum, P., Schulz, S., and Sebastiani, R., "The MathSAT 3 System," *Proceedings of CADE*, Vol. 3632, 2005, pp. 315–321.

[9] Bruttomesso, R., Cimatti, A., Franzen, A., Griggio, A., and Sebastiani, R., "The MathSAT 4SMT Solver," *Lecture Notes in Computer Science*, Vol. 5123, 2008, pp. 299–303.

[10] Cimatti, A., Griggio, A., and Sebastiani, R., "Efficient Interpolant Generation in Satisfiability," *Proceedings of TACAS*, Vol. 7, 2008, pp. 397–412.

[11] Clarke, E. M., Kurshan, R. P., and Veith, H., "The Localization Reduction and Counter-example Guided Abstraction Refinement," *Essays in Memory of A. Pnueli*, 2010, pp. 61–71.

[12] Clarke, E. M., Fehnker, A., Han, Z., Krogh, B. H., Stursberg, O., and Theobald, M., "Verification of Hybrid Systems Based on Counterexample-Guided Abstraction Refinement," *Proceedings of TACAS* Vol. 2619, 2003, pp. 192–207.

[13] Brafman, R. I., and Hoffmann, J., "Conformant Planning via Heuristic Forward Search: A New Approach," *Proceedings of the 14th International Conference on*

Automated Planning and Scheduling, British Columbia, Canada, June 2004, pp. 568–580.

[14] Hoffmann, J., and Brafman, R. I., "Conformant Planning via Heuristic Forward Search: A New Approach," *Proceedings of the 14th International Conference on Automated Planning and Scheduling*, British Columbia, Canada, June 2004, pp. 421–431.

[15] Steel, R., Niezette, M., Cesta, A., Fratini, S., Oddi, A., Cortellessa, G., Rasconi, R., Verfaille, G., Pralet, C., Lavagna, M., Brambilla, A., Castellini, F., Donati, A., and Policella, "Advanced Planning and Scheduling Initiative: MrSPOCK AIMS for XMAS in the Space Domain," *6th International Workshop on Planning and Scheduling for Space*, IWPSS-09, 2009.

[16] "OMCARE: On Board Model Checking Autonomous Reasoning Engine," Oct. 2010, https://es.fbk.eu/projects/esa_omc-are/ [retrieved 29 October 2011].

CHAPTER 31

Integral: Investigation into Van Allen Belt and Geotail Crossings

M. J. H. Walker*
SCISYS Deutschland GmbH, Darmstadt, Germany

J. B. Palmer[†]
Logica GmbH, Darmstadt, Germany

I. INTRODUCTION

The Integral spacecraft, a mission to research gamma ray sources, was launched on 17 October 2002 into a highly elliptical near-polar orbit with a period ("revolution") of three sidereal days. The scientific payload consisted of a gamma-ray spectrograph (SPI), a gamma-ray imager (IBIS), two X-ray imagers (JEMX 1 & 2), and an optical monitor (OMC). Because all these instruments are sensitive to and are impacted by high radiation levels, a radiation environment monitoring device (IREM) [1] was also included as part of the payload. To protect the instruments from excessive radiation, they are commanded to a safe mode during periods of known high particle density, for example, during Van Allen Belt transits or when the IREM detects other high-level radiation environments such as during solar flares.

Because the satellite is periodically outside ground-station contact, use is made of onboard autonomy. This is governed by the spacecraft's Central Data Management Unit, which issues environmental information in the form of a telecommand packet to each instrument every 8 s. This packet includes, amongst other parameters, both ground-specified events such as predictions of entry/exit times of the radiation belts and eclipses, as well as three real-time radiation environment parameters directly from the IREM. The instruments react autonomously to the entry and exit times, ensuring an orderly shutdown of the instruments before the predicted entry into hostile environments. Without such orderly shutdowns, the instruments would undergo emergency shutdowns based on the IREM measurements and subsequently have to undergo lengthy reactivation procedures.

With this in mind, it is clear that the planned duration of scientific operations will be influenced by the accuracy of the Van Allen Belt predictions. The more accurate the prediction, the closer to the predicted times and therefore the longer the instruments can be safely operated, avoiding emergency switch offs.

*Spacecraft Operations Engineer, ESA/ESOC HSO-OAI; mike.walker@scisys.de
†Space Flight Dynamics Engineer, ESA/ESOC Flight Dynamics; jpalmer@esa.int

II.   IREM AND INSTRUMENT SWITCH-OFF LOGIC

The onboard IREM device measures 15 radiation types, which are telemetered to ground as part of the science data (Table 31.1).

Three radiation environment parameters, supplied by IREM, were selected to be distributed to the payload instruments:

1.   TC3: $E(e) > 0.5$ MeV (soft electrons & protons);

2.   S14: 27 MeV $> E(p) > 20$ MeV (protons);

3.   Dose: a derived quantity based on the above values, plus dead times.

Note that these are not the actual counts but a scaled measure of the environment (number of counts per 10 s divided by 256). The actual count rates would require more telemetry and command bandwidth, and such accuracy is not required for autonomous onboard protection.

Each instrument has three threshold settings for each of the above parameters, which, when exceeded, will cause the unit to switch immediately to safe mode. In practice, it has been observed that the TC3 and S14 parameters are the first to hit the limit, so subsequent analysis has been based only around these quantities. The

TABLE 31.1   MEASURED RADIATION TYPES

Parameter	Description	Particle types
TC1	$E(p) > 20$ MeV	Proton
S12	550 MeV $> E(p) > 20$ MeV	Proton
S13	120 MeV $> E(p) > 20$ MeV	Proton
S14	27 MeV $> E(p) > 20$ MeV	Proton
S15	34 MeV $> E(p) > 20$ MeV	Proton
TC2	$E(p) > 39$ MeV	Proton
S25	185 MeV $> E(p) > 150$ MeV	Ions
C1	50 MeV $> E(p) > 40$ MeV	Coincident protons
C2	70 MeV $> E(p) > 50$ MeV	Coincident protons
C3	120 MeV $> E(p) > 70$ MeV	Coincident hard protons
C4	$E(p) > 130$ MeV	Coincident protons
TC3	$E(e) > 0.5$ MeV	Soft electrons & protons
S32	2.3 MeV $> E(e) > 0.55$ MeV	Soft electrons & protons
S33	90 MeV $> E(p) > 11$ MeV	Soft proton
S34	30 MeV $> E(p) > 11$ MeV	Proton

TABLE 31.2   INSTRUMENT RADIATION THRESHOLDS FOR TRIGGERING SAFE MODE

Instrument	Electron threshold TC3	Proton threshold S14
SPI	Not used	Not used
IBIS	200	60
JEM-X	64	20
OMC	2048	16

thresholds specified for each instrument in terms of IREM counts are given in Table 31.2.

III.   ORBIT GEOMETRY AND INSTRUMENT OPERATIONS

It was decided for the purposes of this investigation to use the Integral Science Data Centre [2] definition of 600 counts per minute to define belt entry/exit, as this permits the analysis of data stored onboard during the perigee passage that is not available in real time. These times were recorded and, from them, spacecraft altitudes were calculated. Figure 31.1 shows the resulting plot of the altitudes for threshold crossings during about 1215 orbital revolutions, that is, about 10 years. These values are taken for the purpose of defining Van Allen Belt entry and exit altitudes.

In trying to interpret the data, a first attempt was to separate the entry and exit curves and overlay them with a curve showing the eclipse season times, giving an

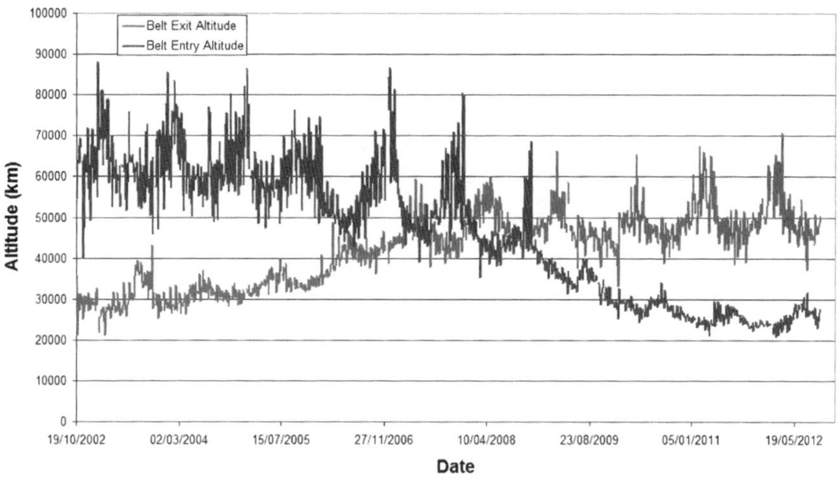

Fig. 31.1   Integral radiation belt entry/exit altitudes.

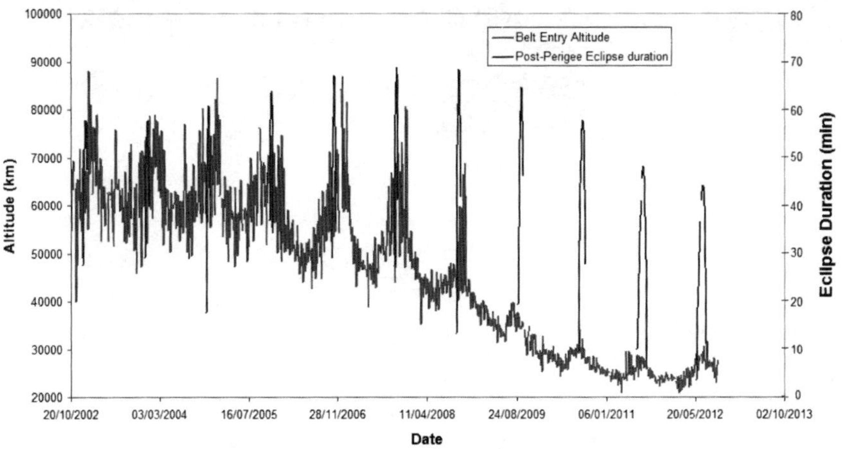

Fig. 31.2   Integral radiation belt entry with post-perigee eclipse duration.

indication of when the Sun crosses the spacecraft orbital plane. The eclipses fall into two categories, pre and post perigee, separated by about six months. The belt entry curve was overlaid with the post-perigee eclipse, and the belt exit with the pre-perigee eclipse (Figs. 31.2 and 31.3).

From Figs. 31.2 and 31.3, it is immediately apparent that the spikes on the belt entry/exit altitudes are correlated to the eclipse periods and therefore to the Sun crossing the orbital plane. It is at these times that the spacecraft is passing behind the Earth through the geotail, a region where charged particles are

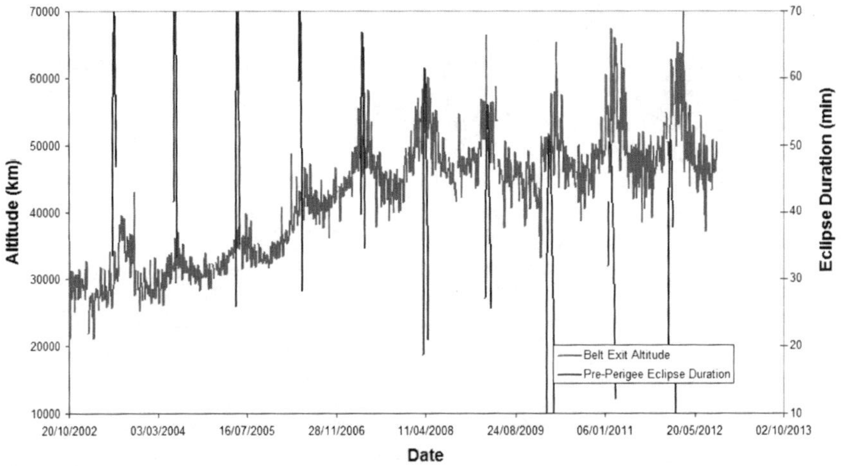

Fig. 31.3   Integral radiation belt exit with pre-perigee eclipse duration.

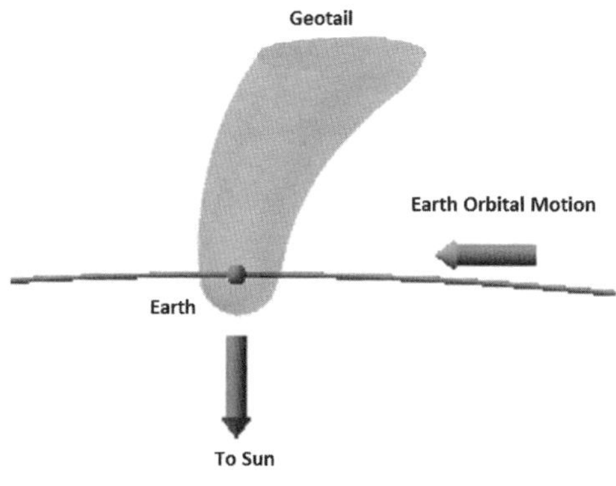

Fig. 31.4   The geotail modeled as a curved paraboloid.

concentrated and shaped much like a comet's tail, and hence the higher spacecraft altitude where this region is encountered.

Kalegaev et al. [3, 4] and others [5] model the geotail as a paraboloid, with the Earth at the focus and the tail extending away from the Sun. However, the data collected by Integral indicate that the geotail does not extend directly away from the Sun, but may be curved, again like a comet's tail, as in Fig. 31.4. This would agree with the east–west asymmetry of He^{2+} ions observed by Stubbs et al. [6]. The AP8–AE8 models of the radiation belts do not include any parameters that could explain any such east–west asymmetry [7, 8]. However, more recent models of the magnetosphere include a dawn–dusk asymmetry (e.g., Tsyganenko [9]), which may help to explain the effect. Bearing in mind the orbital motion of the Earth it seems reasonable that the geotail is asymmetric (also shown in Fig. 31.4).

By overlaying the annual plots of the belt entry/exit and aligning them using the maximum eclipse duration (vertical line), the shearing of these curves with respect to the eclipse becomes obvious. The belt entry altitudes for the post-perigee eclipse seasons and the belt exit for the pre-eclipse seasons are shown in Figs. 31.5 and 31.6, respectively. Note that in Fig. 31.5, in later years, when the entry altitude was lower, the shearing effect is not as pronounced compared with the early years, which have a higher entry altitude. The same effect can be seen in Fig. 31.6; namely, the shearing effect is much more pronounced at higher exit altitudes.

The cause of this shearing is more of a riddle. A possible explanation for the effect may lie with the nature of Integral's orbit. For ground-station visibility reasons, this was chosen to have a period of three sidereal days, which gives a three-day repeating ground track. Consequently, Integral will pass through the Van Allen belts at more or less the same location with respect to the Earth's magnetic field every orbit, in a kind of orbital strobe. Other spacecraft in nonsynchronous orbits would fly through the belts at a wide variety of geocentric longitudes, making the observation of a regular pattern more difficult. It should be noted that the motion of the Integral orbit was uncontrolled during 2006, at the end of which an orbit maneuvre was performed to halt the eastward drift and maintain the new orbital "station." Thus, it has been possible to collect data for slowly varying belt entry/exit locations with respect to a geocentric coordinate system. Figures 31.7 and 31.8 show the location of belt entry and exit expressed in terms of latitude

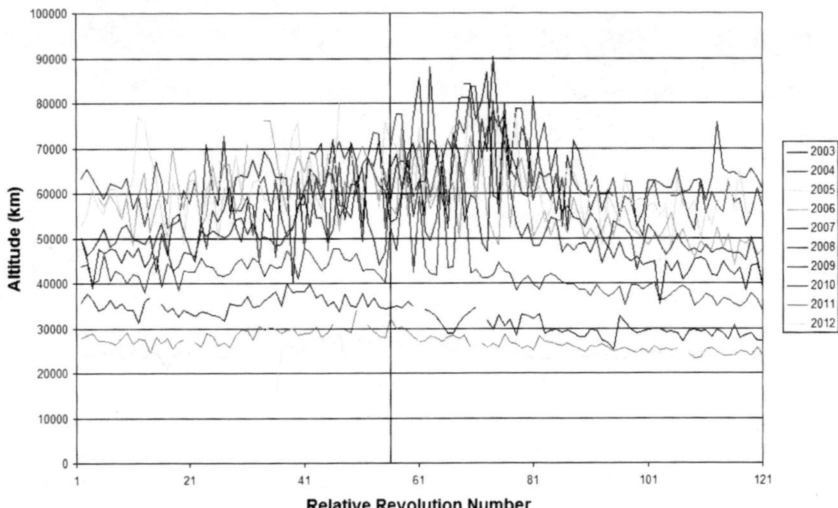

Fig. 31.5 Overlaid belt entry altitudes (the vertical black line marks the maximum eclipse duration).

and longitude. However, if the Earth's magnetic field is responsible for the asymmetric nature of the geotail, it should be expected, at different times of day, to point forward as well as backward. Using the data measured by Integral's IREM instrument, this has not been observed. Analysis of data collected by other spacecraft is necessary to test such a theory.

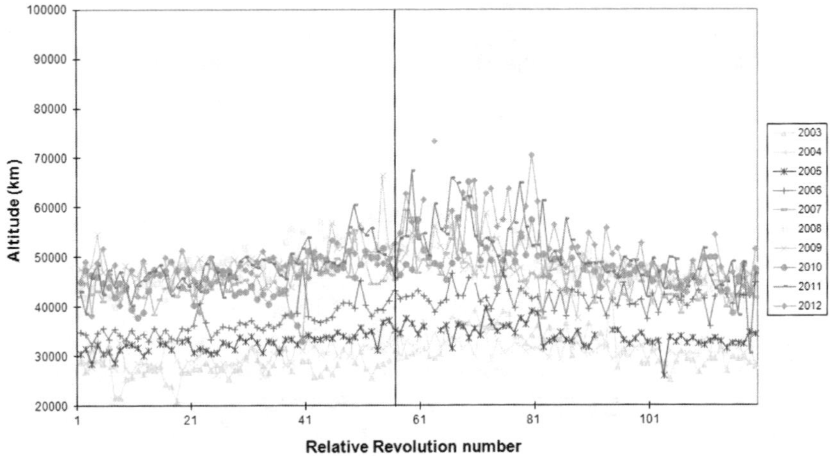

Fig. 31.6 Overlaid belt exit altitudes (black line marks the maximum eclipse duration).

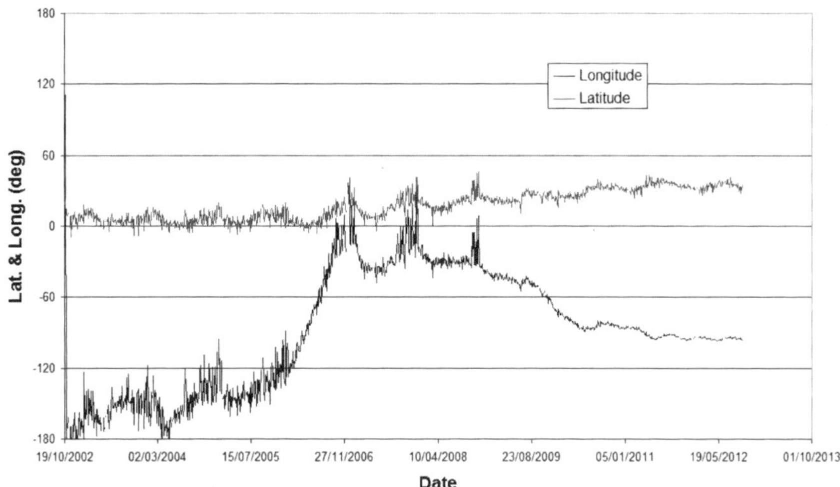

Fig. 31.7 Geocentric location of the radiation belt entry.

As input to the onboard autonomy, Integral requires specific times to be loaded with the predicted radiation belt entry and exit times, as well as eclipse entry and exit times. These inputs are known as "critical altitude descending," which define the radiation belt entry, and "critical altitude ascending," which in turn define the belt exit. Plots showing the history of these two parameters are

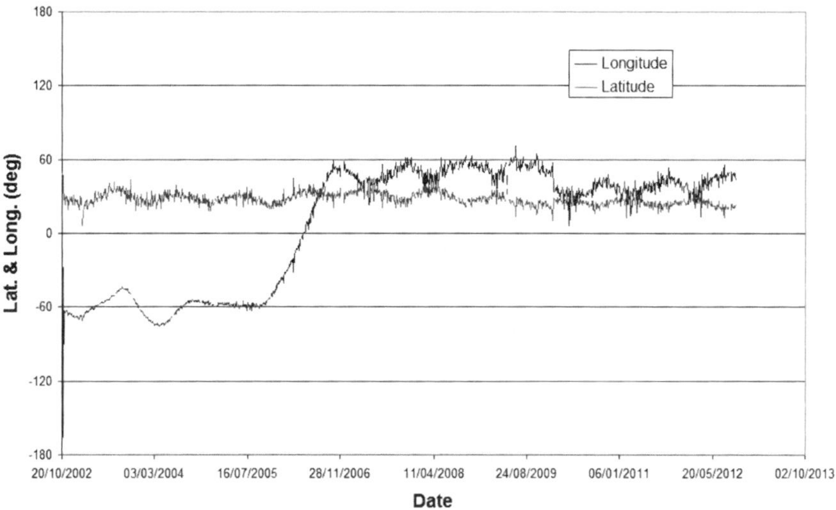

Fig. 31.8 Geocentric location of the radiation belt exit.

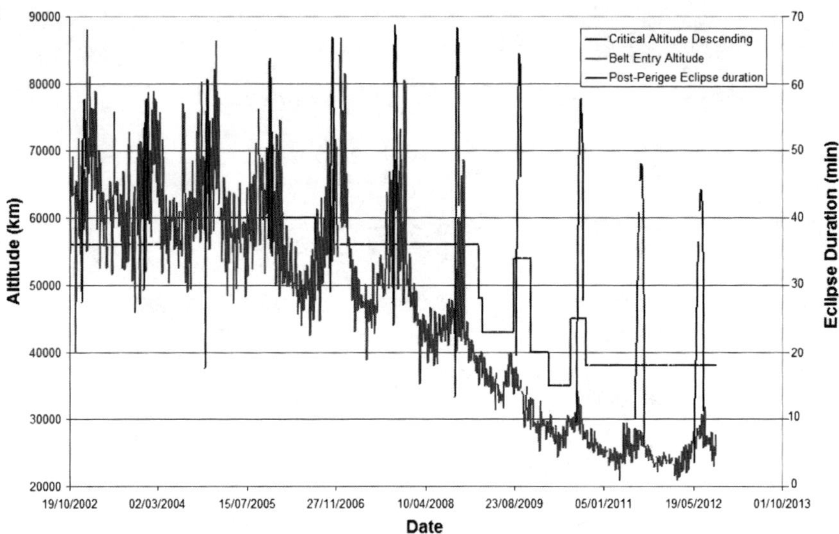

Fig. 31.9   Radiation belt entry with eclipse times and the "critical altitude descending."

given in Figs. 31.9 and 31.10. It should be noted that it is possible to commence payload activation operations before the belts exit, as it takes about 20 min to ramp up the high voltages, and in this time the radiation environment will naturally fall to tolerable levels. It was not possible to apply a similar method to

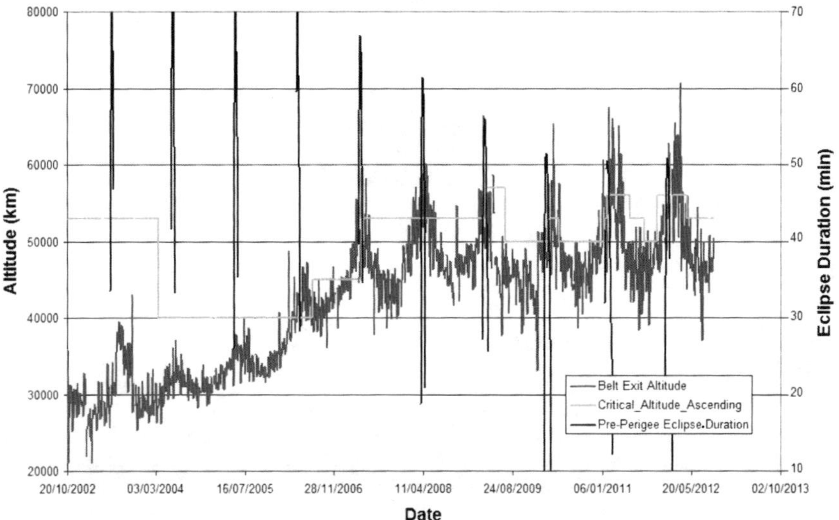

Fig. 31.10   Radiation belt exit with eclipse times and the "critical altitude ascending."

radiation belt entry due to the operational requirement to provide sufficient margin for manual intervention should a problem prevent the onboard autonomy from working.

IV. ORBIT EVOLUTION

The other pronounced feature of Figs. 31.2 and 31.3 is the trend of the belt entry altitude to reduce over time, whereas the belt exit altitude increases. This is explained by the orbit evolution experienced by Integral since launch. As the line of apsides rotates, this affects the measured belt entry and exit altitudes. With an argument of perigee around 270°, the orbit is "upright," but as this value decreases, the belt exit altitude increases and, correspondingly, the belt entry altitude decreases. As an example, Fig. 31.11 shows the belt entry and exit curves for revolution 1220 (left- and right-hand tracks, respectively). Note the very low belt entry altitude compared with that for the belt exit. However, it is expected from orbital perturbations that this evolution will reverse with time. As of September 2012, the belt entry was deep within the cusp of the Earth's magnetosphere. Owing to the nature of the orbit being an integral number of sidereal days, this cusp crossing will always be in the same place on the geoid.

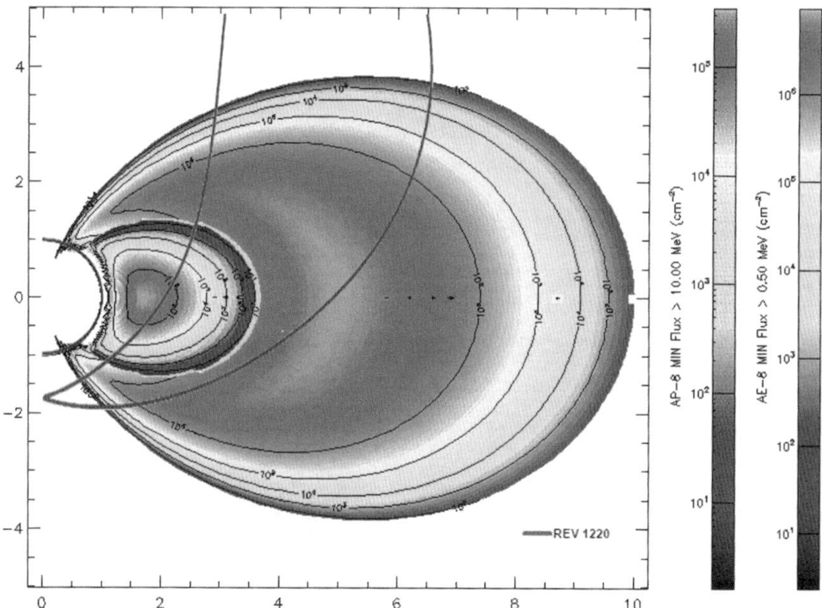

Fig. 31.11 'Fish-eye' plot showing a schematic of belt entry and exit trajectories (data from the AP8/AE8 programs [6, 7]).

At launch (17 October 2002), the Integral orbit had the following characteristics:

1. Semi-major axis 86,791.02 km;

2. Eccentricity 0.83252;

3. Inclination 52.17°;

4. Longitude of ascending node 103.29°;

5. Argument of perigee 301.73°.

This corresponds to an apogee height of 153,666.7 km and a perigee height of 9059.3 km, with a period of three sidereal days. The first four graphs in Fig. 31.12 show how the relevant orbital parameters evolved and are predicted to evolve. The plots show the actual evolution during Integral's lifetime and a prediction of their

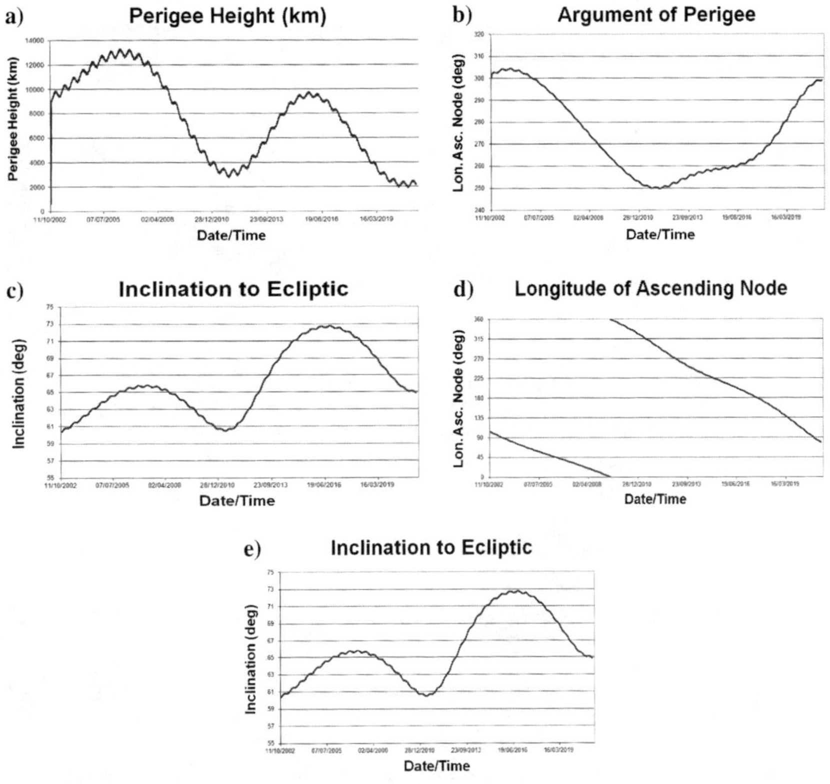

Fig. 31.12   Orbital evolution.

evolution to February 2021. Note that the inclination to the ecliptic is derived using the inclination and longitude of the ascending node and also that the semi-major axis has not been included, as active orbital control is exercised to maintain the period to three sidereal days.

Thus, applying the geometry observed during the first 10 years of operations, it is expected that the effect of the proton belts will gradually reduce as the perigee height rises, reaching a minimum around December 2015. After that the perigee height falls again, and the proton counts will rise once more. This will continue until late 2020, after which orbital perturbations will again cause the perigee height to rise. The effect of the rotation of the line of apsides will mean that the Van Allen Belt (electron and where appropriate proton) entry altitude will start to rise, reaching levels similar to those seen near the start of the mission. However, in compensation the belt exit altitude will reduce, maintaining science time. The orbit inclination with respect to the ecliptic will also oscillate considerably, but will remain high enough that the eclipse periods will remain seasonal and not extend over large parts of the year. Thus, the times when Integral passes through the geotail will remain relatively short.

V. SOLAR FLARES

In addition to the science operation restrictions due to Van Allen Belt entry and exit, solar flares occasionally hinder science operations. Although these solar flares are an order of magnitude less intense than the radiation encountered during Van Allen Belt passage, they can still cause the instruments to automatically go to safe mode. This occurs with a frequency following the 11-year solar cycle. During the solar minimum it was a rare occurrence, whereas during the solar maximum there could be about 10 incidents a year. A typical belt passage will cause the electron count to exceed 65,535 IREM counts in 8 s and go off scale, whereas the peak of the most intense solar storm for five years reached only 11,715 IREM counts, although with a much longer duration. It should also be noted that the proton belt, although more damaging than the electron belt, always lies under the latter. Therefore, the proton counts themselves are only rarely the limiting factor for instrument switch-off. When this occurs it is usually during the dissipation phase following a flare. For this study, however, such flares mask the belt entry and exit altitudes for that revolution. The effect of the solar cycle is to increase the frequency of such events; the belts themselves seem to remain more or less unaffected. As an example, Fig. 31.13 shows the radiation counts experienced during a solar flare where operations were interrupted for an entire orbit.

VI. LONG-TERM PREDICTIONS OF BELT ENTRY AND EXIT ALTITUDES

An initial evaluation of the future trend of belt entry/exit altitudes was made based on the observations made since launch. The major component of this

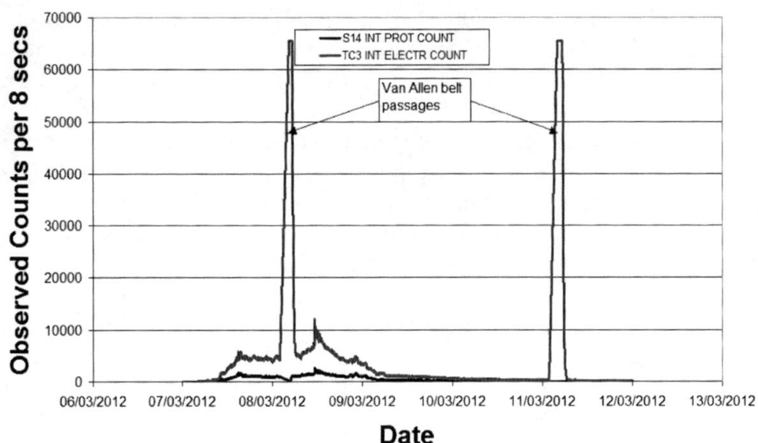

Fig. 31.13   Observed radiation levels during the storm of 8–9 March 2012.

trend is the evolution of the argument of perigee (perigee height does not contribute that much owing to the radial nature of the belts). By comparing the belt entry/exit altitude values observed earlier in the mission and projecting these forward using the argument of perigee as a reference, a prediction was prepared (see Figs. 31.14 and 31.15).

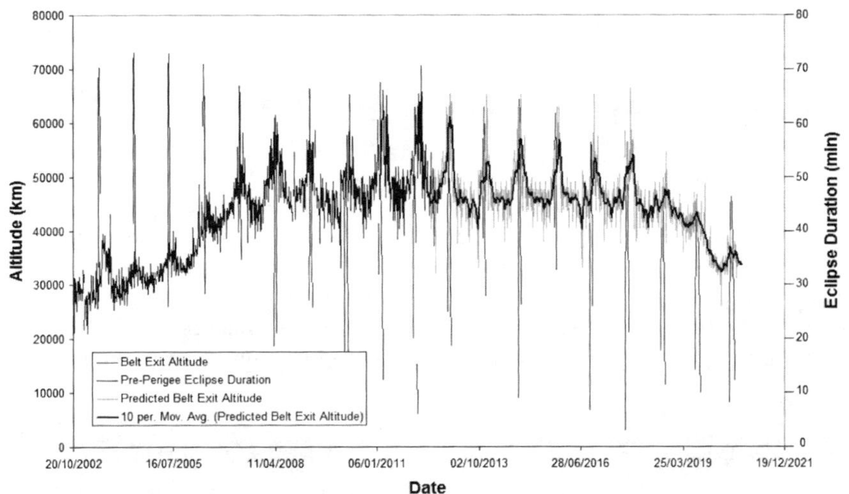

Fig. 31.14   Predicted evolution of the radiation belt exit altitude.

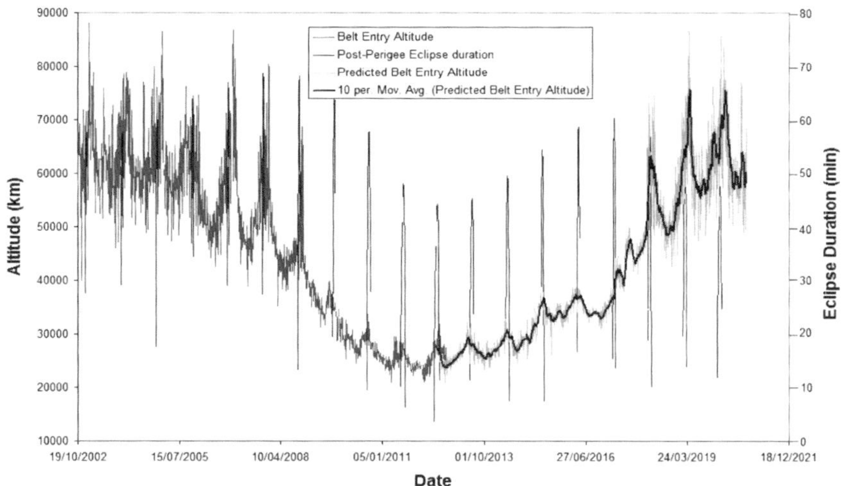

Fig. 31.15   Predicted evolution of the radiation belt entry altitude.

VII.   CONCLUSION

By making use of the findings in this chapter, Integral Science Operations achieved a gain of up to 40 min science time per orbit. However, this was achieved during periods of relatively low radiation at belt exits (i.e., not near the geotail). These gains were not applied at radiation belt entries due to the operational requirement to provide adequate time for manual commanding should an instrument or spacecraft problem prevent the onboard systems from switching the instruments to safe mode. The long-term prediction of radiation belt entries and exits may assist in strategic planning of science operations into the future.

Despite a distinct difference between the geocentric locations of the belt transitions in the early phase of the mission and those in the later phase, it has not been possible to demonstrate a correlation between the bending of the geotail and Earth's magnetic field, and the cause of this effect remains unknown. Further analysis using other spacecraft in different orbits may give results that enable conclusions to be drawn.

REFERENCES

[1]   Hajdas, W., Buhler, P., Eggel, C., Favre, P., Mchedlishvili, A., and Zehnder, A., "Radiation Environment Along the INTEGRAL Orbit Measured with the IREM Monitor," *Astronomy & Astrophysics* Vol. 411, 2003, pp. L43–L48.

[2]   Integral Science Data Centre, http://www.isdc.unige.ch/integral/ [last accessed 02 September 2012].

[3]   Alexeev, I., and Kalegaev, V., "Global Modeling of the Magnetosphere in Terms of Paraboloid Model of Magnetospheric Magnetic Field," *Journal of Geophysical Results*, Vol. 106, January 2001, pp. 245–257.

[4]   Kalegaev, V. V., Alexeev, I. I., and Feldstein, Ya. I., "The Geotail and Ring Current Dynamics Under Disturbed Conditions," *Journal of Atmospheric and Solar-Terrestrial Physics*, Vol. 63, No. 5, 2001, pp. 102–119.

[5]   Garrett, H., "Guide to Modeling Earth's Trapped Radiation Environment", AIAA Technical Standard Q11 G-083-1999. AIAA, Washington, DC, July 1999, pp. 34–36.

[6]   Stubbs, T. J., Lockwood, M., Cargill, P., Fennell, J., Grande, M., Kellett, B., Perry, C., and Rees, A., "Dawn–Dusk Asymmetry in Particles of Solar Wind Origin Within the Magnetosphere," *Annales Geophysicae*, Vol. 19, 2001, pp. 1–9. © *European Geophysical Society*, 2001.

[7]   Sawyer, D. M., and Vette, J. I., "AP-8 Trapped Proton Environment for Solar Maximum and Solar Minimum," Report number 76–06, NSSDC/WDC-A-R&S, December 1976, pp. 79–80.

[8]   Vette, J. I., "The AE-8 Trapped Electron Model Environment," Report number 91–24, NSSDC/WDC-A-R&S, June 1991, pp. 83–86.

[9]   Tsyganenko, N. A.," A Model of the Near Magnetosphere with a Dawn–Dusk Asymmetry. 1. Mathematical Structure," *Journal of Geophysical Research*, Vol. 107, 2002, p. 1179.

Note: Page numbers are followed by *f* or t (indicating figures or tables).

COLOR PLATES

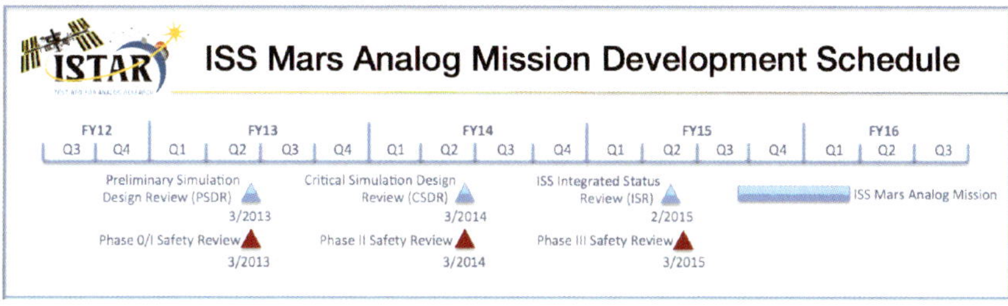

Fig. 1.9   Representative (notional) ISS Mars Analog mission development schedule.

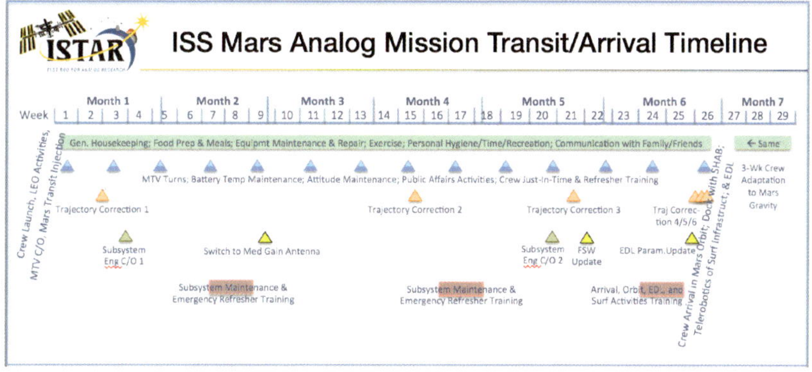

Fig. 1.10   ISS Mars Analog mission transit/arrival timeline (notional).

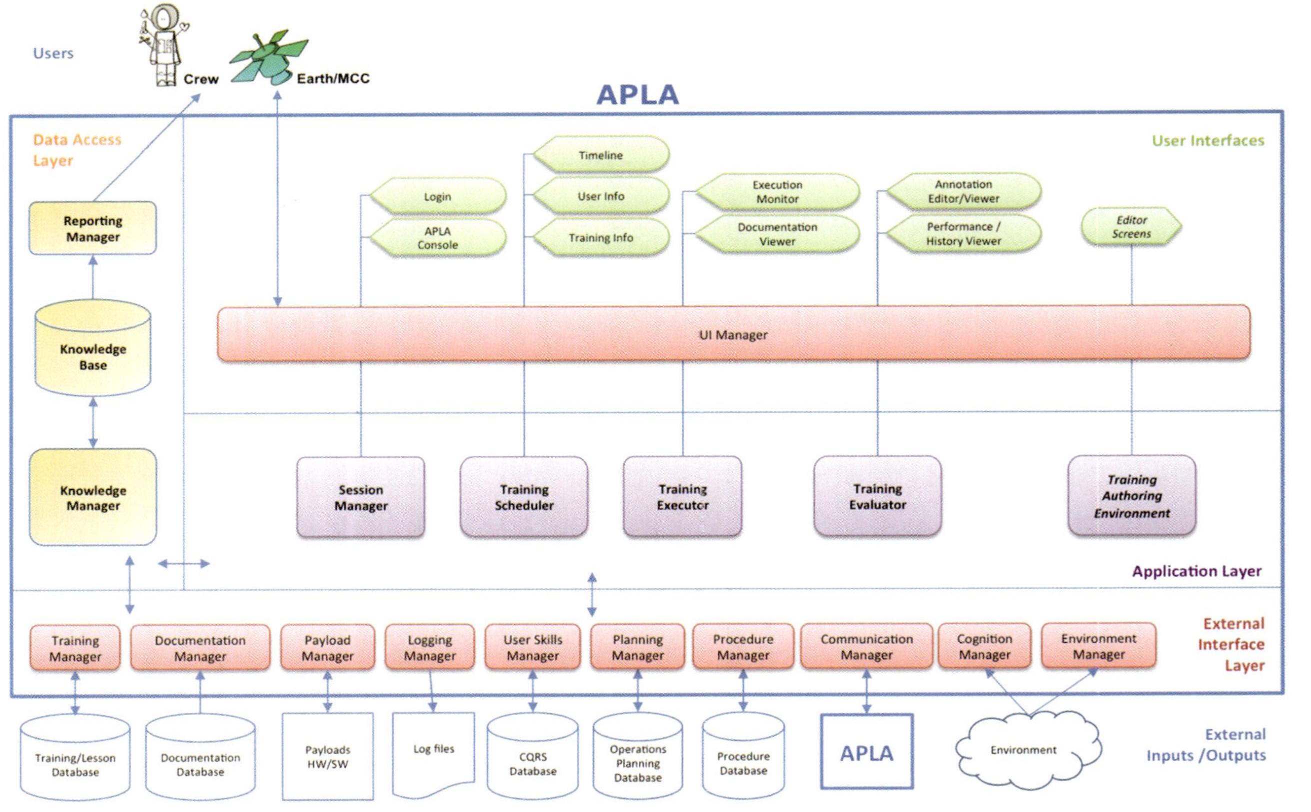

Fig. 6.1   The APLA architecture.

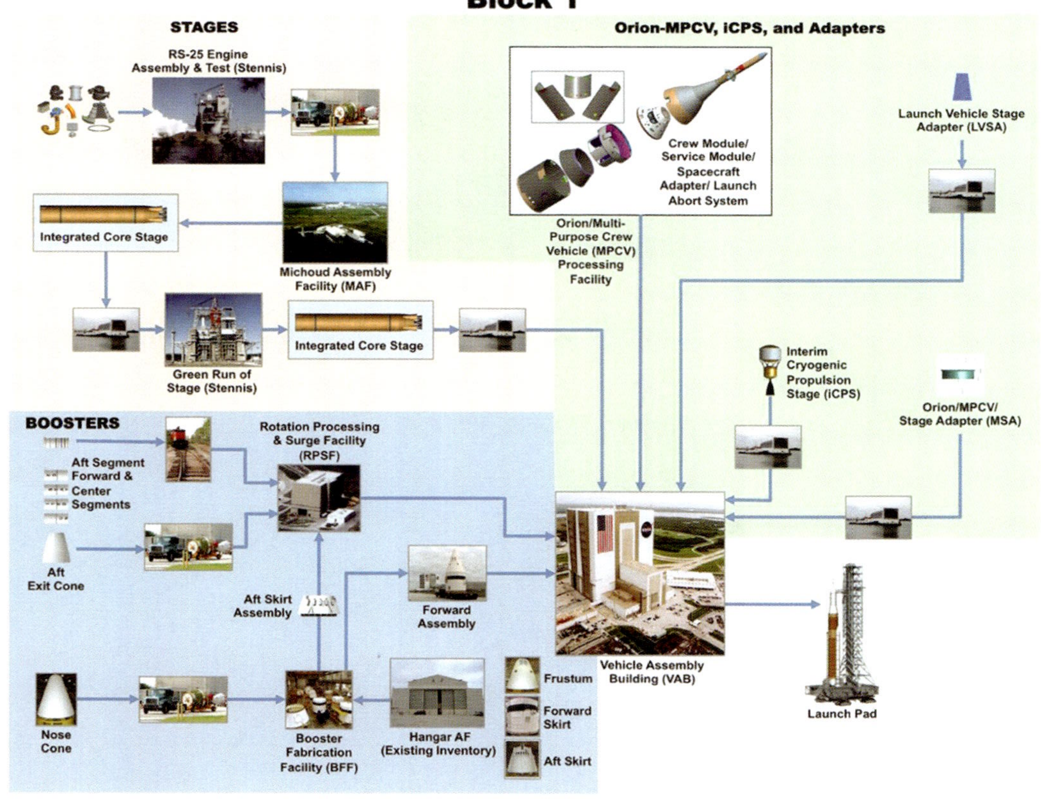

Fig. 9.9   A typical process flow for the SLS Block 1 (70 t) configuration.

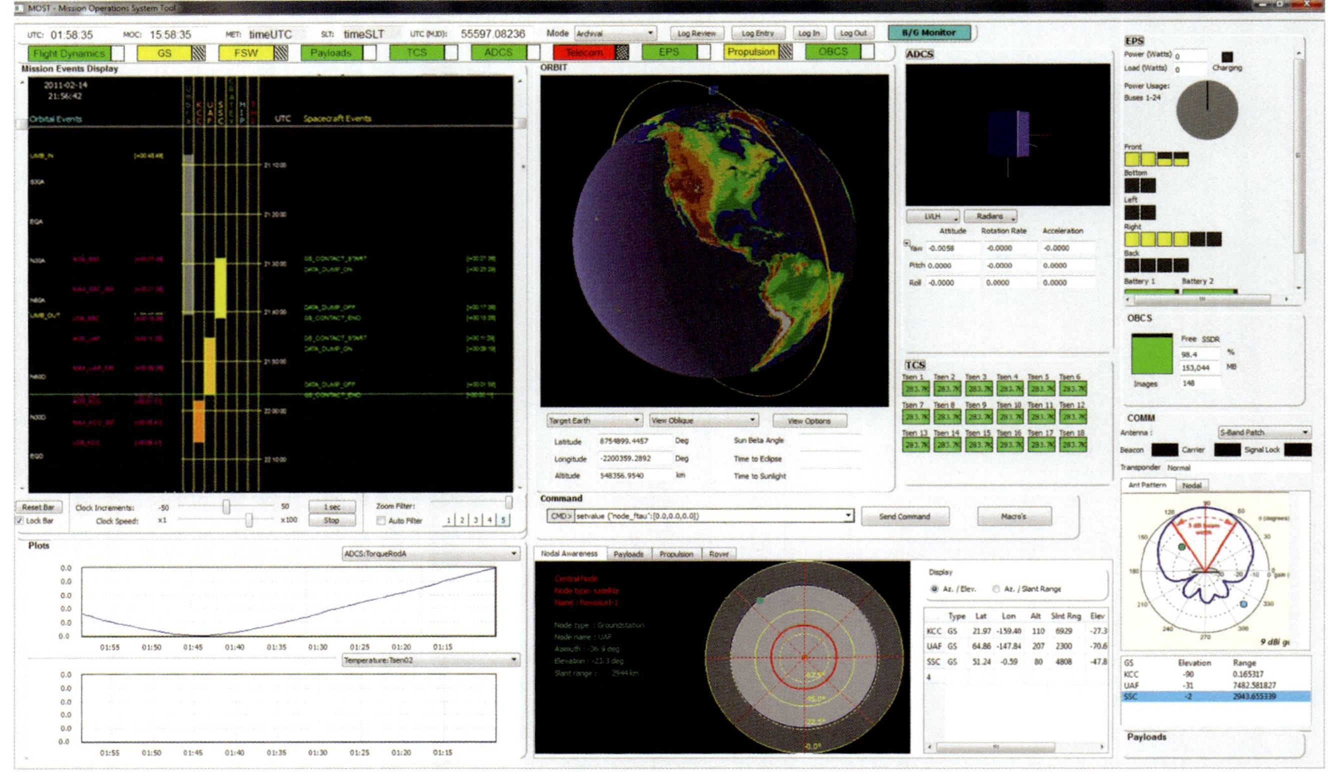

Fig. 10.2   MOST overview display for 3-U CubeSat.

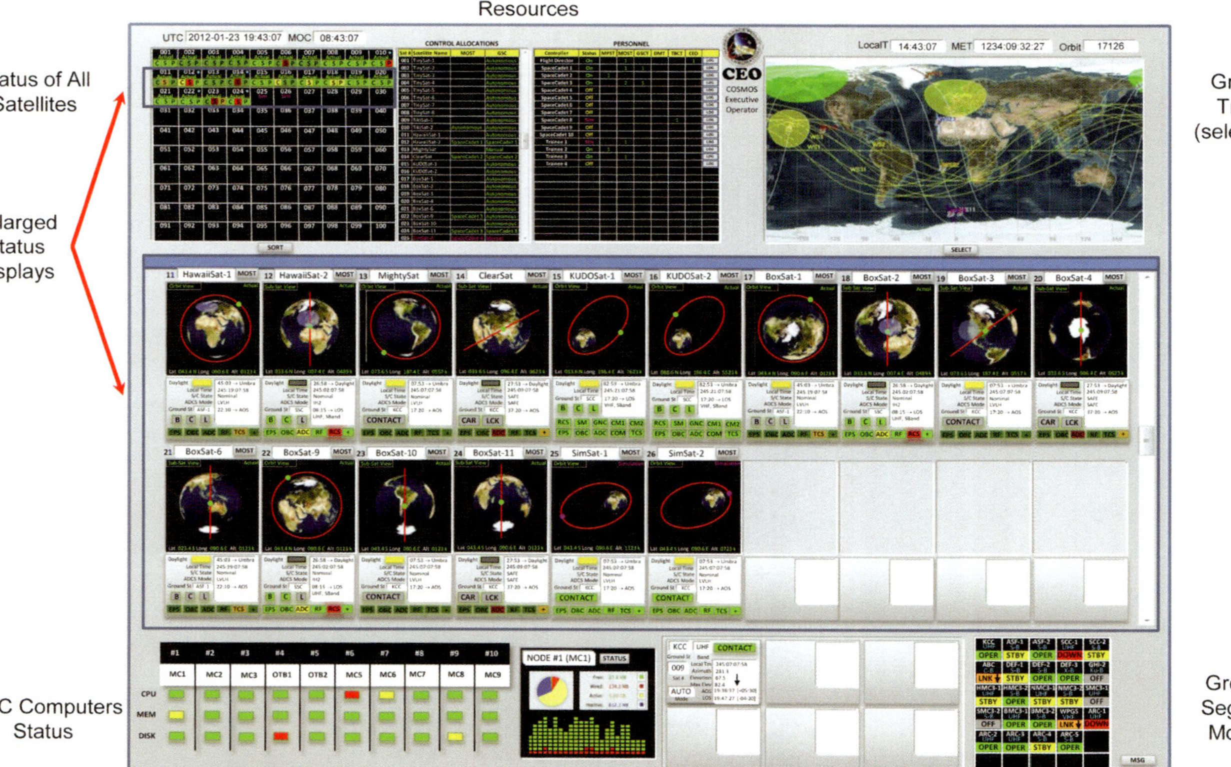

Fig. 10.11   Mockup CEO display, showing 20 satellites and 24 ground station antennas.

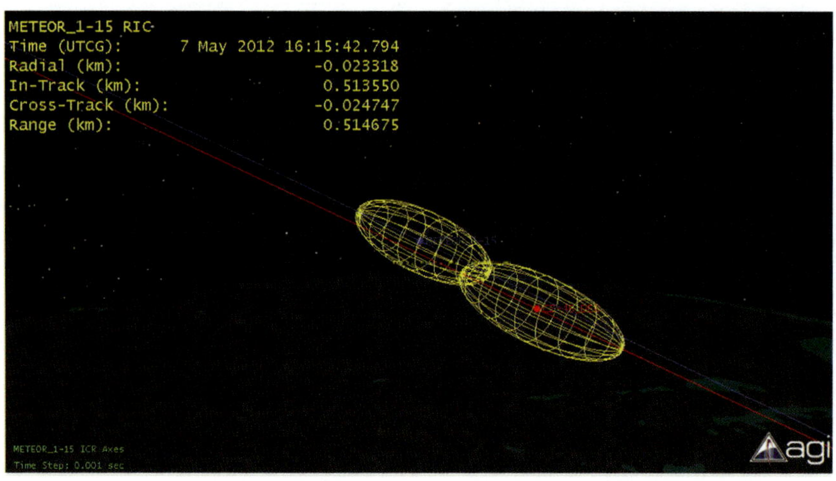

Fig. 11.4   Conjunction encounter geometry. (STK image courtesy of AGI.)

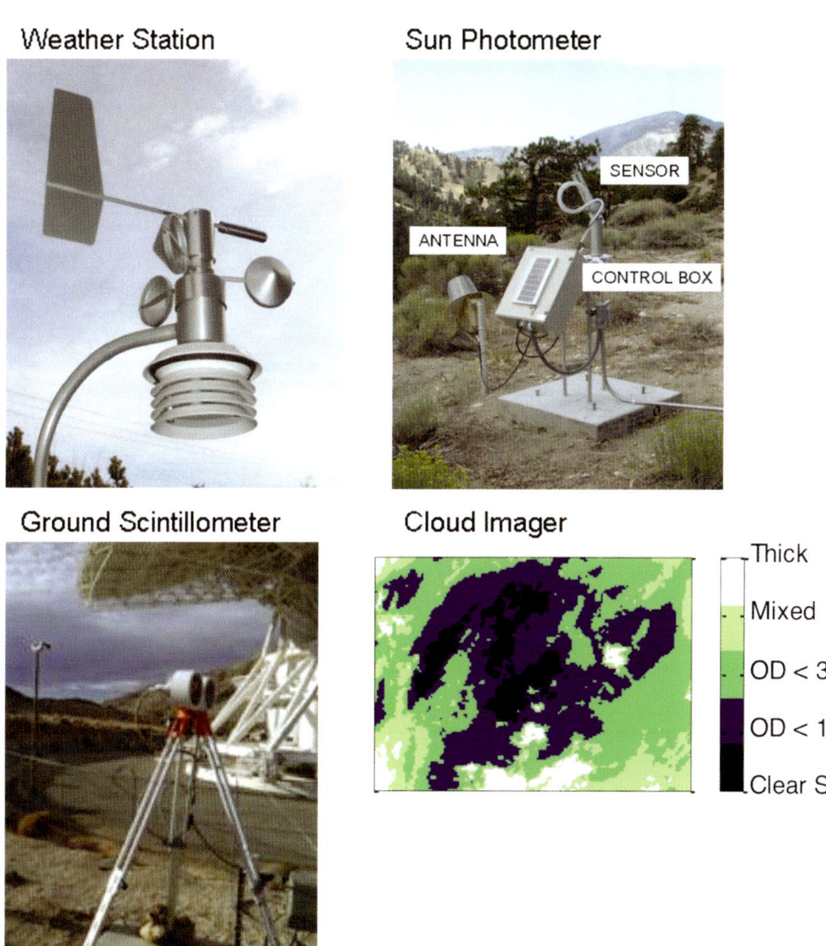

Fig. 12.4 Suite of atmospheric monitoring instruments to characterize the optical channel.

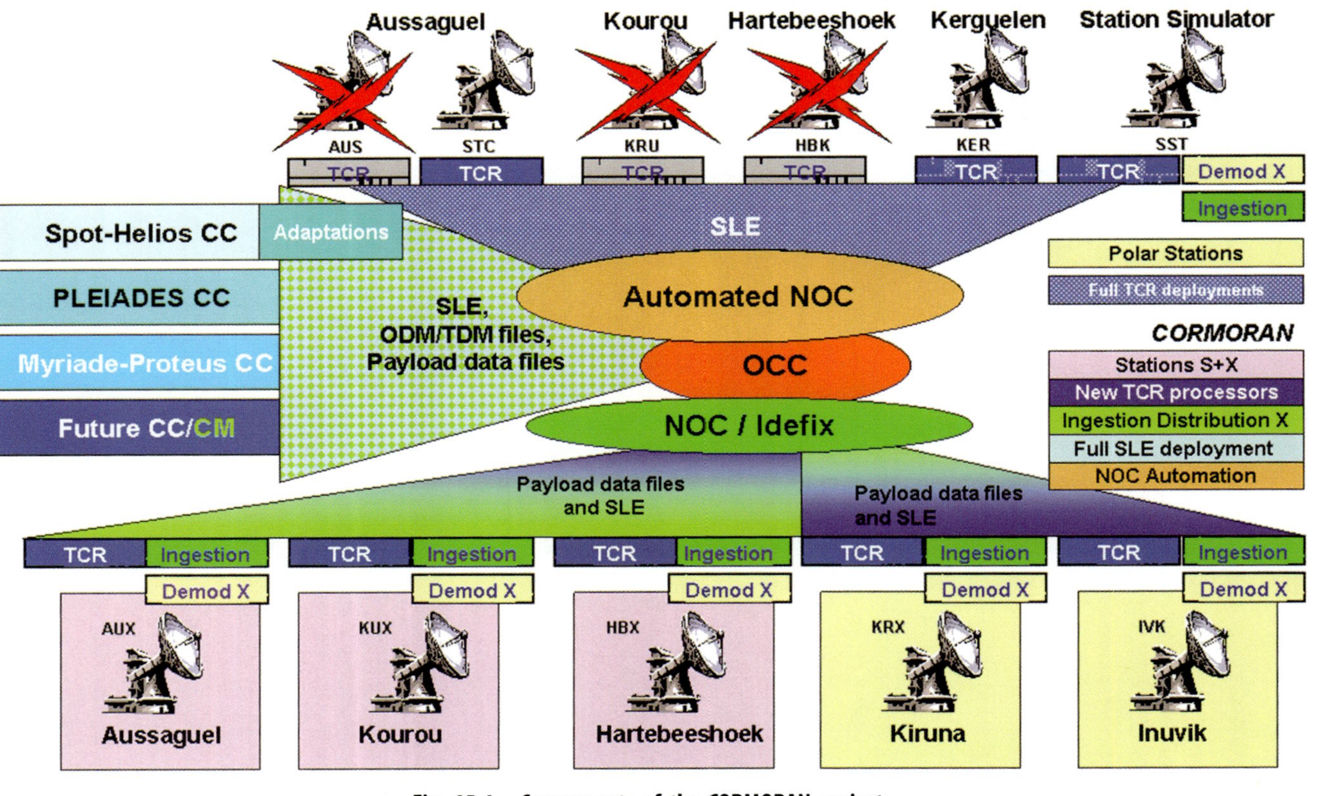

Fig. 15.4   Components of the CORMORAN project.

 ....

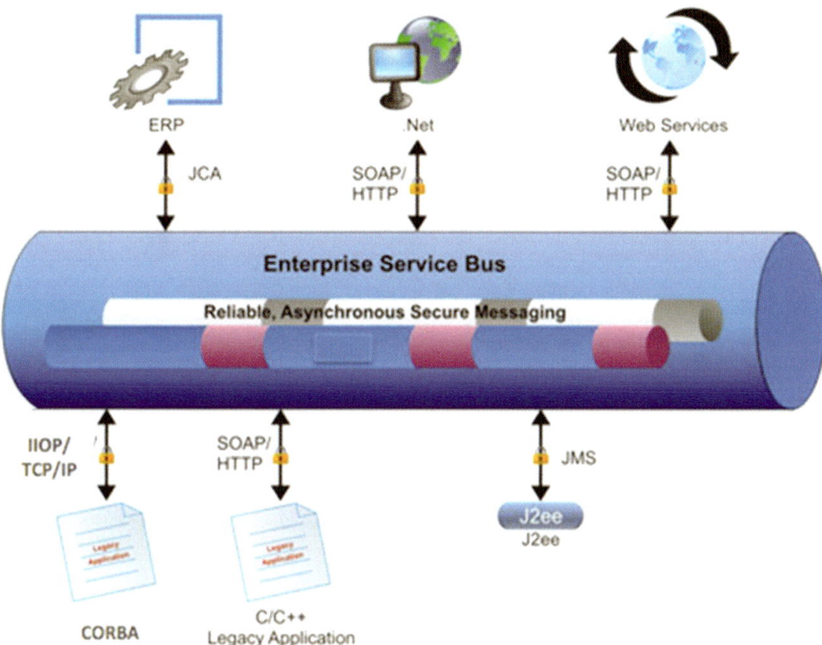

Fig. 19.2 Enterprise Service Bus as integration middleware.

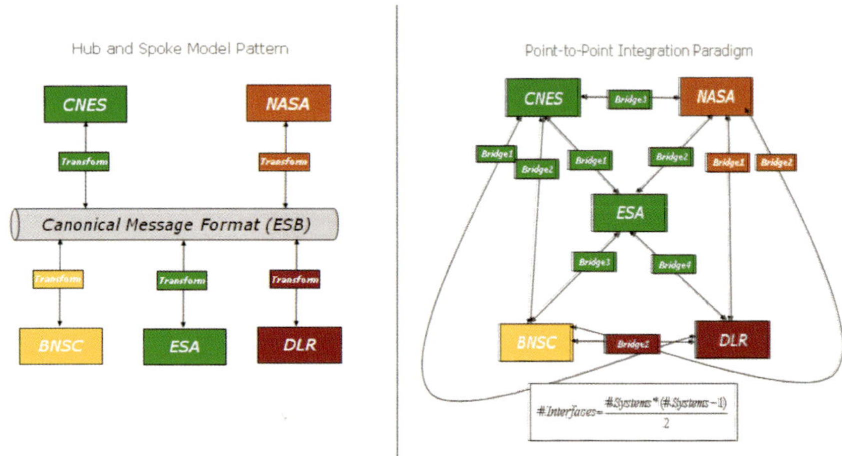

$$\#Interfaces = \frac{\#Systems * (\#Systems - 1)}{2}$$

Fig. 19.5   The hub-and-spoke pattern vs. point-to-point integration.

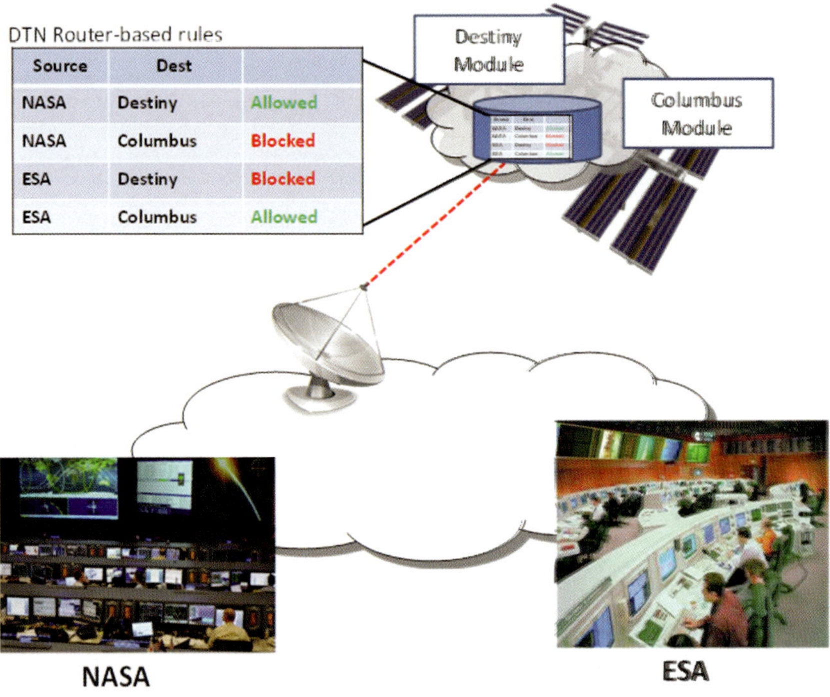

Fig. 20.6   Future test scenarios will incorporate fine-grained security and flight assets, such as the ISS.

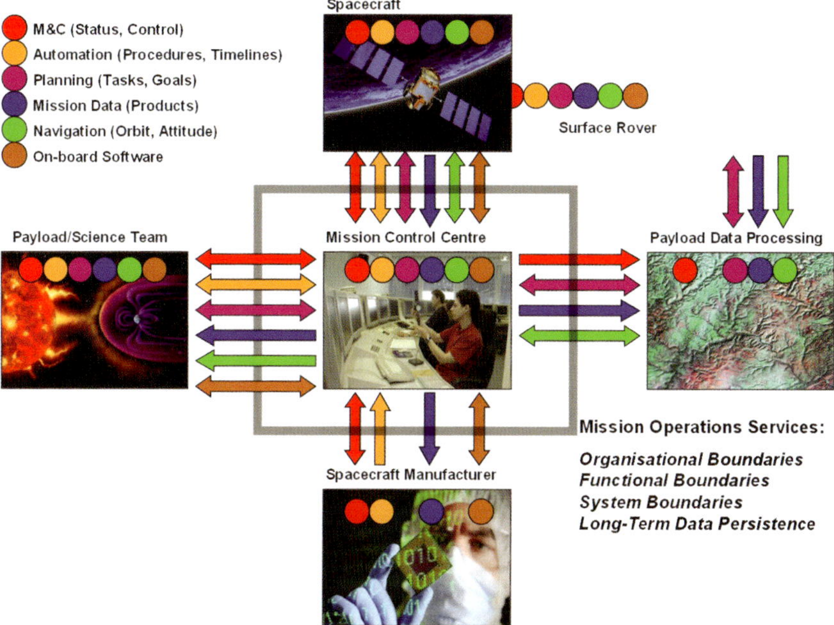

Fig. 21.2   Distributable MO functions.

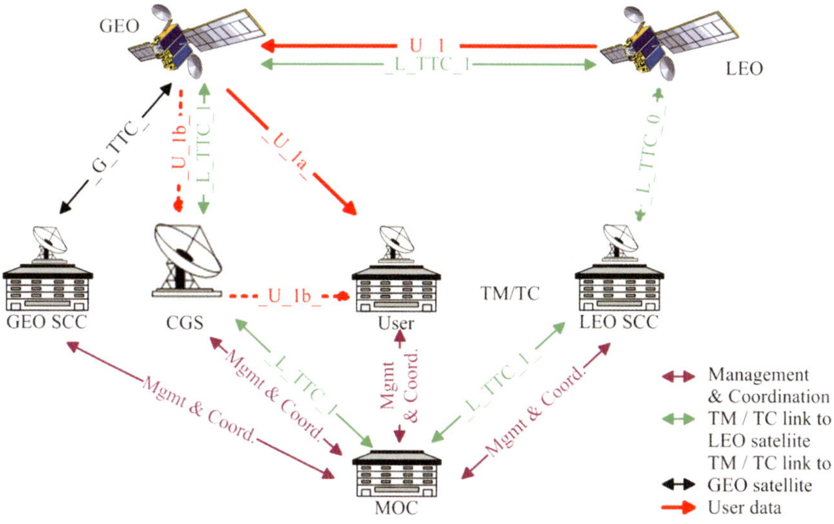

Fig. 22.1   Concept of a GEO relay system for Earth observation data.

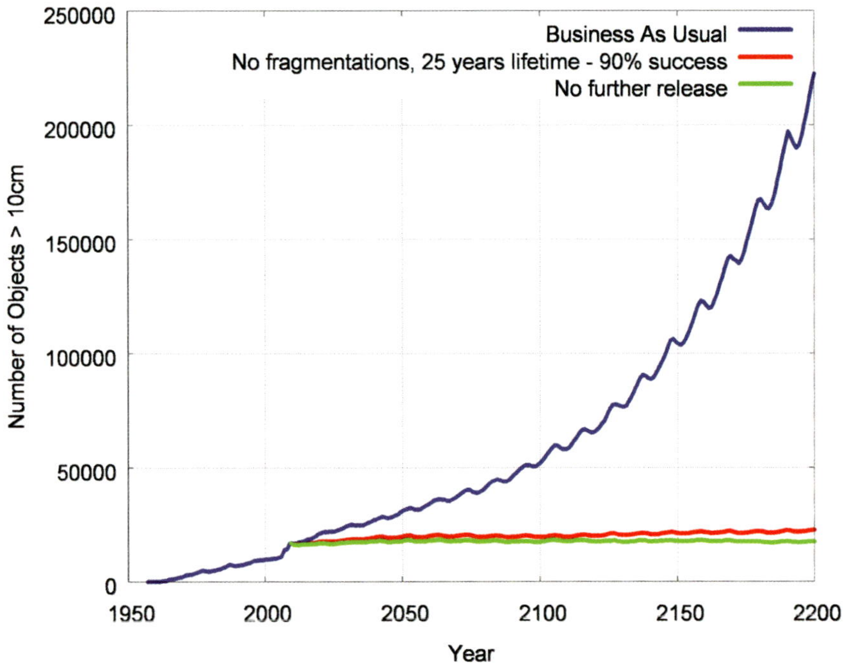

Fig. 23.1   ESA predictions for the evolution of the number of objects >10 cm in LEO.

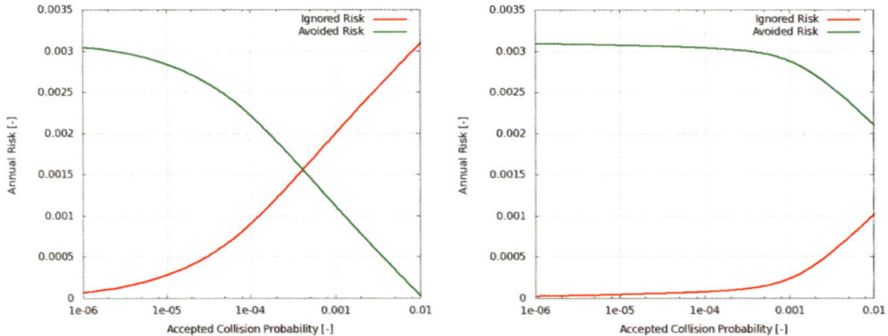

Fig. 23.2   Ignored and avoided risk for collision avoidance procedures based on collision probability levels, for estimated orbit covariances for a typical TLE and CSM.

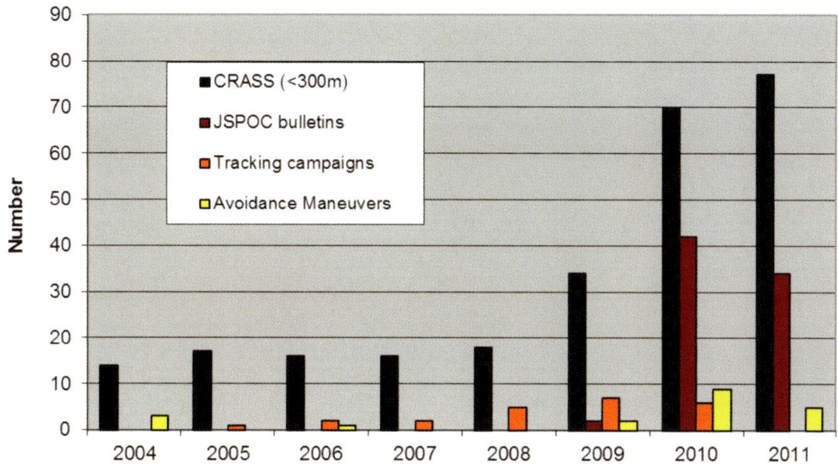

Fig. 23.5   Statistics on collision avoidance for ERS-2, Envisat, and Cryosat (since 2010).

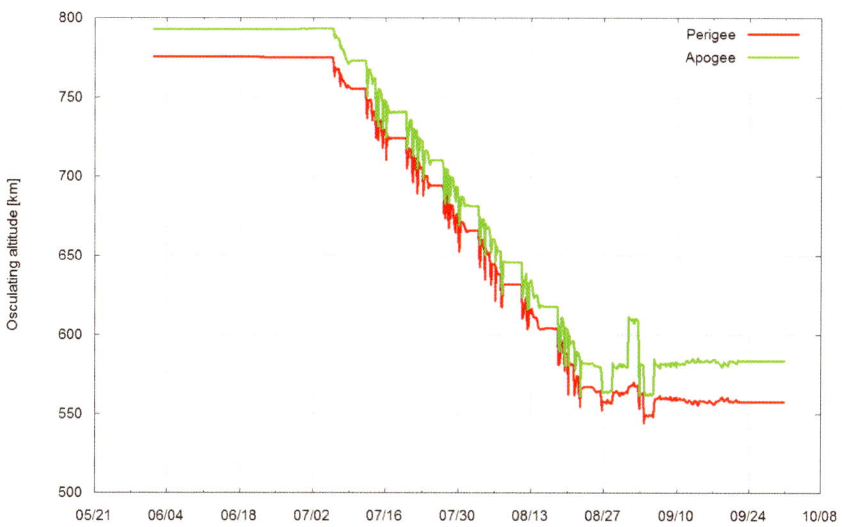

Fig. 23.7   Deorbiting of ERS-2 during July, August, and September 2011.

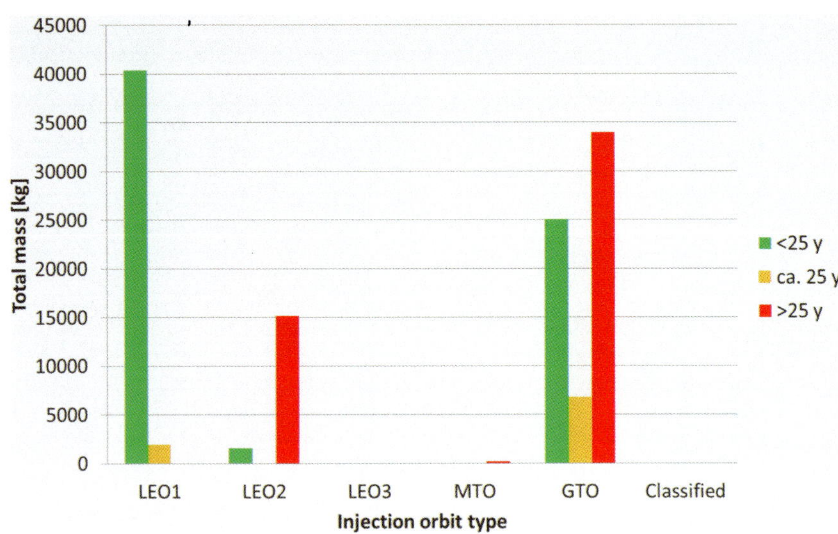

Fig. 23.10 Accumulated mass of upper stages injected in 2010.

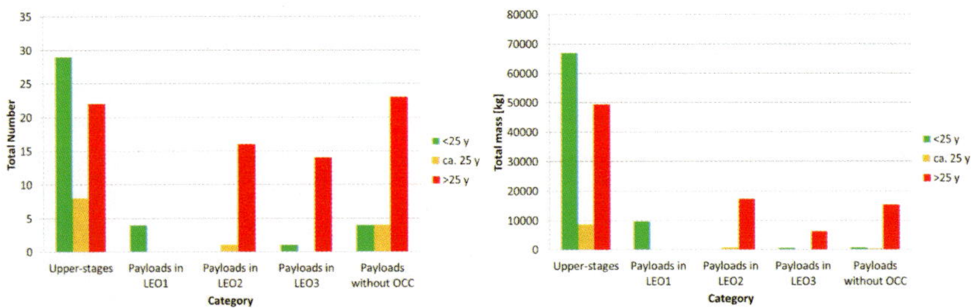

Fig. 23.12 Cumulated numbers (left) and mass (right) of objects that ceased operations in 2010, as a function of type and orbital regime and as a function of remaining orbital lifetime.

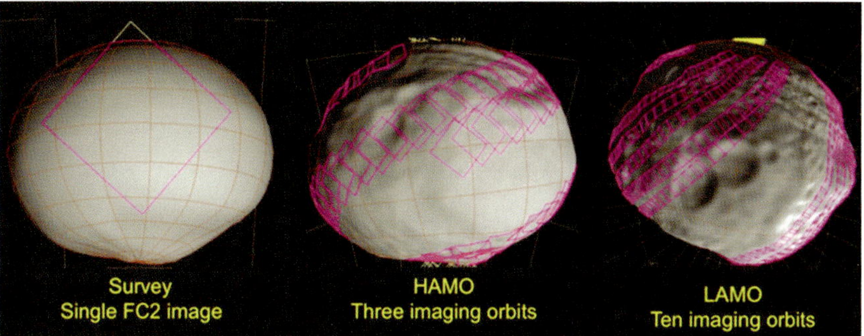

Fig. 27.2 FC2 footprints on Vesta are shown for each science orbit (Vesta is represented by more mature versions of the shape model at each science orbit).

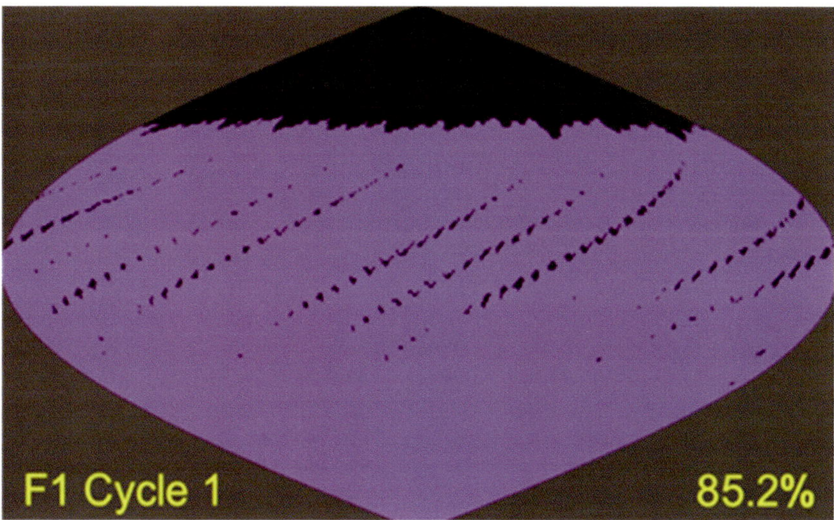

Fig. 27.3 FC2 planned imaging coverage for HAMO-1, Cycle 1, clear filter.

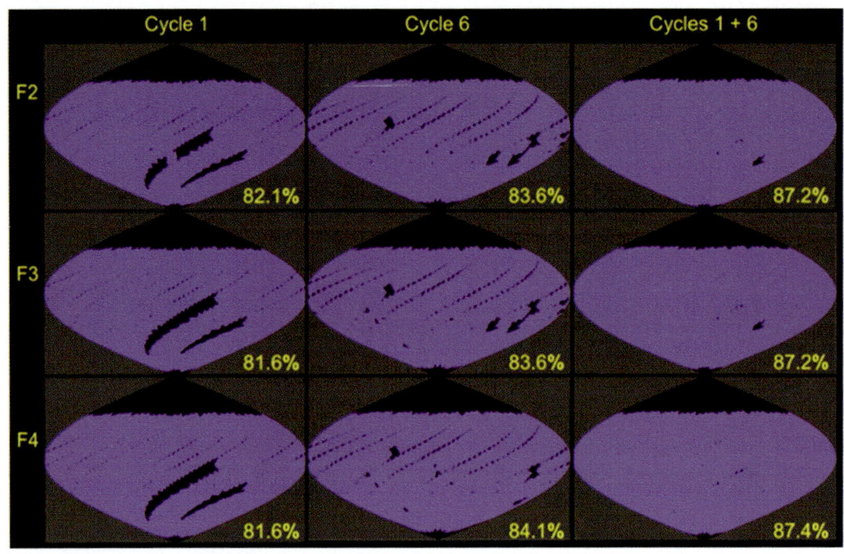

Fig. 27.4   FC2 actual coverage for HAMO-1 three color filters in Cycles 1 and 6.

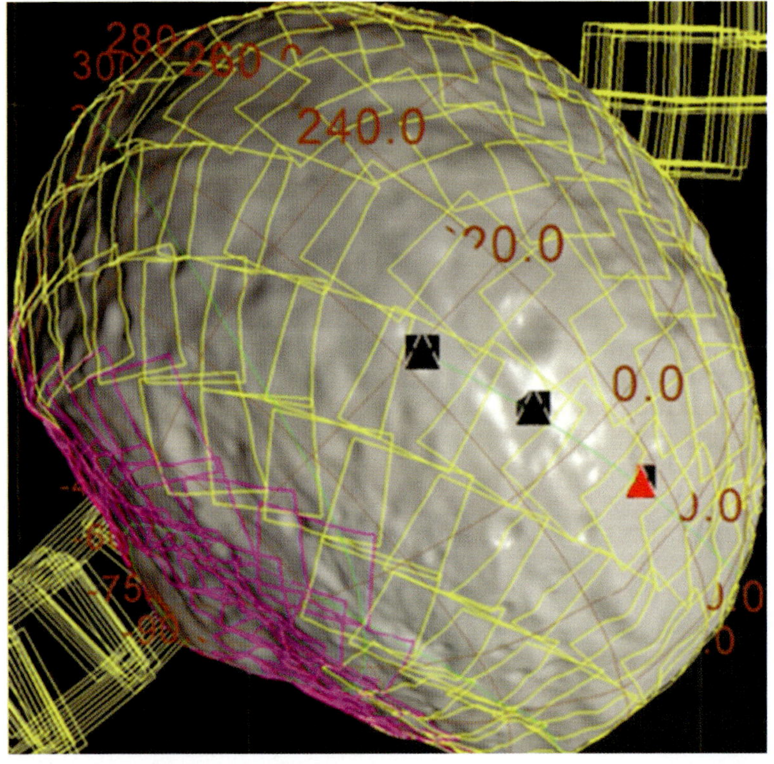

Fig. 27.5   FC2 planned color imaging for HAMO-2.

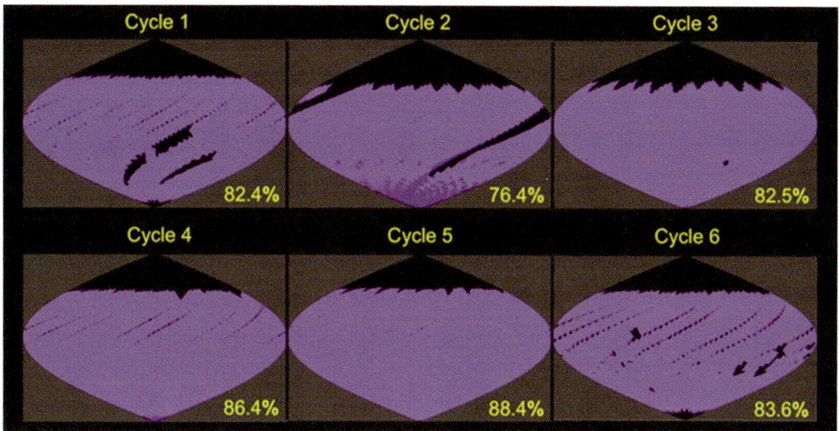

Fig. 27.6   Actual FC2 coverage for HAMO-1 clear imaging in all six cycles.

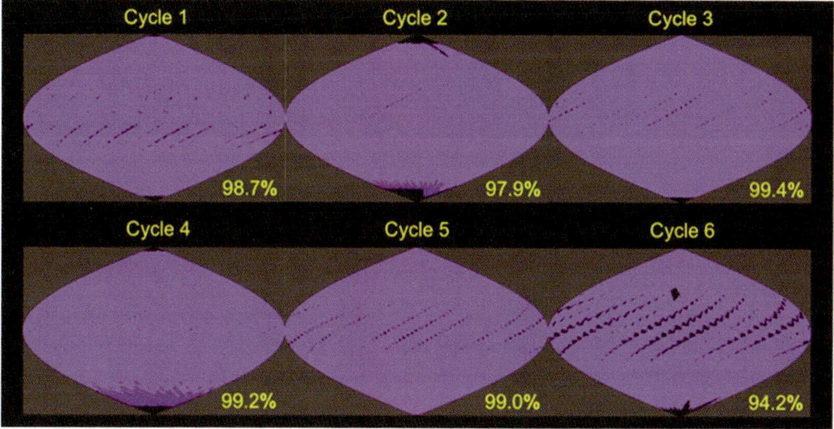

Fig. 27.7   Actual FC2 coverage for HAMO-2 clear imaging in all six cycles.

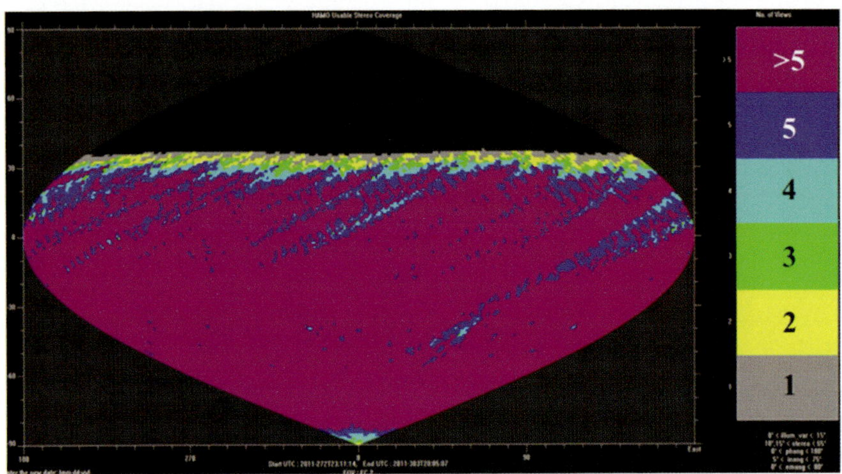

Fig. 27.8   Actual FC2 usable stereo coverage for topography analysis during HAMO-1 (incidence angle cutoff is <75°; colors display the number of available stereo views).

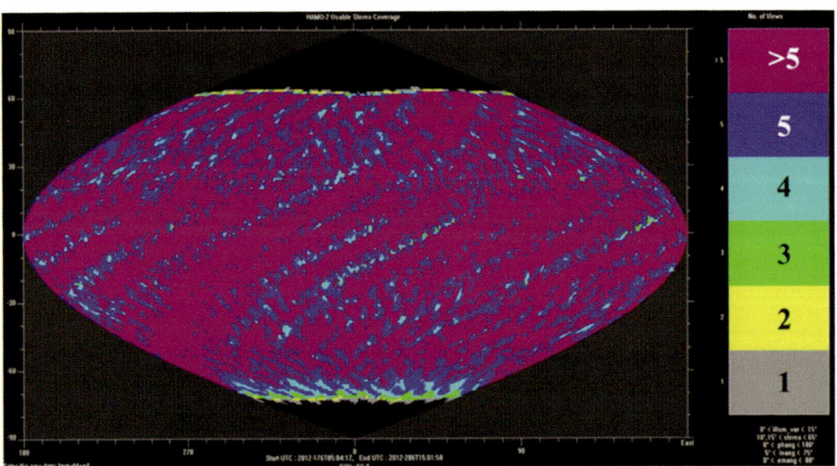

Fig. 27.9   Actual FC2 usable stereo coverage for topography analysis during HAMO-2 (as in HAMO-1, the incidence angle cutoff is <75°).

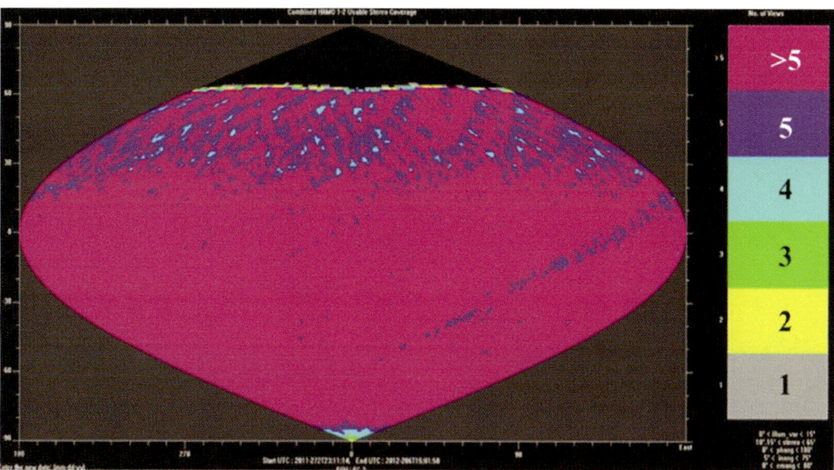

Fig. 27.10   Combined topography imaging coverage using data from HAMO-1 and HAMO-2.

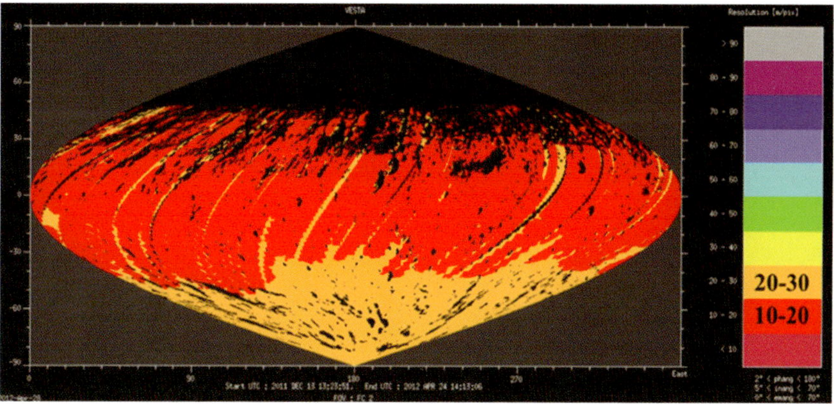

Fig. 27.11   Actual FC2 coverage obtained during the LAMO phase.

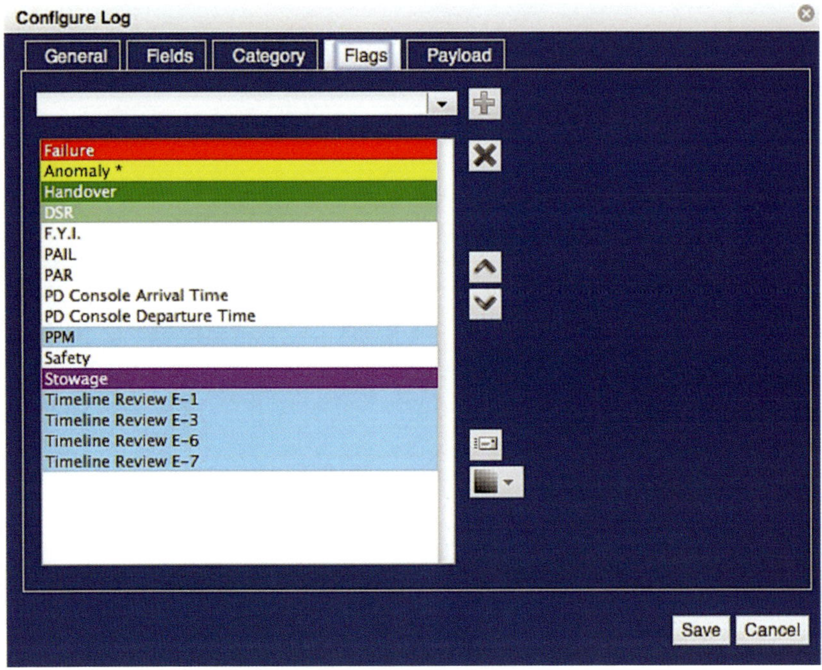

Fig. 28.6   Defining flags.

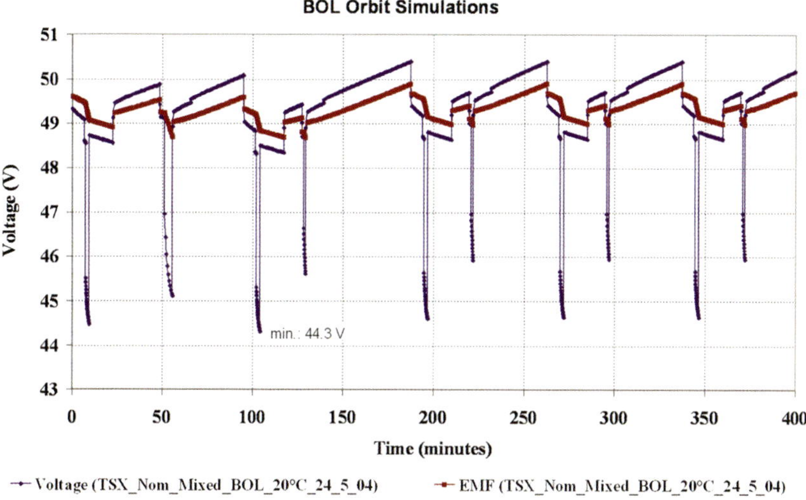

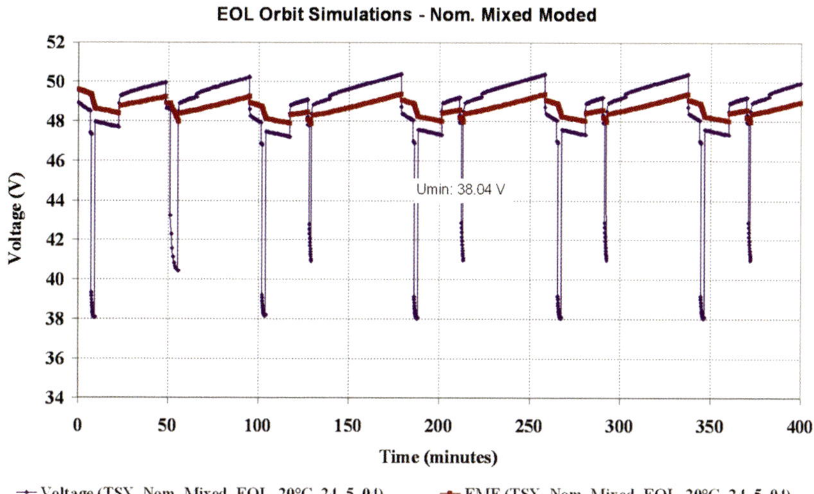

Fig. 29.2 Worst-case primary voltage behavior at BOL and EOL for a nominal operation scenario (design phase).

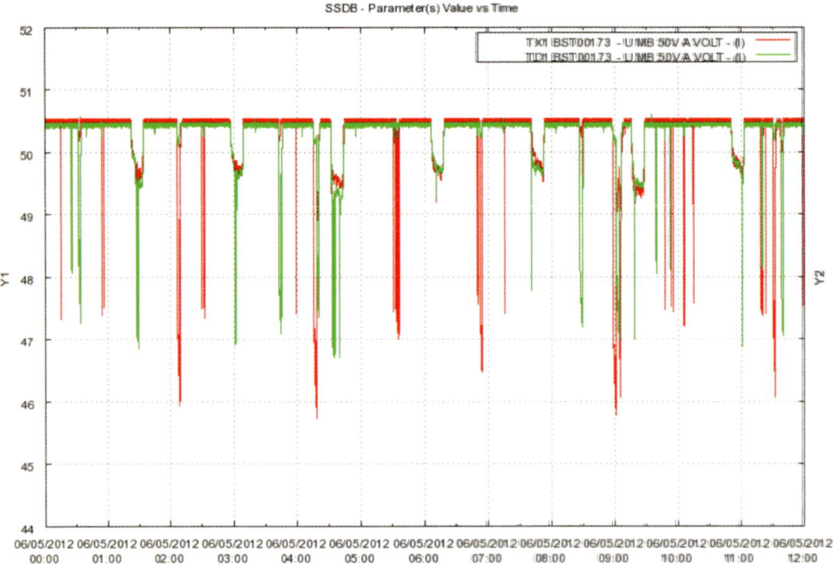

Fig. 29.12   Battery main bus voltage during SAR operation.

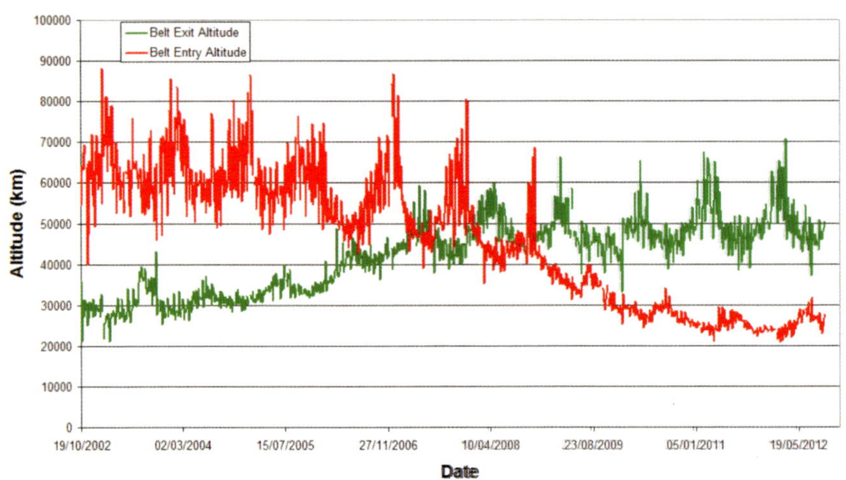

Fig. 31.1   Integral radiation belt entry/exit altitudes.

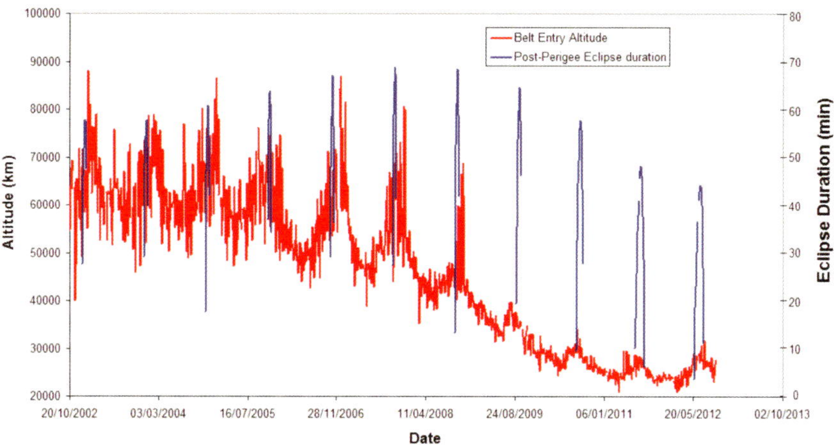

Fig. 31.2   Integral radiation belt entry with post-perigee eclipse duration.

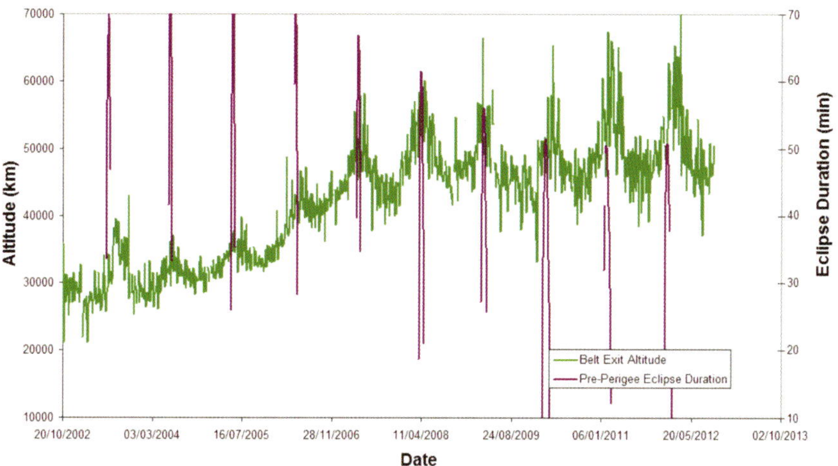

Fig. 31.3   Integral radiation belt exit with pre-perigee eclipse duration.

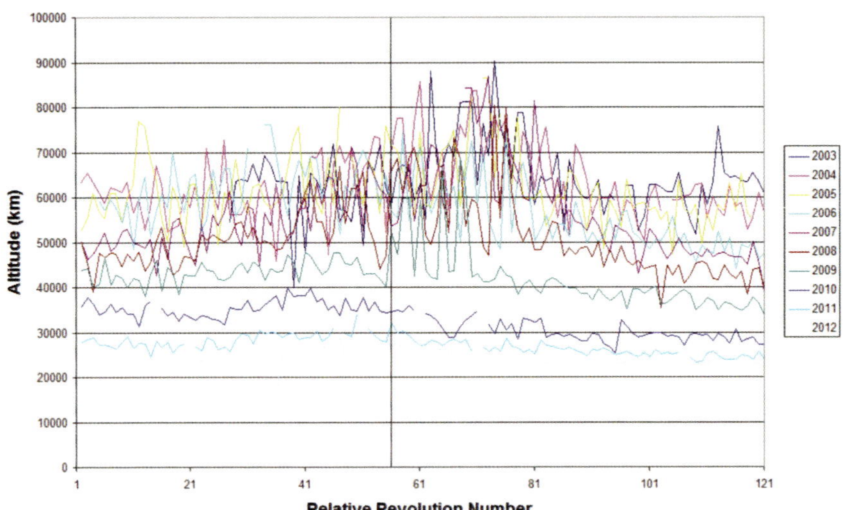

Fig. 31.5   Overlaid belt entry altitudes (the vertical black line marks the maximum eclipse duration).

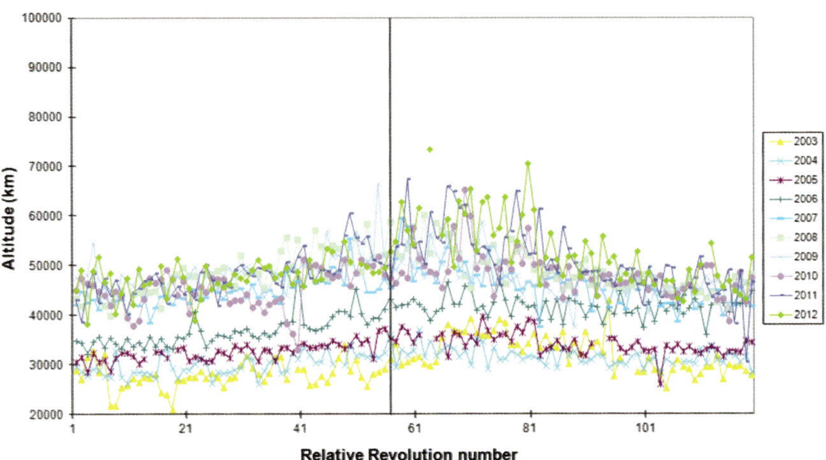

Fig. 31.6   Overlaid belt exit altitudes (black line marks the maximum eclipse duration).

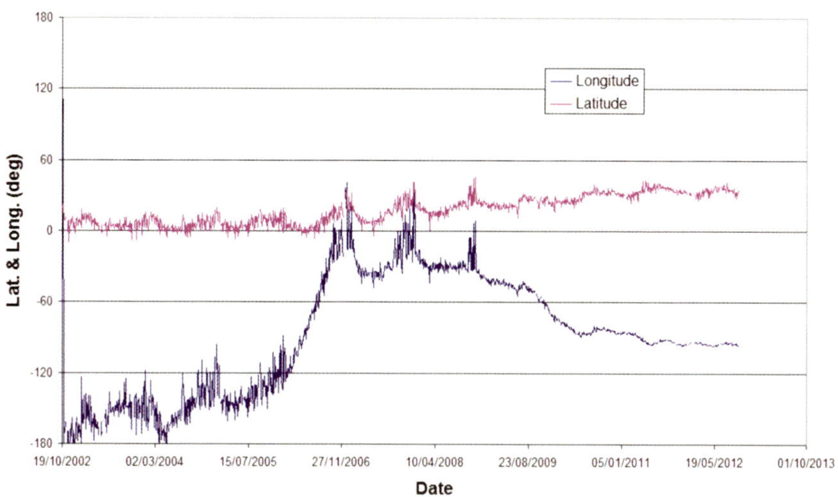

Fig. 31.7   Geocentric location of the radiation belt entry.

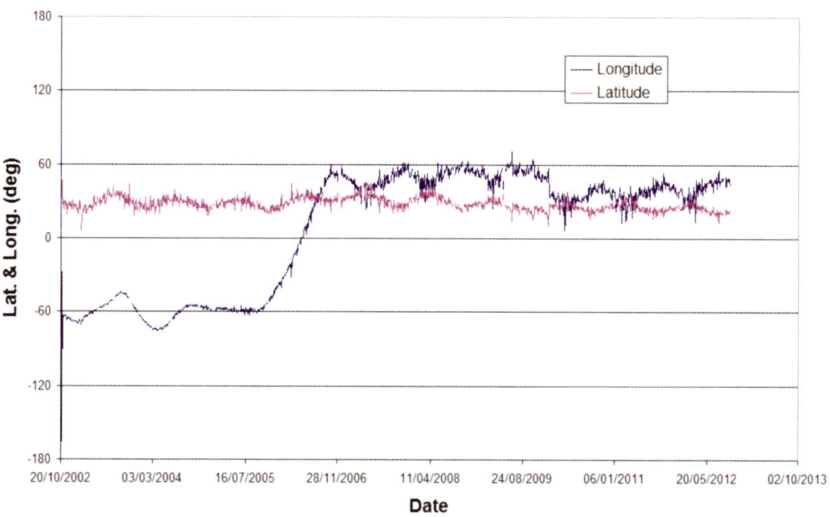

Fig. 31.8   Geocentric location of the radiation belt exit.

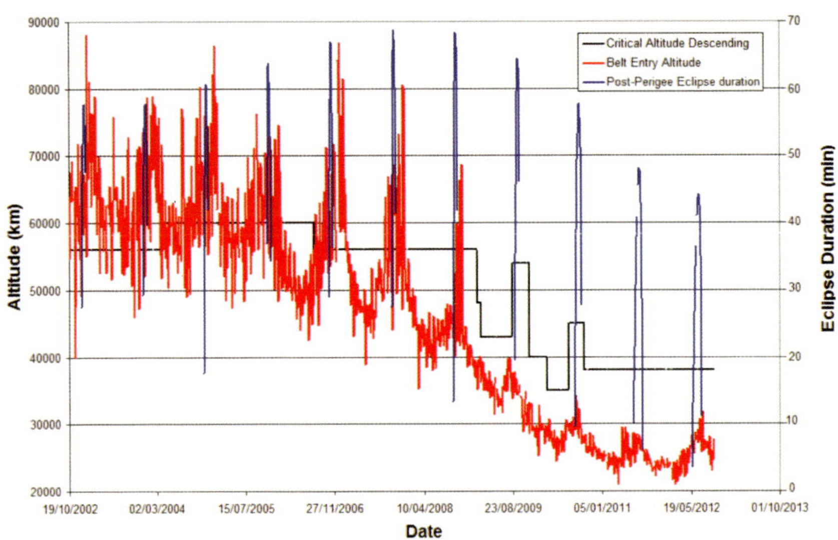

Fig. 31.9   Radiation belt entry with eclipse times and the "critical altitude descending."

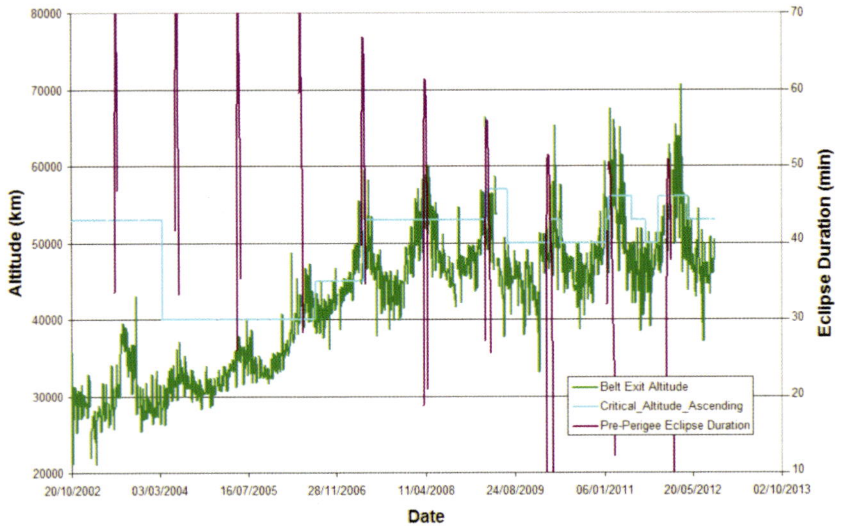

Fig. 31.10   Radiation belt exit with eclipse times and the "critical altitude ascending."

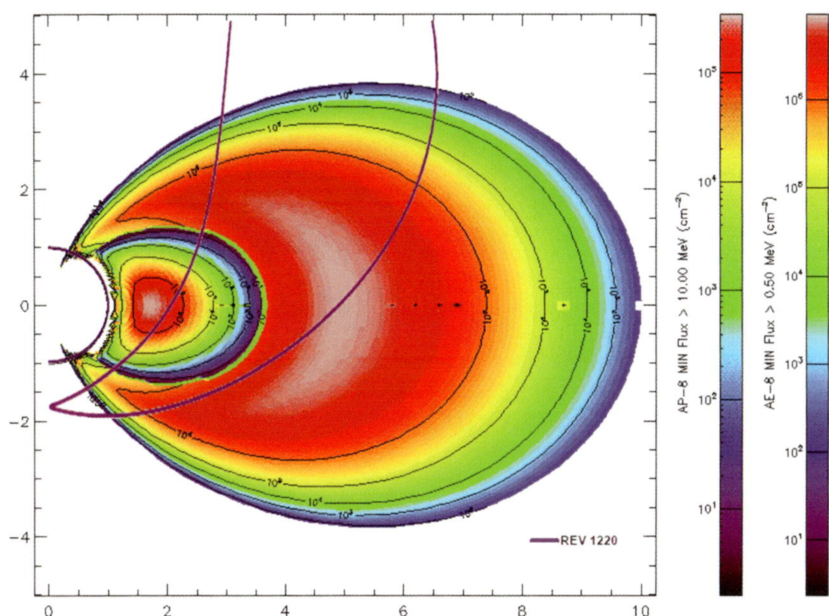

Fig. 31.11 'Fish-eye' plot showing a schematic of belt entry and exit trajectories (data from the AP8/AE8 programs [6, 7]).

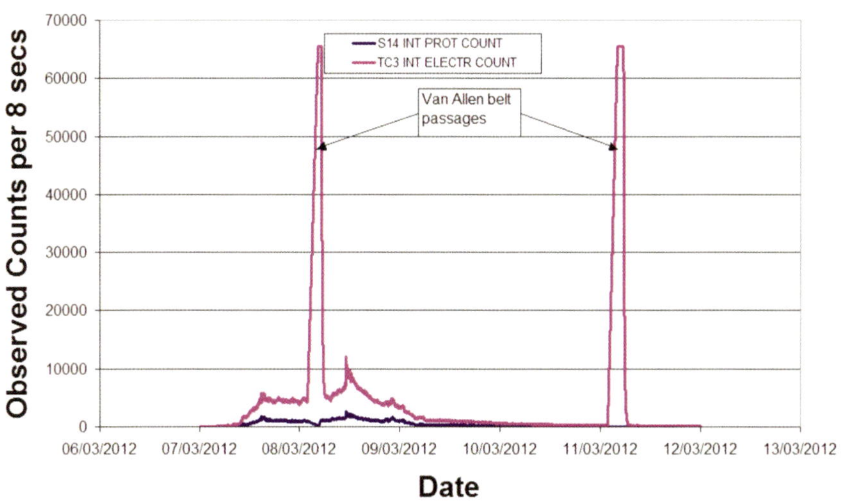

Fig. 31.13 Observed radiation levels during the storm of 8–9 March 2012.

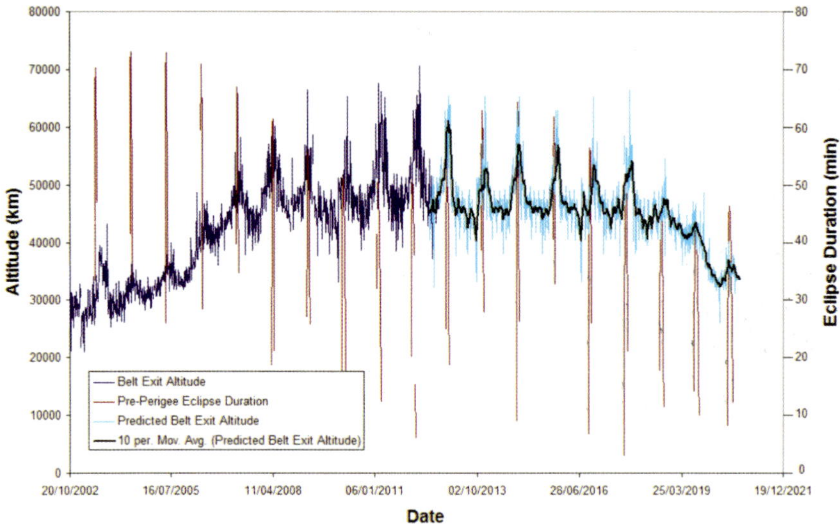

Fig. 31.14 Predicted evolution of the radiation belt exit altitude.

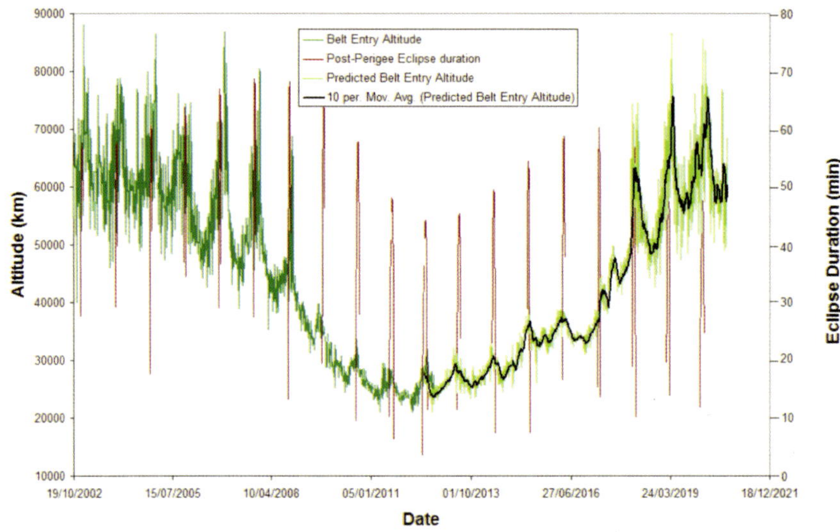

Fig. 31.15 Predicted evolution of the radiation belt entry altitude.